36-50

Fundamentals of Three-Dimensional Descriptive Geometry

Board of Advisors, Engineering

Fundamentals of Three-Dimensional Descriptive Geometry

Second Edition

Steve M. Slaby
Princeton University

JOHN WILEY & SONS, New York • Chichester • Brisbane • Toronto

Library of Congress Cataloging in Publication Data

Slaby, Steve M
Fundamentals of three-dimensional descriptive geometry.

Bibliography
Includes index
1. Geometry, Descriptive I. Title.
QA501.S448 1976 516'.6 76-18152
ISBN 0-471-79621-2

Printed in the United States of America

10 9 8 7 6 5 4

TO

Karin

Stefan

Kristin

Projective Graphics—
Technical Design and Analysis

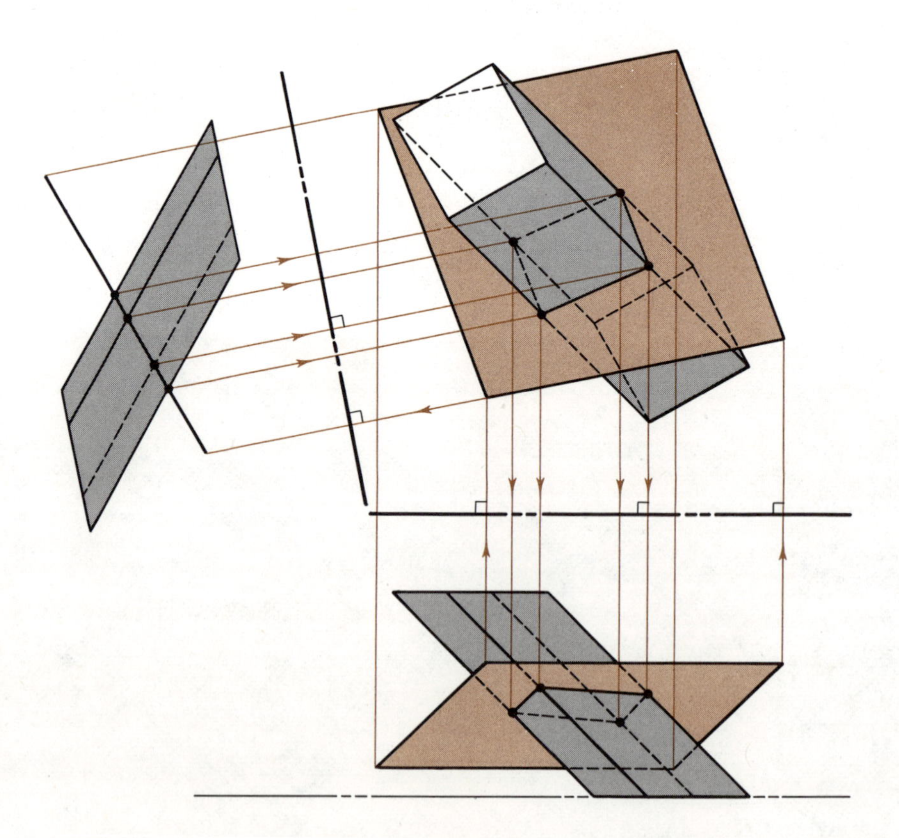

Preface to the Second Edition

The first edition of this book was published by Harcourt, Brace & World, Inc. This second edition, published by John Wiley & Sons, Inc., in addition to making corrections in the first edition, includes a new chapter (XI) entitled "Principles of Descriptive Geometry Applied to Practical Problems." Chapter XI is devoted entirely to problems for solution by the student. These problems are related to real life situations in various areas of engineering and architecture. In addition, this new material introduces the International System of Units, which is officially designated as the "SI" system. This system is based on a modern version of the metric system.

I want to thank the publisher, John Wiley & Sons, Inc. and its staff for making this second edition of my book possible.

Princeton, N.J. Steve M. Slaby

Preface to the First Edition

Three-dimensional descriptive geometry is the graphical method of solving space or solid analytic geometry problems. It is also the theory of engineering drawing. This book is designed for teachers and students of three-dimensional descriptive geometry in technical institutions, colleges, engineering schools, and universities. It can also be used in a self-study education program.

The fundamental concepts of three-dimensional descriptive geometry that are presented in this book are based on classical geometric reasoning and relations. These concepts are presented in such a manner that the student will readily understand descriptive geometry and be able to relate it to engineering, science, architecture, and technology.

One main objective of this book is to encourage the development of a "graphic mind" in the student, through a carefully organized presentation of the fundamental material in a logical and sequential manner. Color is used in most of the illustrations to aid the learning (and teaching) process. After each discussion of the fundamental concepts, orthographic examples are presented with given and required data. Included are detailed construction programs that identify the procedure to be used for the complete understanding of the fundamental concepts. At the ends of Chapters I–IX are sample quizzes and practice problems, accompanied by illustrations on graph paper showing the given data. These exercises test the student on what he has learned in the chapter. Sample examinations are given at the ends of Chapters II and V; the answers to these tests are in the appendix. A sample final examination is also presented (with answers) in the appendix.

I want to thank the editors of Barnes & Noble, Inc., for their cooperation in the development of this book, which is based, in part, on my *Engineering and Descriptive Geometry,* 1st ed. (New York: Barnes & Noble, 1956). I also want to thank Mrs. Suzanne Fisher for the wonderful job she did in typing the manuscript. Finally, I express my special thanks to Bert Schneider for doing an outstanding job in preparing the finished artwork for my book.

Princeton, N.J. Steve M. Slaby

Contents

Fundamentals of Three-Dimensional Descriptive Geometry

Principles and Basic Concepts of Three-Dimensional Descriptive Geometry

1 Introduction

Descriptive geometry deals with physical space. In this book we shall be concerned with the three-dimensional space with which we have had direct personal contact since birth. The use of descriptive geometry occurs in various forms throughout the disciplines of engineering, science, and architecture. For example, the relation of parts of a structure to another structure; the relation of a component of a machine to another component in the same machine, and how they effect each other in terms of motion, strength, and total effect on the machine; and the relation of molecules in a crystal or chemical compound are all examples of *spatial relations.* The relationships of these components are all important in the proper operation of mechanical or scientific apparatus in terms of gaining experimental data. The relationship between various segments and components of a structural system is also important, where, if the geometric relation of one component to another does not follow basic geometric and physical laws, the system might collapse under loading. In other words, in terms of engineering, science, and architecture, spatial relations that are expressed and analyzed through descriptive geometry are one of the most important factors in the design of scientific apparatus, engineering systems, and architectural structures.

The history of descriptive geometry started about 1790 with Gaspard Monge, a French military engineer. Monge was a designer of fortifications as well as a mathematician. In those days, fortifications usually consisted of stone walls and turrets. The stones used in the construction of this type of fortification had to be cut accurately to fit one stone to another so that the wall or turret was self-supporting and strong enough to withstand bombardment. In order to determine the angles at which the stones had to be cut, laborious arithmetical calculations were necessary. Monge conceived of the idea of determining the necessary "design" information by means of a graphic analysis. The graphic analysis resulted in the discipline and area of knowledge known today as *descriptive geometry*. (In his system of graphic analysis Monge used the concept of the "trace" of a plane, that is, the line of intersection between two planes.) Using this new method of analysis, Monge completed the design of a fortification in record-breaking time, handing over his completed plans to an incredulous commanding officer, who refused to believe that they were complete and accurate. Monge insisted on the accuracy of the plans, and the officer reluctantly agreed to build the fortification. Immediately, Monge's descriptive geometry system was declared "classified" and became a "military secret" for a number of years. The system was eventually "declassified," and was immediately adopted in French engineering schools as a basic engineering discipline.

Throughout the years, the Mongean system of descriptive geometry has evolved into the form that is presented in this book. We shall deal with the trace-of-plane-method only partially, when we present the intersection of geometric solids and surfaces in Chapter VIII.

2 Generation of a three-dimensional space

When you speak of space you are actually speaking of "dimensions." For example, consider a point mathematically. You can consider a point in the mathematical sense, by analogy, to be a sphere whose diameter is equal to zero. Obviously, to represent such a point is impossible, and yet when you solve problems by using graphics and descriptive geometry you must represent this point as a mark on your paper. Since you can see this mark or point on your paper, it actually does have a dimension, however small it may be.

Figure 1 shows a dot which "represents" a point *P* that has a diameter equal to zero. This means that the point *P* is an ideal point that has no dimension and occupies zero space. You can therefore say that an ideal point is a *space of zero dimension.*

FIGURE 1

A space of zero dimension—point

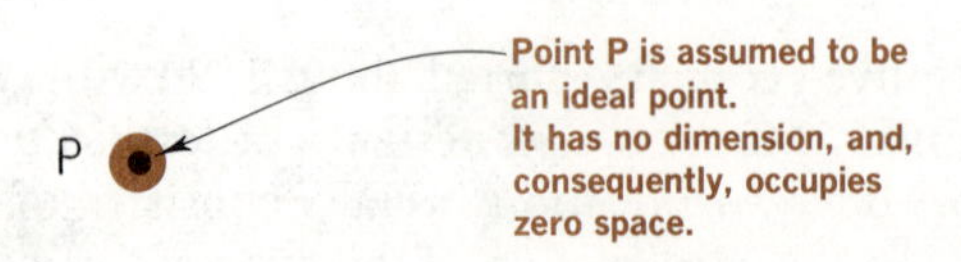

If you take the point P and move, or "translate," it in a straight-line direction to another location, the point P will have *generated* a line that can be considered to be a *one-dimensional space,* since one dimension can be measured along the line. No other measurement is possible except along the line generated by point P, since the line generated by an ideal point has no thickness. Therefore, you can consider an ideal line to be a *space of one dimension* (see Fig. 2).

FIGURE 2

A space of one dimension—line

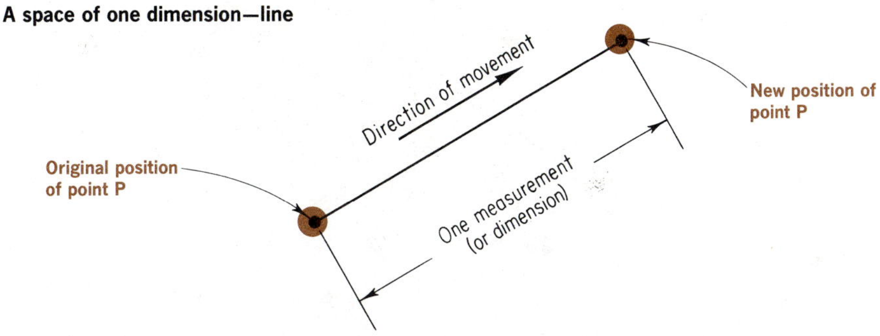

If you now take the ideal line and translate it parallel to itself from one given position to a second given position, the ideal line will generate a *plane* on which *two measurements* can be made—one along the line itself and one in the direction of the translation of the line's movement (see Fig. 3). Therefore, you can consider a plane to be a *space of two dimensions.*

FIGURE 3

A space of two dimensions—plane

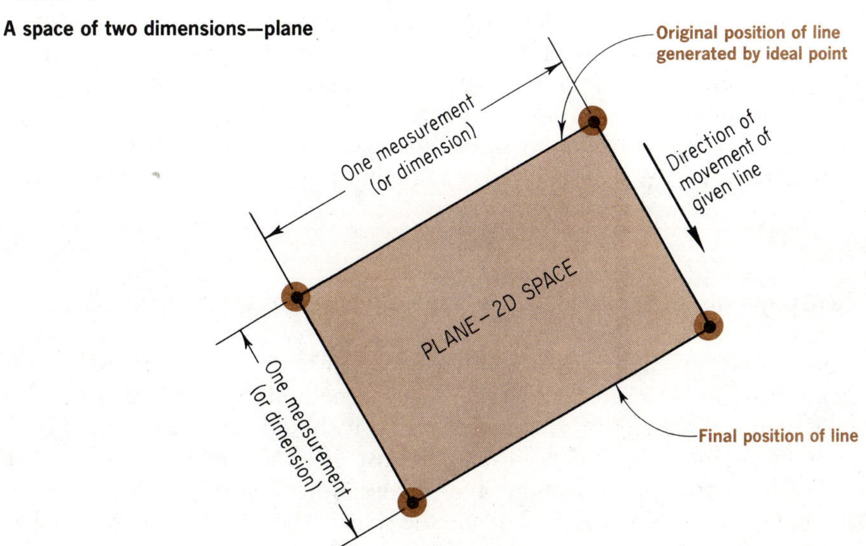

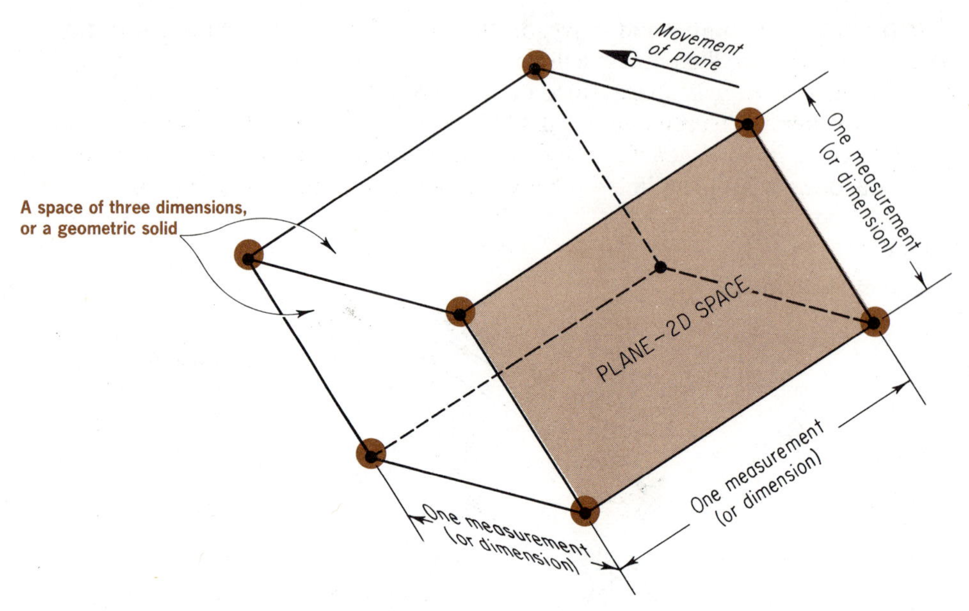

FIGURE 4

A space of three dimensions—geometric solid

Now if you take a *plane,* which is a *space of two dimensions,* and move or translate it in a direction, for instance, parallel to itself from one given position to a final position, the plane will generate a geometric solid that encloses a *space of three dimensions* (see Fig. 4).

In Fig. 4 you see that the geometric solid has a *first dimension,* determined originally by the translation of the *ideal point* from one given location to another; a *second dimension,* determined by the translation of the *line* from one given location to another; and a *third dimension,* generated by the *two-dimensional space* or plane being moved from one given location to another. A total of *three dimensions,* measured in different directions, has now been established. What you have derived is that space which is as familiar to you as the inside of a given room.

3 Points and projection planes in a three-dimensional space

Since descriptive geometry deals with space it can be defined as *the science of spatial relations and analysis.* In order to develop methods by which this analysis can take place on a sheet of drawing paper, you should represent spatial relations among geometric elements through the use of *orthographic projection.* The concept of orthographic projection involves the projection of points in a *three-dimensional space* onto planes that define the space. These planes are referred to as *projection planes* (see Figs. 5 and 6).

FIGURE 5

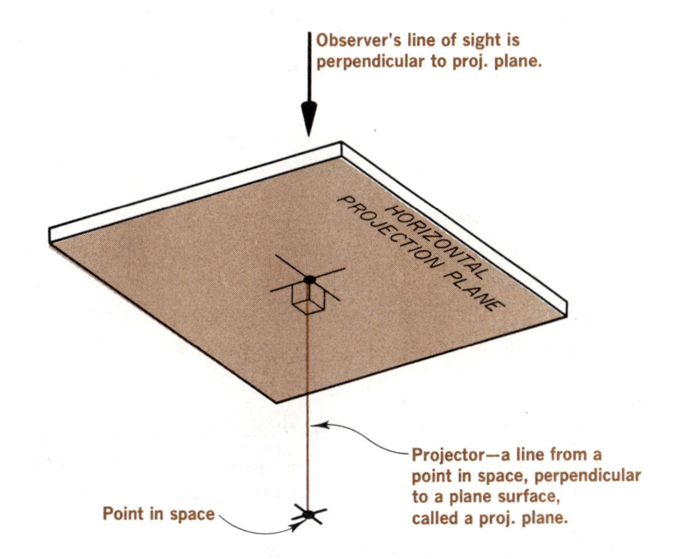

From section 2 you can see that two intersecting planes determine a three-dimensional space. If you place a point within this three-dimensional space and *project* it onto the perpendicular planes that define the space, you then obtain the projections of the point on each of the space-defining planes (see Fig. 6).

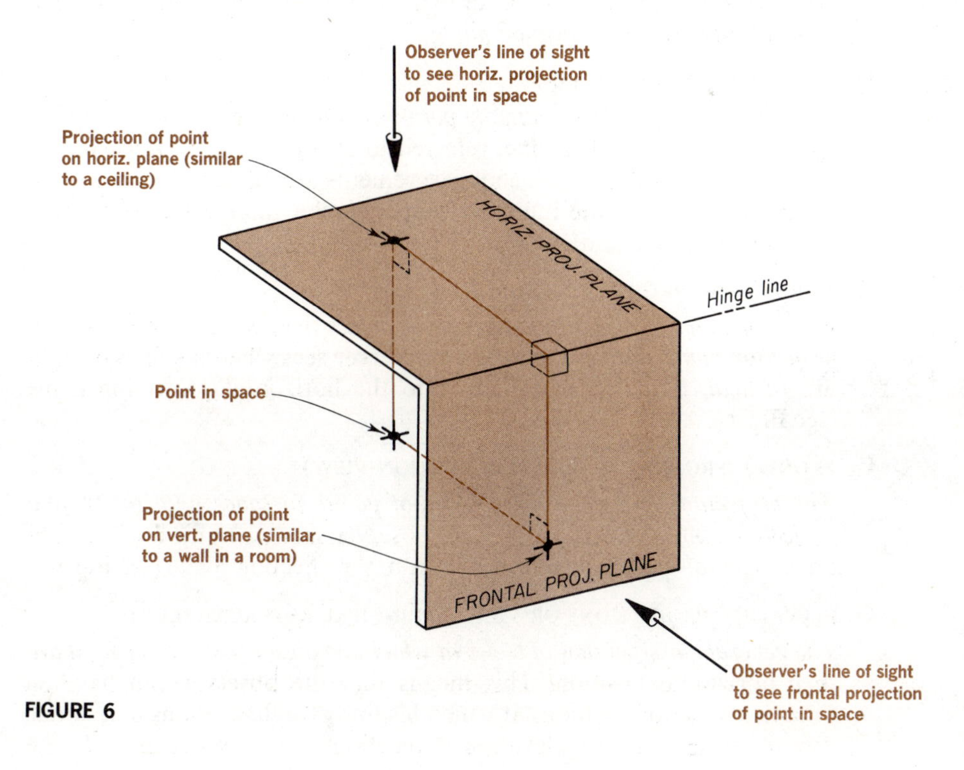

FIGURE 6

4 Definition of terms used in orthographic projection

In section 3 we referred to projections and projection planes. Following is a list of basic definitions related to the theory of orthographic projection:

A. LINES OF SIGHT

In the system of orthographic projection, the lines of sight *of an observer who is viewing a point or a series of points.* They are assumed to be *parallel straight lines.* The observer's position relative to the point or series of points he is viewing is considered to be at *infinity.*

B. PROJECTION PLANES

A plane surface that has no thickness, is transparent, and is assumed to be placed between the observer and the point or series of points that he is viewing. The position of the *projection plane* is such that it is *perpendicular* to the observer's *lines of sight.*

C. PROJECTORS

Lines that can be considered as extensions *of the observer's* lines of sight *from the projection planes to a point in space relative to the projection planes or from a series of points in space relative to the projection planes* (see Fig. 6). Actually, in solving problems graphically, the observer assumes the projectors to be drawn *from* the given point (or points) *perpendicular* to the *projection planes.*

D. REFERENCE, OR HINGE, LINES

The line common to two mutually perpendicular projection planes at their point of intersection. This line, referred to as a *reference* or *hinge line,* is used as a base line from which measurements are made to locate a point in space relative to the mutually perpendicular intersecting projection planes (see Figs. 6 and 10).

E. HORIZONTAL PROJECTION (TOP, OR PLAN, VIEW)

The projection of a point or a series of points in space onto a horizontal *projection plane.* This is the view an observer sees when his lines of sight are *vertical,* that is, perpendicular to the horizontal projection plane (see Fig. 6).

F. FRONTAL PROJECTION (FRONT ELEVATION VIEW)

The projection of a point or a series of points in space onto the frontal *projection plane.* This is the view an observer sees when his lines of sight are *horizontal* (perpendicular to the frontal projection plane [see Fig. 6]).

G. ELEVATION PROJECTIONS OR VIEWS (PRINCIPAL AND AUXILIARY)

The general classification of views at which the observer's lines of sight are in a horizontal position. This means that the observer can have an *infinite number* of positions at which his line of sight remains horizontal. Two classifications of elevation projections, or views, exist: (1) the principal elevation projections, which consist of the front, right and left sides, and rear views; and (2), auxiliary elevation projections, which include all elevation projections *other than* the principal ones. Note that elevation projection planes are *always perpendicular* to the *horizontal projection plane* (see Figs. 6 and 15).

H. AUXILIARY INCLINED PROJECTIONS OR VIEWS

The general classification of views at which the observer's lines of sight are neither vertical *nor* horizontal. The auxiliary inclined projection planes are *infinite* in number and may be *perpendicular* to other inclined projection planes or to any desired elevation projection planes (see Figs. 18 and 20).

Detailed discussions of each of the foregoing definitions follow. It is very important at this early stage that you make a conscientious effort to visualize these definitions and concepts so that they become an integral part of your thinking as you continue the study of descriptive geometry. As an aid to this visualization, draw freehand, pictorial sketches to illustrate each definition.

5 Horizontal projection plane (top or plan view)

Figure 5 shows a point in space located at a *fixed perpendicular distance* from a *horizontal projection plane.* This plane may be considered to be similar to a ceiling in a room. The distance from the ceiling—in this case, the horizontal projection plane—to the point in space has been arbitrarily fixed. This *perpendicular distance* is called a *projector,* the "foot" (or point of intersection) of which, on the horizontal projection plane, represents the projection of the point in space on the plane.

Because the given point exists in a three-dimensional space, it has *three degrees* of movement, or freedom: up and down, right and left, back and forth. The point in space in Fig. 5 has *one degree* of freedom fixed relative to a given horizontal projection plane. Two other directions, or distances, remain to be established relative to other projection planes so that this unique point in the given three-dimensional space may be identified.

6 Frontal projection plane (front or front elevation view)

To fix the given point relative to a back and forth position we introduce the *frontal projection plane,* which is *perpendicular* to the *horizontal projection plane.* We consider this plane to be similar to one wall of a room that is perpendicular to the ceiling of the same room. Referring to Fig. 6, we see that the point in space is located a specific *perpendicular distance* from the *frontal projection plane.* This means that a given point is *simultaneously* located at a *definite perpendicular distance* from the *horizontal projection plane* and at a *definite perpendicular distance* from the *frontal projection plane,* and that its only possible movement now is right or left.

7 Profile projection planes (right- and left-side views of elevations)

When we introduce a *third projection plane, simultaneously perpendicular* to the *horizontal* and *frontal projection planes,* the third possible movement of the point in space is then fixed relative to this third plane. We refer to this third plane as a *profile projection plane,* which may be placed in two positions—either to the left or right of the point as we look at it in the frontal projection. These profile projection planes may be considered similar to two opposite walls in a room which are *perpendicular* to the ceiling and to a third wall (the horizontal and frontal projection planes, respectively). It will be noted in Fig. 7 that the observer's *lines of sight* are always *perpendicular* to each projection plane. This means that the observer *moves* from one position to the other to view the horizontal, frontal, and profile projections of the points in space.

FIGURE 7

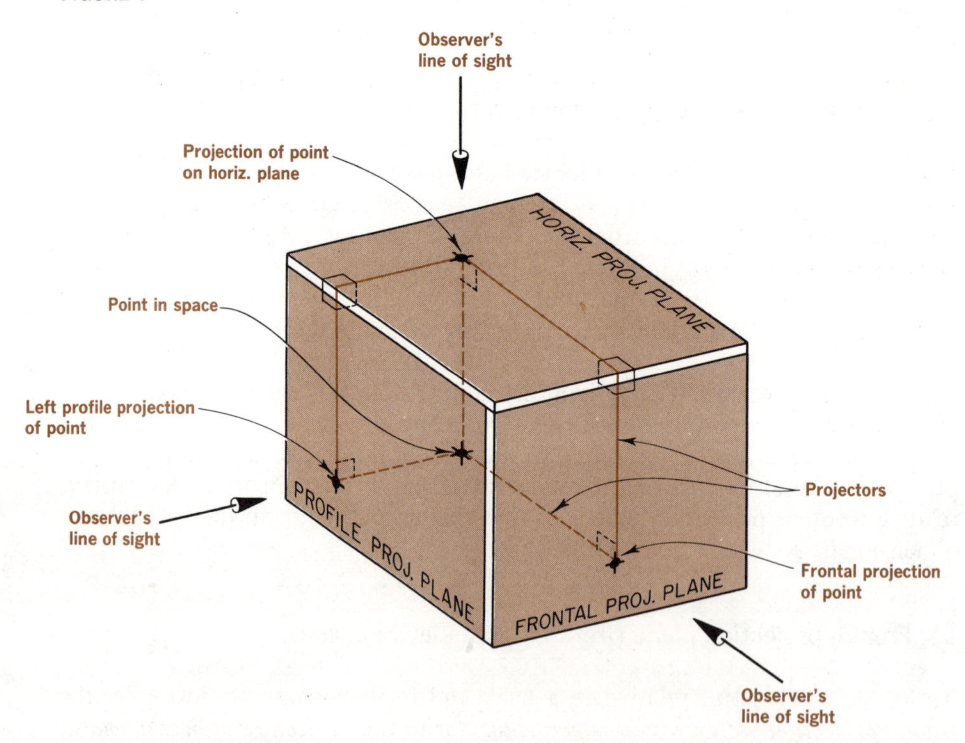

The *three mutually perpendicular planes,* consisting of horizontal, frontal, and profile projection planes, and the projectors that are drawn from a point in space *perpendicular* to each of these planes constitute the basic notion of orthographic projection on which descriptive geometry is based.

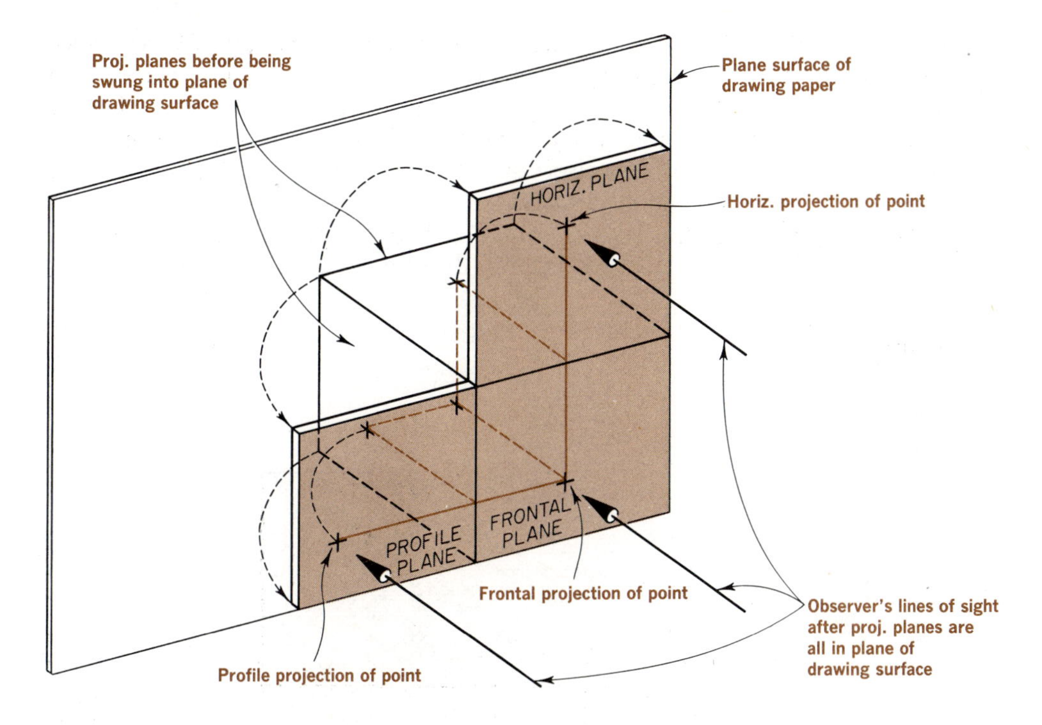

FIGURE 8

8 Basic principles of orthographic projection applied to descriptive geometry

Since *measurement* is involved in descriptive geometry, we should devise some means by which accurate perpendicular measurements from a point in space to the respective projection planes, as represented on the plane surface of a sheet of drawing paper, are possible. This is done by swinging or rotating the projection planes that are *common* to each other so that they *all* end up being in *one* plane. For example, if we imagine the horizontal projection plane to be *hinged* with the frontal projection plane at their line of intersection, and the profile projection plane also to be hinged with the frontal projection plane at their line of intersection, and if *both* these planes are swung into the *same plane* as the frontal projection plane, the result would be what is shown in Fig. 8.

Note that the observer's lines of sight remain perpendicular to the respective projection planes. Remember that after the horizontal projection plane has been swung into the same plane as the frontal plane, the observer continues to consider the horizontal projection plane as being *perpendicular* to the frontal projection plane. Also, remember that the profile projection plane is *perpendicular* to the frontal projection plane. This is basic since it can be seen that when the observer first looks at the horizontal projection plane (before the planes are swung into one plane), he sees the frontal projection plane as an *edge;* and when he looks at the profile projection plane, the frontal projection plane again appears as an *edge.* Likewise, when he looked at the frontal projection plane he saw *both* the horizontal and profile projection planes as edges. By seeing these projection planes as edges he also can see the *perpendicular distance* from the respective planes to the point in space.

Figure 9 shows the horizontal, frontal, and profile projection planes as they actually appear in the plane of a sheet of drawing paper.

FIGURE 9

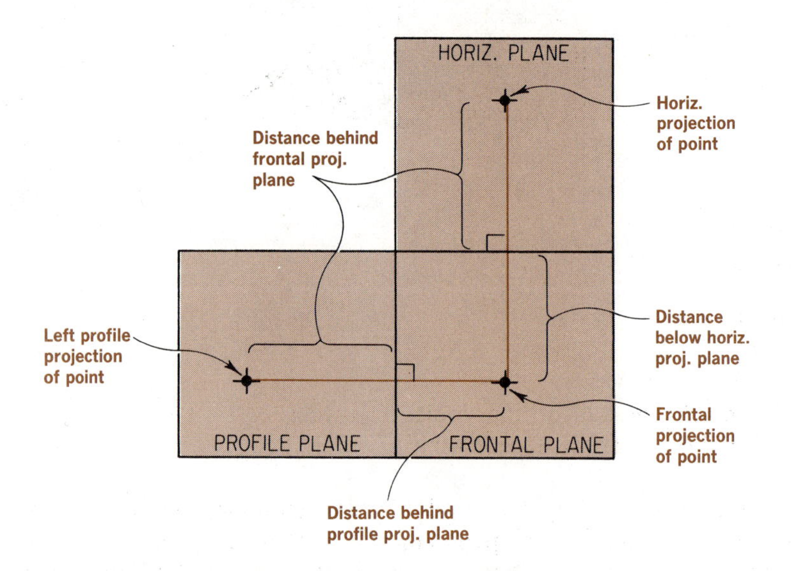

Note that the observer's lines of sight, when he is looking at each individual projection plane, appear as *points.* He can visualize this by imagining himself to be standing *behind* the lines of sight that view the horizontal, frontal, and profile projections, respectively, of the point in space. Therefore, when the observer views the frontal projection of the point, he can see the point's distance *below* the horizontal projection plane. When he views the horizontal projection of the point in space, he sees the point's distance *behind* the frontal projection plane. When he views the profile projection of the point, he again sees its distance *behind* the frontal projection plane. When he views the frontal projection of the point, he also sees the point's distance *behind* the profile projection plane, since the profile projection plane appears as an *edge* in *the frontal view.*

When the observer looks at the individual views as indicated in Fig. 9, what he visualizes is a picture of the projection planes as illustrated in Fig. 10.

FIGURE 10

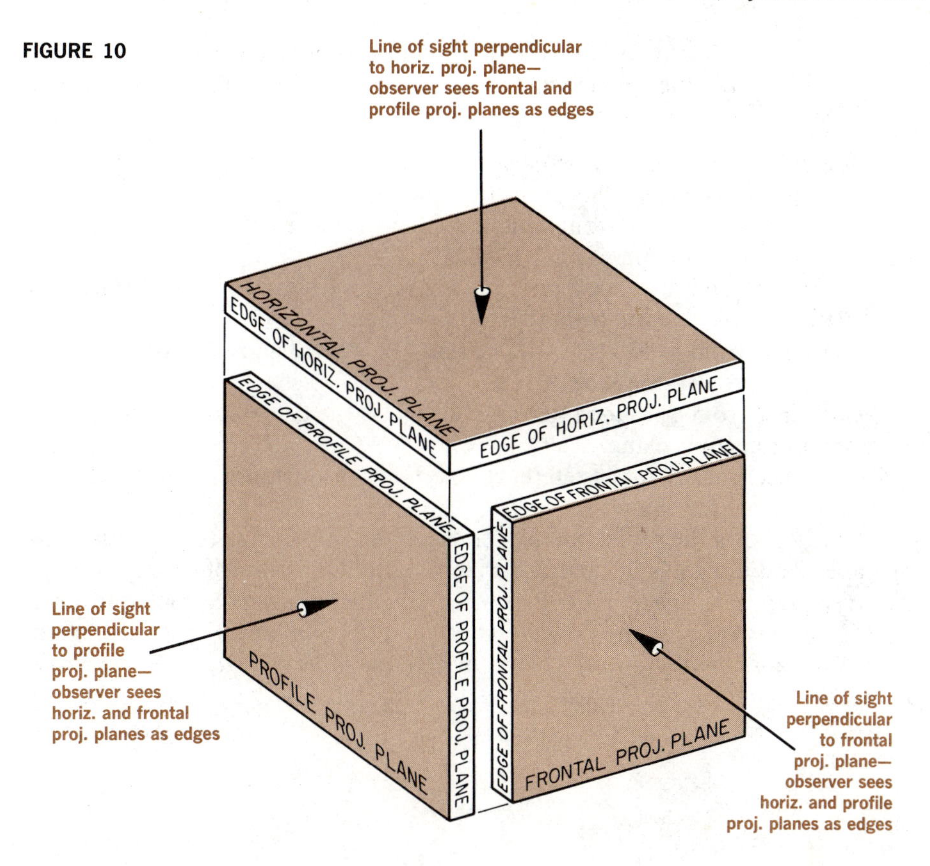

9 A system of notation used in descriptive geometry

In order to efficiently solve and analyze problems using the methods and principles of orthographic projection, we must adopt an organized system of notation for identifying points in space, the projection of points in space, projection planes, and reference lines. In this book we shall use the following notations:

1. *Actual points, lines, planes,* or *geometric solids* in space are denoted by *capital* (upper-case) letters. This type of notation is used in the pictorial representations (see Figs. 11 and 20).
2. *Projections* of *points, lines, planes,* or *geometric solids* in space are denoted by small (lower-case) letters with subindices that indicate the projection planes to which they belong (see Figs. 11 and 20).
3. *Projection planes* are identified by their lines of intersection with mutually perpendicular projection planes. These lines are referred to as *reference* (or hinge) *lines* and are noted by numerals placed on either side of them as shown in Figs. 11 and 12. Note, for example, that the horizontal projection plane receives the number 1 and the frontal projection

plane receives the number 2. This will be kept standard throughout the book. Any other plane will receive the number 3 and additional planes will proceed in numerical order.

Figure 11 shows a point *P* in space projected onto horizontal, frontal, and profile projection planes. The projection on the horizontal plane is identified as p_1, the projection on the frontal plane is identified as p_2, and the projection on the profile projection plane is identified as p_3.

The perpendicular distance from p_1 to the reference line 1–2 is the distance of point *P* in space *behind* the frontal projection plane.

The distance from p_2 to the reference line 1–2 is the distance of point *P below* the horizontal projection plane.

The distance from p_3 to the reference line 2–3 is the distance of point *P behind* the frontal projection plane.

The distance from p_2 to the reference line 2–3 is the distance of point *P* behind the profile projection plane.

Figure 12 shows the projection planes swung into one plane (that of the drawing paper) and indicates the actual horizontal, frontal, and profile projections of the given point *P* in space. Note that when you are an observer and your line of sight is *perpendicular* to the *frontal plane,* the reference line 1–2 represents the *edge* of the *horizontal projection plane.* When your line of sight is *perpendicular* to the *horizontal projection plane,* the reference line 1–2 represents the *edge*

FIGURE 11

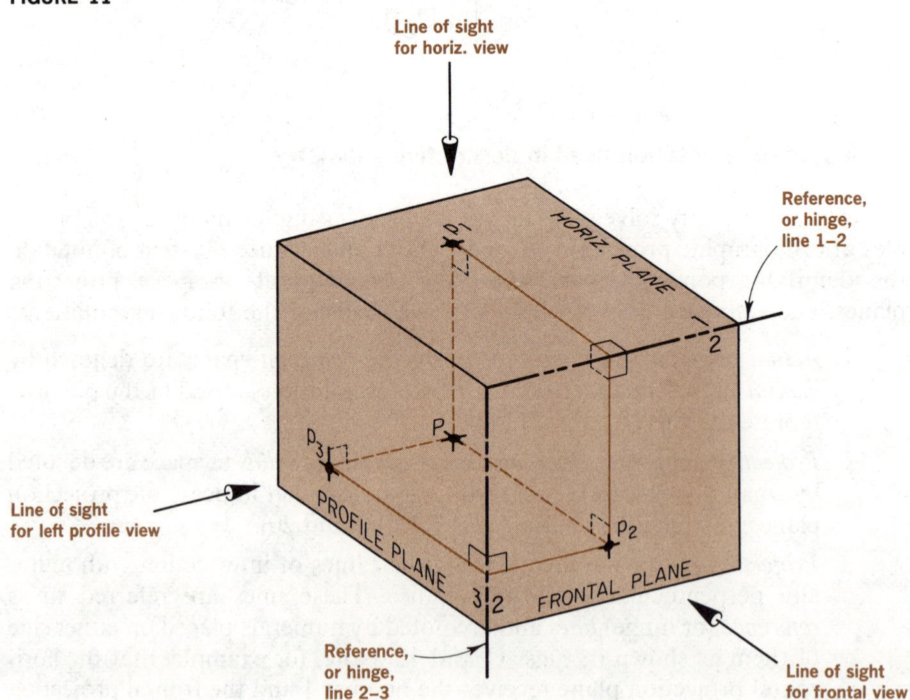

FIGURE 12

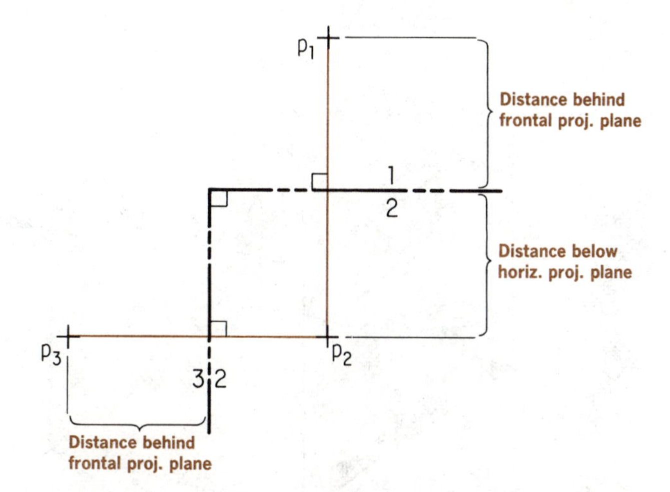

of the *frontal projection plane.* When your line of sight is *perpendicular* to the *profile projection plane,* the *frontal projection plane* appears as an edge. Also note that the outline of the projection planes has been eliminated, since actually these planes can have any size or shape (always remaining flat, of course). As far as solving actual problems is concerned, the actual shape and size of the planes is immaterial as long as their *line of intersection* (the reference line*) is known and located.

10 Principal elevation views (or projections)

Figure 13 shows the principal (or common) elevation views, which consist of the front, right and left sides, and rear elevations.

The main characteristics of elevation views are as follows:

1. The projection planes of elevation views are always *perpendicular* to the *horizontal* projection plane.
2. The lines of sight that are *perpendicular* to the *elevation projection planes* are therefore always *horizontal.*
3. Since the lines of sight at elevation views are always *horizontal,* upward and downward distances, or "*elevations,*" are seen. This means that *perpendicular* distances below (or above) the horizontal projection plane are *always seen* in *elevation views.*

Figure 14 shows the principal elevation projection planes swung into the plane of the drawing paper (see the RL notations).

---

* For the rest of this book, when referring to a particular reference line, the abbreviation RL will be used.

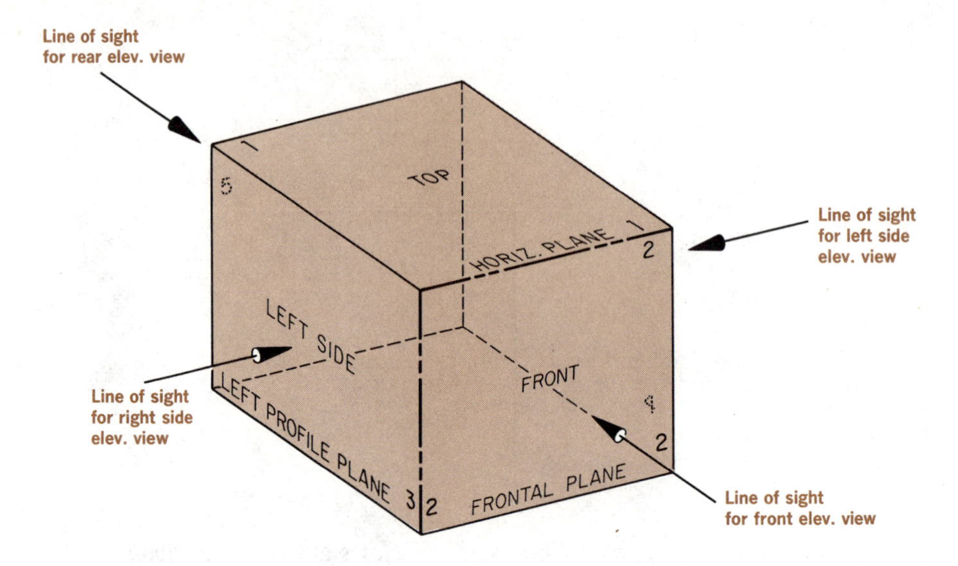

FIGURE 13

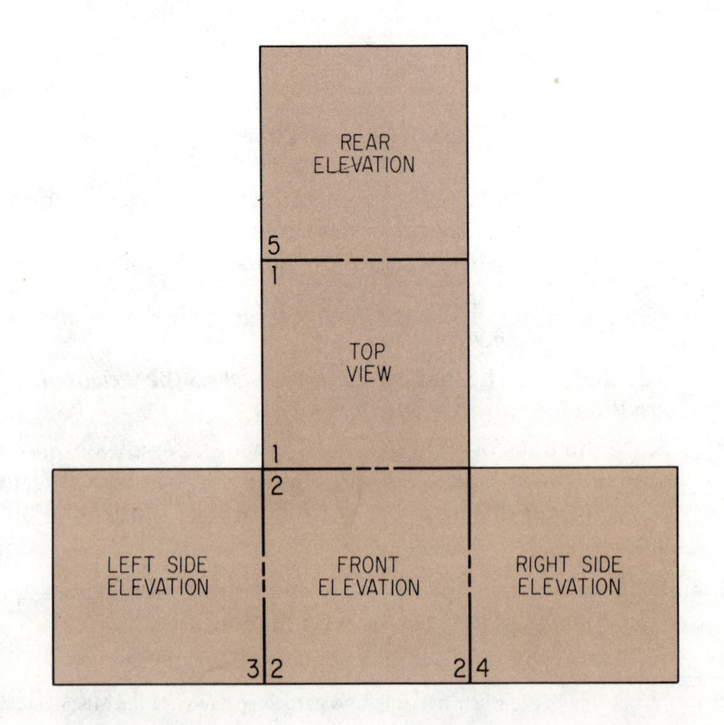

FIGURE 14

11 Auxiliary elevation views (or projections)

Figure 15 shows a series of projection planes that are *perpendicular* to the horizontal projection plane. By definition, then, these planes are all *elevation projection planes.* One plane, the front elevation or frontal projection plane, is denoted as such, which means that automatically all the other elevation projection planes in this illustration are considered to be auxiliary elevation projection planes since they are different from the principal elevation views or projection planes. Note that the observer's lines of sight are perpendicular to all the elevation projection planes and, therefore, they are all *horizontal.* This means that an observer can take an *infinite number* of positions in space at which his line of sight remains horizontal and at which he will see an *infinite number of elevation views,* all of which will be auxiliary elevation views except those that have been defined as principal elevation views. Note also that whenever the observer looks at an elevation projection plane, he will *always see the horizontal projection plane as an edge.*

FIGURE 15

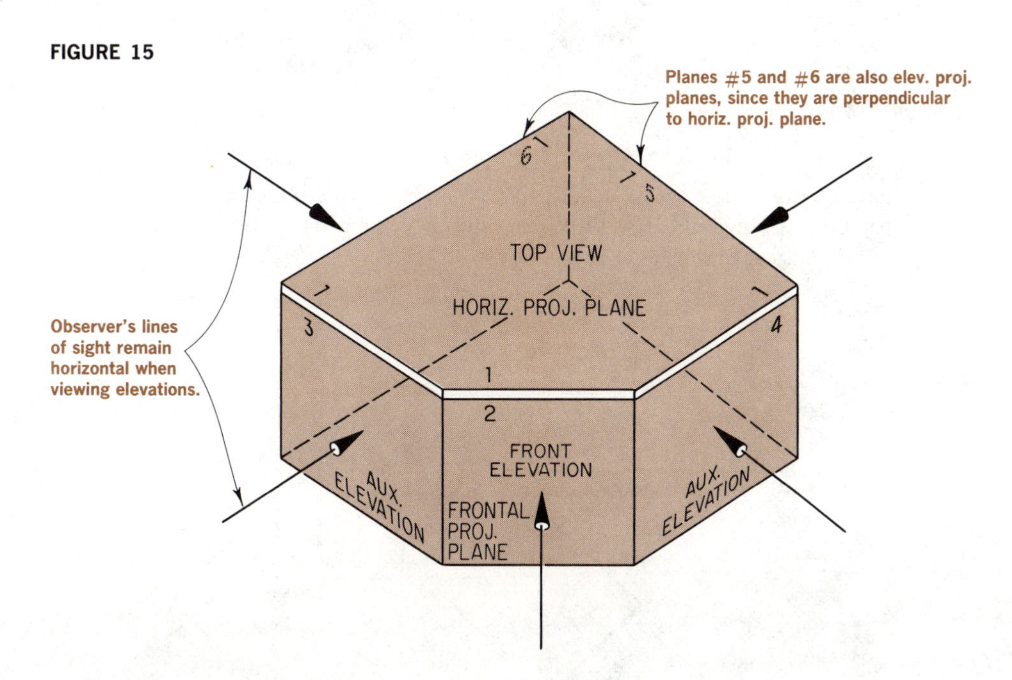

Figure 16 shows all the elevation projection planes swung into a plane common with the horizontal projection plane, which may be considered to be the plane of the drawing paper. The planes are hinged at their lines of intersection (the reference, or hinge, lines), and the lines of intersection for each elevation projection plane and for the horizontal projection plane are identified by numbers on either side.

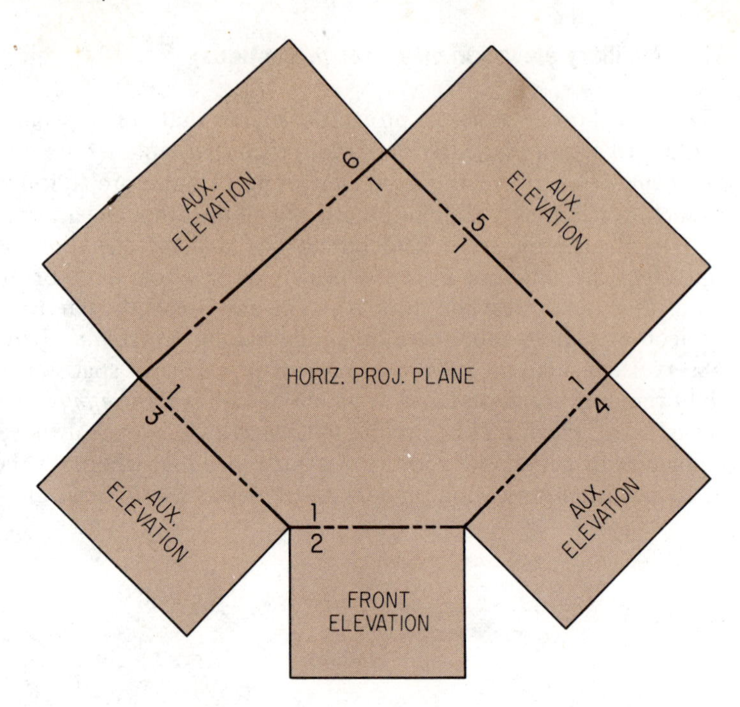

FIGURE 16

When solving problems in descriptive geometry, we eliminate the outlines of the projection planes and use only the reference, or hinge, lines to identify the various projection planes (see Fig. 17).

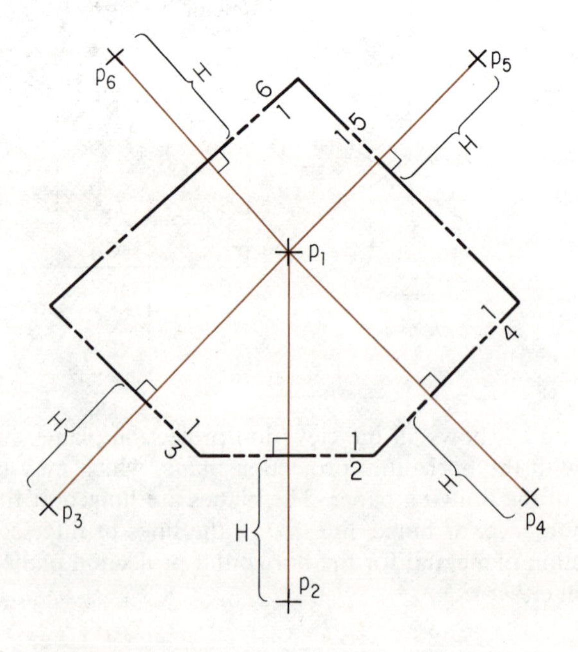

FIGURE 17

Figure 17 shows the horizontal, frontal, and auxiliary elevation projections of a point P. The projectors from one view to another are *perpendicular* to the reference lines. The point P in space is located a perpendicular distance H below the horizontal projection plane. We see this distance in *every* elevation view since we see the horizontal projection plane *as an edge* in *all elevation views.*

Assume that the horizontal projection p_1 and frontal projection p_2 are given and that you are to determine the auxiliary elevation view #5 of the point P. The following procedure is necessary.

a. Construct a projector from p_1 perpendicular to RL 1–5.

b. In the frontal projection view (#2) measure the perpendicular distance H from RL 1–2 to the frontal projection p_2 (measurement is made with dividers).

c. Transfer the distance H (with dividers) to the auxiliary elevation view by measuring it along the perpendicular projector starting from RL 1–5. The point thus located is noted as p_5 and is the required auxiliary elevation view.

12 Auxiliary inclined views (or projections)

Up to this point we have been dealing with lines of sight that have been vertical (lines of sight obtained by looking down on the horizontal projection plane) or horizontal (lines of sight obtained by looking out at elevation projection planes). When the observer's line of sight is *neither* horizontal nor vertical but inclined at

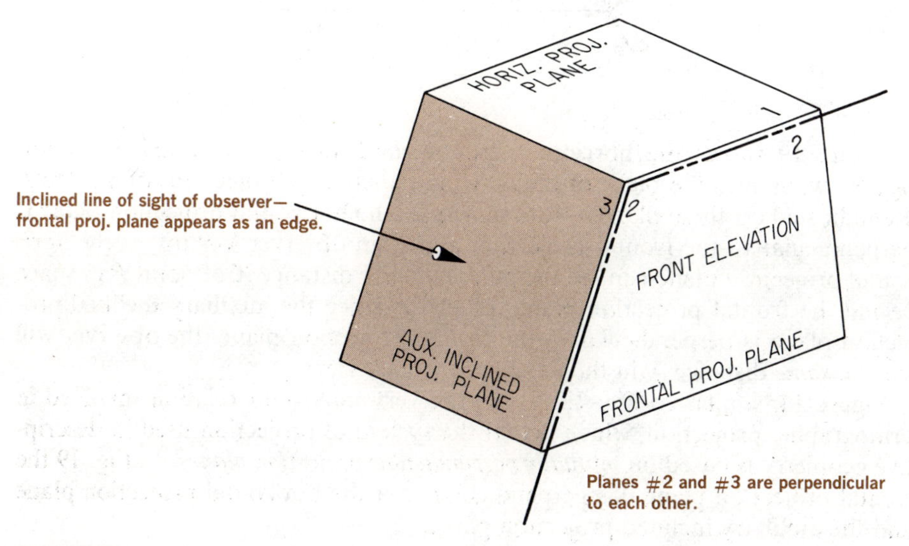

FIGURE 18

some angle, the projection plane, which is placed between the observer and the point he is viewing by orthographic projection principles, makes 90° with the line of sight and therefore is also inclined relative to the horizontal and frontal projection planes (see Fig. 18).

In Fig. 18 you see an auxiliary inclined projection plane that is perpendicular to the frontal projection plane. This means that when you look in a direction perpendicular to the auxiliary inclined projection plane, you will see the *frontal projection plane as an edge.* Therefore, if you place a point behind the frontal projection plane, you will be able to see the perpendicular distance from the point to the frontal projection plane (see Fig. 19).

FIGURE 19

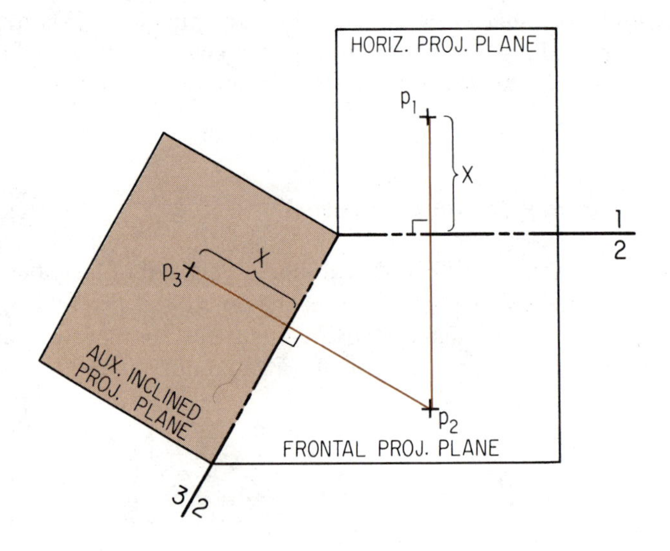

Figure 19 shows the horizontal, frontal, and auxiliary inclined projection planes swung into the plane of the drawing paper. The projections of a point P are indicated on these planes. Note that since the horizontal projection plane is perpendicular to the frontal projection plane, an observer looking at the horizontal projection plane can see the perpendicular distance X of point P in space *behind* the frontal projection plane. Likewise, since the auxiliary inclined projection plane is perpendicular to the frontal projection plane, the observer will see the *same* distance X in the auxiliary inclined view.

Figure 14, similar to Fig. 12, shows the very important relation involved in orthographic projection, which is that the system of projection used in descriptive geometry is based on *mutually perpendicular projection planes.* In Fig. 19 the frontal projection plane has perpendicular to it the horizontal projection plane and the auxiliary inclined projection plane.

13 Key to the system of orthographic projection used in descriptive geometry

The system of orthographic projection used in descriptive geometry depends upon mutually perpendicular projection planes and upon projectors that are perpendicular to the projection planes. Figure 20 pictorially presents the key to this system. It is vital that you study and visualize Fig. 20 and also Fig. 21. Visualize *each plane* and *each projection* and fix in your mind the ways in which they relate to each other. Once you understand the system you should be able to proceed rapidly through the basic material in this book.

Figure 20 shows a series of five consecutive projection planes in which two projection planes are *always* perpendicular to a third. In analyzing this figure we see that we can progress from one projection plane to another in directions that are *perpendicular* to these planes. For example, the familiar frontal projection plane has perpendicular to it the familiar horizontal projection plane and an auxiliary inclined projection plane. The auxiliary elevation projection plane has perpendicular to it the horizontal projection plane and another auxiliary inclined projection plane #4. We see from the figure that point P in space is located at a perpendicular distance H *below* the horizontal projection plane. We also see this distance H when observing the frontal projection plane in order to locate the frontal projection p_2 of point P in space.

FIGURE 20

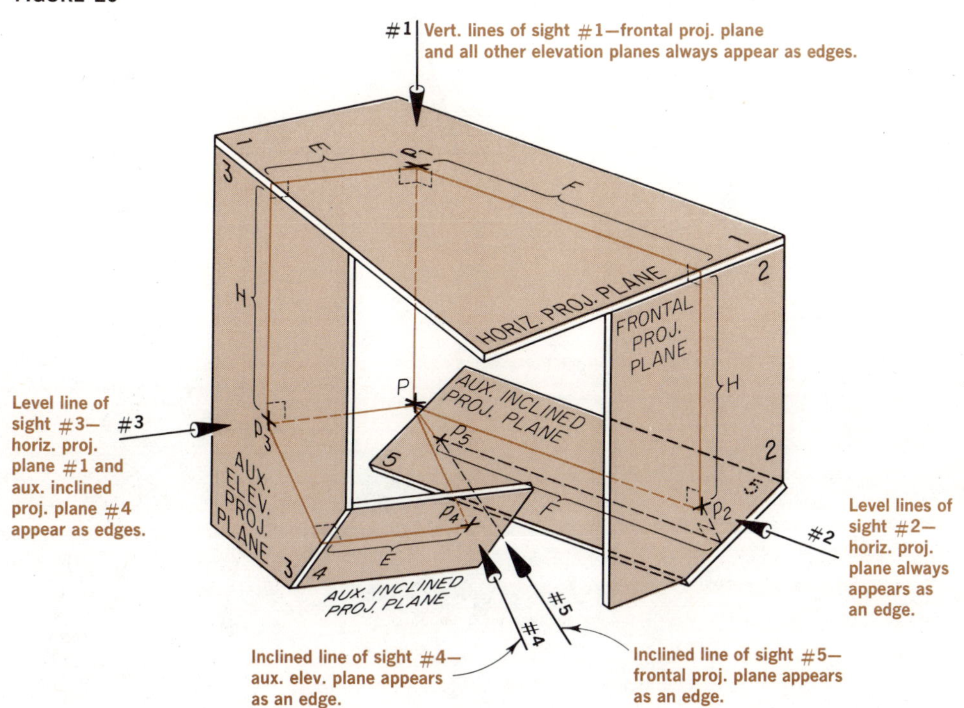

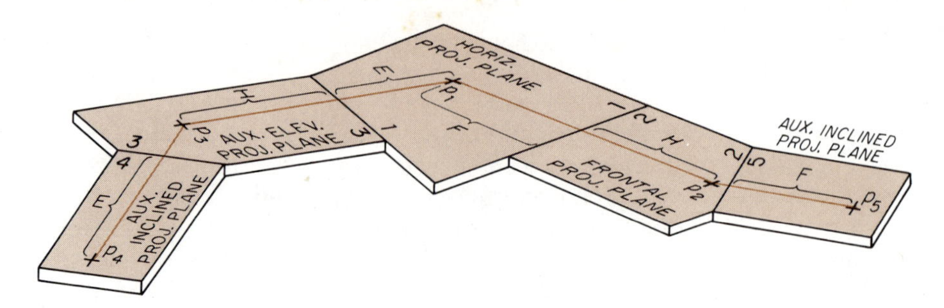

FIGURE 21

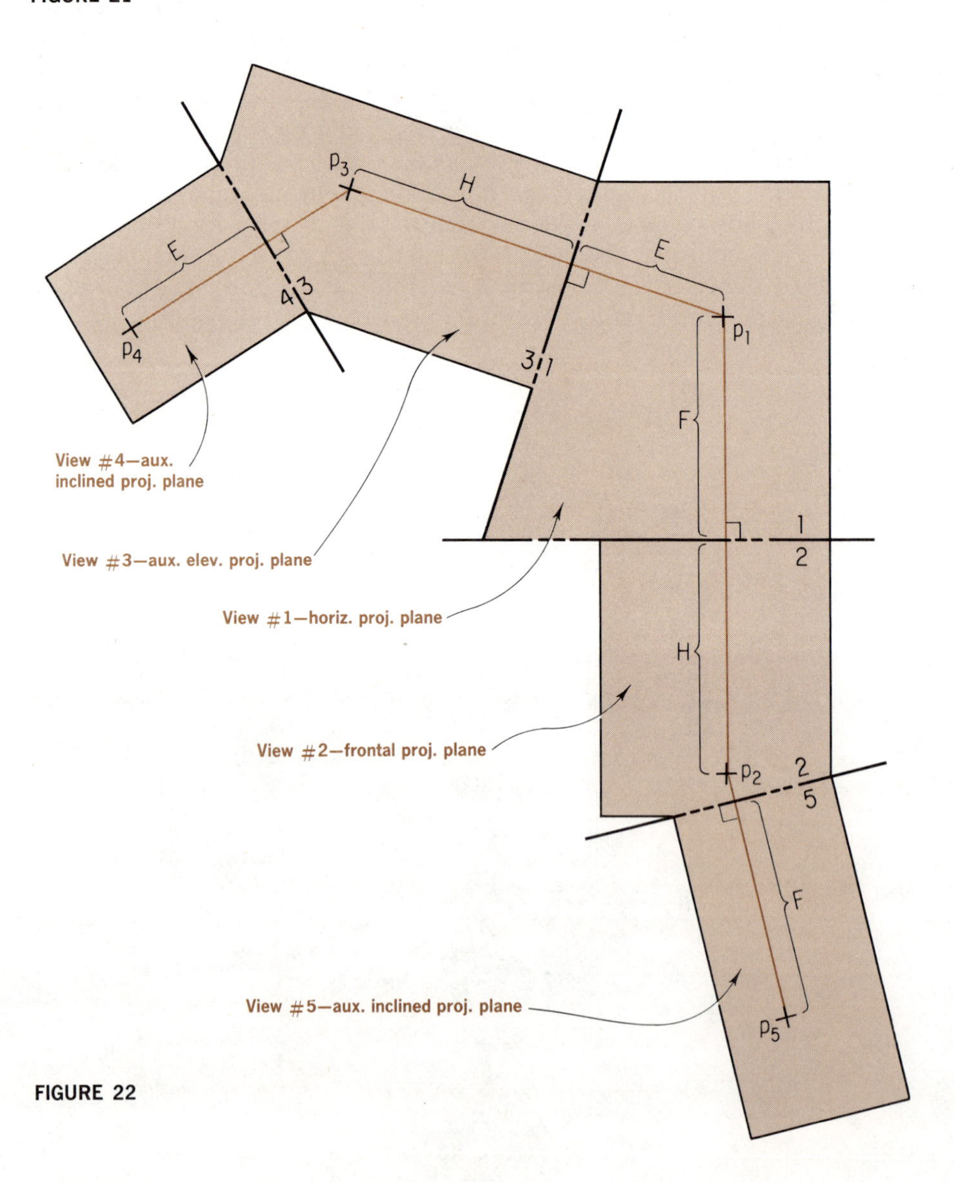

FIGURE 22

We note that point P in space is located a perpendicular distance E *behind* the auxiliary elevation plane #3. We see distance E when viewing the horizontal projection plane and the auxiliary inclined projection plane #4.

Further examining Fig. 20 we see that point P in space is located at a perpendicular distance F *behind* the frontal projection plane. We see the distance F when viewing the horizontal projection plane and the auxiliary inclined projection plane #5.

Figure 21 shows all the projection planes swung into one plane; each plane is hinged at the reference line, which is determined by the line of intersection of adjacent planes. Notice the *consecutive nature* of the system. Notice again that the frontal projection plane has on either side of it the horizontal projection plane and the auxiliary inclined projection plane, while the auxiliary elevation projection plane #3 has on either side of it the horizontal projection plane #1 and the auxiliary inclined projection plane #4.

Figure 22 shows the consecutive projection planes as they appear in the plane of a sheet of drawing paper. Let us analyze this figure and in so doing continue to visualize the pictorial presentation shown in Fig. 20, which is a picture of the actual space condition. From Fig. 22 we see distance H of point P *below* the horizontal projection plane (view #2—frontal elevation projection plane—and view #3—auxiliary elevation projection plane). This occurs in both views because the reference lines represent the *edge view* of the *horizontal projection plane.*

Note the perpendicular distance F of P *behind* the frontal projection plane in views #1 (horizontal projection plane) and #5 (auxiliary inclined projection plane). RL 1–2 represents the *edge view* of the frontal projection plane in view #1 and RL 2–5 represents the *edge view* of the frontal projection plane in view #5.

The perpendicular distance E of P *behind* the auxiliary elevation projection plane #3 is seen in view #1 (horizontal projection plane) and view #4 (auxiliary inclined projection plane). RL 1–3 represents the *edge view* of the auxiliary elevation projection plane in view #1 and RL 3–4 represents the edge view of the auxiliary projection plane in view #4.

If the horizontal and vertical projections p_1 and p_2 of a point P in space were given and if it were necessary to locate the auxiliary elevation projection (view #3) and the auxiliary inclined projection (view #4) of the point, you would apply the following procedure (refer to Fig. 23):

a. Draw RL 1–3 in the same position as shown in Fig. 22.
b. Construct a *perpendicular* projector from p_1 to RL 1–3 and extend it beyond the reference line.
c. Measure the perpendicular distance H (with dividers) that the point P is below the horizontal projection plane. (This distance is seen in the frontal projection plane, which shows p_2.)
d. Transfer this distance H to projection plane #3 (auxiliary elevation), measuring from RL 1–3 along the projector as shown in Fig. 23.

e. Note the projection of point P in view #3 as p_3. Notice that when "information" in terms of a distance H below the horizontal plane was required in view #3, it was necessary to go *back* to view #1 and then to view #2 where the measurement of distance H was made. In other words, looking at view #3 you had to go back *two views* to get the measurement for H.

f. Draw RL 3–4 in the same position as shown in Fig. 22.

g. Construct a perpendicular projector from p_3 to RL 3–4 and extend it beyond this reference line (refer to Fig. 23).

h. The perpendicular distance that the point P is behind the auxiliary elevation plane #3 is equal to E and is seen in view #1. Measure distance E with dividers and transfer this distance to view #4 measuring distance E along the perpendicular projector from RL 3–4 into view #4.

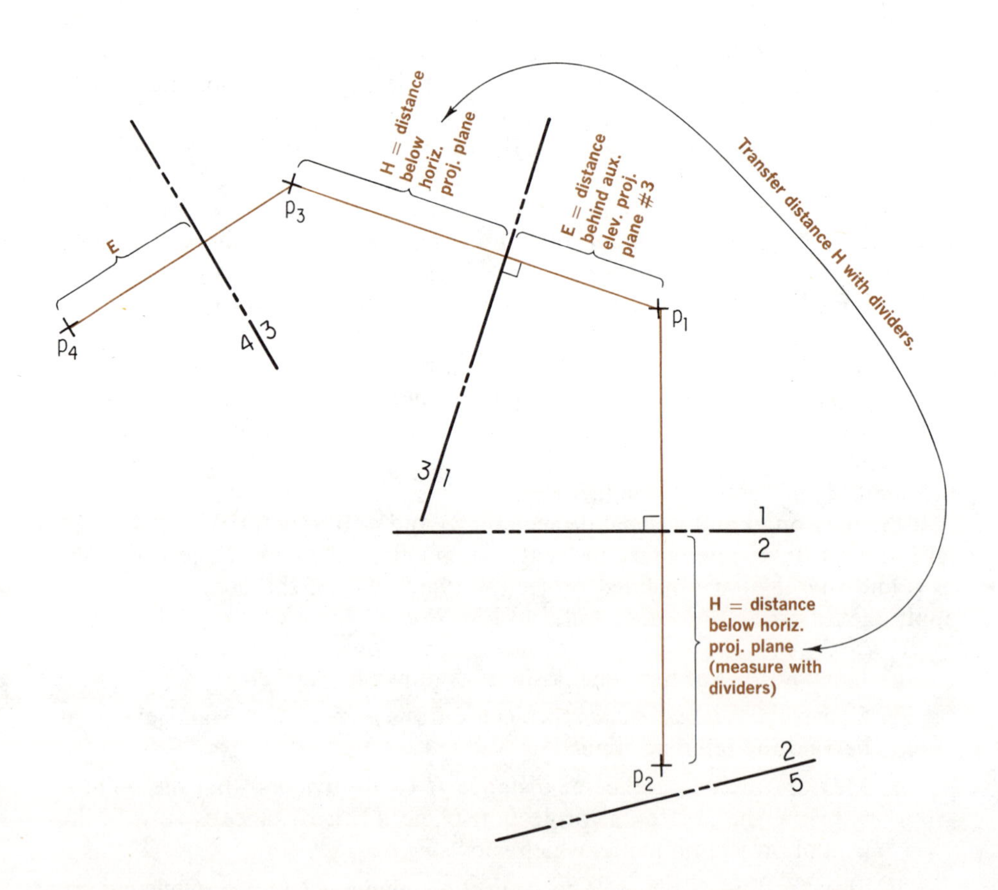

FIGURE 23

i. Properly note or record the projection of P in view #4 as p_4. Notice that again you were interested in getting a measurement for view #4 and had to *go back* to views #3 and #1. In other words, go *back two views* in order to get the distance E.

(NOTE: The procedure of working in one view and going back two consecutive views to determine a distance applies to all problem-solving in descriptive geometry.)

Keeping the foregoing procedure in mind, determine how you locate the projection of P in view #5 (see Fig. 23).

Sample Quizzes and Practice Problems

At the end of each chapter there are a number of "Sample Quizzes" and "Practice Problems" to be answered and solved. The quizzes are based directly on the material in each chapter, so check your answers against the presentation in the text. A number of quizzes require freehand approximate "solutions" which are used principally as a means for you to become familiar with the methodology of descriptive geometry.

The practice problems are also based on the text in each chapter (and combinations of the contents of previous chapters). Usually these problems require an individual approach on your part to determine the best method of attacking the problem. You will need actual accurate graphical constructions, using drawing instruments, to solve these problems. Answers to a representative number of practice problems are given. When numerical answers are required you should allow a discrepancy of 1 or 2% between your answer and the given answer since the accuracy of an answer depends on the accuracy with which a problem is reproduced and the quality and correctness of the graphical construction done.

Practice problems have been reduced (scale: 1 in. = $\frac{7}{16}$ in.) but they may be reproduced on 8 in. × 10 in. cross-section paper having $\frac{1}{8}$-in. squares. By counting the squares in each given problem setup in the textbook, you can accurately reproduce these problems on the cross-section paper.

Location of given data in all practice problems is based on 8 squares per inch with the *left end* of the RL 1–2 located relative to the lower left-hand corner (as origin O) of the cross-section paper (see the example of the practice problem figure). The left-hand end of RL 1–2 in this figure is at $X = 2$ and $Y = 5$. The remaining data (in this case points of plane ABC in the horizontal and frontal projections) are located relative to the left-hand end of RL 1–2 counting squares. For example, point A is 8 squares to the right of, 8 squares below, and 8 squares behind the left-hand end of RL 1–2. Points B and C of plane ABC are located in the same manner. This data location procedure is typical for all problems presented in this book. (Partial guides are generally indicated at the end of each chapter.)

EXAMPLE OF A PRACTICE PROBLEM LAYOUT OF GIVEN DATA

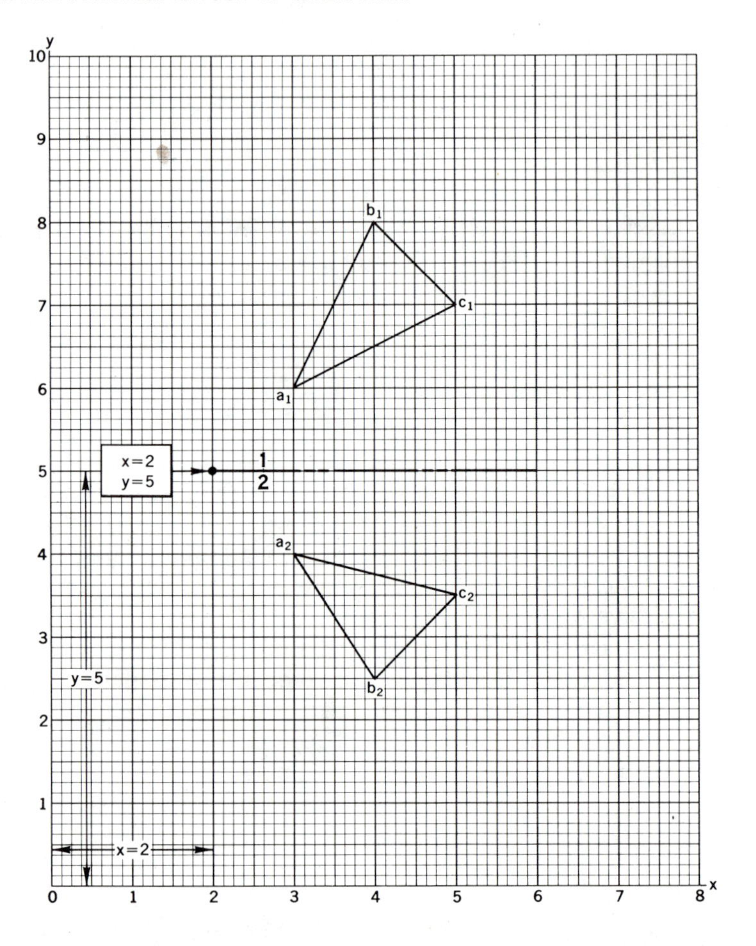

Problems should be laid out accurately using well-sharpened 6H pencils for construction and projection lines. These lines should be light in weight but visible. Given or constructed points, lines, planes, and solids should be dark and sharp in all views. It is recommended that 2H or 3H pencils be used for these lines. *All* given and constructed information must be carefully noted according to the system of notation adopted in this book. Remember, you are not only solving these problems for yourself; you are also "communicating" your solution to your instructor.

At two places in the book "Sample Tests" are presented for you to solve. These tests in each case cover groups of topics. Solutions to these tests are given in the Appendix. Also in the Appendix is a sample final examination (with solutions), designed to cover most of the basic principles of descriptive geometry presented in this book in a comprehensive, challenging manner. Set a time limit of three hours in which to complete this sample final examination and then check your results with the solutions given in the Appendix.

SAMPLE QUIZZES

1. Briefly define the following:

a. Projection plane.
b. Reference, or hinge, line.
c. Projector.
d. Projection.

2. A. What are the relative positions of observer, object, and projection planes in orthographic projection?

B. What is the direction of the line of sight for:

a. Horizontal or top view?
b. Frontal or front elevation view?
c. Any elevation view?

3. How does an inclined view differ from horizontal and elevation views?

4. A. Inclined views (or projections) are taken directly from what projection planes?

B. Inclined views (or projections) are never taken directly from which projection planes? Why?

5. State in your own words the basic concept of orthographic projection.

PRACTICE PROBLEMS

1. Given Data: RL 1–2, RL 1–3, and RL 2–4.

Required: A. Referring to RL 1–2, RL 1–3, and RL 2–4, answer the following questions:

a. When an observer looks perpendicular to frontal plane #2, what does RL 1–2 represent?

b. When an observer looks perpendicular to horizontal plane #1, what does RL 1–2 represent?

c. When an observer looks perpendicular to auxiliary elevation plane #3, what does RL 1–3 represent?

d. When an observer looks perpendicular to horizontal plane #1, what does RL 1–3 represent?

e. When an observer looks perpendicular to frontal plane #2, what does RL 2–4 represent?

f. When an observer looks perpendicular to profile elevation plane #4, what does RL 2–4 represent?

B. Make a pictorial freehand sketch of planes #1, #2, #3, and #4 as they would appear in space before being unfolded into one plane.

PROBLEM 1

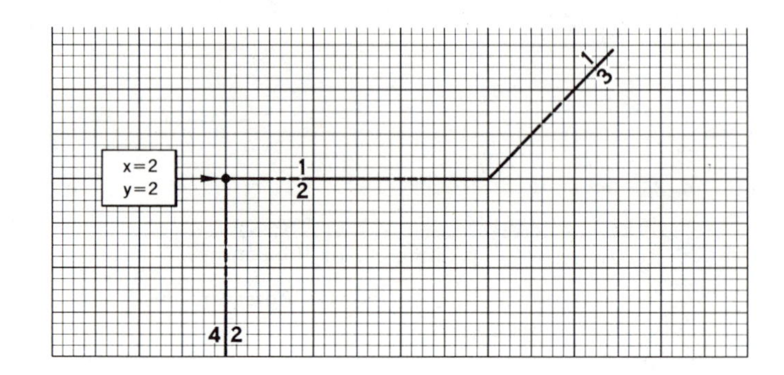

2. Given Data: RL 1–2, RL 2–3, and RL 3–4.

Required: A. Referring to RL 1–2, RL 2–3, and RL 3–4, answer the following questions:

a. When an observer looks perpendicular to plane #2, what does RL 1–2 represent?

b. When an observer looks perpendicular to plane #1, what does RL 1–2 represent?

c. When an observer looks perpendicular to plane #2, what does RL 2–3 represent?

d. When an observer looks perpendicular to plane #3, what does RL 2–3 represent?

e. When an observer looks perpendicular to plane #3, what does RL 3–4 represent?

f. When an observer looks perpendicular to plane #4, what does RL 3–4 represent?

PROBLEM 2

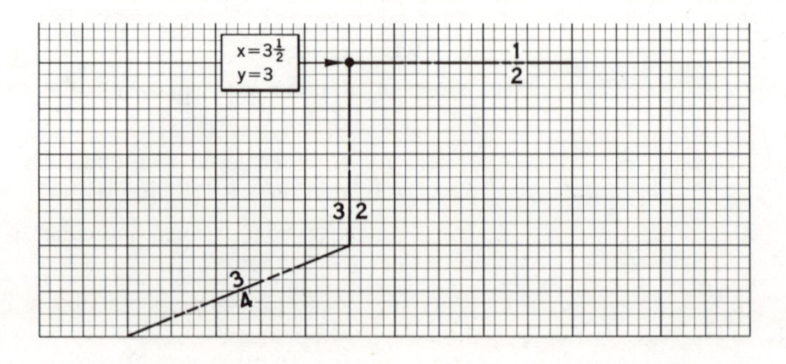

B. Identify by name planes #1, #2, #3, and #4.

C. Make a pictorial freehand sketch of planes #1, #2, #3, and #4 as they appear in space before being unfolded into one plane.

D. Make a paper model of planes #1, #2, #3, and #4 as they appear in space before being unfolded into one plane.

3. Given Data: A pictorial representation of four projection planes (#1, #2, #3, and #4).

Required: A. Represent (approximately) the given projection planes by their reference lines as they would appear unfolded into one plane.

B. Identify all the given projection planes by name.

PROBLEM 3

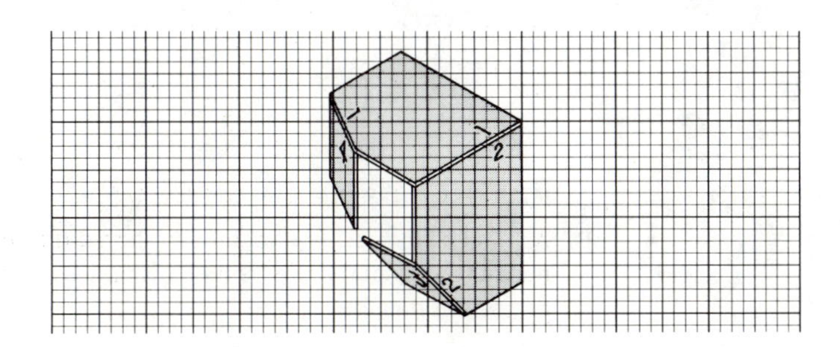

4. Given Data: RL 1–2, RL 1–3, RL 3–4, RL 2–5, RL 5–6, and RL 6–7.

Required: A. Indicate what each reference line represents when an observer looks perpendicular to planes #1, #2, #3, #4, #5, #6, and #7, respectively.

B. Identify by name planes #1, #2, #3, #4, #5, #6, and #7.

PROBLEM 4

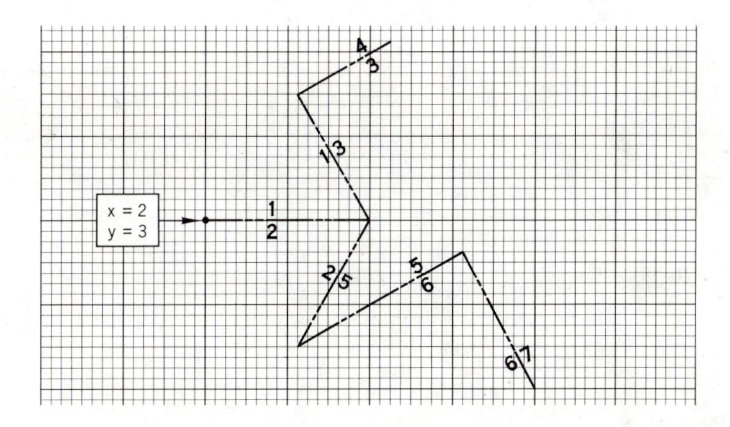

5. Given Data: Frontal and left profile projections of point *P* and RL 1–2, RL 2–3, and RL 1–4.

Required: The horizontal projection p_1 and rear elevation p_4 of point *P*.

PROBLEM 5

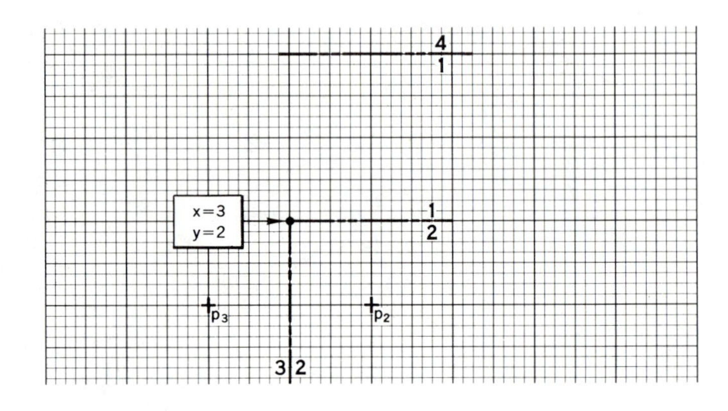

6. Given Data: A. RL 1–2, RL 1–3, RL 2–4, and RL 4–5.

B. A point *P* located $\frac{1}{2}$ in. below the horizontal projection plane, 1 in. behind the frontal projection plane, and 1 in. behind the right profile projection plane.

Required: The right profile projection p_4, auxiliary elevation projection p_3, and the auxiliary inclined projection p_5 of the given point *P*.

PROBLEM 6

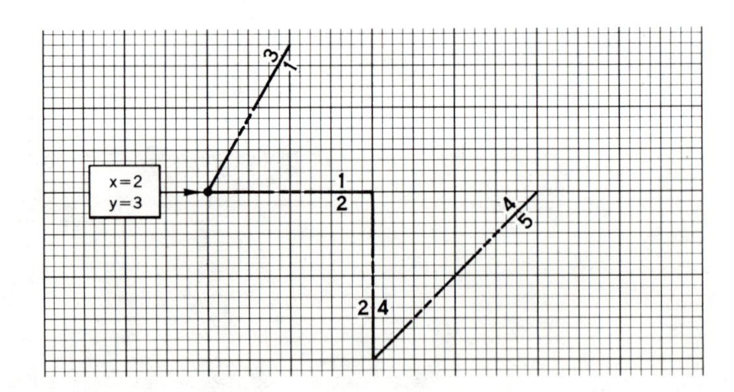

7. Given Data: A. RL 1–2, RL 1–3, and RL 2–4.

B. The frontal projection p_2 and auxiliary elevation projection p_3 of point *P*.

Required: The horizontal projection p_1 and left profile projection p_4 of point *P*.

PROBLEM 7

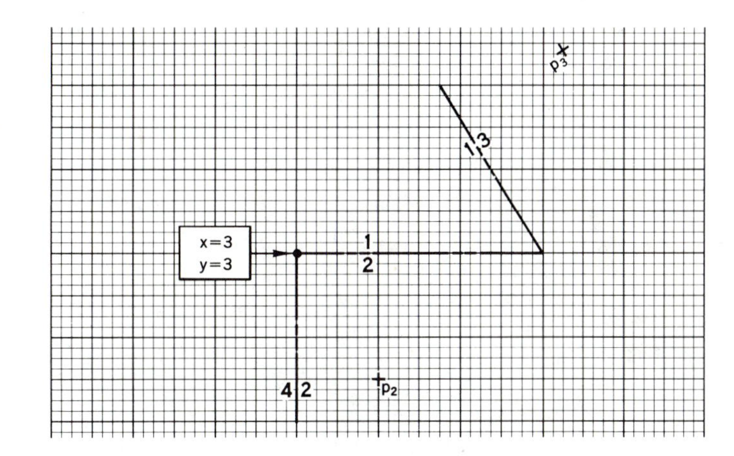

8. Given Data: A. RL 1–2, RL 1–3, RL 3–4, and RL 2–5.

B. Frontal projection p_2 and auxiliary inclined projection p_4 of point P.

Required: The horizontal projection p_1, auxiliary elevation projection p_3 and right profile projection p_5 of point P.

PROBLEM 8

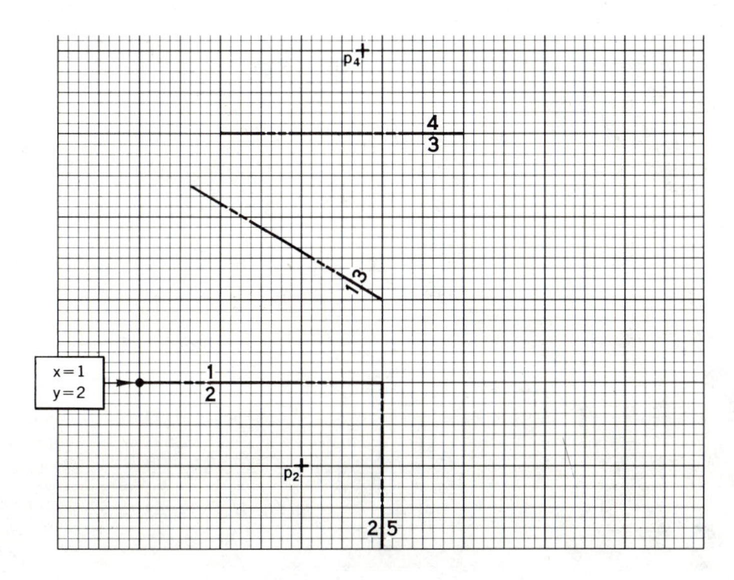

9. Given Data: A. RL 1–2, RL 1–3, RL 3–4, and RL 2–5.

B. The following spatial descriptions of points P and Q:

a. Point P is located $\frac{1}{2}$ in. below the horizontal projection plane, $1\frac{1}{2}$ in. behind the frontal projection plane, and $1\frac{1}{4}$ in. behind the left profile projection plane.

b. Point Q is located $1\frac{1}{2}$ in. below the horizontal projection plane, $\frac{1}{4}$ in. behind the frontal projection plane, and 2 in. behind the left profile projection plane.

Required:

A. The left profile projections (p_5 and q_5) of points P and Q.

B. The auxiliary elevation projections (p_3 and q_3) of points P and Q.

C. The auxiliary inclined projections (p_4 and q_4) of points P and Q.

D. The distance of point Q in front of and below point P.

E. What is the actual shortest distance between points P and Q? Do any of the views show this distance directly?

PROBLEM 9

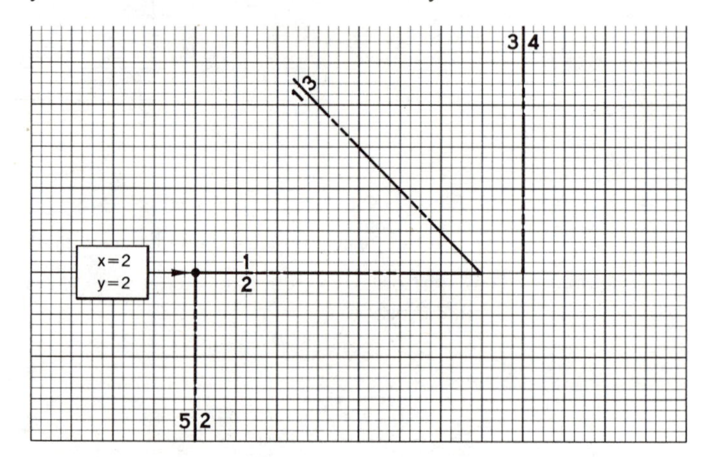

10. Given Data: A. RL 1–2, RL 1–3, and RL 2–4.

B. The following spatial description of point A:

Point A is $1\frac{1}{2}$ in. in front of the frontal projection plane, 1 in. above the horizontal projection plane, and 2 in. behind the left profile projection plane. (NOTE: Carefully indicate proper notation numbers for each given projection.)

Required: The auxiliary elevation projection a_3 and left profile projection a_4 of point A.

PROBLEM 10

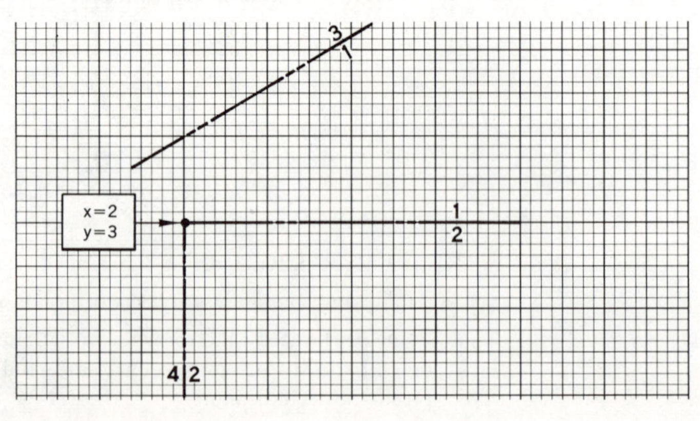

Lines in a Three-Dimensional Space

14 General case of lines parallel to projection planes

When a line in space is oriented so that it is *parallel* to a given projection plane, its projection on that plane represents the *true length* (TL) of the given line. Figure 24 shows a line AB in space with three planes (#1, #2, and #3) parallel to it. The projection of line AB onto each of these planes can be seen as true length since every point on the line AB is equidistant from each plane. An *infinite* number of *parallel projection planes* can "surround" AB, and on each of these planes a true length projection of line AB will be seen.

15 Principal lines

There are three principal lines that are basic to the solution of many three-dimensional descriptive geometry problems. These are: frontal line, horizontal line, and profile line.

A. FRONTAL LINE

A line parallel to the frontal *projection plane.* The projection of such a frontal line appears as true length on the frontal projection plane. Figure 25 pictorially presents a line AB that is parallel to the frontal projection plane. This means that when you view the frontal projection of the line AB, you see its true length noted as a_2b_2.

Figure 26 orthographically shows the horizontal and frontal projections of the line AB. Note that the horizontal projection a_1b_1 of line AB is parallel to the frontal projection plane, which you see as an *edge*

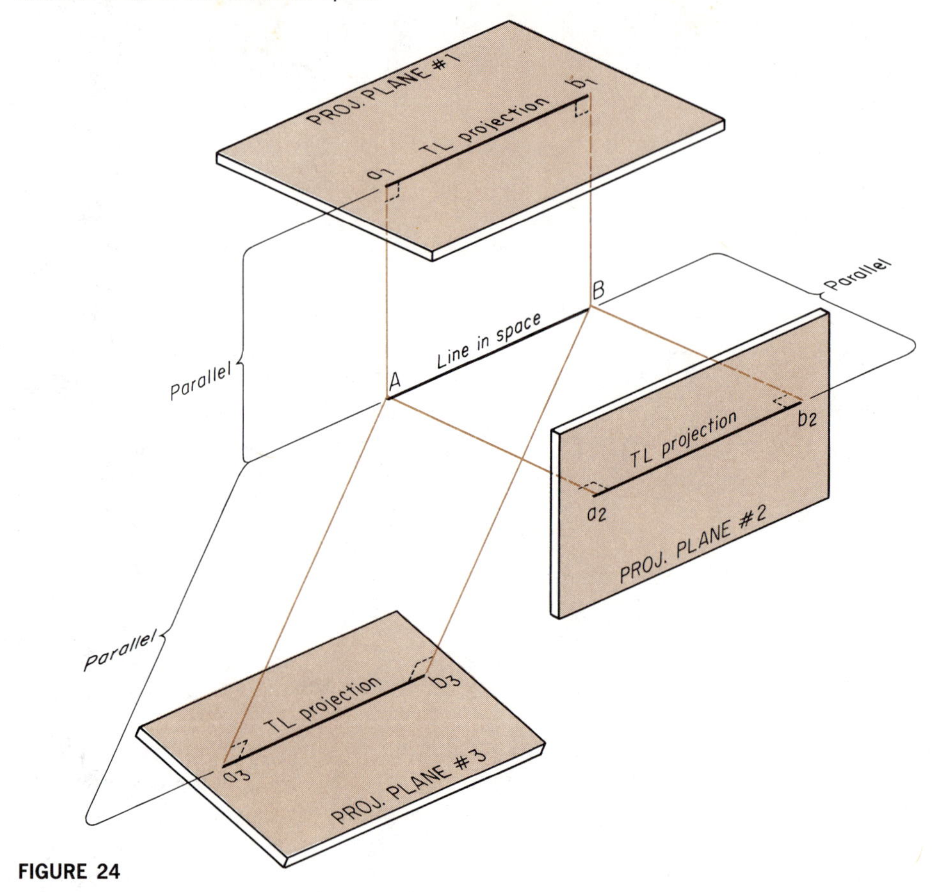

FIGURE 24

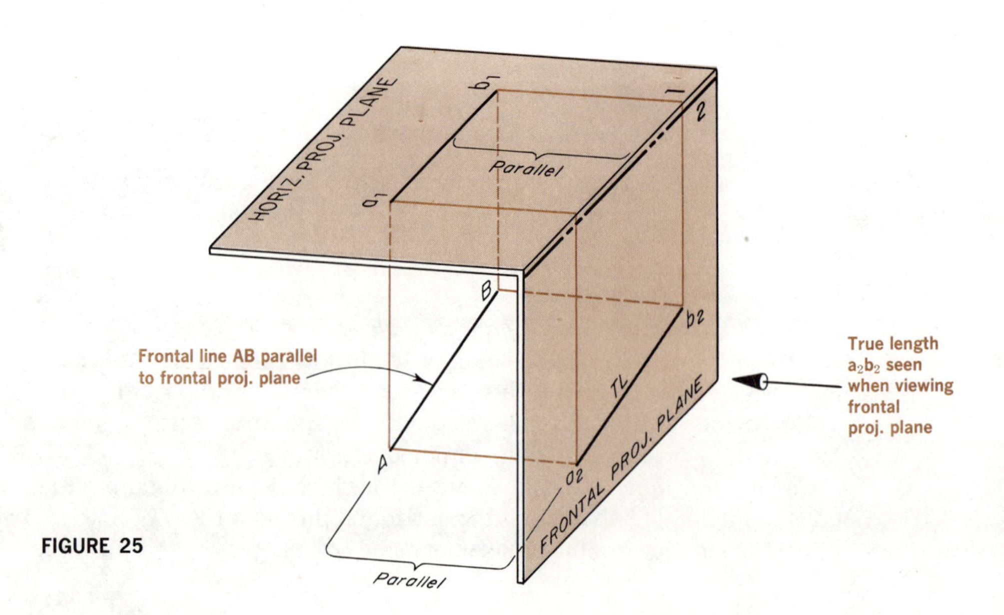

FIGURE 25

when you view the horizontal projection (top view) of the line. RL 1–2 in the top view represents the edge view of the frontal projection plane. Since you see the frontal projection plane as an edge in the top view, line *AB* and the frontal projection plane are parallel.

FIGURE 26

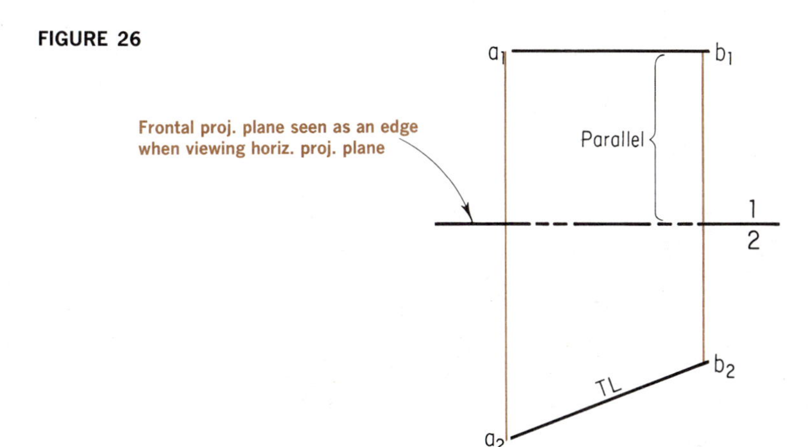

B. HORIZONTAL LINE

A line parallel to the horizontal *projection plane.* The projection of a horizontal line appears as true length on the horizontal projection plane. Figure 27 pictorially presents a horizontal line.

FIGURE 27

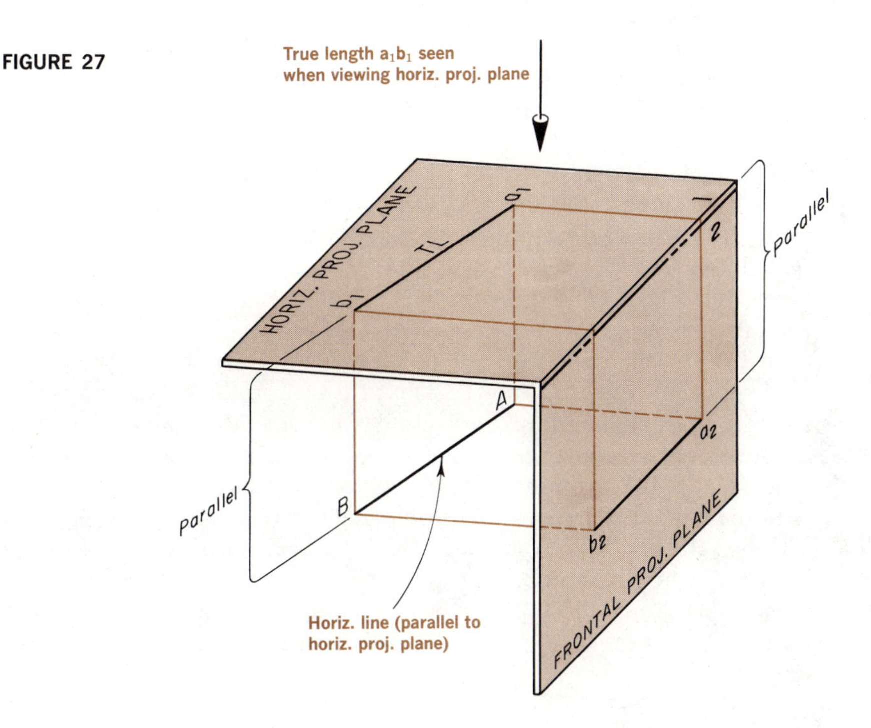

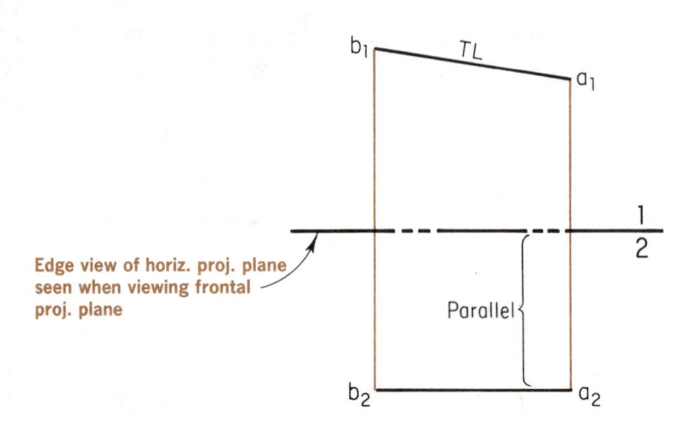

FIGURE 28

Figure 28 orthographically shows the horizontal and frontal projections of a horizontal line *AB*. Note the true length a_1b_1 in the horizontal projection plane (top view). Parallelism exists in the frontal projection where RL 1–2 represents the edge view of the horizontal projection plane. Note also that a horizontal line always remains parallel to the horizontal projection plane, but it can have an infinite number of positions relative to the frontal projection plane. As long as a line in space remains parallel to the horizontal projection plane, you can *always* see its true length in the horizontal projection plane (top view).

C. PROFILE LINE

A line parallel to a profile *projection plane.* A profile line always projects true length onto a profile plane. Figure 29 is a pictorial representation of a profile line *AB* and its true length projection a_3b_3 onto profile projection plane #3.

Figure 30 orthographically shows the horizontal, frontal, and profile projections of a profile line *AB*. Note that when viewing the *frontal* projection plane, the *profile* projection plane is seen as an *edge* and is represented by RL 2–3. This means that you can determine parallelism between the line *AB* and the profile projection plane in the frontal projection view. Note also the true length a_3b_3 of line *AB* in view #3.

In Fig. 30 the distance *X* of point *A* behind the frontal projection plane as seen in the top view (#1) is also seen in the profile projection view (#3), since the frontal projection plane appears as an edge in each of these views. Similarly, in views #1 and #3, you can see the distance *Y* of point *B* (on the line *AB*) behind the frontal projection plane.

In the frontal projection the profile line normally *appears* to be vertical when the line itself is not necessarily vertical. Therefore, its position in space, when determined by its horizontal and frontal projections, must be analyzed carefully. In order to eliminate any doubt as to the position of a profile line *always* construct the profile projection view.

FIGURE 29

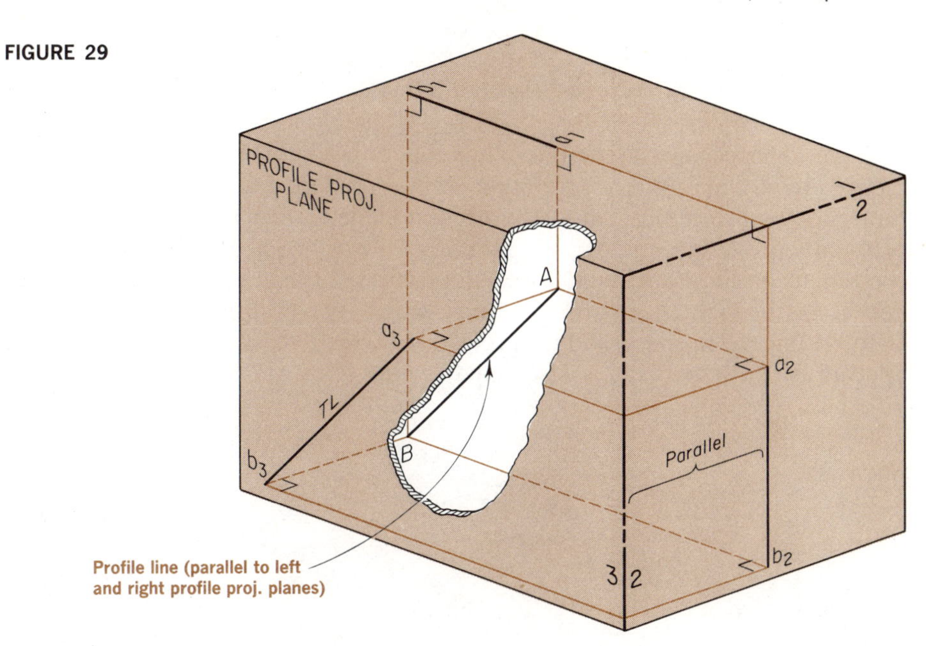

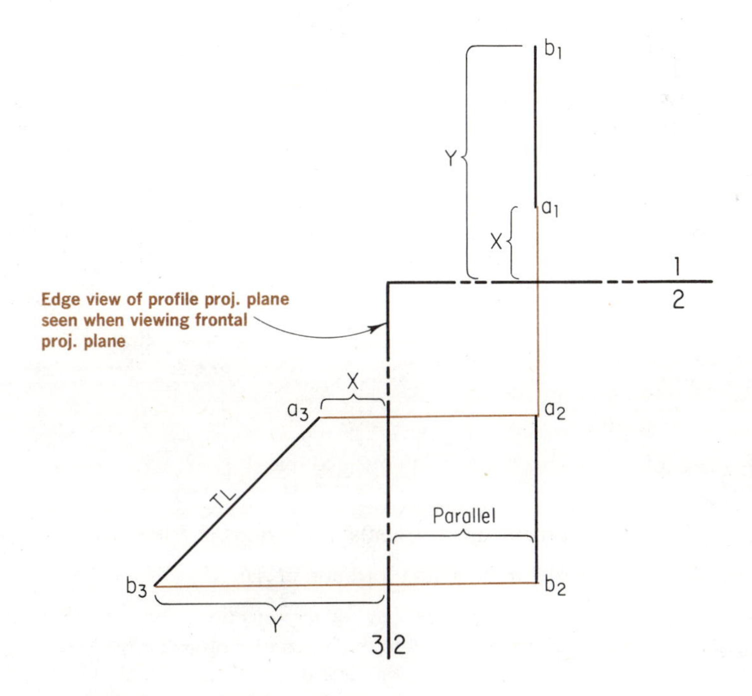

FIGURE 30

16 General case of the true length of a line

If a line in space is neither frontal, horizontal, nor profile, but is a line that is inclined to all the principal projection planes, you can determine its true length projection by projecting it onto a projection plane *parallel* to the line. Figure 31 pictorially presents a line *AB*, *not* parallel to the principal planes. Therefore, its projections on the frontal and horizontal projection planes, for example, do not appear as true (actual) lengths of the line. Auxiliary elevation plane #3 is *parallel* to the line *AB* in space, and the projection of line *AB* onto this auxiliary plane appears as true length and is designated as a_3b_3.

FIGURE 31

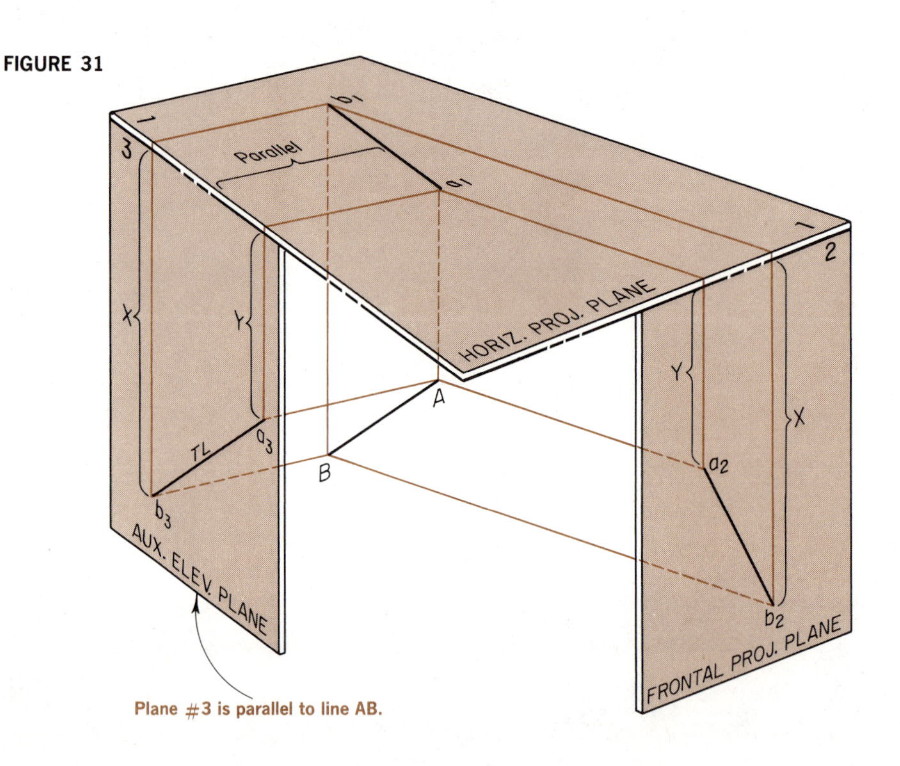

Orthographic Example: True length of an inclined line (Fig. 32).

Given Data: Frontal and horizontal projections of a line *AB*.

Required: The true length projection of *AB*.

Construction Program:

a. Draw the edge view of a projection plane represented by RL 1–3, which is parallel to the horizontal projection a_1b_1 of line *AB*. (RL 1–3 represents the edge view of plane #3, which is seen in the top view and is therefore an auxiliary elevation projection plane.)

b. Construct projectors from points a_1 and b_1 *perpendicular* to RL 1–3. Extend these projectors into view #3.

c. Starting from view #3 go back *two views* to determine the distances that will locate points a_3 and b_3. (This is the application of the key to the System of Orthographic Projection used in descriptive geometry—see Figs. 21 and 22.)

d. The perpendicular distance Y of point A below the horizontal projection plane and the perpendicular distance X of point B below the horizontal projection plane are both seen in the frontal view #2. Transfer these distances with dividers to view #3 along their respective projectors to locate a_3 and b_3.

e. Connect the newly determined points a_3 and b_3 with a straight line. Measure and note the true length (TL) of line AB in auxiliary elevation view #3.

FIGURE 32

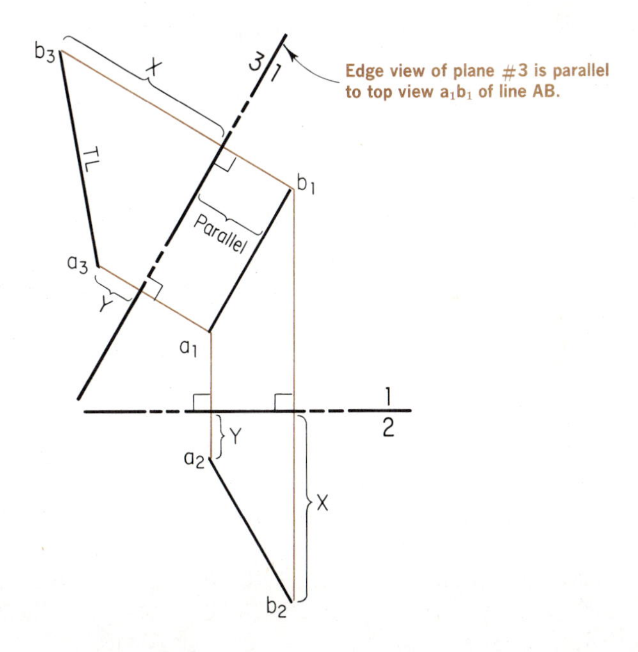

You can also determine the true length projection of line AB by constructing a plane parallel to the frontal projection a_2b_2 of the given line. In Fig. 33 an inclined projection plane #4 appears as an edge in the frontal projection view (RL 2–4) and is parallel to the frontal projection a_2b_2 of line AB. Note that the distance L of point A behind the frontal projection plane and the distance M of point B behind the frontal projection plane are seen in the horizontal view. Transfer these two distances with dividers to view #4 to locate points a_4 and b_4

as shown in Fig. 33. The true length you found in view #4 must be the same as that which you found in view #3.

FIGURE 33

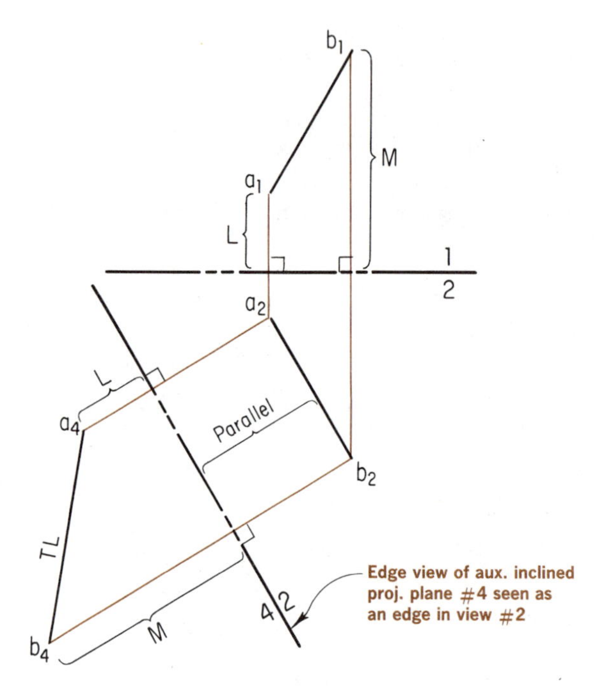

17 Specifications of the position of a line segment in space

In order to completely specify the position of a line segment in space, we must establish either (1) two points on the line segment or (2) one point on the line as well as the angular position of the line relative to a frame, or system, of reference. In our case the projection planes comprise the system of reference. The "fixing" of the position of a line must be "described" in such a manner that there is no question about the particular line segment with which we are concerned. Therefore, applying orthographic projection concepts, we can locate two points on a line by determining the perpendicular distances from the horizontal projection plane and the frontal projection plane. These perpendicular distances define exactly the position of a given line segment relative to the horizontal and frontal projection planes. This means that the distances of the two points *below* the horizontal projection plane can be seen in the frontal projection view, and the distances of the two points *behind* the frontal projection plane can be seen in the horizontal projection view.

Figure 34 pictorially presents a line segment AB in space related to the horizontal and frontal projection planes. (The projection planes in this figure have been oriented so that you are looking "behind" the frontal projection plane and "underneath" the horizontal projection plane in order to more clearly illustrate the distances for points A and B.) From this figure you can see that the distances of A and B behind the frontal projection plane are R and S, respectively, and that the distances of A and B below the horizontal projection plane are T and U, respectively. Figure 35 shows the orthographic horizontal and vertical projections of line AB.

A profile projection plane was not included in Fig. 34 so that the pictorial representation would not become too cluttered. A profile projection plane has been added in Fig. 35 and you see two other distances (L and M) from points A and B perpendicular to the profile projection plane. This now completely defines the position of the line AB relative to the principal projection planes.

The other method of specifying the position of a line segment in space involves (1) locating one point on the line segment and (2) defining the angle the line segment makes with the *horizontal projection plane* (*slope of a line*) and (3) the direction the line takes relative to a *compass reading* (*bearing of a line*). If one point on the line is specified relative to the horizontal and frontal projection planes, then the rest of the line can be positioned by using the concepts of slope and bearing simultaneously.

FIGURE 34

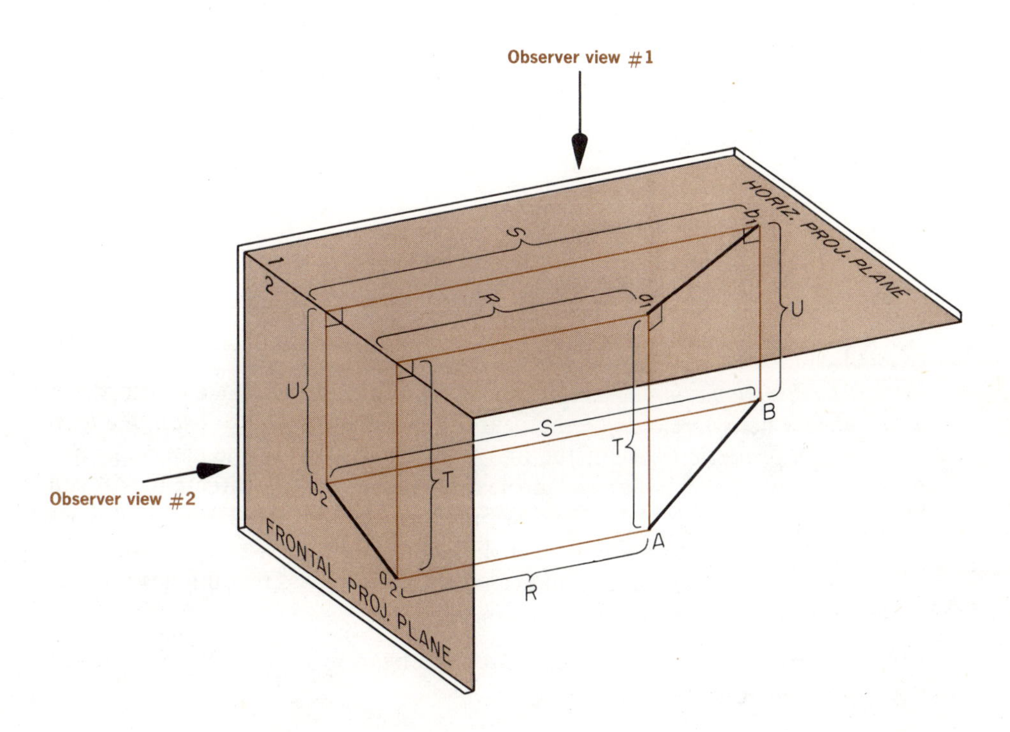

FIGURE 35

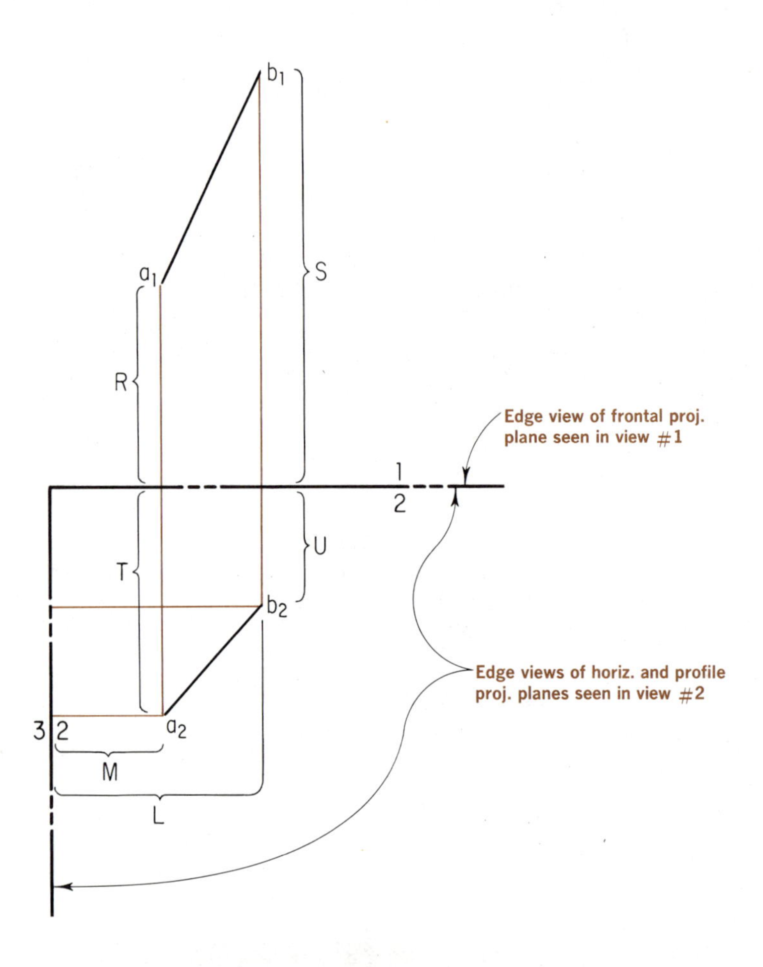

A. BEARING OF A LINE

The direction a line has relative to a compass reading. Since a compass is *always* held *horizontally,* the bearing of a line is *always seen in the horizontal projection view* of the line. This means that a line can make any angle with the horizontal projection plane and its direction will still always be seen in the top view of the line (see Fig. 36). This is analogous to a person with a compass in his hand climbing a steep incline. The *direction* of the incline as indicated by the compass reading has nothing to do with the angle of the incline.

Figure 36 shows a line AB that has a bearing of N60°W (the compass is assumed to be at point a_1).

FIGURE 36

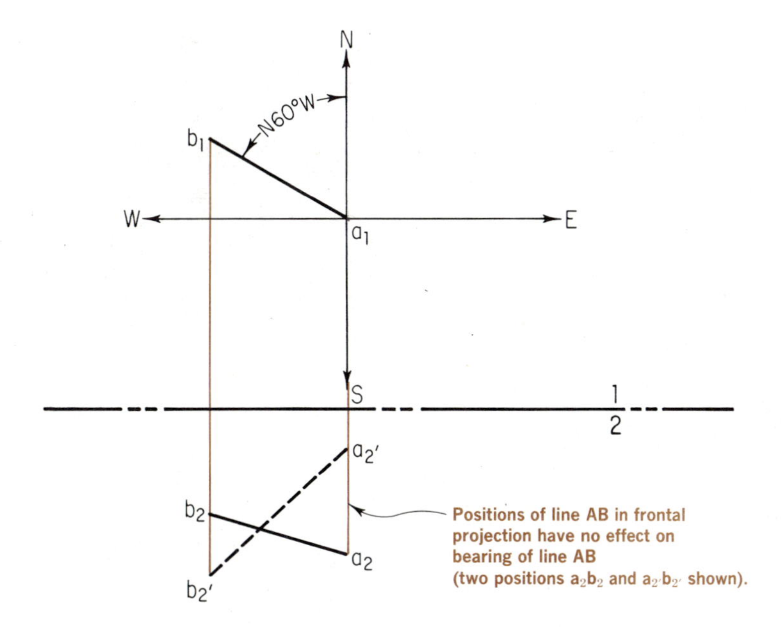

B. SLOPE OF A LINE

The angle of inclination a line has with the horizontal projection plane. (This angle is the same angle a line makes with any horizontal plane.) The *slope angle,* that is, the angle between a line and the horizontal projection plane, can be seen *only* when *two conditions* are satisfied: the line must appear as a *true length projection* in a view where the *horizontal projection plane* appears as an edge. Only then can the *true slope angle* of a line be seen. Unless these conditions exist, the true slope angle cannot be seen (see Figs. 37 and 38). Note in Fig. 38 that the auxiliary elevation plane #3 is seen as an edge in the top view and is parallel to the top view a_1b_1 of line AB. Therefore, you see the *true length* of AB in the auxiliary elevation view #3 where at the *same time* you see the *edge* of the *horizontal projection plane.* This satisfies the two requirements for seeing the true slope angle of line AB.

The slope of a line segment can also be measured in terms of *percent grade.* Percent grade is the civil engineer's way of indicating, for example, the slope or incline of a road. Percent grade is defined as *the number of units of vertical rise*

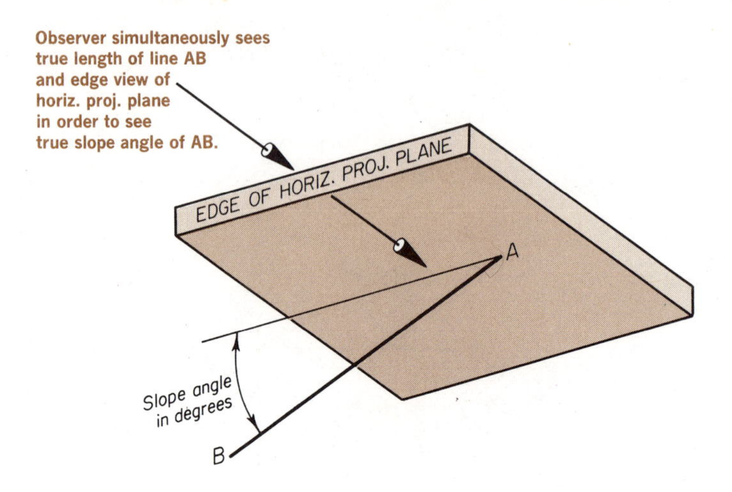

FIGURE 37

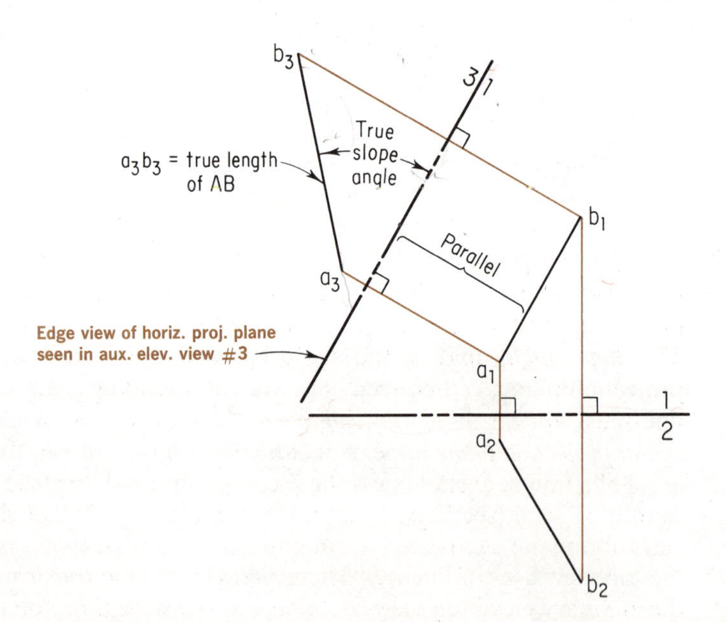

FIGURE 38

for each one hundred units of horizontal distance. Figure 39 shows a line segment AB that has a grade of 50 percent.

Combining the concepts of the true length, bearing, and slope of a line, we can now present an illustrative example involving these concepts and definitions.

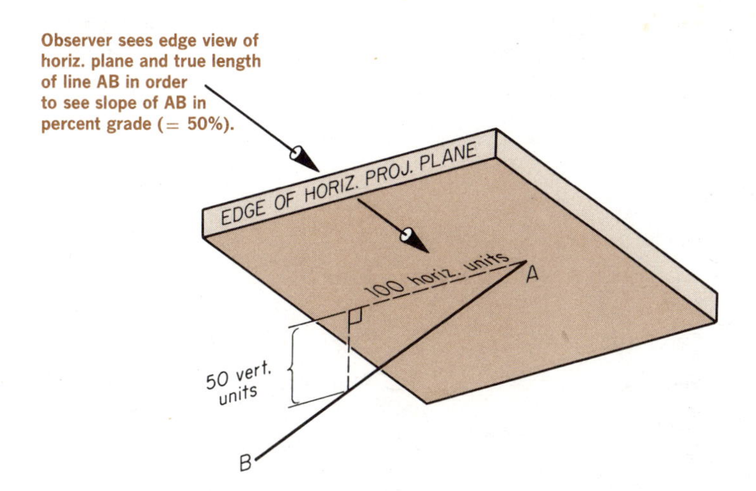

FIGURE 39

Orthographic Example: True length, bearing, and slope of a line segment (Fig. 40).

Given Data: The horizontal and frontal projections of a point A.

Required: The horizontal and frontal projections of a line segment AB, which has a bearing of N30°E, a slope angle of 45° down from A, and a true length of 1.5 in.

Construction Program:

a. Establish the bearing of the line in the top view from a_1. Assuming the top of the paper to be north, draw a line having a bearing of N30°E from a_1 to an assumed point x_1. At this stage, a_1x_1 is an *arbitrary* length.

b. Determine a true length projection of line AX in an elevation view by drawing the edge view of an elevation projection plane represented by RL 1–3, which is *parallel* to the horizontal projection of line a_1x_1.

c. Project point a_1 into the auxiliary elevation view #3, locate a_3 by measuring the distance D of a_2 below the horizontal projection plane as seen in view #2, and transfer this distance with dividers to view #3 and locate a_3.

d. From a_3 construct a line that makes a slope angle of 45° with the edge of the horizontal projection plane as seen in elevation view #3. Extend this line until it intersects with the projector from x_1, thereby locating x_3.

e. Using a convenient scale, measure 1.5 in. on a_3x_3 starting from a_3. This locates point b_3 and defines the required line segment.

f. Project the point b_3 back to horizontal projection plane #1 and frontal projection plane #2, thereby locating b_1 and b_2. (The distance H of b_2 below the horizontal projection plane is obtained in auxiliary elevation view #3.)

g. Connect a_1b_1 and a_2b_2 with straight lines to obtain the required data.

FIGURE 40

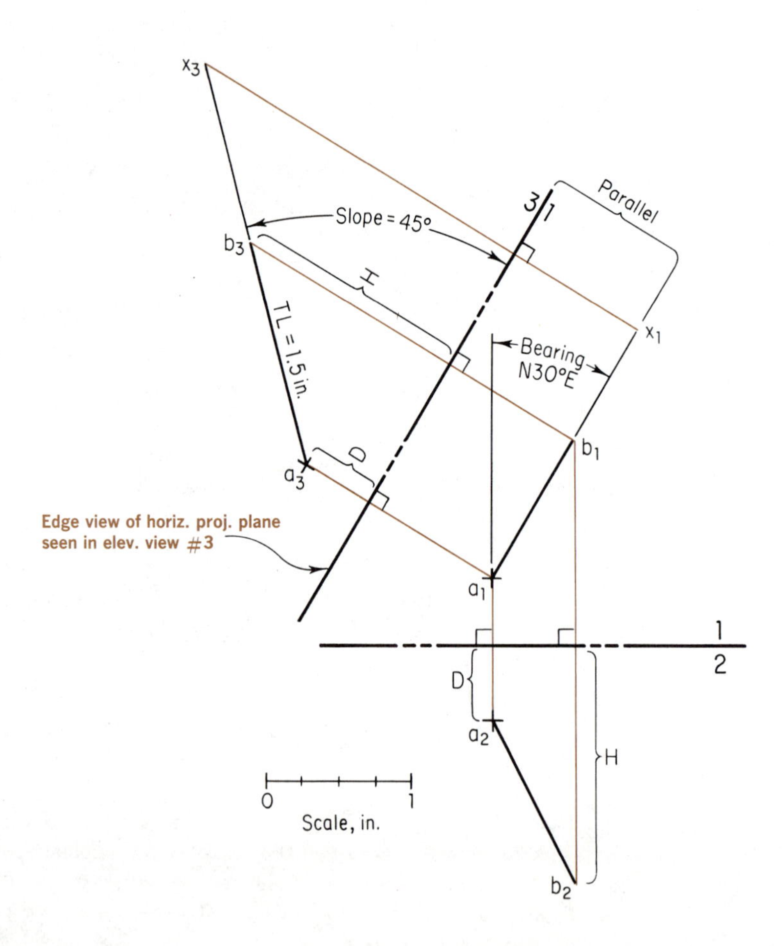

18 Point view of a line

You see the point view of a line when you view a plane that is *perpendicular* to the line. Figure 41 shows a line AB in space with an inclined auxiliary plane #4 perpendicular to it. Note that auxiliary projection plane #4 must be perpendicular to the true length of line AB in order for it to project as a point onto plane #4.

FIGURE 41

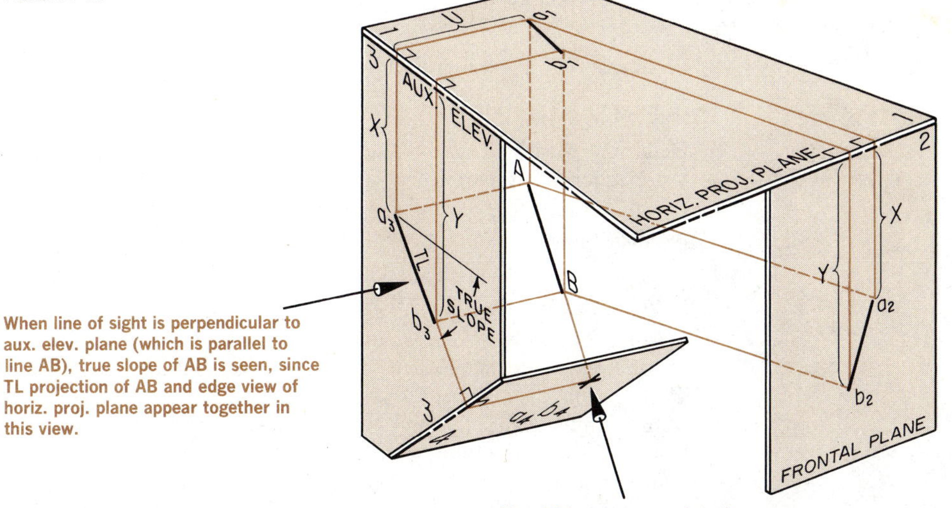

FIGURE 42

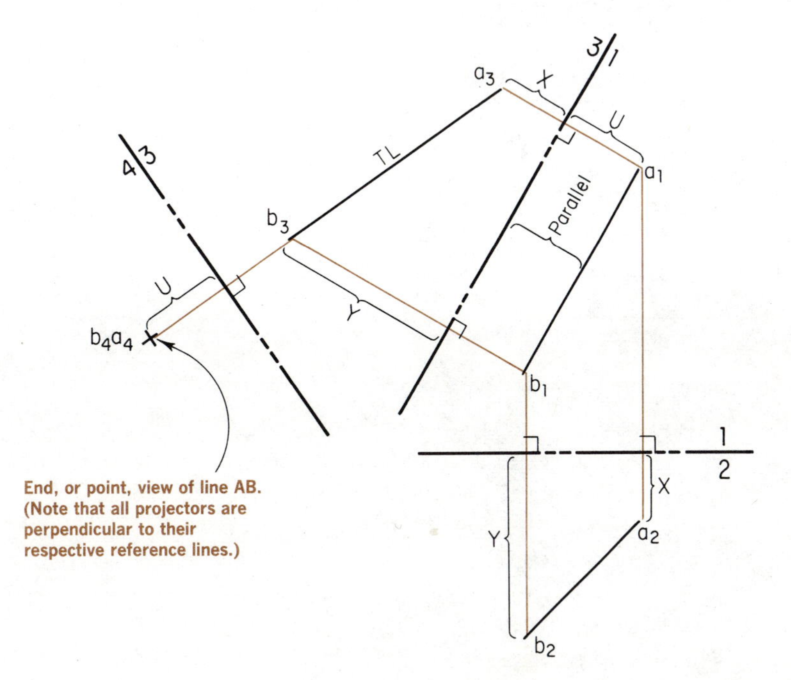

Orthographic Example: Point view of an inclined line (Fig. 42).

Given Data: Frontal and horizontal projections of a line *AB*.

Required: A point view (projection) of the given line *AB*.

Construction Program:

a. Determine the true length of line *AB* by projecting the given line onto a projection plane parallel to one of the given views of the line. (In Fig. 42 an auxiliary elevation projection plane #3 has been drawn parallel to the horizontal projection a_1b_1 of line *AB*.)

b. Draw the edge view of an auxiliary projection plane #4 *perpendicular* to the *true length* (a_3b_3) of *AB* as seen in view #3.

c. Starting from view #4 and going back two views to view #1, measure the perpendicular distance *U* of points a_1 and b_1 from RL 1–3. Transfer this distance to view #4, measuring along the projectors from a_3 and b_3. Since a_1 and b_1 are equidistant from RL 1–3, the distance *U* falls on one point, which is the required point view of line *AB* and is designated as b_4a_4. (Point b_4 is noted before point a_4, since when viewing the line *AB* in view #4, point *B* would be seen before point *A*.)

19 Characteristics of parallel lines in space

Parallel lines, lines equidistant from each other in all points, are lines that will not intersect no matter how far they are extended. Parallel lines have the prop-

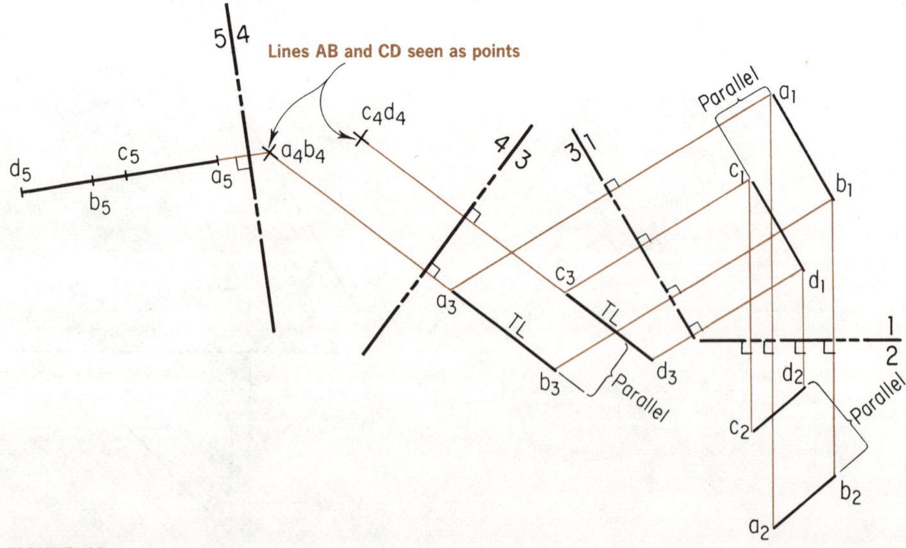

FIGURE 43

erty of appearing parallel in space when viewed from all positions except (1) where the observer sees the lines simultaneously as points and (2) where the lines appear one behind the other. In the first case, in which two lines simultaneously appear as points, the observer has a proof of parallelism, while in the second case, if two lines appear one behind the other, the observer cannot with confidence state that they are parallel until he determines a view that shows them either to be parallel or to appear as points.

Figure 43 shows two parallel lines *AB* and *CD* given by their horizontal and frontal projections. *AB* and *CD* are projected onto an auxiliary elevation view #3, which shows them to be parallel; they are then projected onto an auxiliary inclined view #4, which shows them to be points. Finally, they are projected onto another auxiliary inclined view #5, in which they appear one behind the other. This figure illustrates the possible general views of parallel lines.

The property of parallelism will prove useful in solving certain types of three-dimensional problems and, therefore, it is important that you remember the characteristics indicated above.

20 Perpendicular lines in space—intersecting and nonintersecting

If two intersecting lines are perpendicular to each other, the 90° angle between them will be seen in any view in which *at least one* of the lines appears as *true length.* This is the condition that proves perpendicularity. The only exception to this condition is a view in which one of the lines appears as true length and the other as a *point* on the true length.

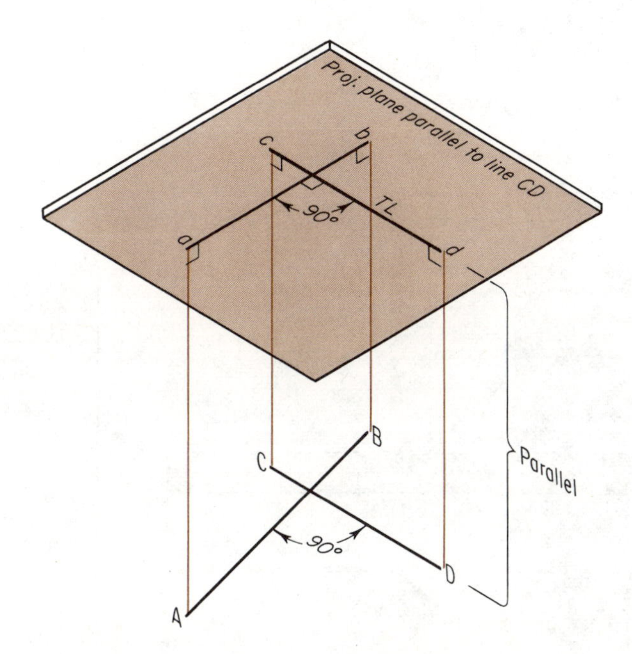

FIGURE 44

Figure 44 shows two intersecting perpendicular lines AB and CD in space, where line CD is parallel to a projection plane. When the lines are projected onto this plane, CD naturally projects as a true length while AB, since it is not parallel to the projection plane, projects as a foreshortened non-true length line. The *projections ab* and *cd* of the lines are at 90° to each other.

FIGURE 45

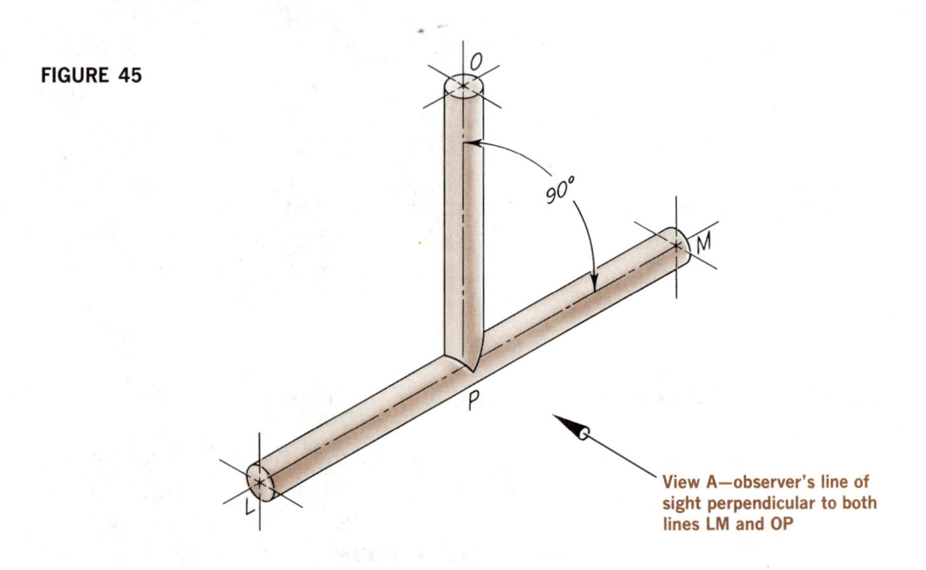

FIGURE 46

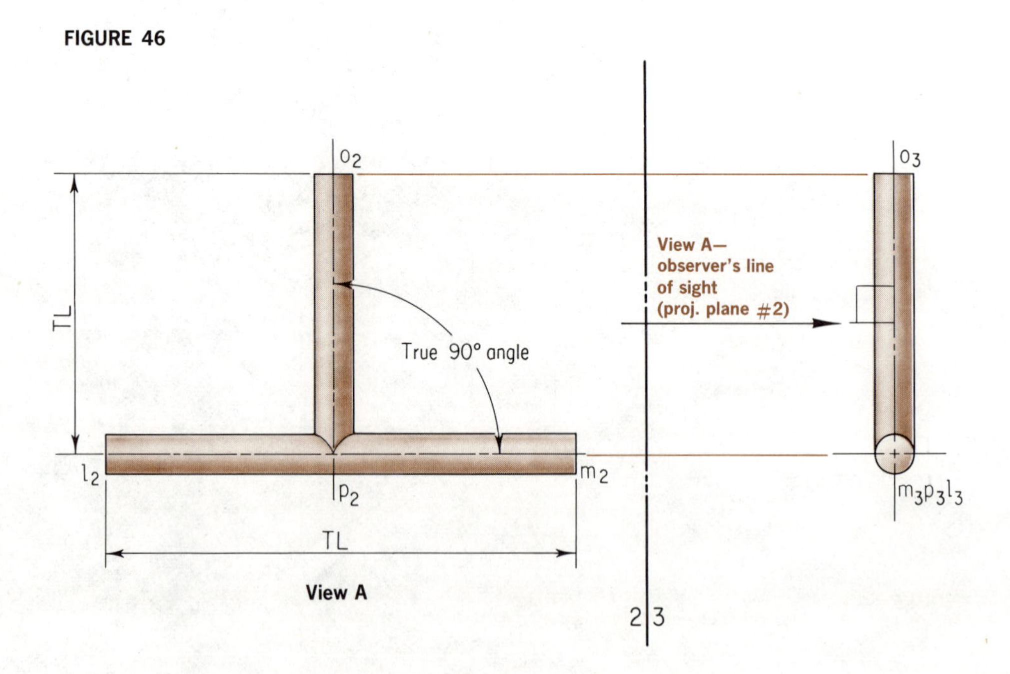

Figure 45 pictorially presents two rods (lines) *LM* and *OP* that are perpendicular to each other. View A is a view in which the observer's line of sight is perpendicular to both rods.

Figure 46 shows the frontal and right profile projections of the rods *LM* and *OP*, and view A as seen by an observer whose line of sight is perpendicular to both rods. [The true 90° angle appears in the frontal projection (view #2).]

Figure 47 pictorially presents the same rods *LM* and *OP*, but this time view B is a view in which the observer's line of sight is perpendicular to line *LM*. This means that *LM* appears as true length while *OP* appears foreshortened.

FIGURE 47

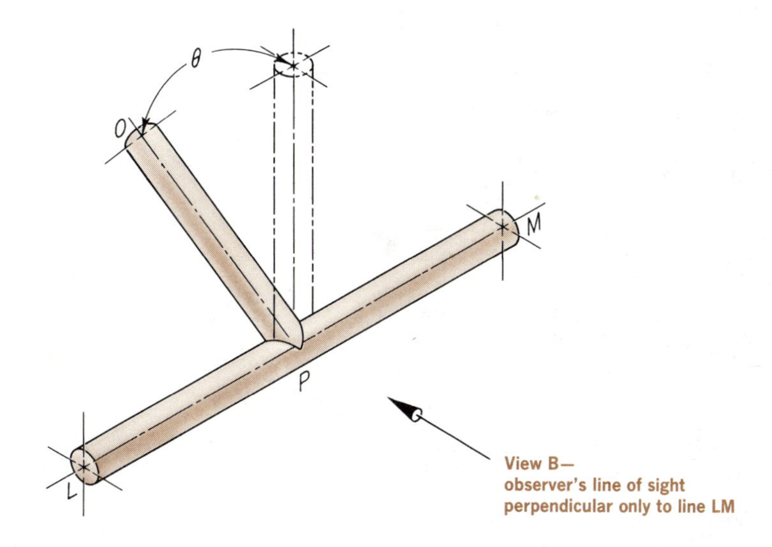

Figure 48 shows the frontal and right profile projections of this new position of *OP*. The frontal projection (view #2) shows *LM* as true length and *OP* as foreshortened; the true 90° angle between the two rods is still seen. This demonstrates the fact that a true 90° angle between two lines will be seen as long as one of the lines appears as true length to the observer.

Figure 49 shows the characteristics of two perpendicular lines *AB* and *CD* in orthographic projection. In the given horizontal and frontal projections we note that 90° between the two lines cannot be seen, since neither of the lines appears as true length. In view #3 you see that line *AB* appears as true length while *CD* does not, yet 90° is defined between them. In view #4 you see line *CD* as true length but line *AB* as a point. Here you *cannot* see 90° between the two lines, but because one line appears as true length and the other as a point on the true length segment, perpendicularity can be proved. In view #5 you see both *CD* and *AB* as true length and, therefore, you also see a true 90° angle between them.

If lines *AB* and *CD* in Fig. 49 were not intersecting perpendicular lines but skew lines (lines that do not intersect and are not parallel) with a direction 90°

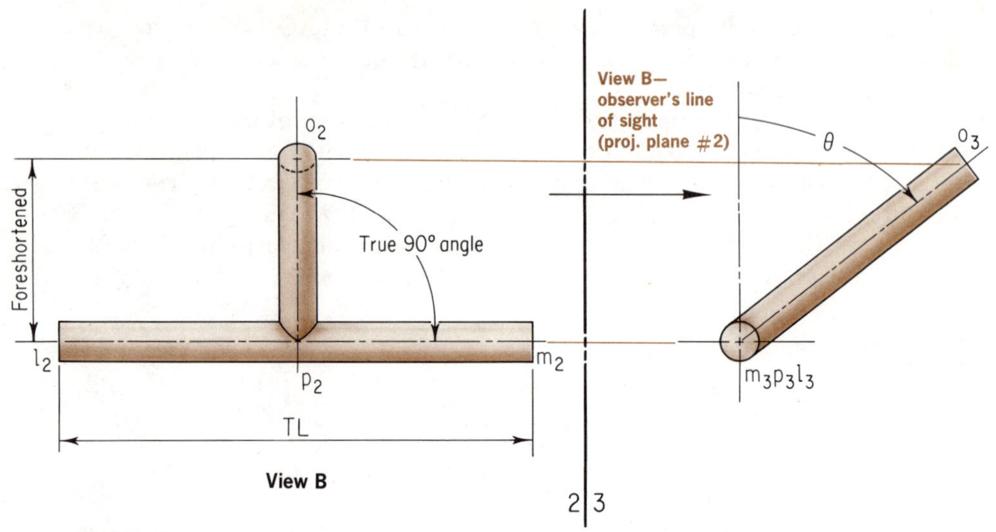

FIGURE 48

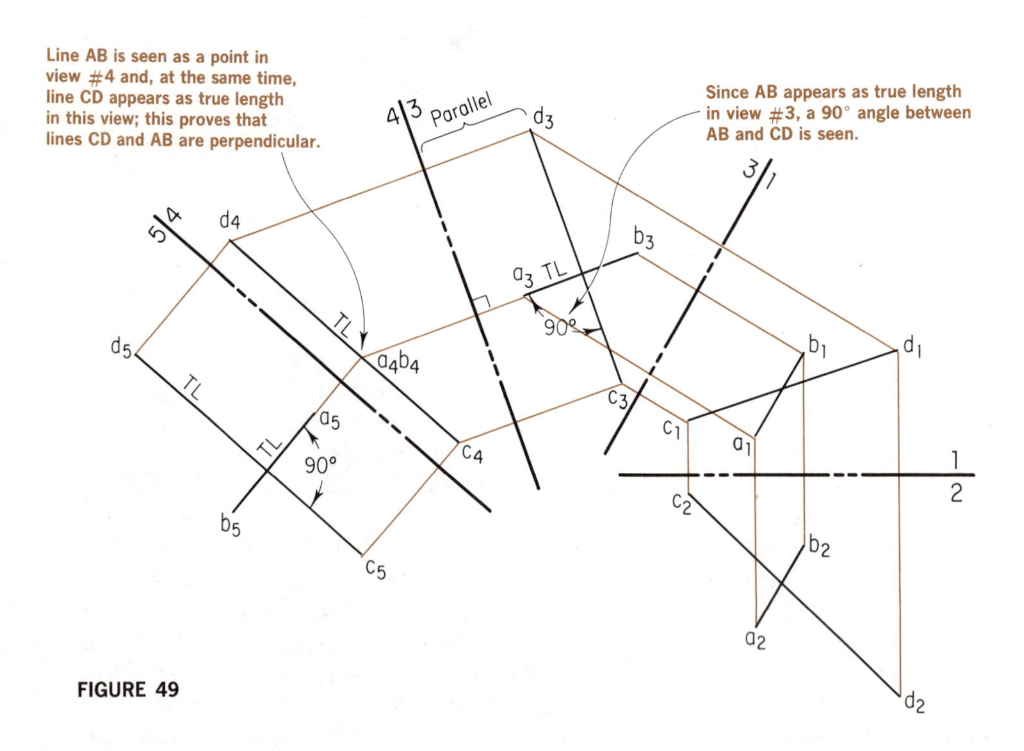

FIGURE 49

to each other, the same reasoning would be applied as was applied to the case of intersecting perpendicular lines.

Figure 50 shows two skew lines in orthographic projection with a direction of 90° to each other.

View #4 shows that lines AB and CD do not intersect but have a direction of 90° to each other, since CD appears as true length and AB as a point in the same view.

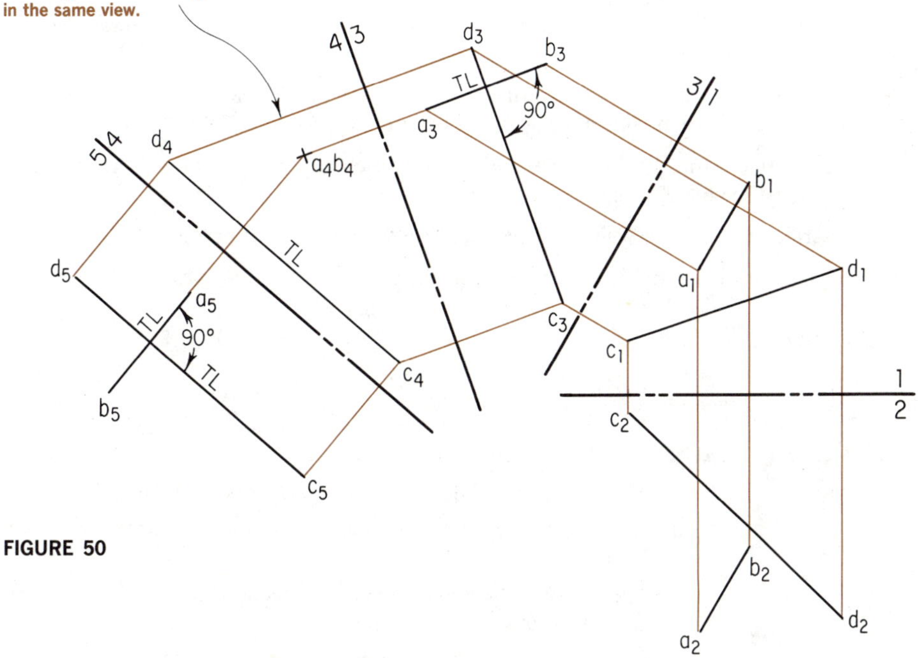

FIGURE 50

SAMPLE QUIZZES

1. Draw freehand and fully note the horizontal, frontal, and profile projections of the following lines:

a. Frontal. b. Horizontal. c. Profile.

2. Draw freehand and fully note the horizontal, frontal, and profile projections of the following:

a. A line that is both horizontal and frontal.
b. A line that is both horizontal and profile.
c. A line that is both frontal and profile.

3. If one leg of a right triangle appears as a point in the front view, and the hypotenuse of the triangle appears as true length in the top view, what is the direction of the third leg of the triangle in relation to the two projection planes? Draw freehand the horizontal, frontal, and profile projections of this triangle, carefully noting all points in all views.

4. Draw freehand the horizontal and frontal projections of a right triangle ABC whose vertical front edge, leg AB, is 1 in. long, and its top edge, leg BC, $1\frac{1}{2}$ in. long, makes 60° with the frontal projection plane, point C to the right. Letter all 3 points in both views.

5. State the two requirements for a view that shows the true slope angle of any line.

6. Draw freehand and fully note the horizontal and frontal projections of the following lines:

a. Frontal line *AB*, length 10 units (a unit is $\frac{1}{8}$ in.), with a true slope of 30° downward from point *A* to the right.

b. Profile line *CD*, length 8 units (a unit is $\frac{1}{8}$ in.), with a true slope of 45° upward from *D* to the front.

7. If a straight line appears as a point in the front view, in which views will it appear as true length? Can a sloping line appear as true length in the top view? Why?

8. If two intersecting lines are perpendicular to each other in space, in what view (or views) can you see the 90° angle between them?

9. Describe two views in which two parallel lines will not appear parallel.

10. In what view can you see the true distance between two parallel lines?

PRACTICE PROBLEMS

1. Given Data: A. RL 1–2, RL 1–3, and RL 2–4.

B. The following spatial description of horizontal line *AB*:

a. Points *A* and *B* are both 1 in. below the horizontal projection plane.

b. Point *A* is $\frac{1}{2}$ in. behind the frontal projection plane and 2 in. behind the right profile projection plane.

Point *B* is 1 in. behind the frontal projection plane and $\frac{1}{2}$ in. behind the right profile projection plane.

Required: A. The horizontal projection a_1b_1, frontal projection a_2b_2, auxiliary elevation projection a_3b_3, and right profile projection a_4b_4 of line *AB*.

B. What must be the position of the auxiliary elevation projection a_3b_3 of line *AB* relative to RL 1–3? (*Ans.:* Parallel)

C. Measure and show where the true length of line *AB* appears. (*Ans.:* TL = $1\frac{9}{16}$ in.)

D. What angle does line *AB* make with the frontal projection plane? (*Ans.:* 19°)

PROBLEM 1

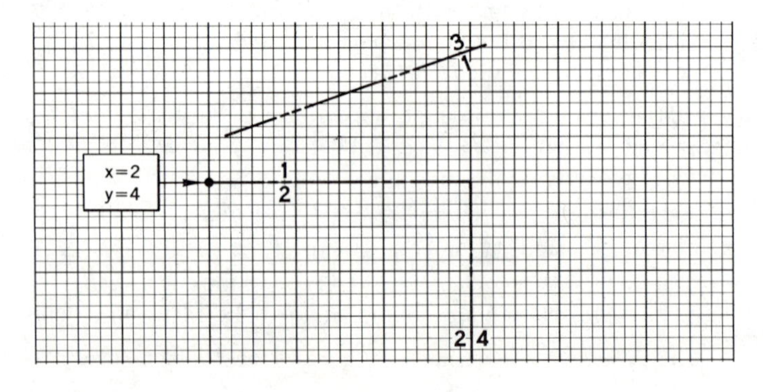

2. Given Data: A. RL 1–2, RL 2–3, and RL 2–4.

B. The following spatial description of frontal line *CD*:

a. Points *C* and *D* are both $\frac{3}{4}$ in. behind the frontal projection plane.

b. Point *C* is $1\frac{1}{2}$ in. below the horizontal projection plane and $\frac{1}{2}$ in. behind the left profile projection plane.

c. Point *D* is $\frac{1}{2}$ in. below the horizontal projection plane and 2 in. behind the left profile projection plane.

Required: A. The horizontal projection c_1d_1, frontal projection c_2d_2, left profile projection c_4d_4, and auxiliary inclined projection c_3d_3 of line *CD*.

B. Measure and show where the true length (TL) of line *CD* appears. (*Ans.:* TL = $1\frac{25}{32}$ in.)

C. What angle does line *CD* make with the horizontal projection plane? (*Ans.:* 34°)

D. What must be the position of the auxiliary inclined projection c_3d_3 of line *CD* relative to RL 2–3? (*Ans.:* Parallel)

PROBLEM 2

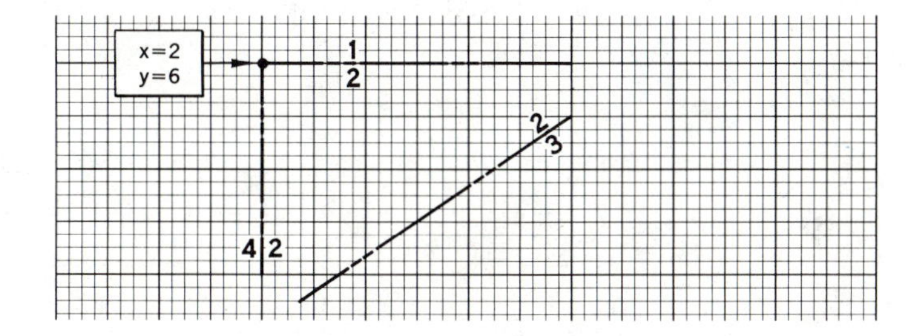

3. Given Data: A. RL 1–2, RL 1–3, and RL 2–4.

B. The following spatial description of profile line *EF*:

a. The true length of line *EF* is $1\frac{3}{4}$ in.

b. Points *E* and *F* are both 1 in. behind the right profile projection plane.

c. Point *E* is $\frac{1}{2}$ in. behind the frontal projection plane and $1\frac{1}{2}$ in. below the horizontal projection plane.

d. Point *F* is 2 in. behind the frontal projection plane.

Required: A. The horizontal projection e_1f_1, frontal projection e_2f_2, right profile projection e_4f_4, and left profile projection e_3f_3 of line *EF*.

B. Measure and show where the true length of line *EF* appears. (*Ans.:* TL = $1\frac{3}{4}$ in., seen in views #3 and #4)

C. Measure and note the angle line *EF* makes with the frontal projection plane. (*Ans.:* $58\frac{1}{2}$°)

D. Measure and note the angle line *EF* makes with the horizontal projection plane. (*Ans.:* 31°)

PROBLEM 3

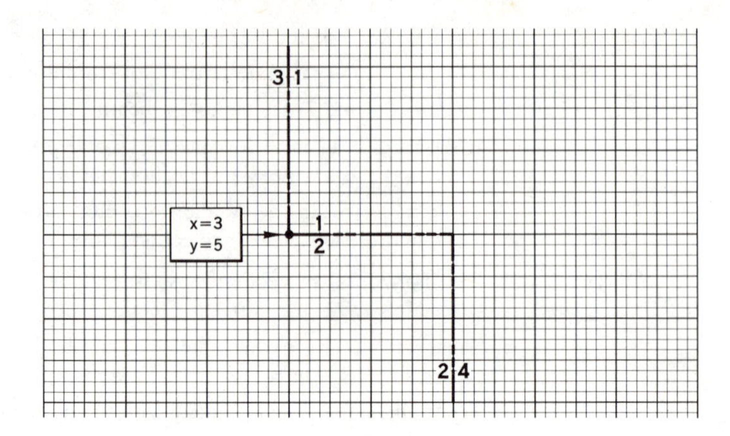

4. Given Data: Horizontal and frontal projections of line AB.

Required: The true length, point view, bearing, and slope of line AB. Note all required data in appropriate views. (*Ans.:* TL = $2\frac{1}{16}$ in., bearing = N56°E from A, slope = 29°)

PROBLEM 4

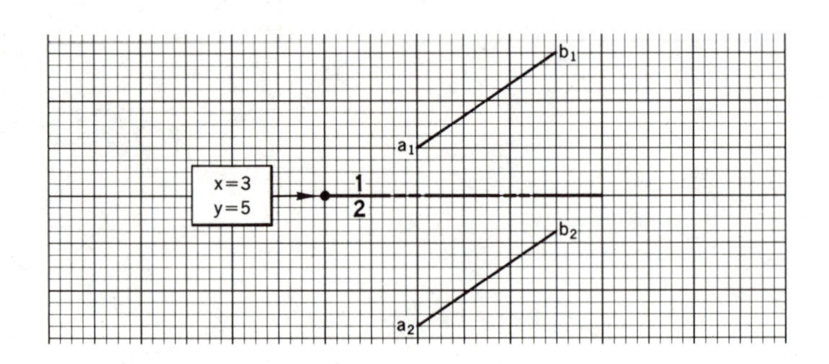

5. Given Data:

A. RL 1–2 and RL 2–4.

B. The following spatial description of line XY:

a. Point X is $\frac{1}{2}$ in. in front of the frontal projection plane, $\frac{3}{4}$ in. above the horizontal projection plane, and 1 in. behind the left profile projection plane.

b. Point Y is 1 in. behind the frontal projection plane, $1\frac{3}{4}$ in. below the horizontal projection plane, and $2\frac{1}{2}$ in. behind the left profile projection plane.

Required:

A. The true length, point view, bearing, and slope of line XY. Note all required data in appropriate views. (*Ans.:* TL = $3\frac{9}{32}$ in., bearing = S45°W from Y, slope = 50°)

B. An approximate freehand pictorial representation of line XY in space relative to the horizontal and frontal projection planes.

PROBLEM 5

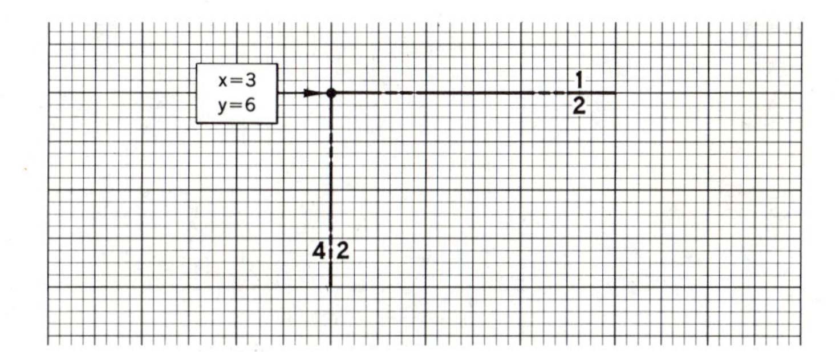

6. Given Data: A. Horizontal and frontal projections of point *B*.

B. The following spatial description of a line *BC*:

a. Line *BC* has a bearing of S30°E from *B*.

b. The slope of line *BC* is 30° upward from *B* (going toward the horizontal projection plane).

c. The true length of line *BC* is 2 in.

Required: A. The horizontal and frontal projections of line *BC*.

B. Label the views in which the true length, bearing, and slope, of line *BC* are measured.

PROBLEM 6

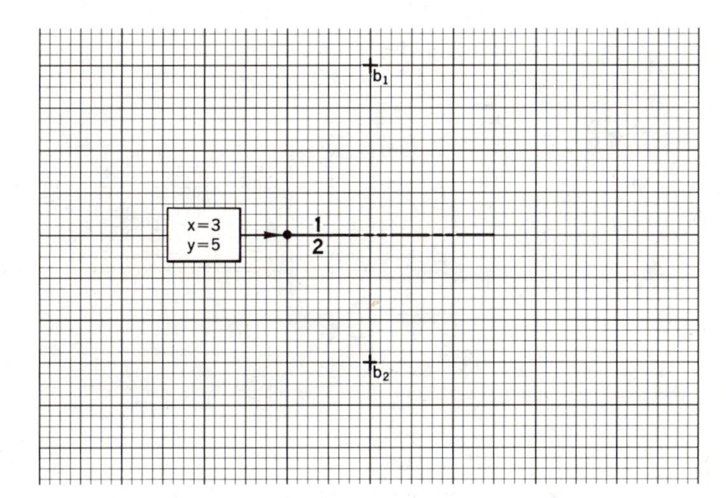

7. Given Data: The frontal projection d_2e_2 and auxiliary elevation projection d_3e_3 of line *DE*.

Required: A. The true length, bearing, slope, and point view of line *DE*. (*Ans.:* TL = $1\frac{3}{16}$ in., bearing = S67°E from *D*, slope = $28\frac{1}{2}$°)

B. Label the views in which the true length, bearing, and slope of line *DE* are measured.

PROBLEM 7

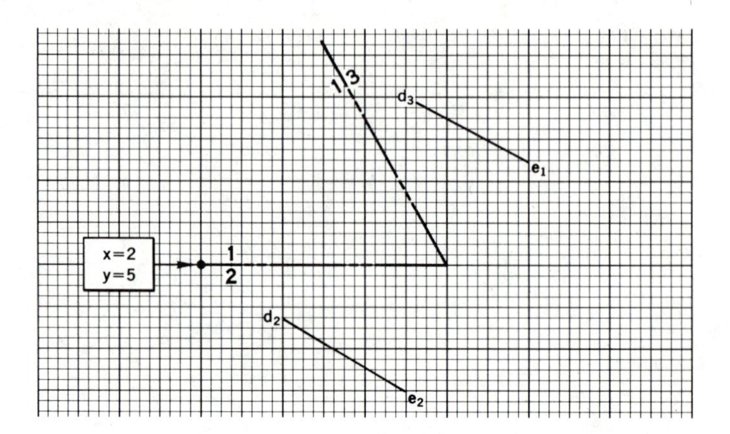

8. Given Data: A. The horizontal and frontal projections of point A.

B. The following spatial descriptions of intersecting lines AC and AD:

a. Line AC has a bearing of S45°W from A, a slope of 30° upward from A, and a true length of $1\frac{1}{2}$ in.

b. Line AD has a bearing of S60°E from point A, a grade of 30% upward from A and a true length of 2 in.

Required: A. The horizontal and frontal projections of intersecting lines AC and AD.

B. Label the views in which the true lengths, bearings, and slopes of lines AC and AD appear.

PROBLEM 8

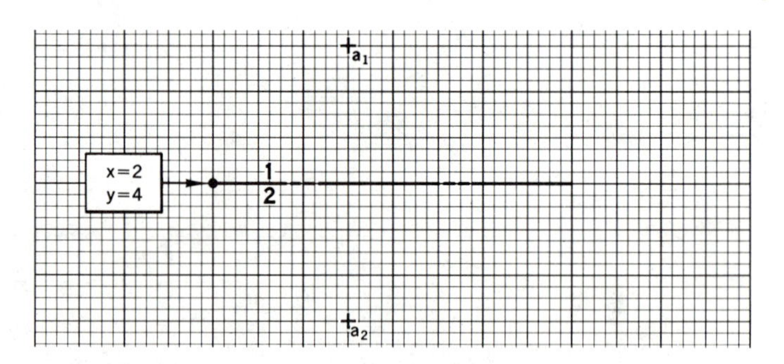

9. Given Data: A. Horizontal and frontal projections of point A.

B. The spatial descriptions of the following lines:

a. A line starting from point A and terminating at point B has a bearing of N45°E, a grade of 50% upward from A, and a true length of $2\frac{1}{4}$ in.

b. A line starting from point B and terminating at point C has a bearing due east, a grade of 30% downward from B, and a true length of $1\frac{1}{2}$ in.

c. A line starting at point C and terminating at point D has a bearing of S30°E, a grade of 100% upward, and a true length of 2 in.

Required:

A. The horizontal and frontal projections of lines AB, BC, CD, and the line connecting points D and A.

B. The true length, bearing, and percent grade of the line connecting points D and A. (*Ans.:* TL of $DA = 4\frac{1}{16}$ in., bearing of DA = N87°E from A, grade = 55%)

C. Label the views in which the bearings, slopes, and true lengths of all lines are measured.

PROBLEM 9

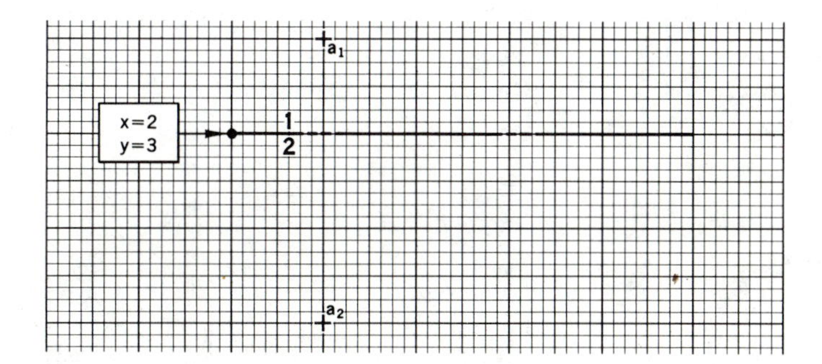

10. Given Data:

A. RL 1–2, RL 1–3, and RL 3–4.

B. The frontal projection a_2b_2 of line AB.

C. The point view a_4b_4 of line AB.

PROBLEM 10

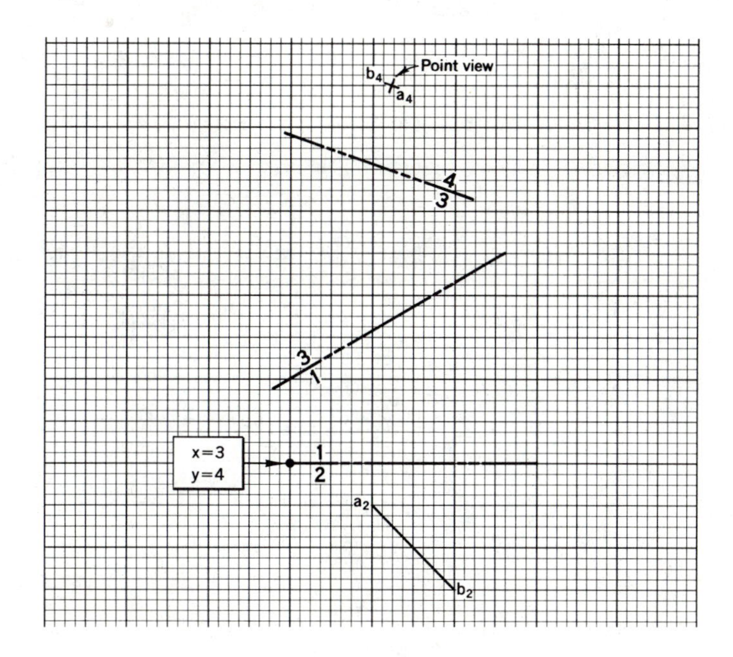

Required: A. The true length, slope, and bearing of line AB. (*Ans.:* TL = $1\frac{1}{2}$ in., bearing = N60°E from A, slope = $41\frac{1}{2}$°)

B. Label the views in which the true length, slope, and bearing are measured.

11. Given Data: Horizontal and frontal projections of skew lines AB and CD.

Required: A. Draw the point view of line AB in an elevation view.

B. In the point view of line AB, measure the perpendicular distance from line AB as a point to line CD. Compare this distance with the perpendicular distance measured from line CD as a point to line AB in view #2. What is your conclusion?

PROBLEM 11

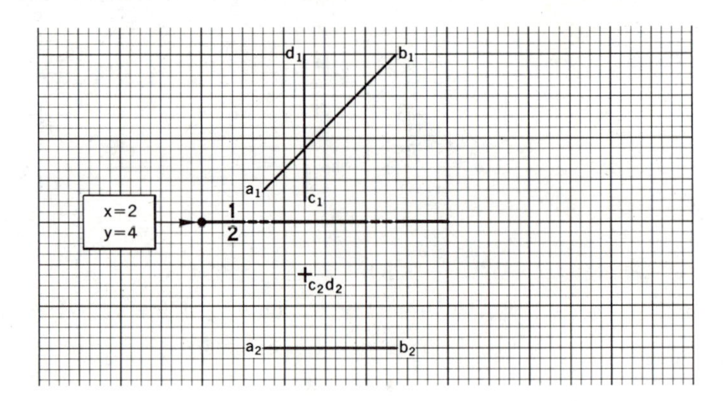

12. Given Data: Horizontal and frontal projections of lines XY and WZ.

Required: Determine by projection whether lines XY and WZ are parallel.

PROBLEM 12

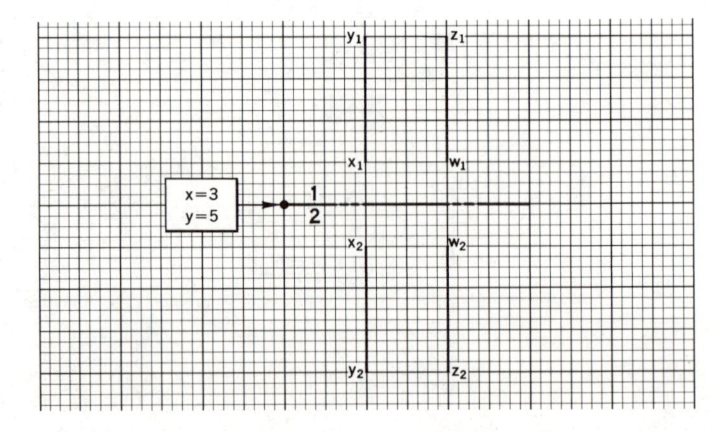

13. Given Data: Horizontal and frontal projections of line AB and a point C outside the line.

Required: A. The horizontal and frontal projections of a line CD which is parallel to line AB.

B. Check the parallelism of lines AB and CD by drawing a right profile projection of these two lines.

PROBLEM 13

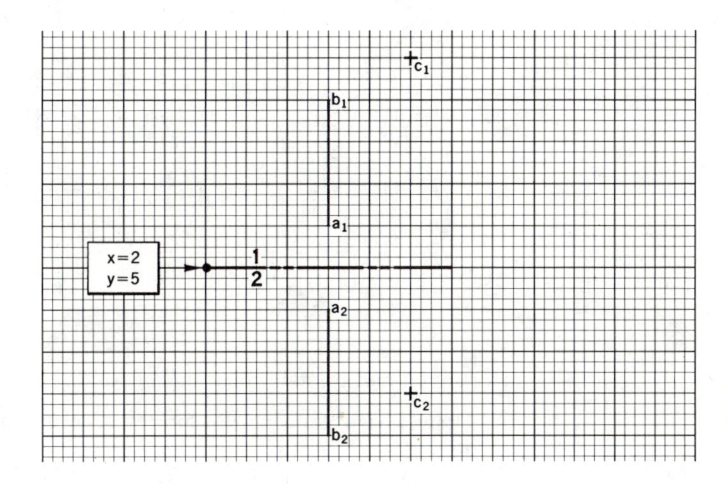

14. Given Data: Horizontal and frontal projections of intersecting lines AB and AC.

Required: Determine by projection whether lines AB and AC are perpendicular to each other.

PROBLEM 14

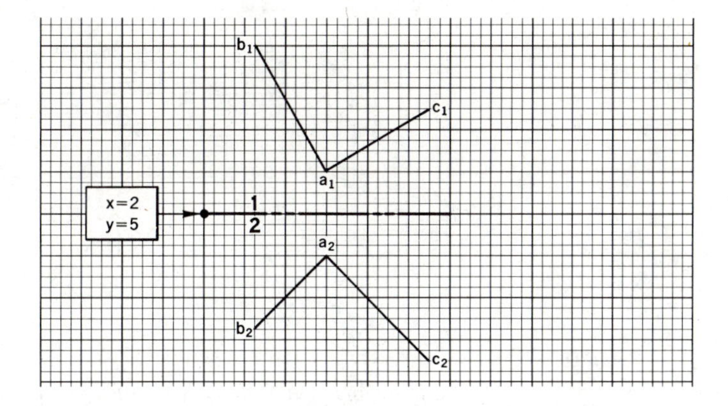

15. Given Data: Horizontal and frontal projections of line AB.

Required: A. The horizontal and frontal projections of a line BC which is perpendicular to line AB at B, and has a bearing of due East and a true length of $1\frac{1}{2}$ in.

B. Label and show all construction.

PROBLEM 15

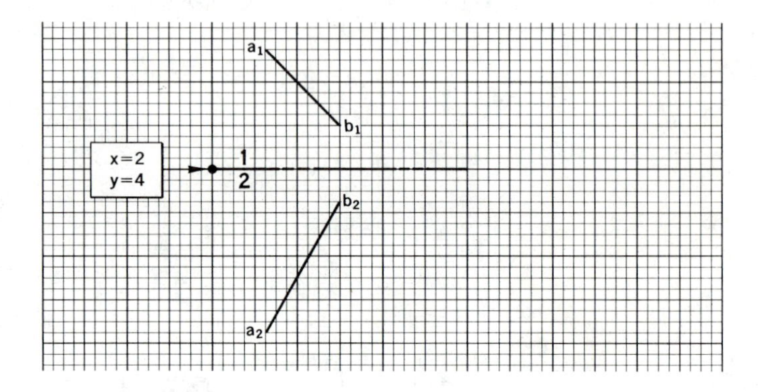

16. Given Data: Horizontal and frontal projections of parallel lines AB and CD.

Required: A. Draw a view showing the shortest true distance between lines AB and CD.

B. Measure and label the shortest distance between lines AB and CD. (*Ans.:* Shortest distance = $\frac{7}{8}$ in.)

PROBLEM 16

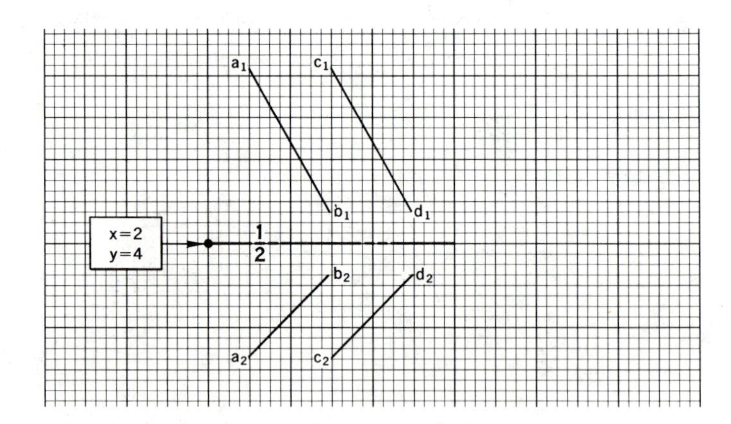

17. Given Data: A. RL 1–2, RL 1–3, RL 2–4, and RL 2–5.

B. The following spatial description of right triangle ABC:

a. The vertical front edge (leg $AB = 1\frac{1}{2}$ in.) of triangle ABC and its top edge (leg $BC = 1$ in.) which makes 60° with the frontal projection plane with point C to the left.

b. Leg BC is $\frac{1}{2}$ in. below the horizontal projection plane, and leg AB is 1 in. behind the right profile projection plane and 1 in. behind the frontal projection plane.

Required: A. The horizontal (#1), frontal (#2), auxiliary elevation (#3), right profile (#4), and auxiliary inclined (#5) projections of plane ABC.

B. Measure and label the true lengths of each side of triangle ABC in the views in which these true lengths appear.

C. Identify the views in which the 90° angle between legs AB and BC appears.

PROBLEM 17

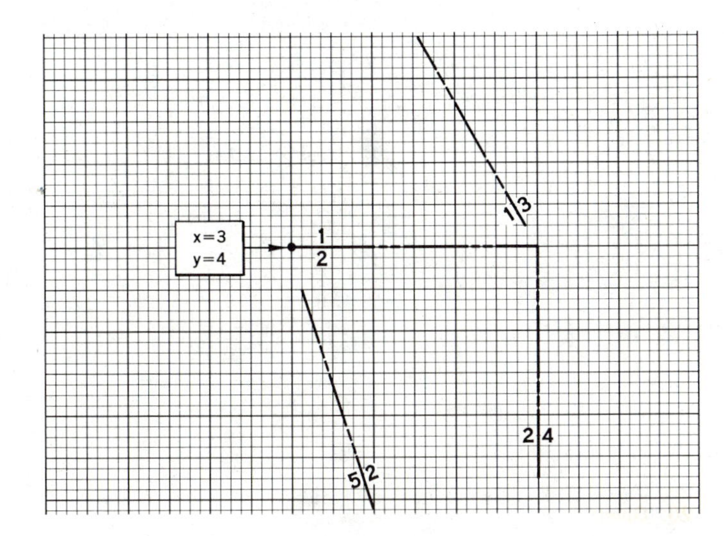

18. Given Data: Horizontal and frontal projections of three nonintersecting lines AB, CD, and EF.

Required: Determine by projection if any of the given lines have directions that are perpendicular to each other.

PROBLEM 18

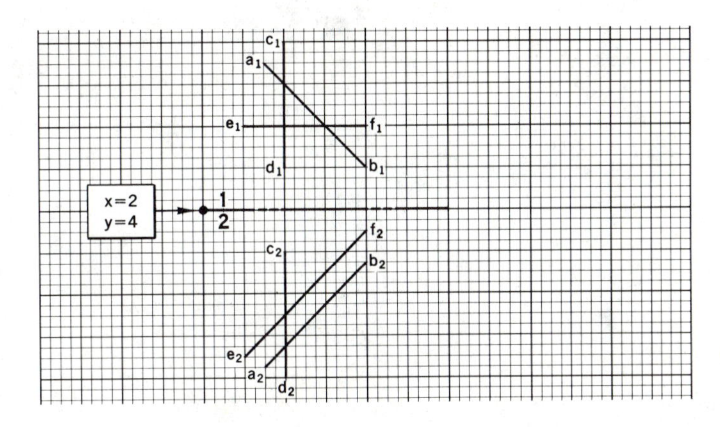

19. Given Data: Horizontal and frontal projections of point *P* and line *AB*.

Required: A. The horizontal and frontal projections of a line *PQ* from point *P* to line *AB*, the bearing of which is N60°E.

B. The true length and slope of line *PQ*. (*Ans.:* TL = $1\frac{3}{4}$ in., slope = 22°)

PROBLEM 19

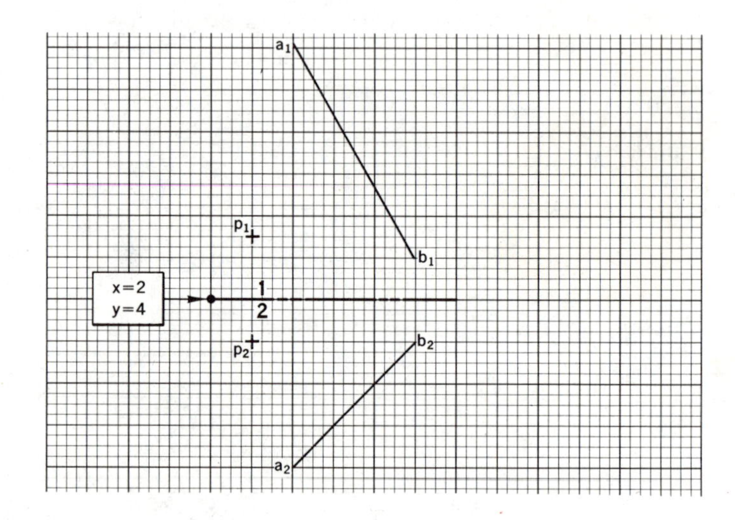

20. Given Data: Horizontal and frontal projections of point *R* and a profile line *CD*.

Required: A. The horizontal and frontal projections of a line *RS* which is perpendicular to line *CD* from *R*.

B. The true length, bearing, and slope of *RS*. (*Ans.:* TL = 2 in., bearing = N38°E, slope = 38°)

PROBLEM 20

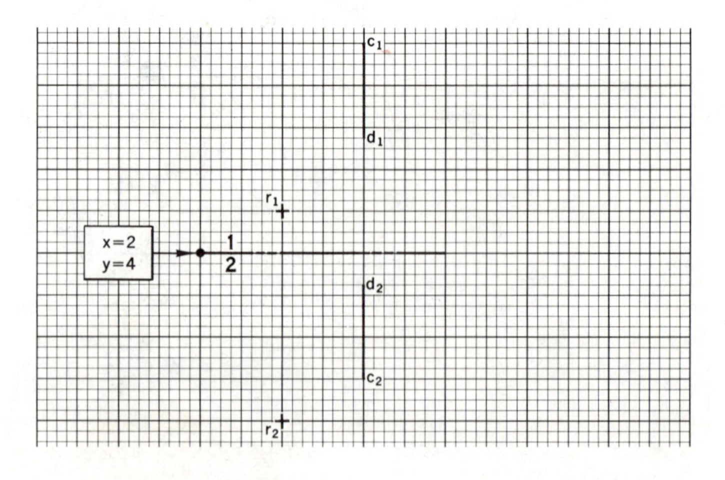

Sample Descriptive Geometry Test #1 (Covering Chapters I and II)

COURSE NAME: __________ COURSE NO.: __________ STUDENT NAME: ________________ DATE: __________

TIME LIMIT: 45 MINUTES

NOTE: Wherever actual construction is necessary in order to solve a problem, the problem can be reproduced on a 8 in. × 10 in. cross-section paper having 8 in. × 8 in. divisions to the inch, horizontally and vertically (see the Appendix for solutions to test problems).

1A. GIVEN DATA: RL 1–2, RL 1–3, RL 2–4, and RL 3–4.

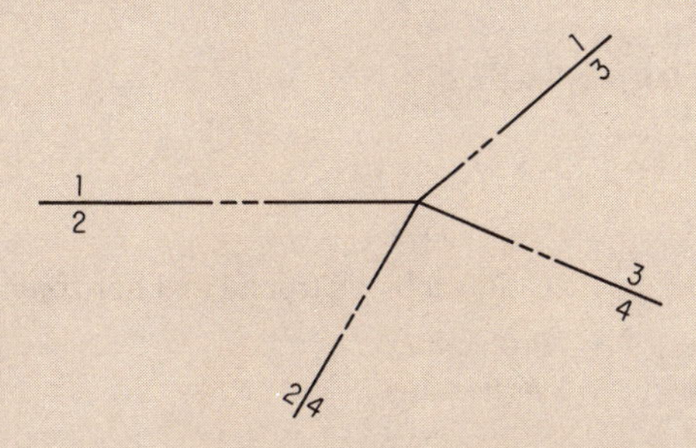

When you are looking at projection plane #1, RL 1–2 represents the edge of __________.

When you are looking at projection plane #2, RL 1–2 represents the edge of __________.

When you are looking at projection plane #3, RL 1–3 represents the edge of __________.

When you are looking at projection plane #1, RL 1–3 represents the edge of __________.

When you are looking at projection plane #4, RL 3–4 represents the edge of __________.

When you are looking at projection plane #2, RL 2–4 represents the edge of __________.

When you are looking at projection plane #4, RL 2–4 represents the edge of __________.

1B. Make a freehand pictorial representation below of the projection planes and their relative positions in space as indicated by the reference lines in 1A. Identify all planes by their respective names and numbers.

2. In orthographic projection the projectors must be

a. Perpendicular to a true length line.
b. Perpendicular to the line of sight.
c. Perpendicular to a projection plane.
d. Parallel to a projection plane.

Ans.: __________

3. To see a view showing a line as true length the projection plane must be

a. Perpendicular to the line.
b. Oblique to the line.
c. Parallel to the line.
d. Parallel to a reference line.

Ans.: __________

4. In the figure below, line *BC* shown by its frontal and horizontal projections is a

a. Horizontal line.
b. Frontal line.
c. Profile line.
d. Vertical line.

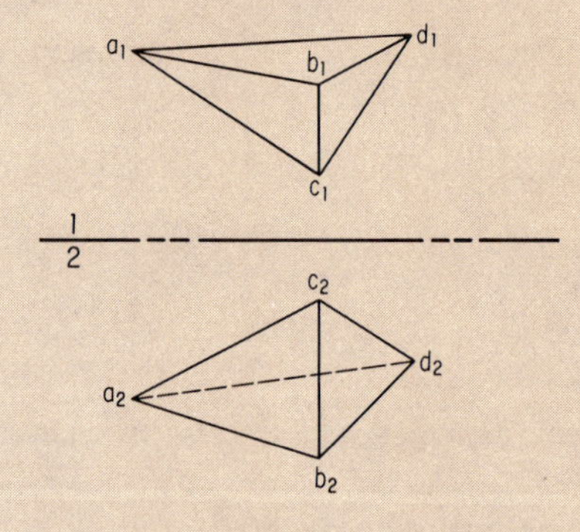

Ans.: __________

5. If two lines are perpendicular to each other in space, what view will show the actual 90° between them?

Ans.: __________

6. GIVEN DATA: Horizontal and frontal projections of point C.

REQUIRED: Draw the top, front, and end views of the line CD, having a bearing of S30°E, a slope of 45° downward from C, and a length of 90 ft. (Use a scale of $\frac{1}{8}$ in. = 10 ft.)

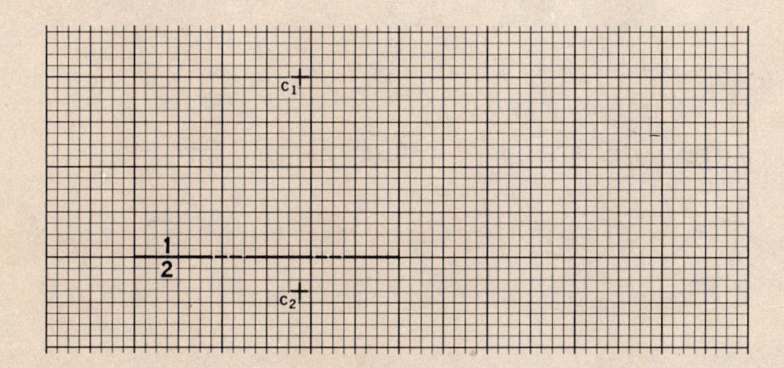

Plane Surfaces in a Three-Dimensional Space

21 Definition and generation of a plane surface

In Chapter I you noted that a *two-dimensional space* was defined as a *plane,* and that it was generated by translating a line parallel to itself. Also, you noted that a plane may be generated by revolving a line about a point on the line so that the line describes a surface having the shape of a sector of a plane circle or, if the revolution goes through 360°, of a complete plane circle [see Figs. 51(a) and 51(b)].

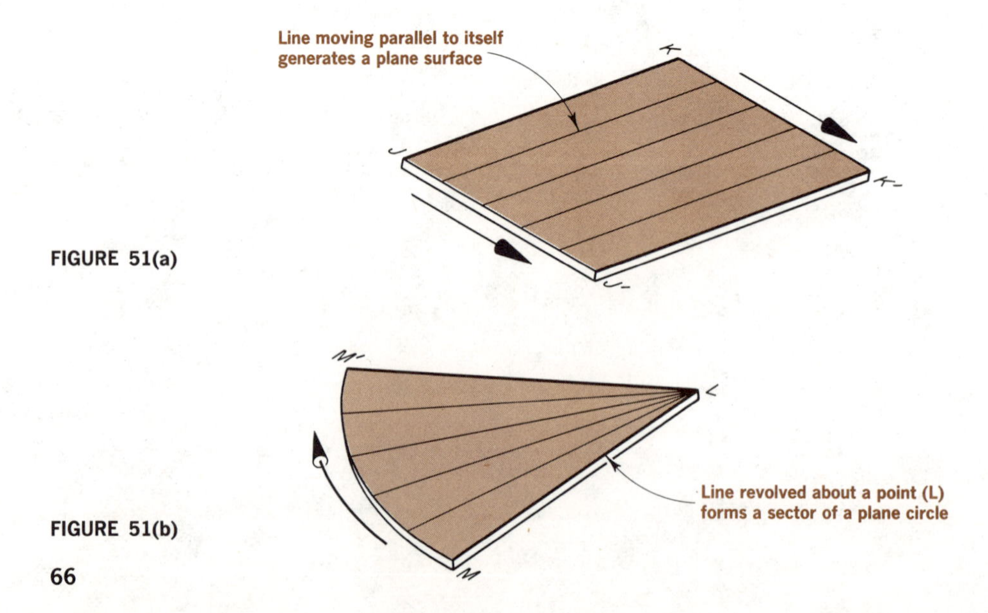

FIGURE 51(a)

FIGURE 51(b)

A plane surface has the characteristic of being a flat surface on which a line or a straight edge may lie in any position. This means that, at every point along its length, the line or straight edge is in contact with the plane surface (see Figs. 52–55).

22 Representation of plane surfaces

Plane surfaces may be represented in four basic ways:

A. PLANE FORMED BY TWO INTERSECTING LINES

Figure 52 shows two intersecting lines *AB* and *CD* which form a plane *ACBD*. A line *XY* on the plane *ACBD* remains in contact with the plane in all positions along its length.

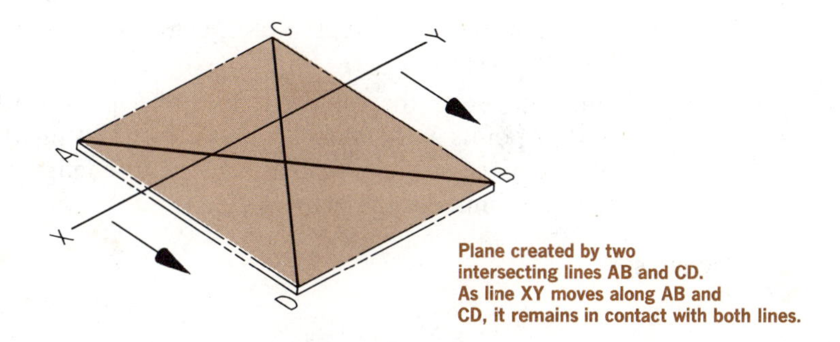

FIGURE 52

Plane created by two intersecting lines AB and CD. As line XY moves along AB and CD, it remains in contact with both lines.

B. PLANE FORMED BY TWO PARALLEL LINES

Figure 53 shows two parallel lines *EG* and *FH* which represent a plane *EFHG*. A line *XY* on this plane remains in contact with the plane in all positions along its length, which is within the confines of the plane limited by *EF* and *HG*.

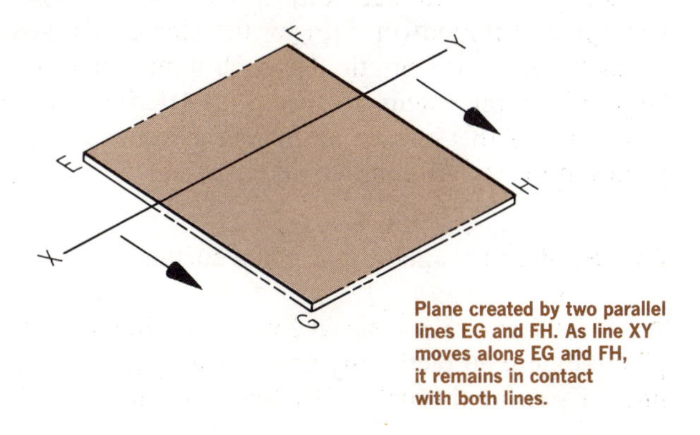

FIGURE 53

Plane created by two parallel lines EG and FH. As line XY moves along EG and FH, it remains in contact with both lines.

C. PLANE FORMED BY A STRAIGHT LINE AND A POINT NOT ON THE LINE

Figure 54 shows a line *AB* and a point *X* outside the line. The plane represented by the point *X* and the line *AB* contains a line *LM* which is in contact with the plane surface *ABX* along its entire length and in any position on the plane.

FIGURE 54

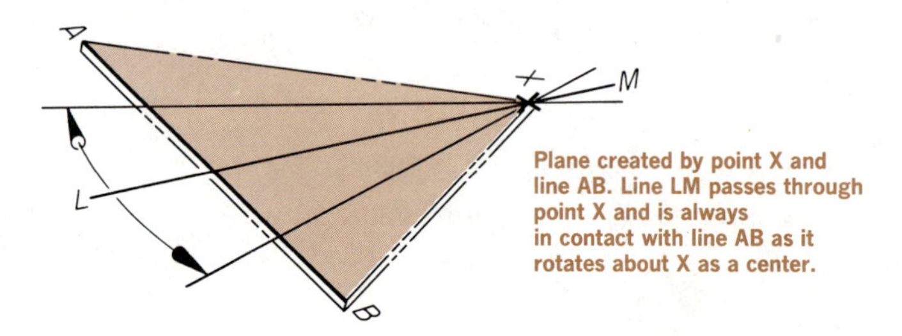

D. PLANE FORMED BY THREE NONCOLLINEAR POINTS (POINTS NOT IN A STRAIGHT LINE)

Figure 55 shows the representation of a plane surface by three noncollinear points *X*, *Y*, and *Z*. A line *LM* on this plane remains in contact with the plane surface *XYZ* along its entire length and in any position within the limits of the plane *XYZ*.

FIGURE 55

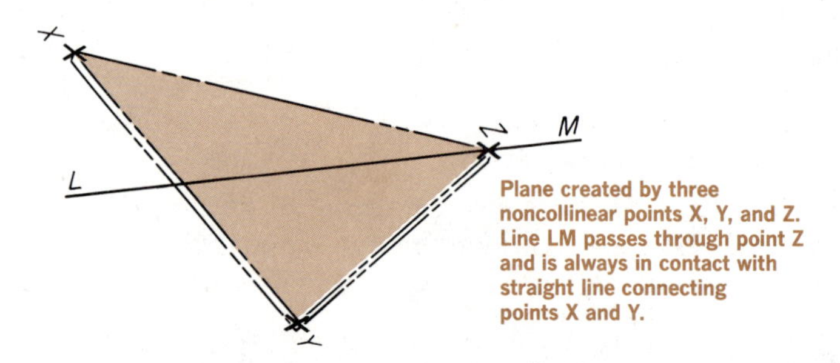

It should be mentioned at this point that the planes represented in Figs. 52–55 were planes limited (in size) by the elements that created the plane surfaces. Actually, the representation in each figure could be considered a "piece" of a larger plane, the extent of which is indefinite. It is convenient in some cases to consider a plane surface as having a limitless size. (For example, projection planes in theory are limitless in size.)

23 Position of a point in a plane surface

A point lies in a plane surface when any line passing through it also lies on the plane. Figure 56 pictorially presents a plane *ABC* that contains a point *P*. The line *XY* passing through *P* also lies on the plane *ABC*.

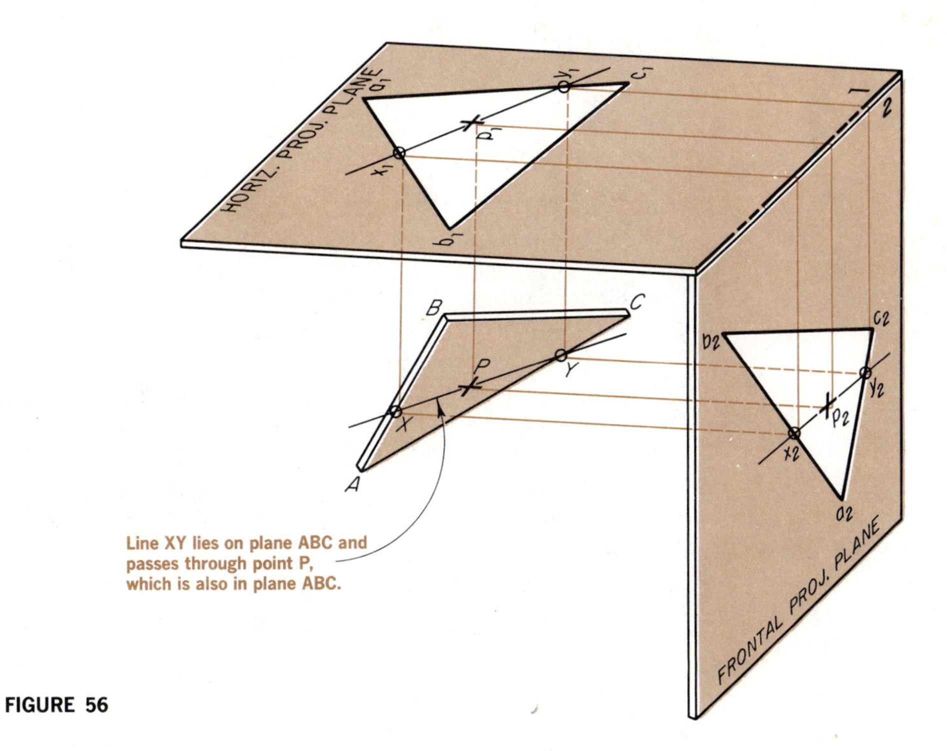

FIGURE 56

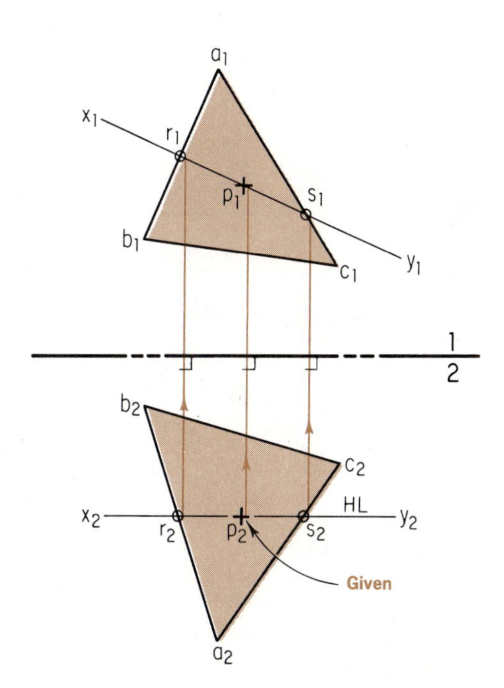

FIGURE 57

Orthographic Example: Determination of the position of a point in a plane surface (Fig. 57).

Given Data: Horizontal and frontal projections of a plane ABC and the vertical projection p_2 of the point P.

Required: The horizontal projection p_1 of point P in the plane ABC.

Construction Program:

a. In the frontal view, draw a line x_2y_2 on the plane projection $a_2b_2c_2$, which passes through p_2. (The line XY used in the example happens to be a horizontal line, but it could be any line.) We know that x_2y_2 is on plane ABC, since it intersects lines AB and AC of the plane ABC at points R and S, respectively.

b. Determine the horizontal projection x_1y_1 of the line XY by constructing perpendicular projectors from points r_2 and s_2, thereby locating r_1 and s_1 in the horizontal projection view.

c. Project p_2 up to view #1. The point at which this projector intersects x_1y_1 is p_1, the required horizontal projection of the point P in the plane ABC.

(NOTE: Figure 58 depicts a line passing through p_2 as any general line on the plane ABC.)

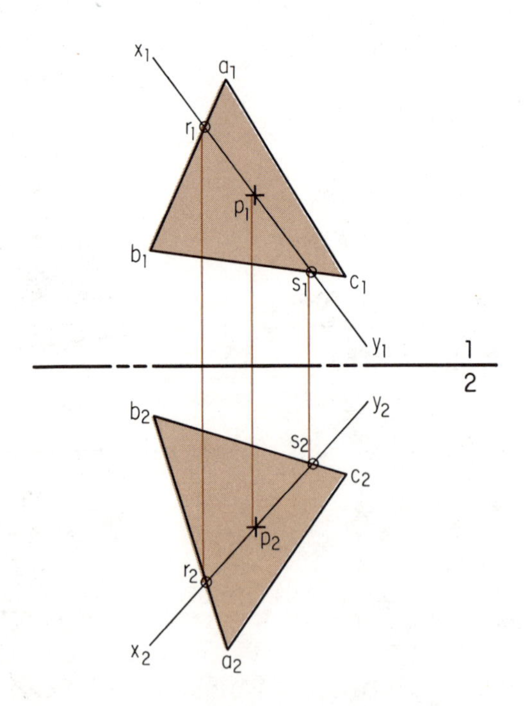

FIGURE 58

24 Edge view of an oblique plane surface

In order to see an oblique plane as an edge, the observer must assume a position in space so that a line in the given plane appears as a point. Figure 59 shows an observer in three positions in which three different lines AX, CY, and BZ appear as points. From each of these positions the plane ABC will appear as an edge.

Because a plane may contain an infinite number of lines, there are an infinite number of positions available from which you may view the plane to see it as an edge.

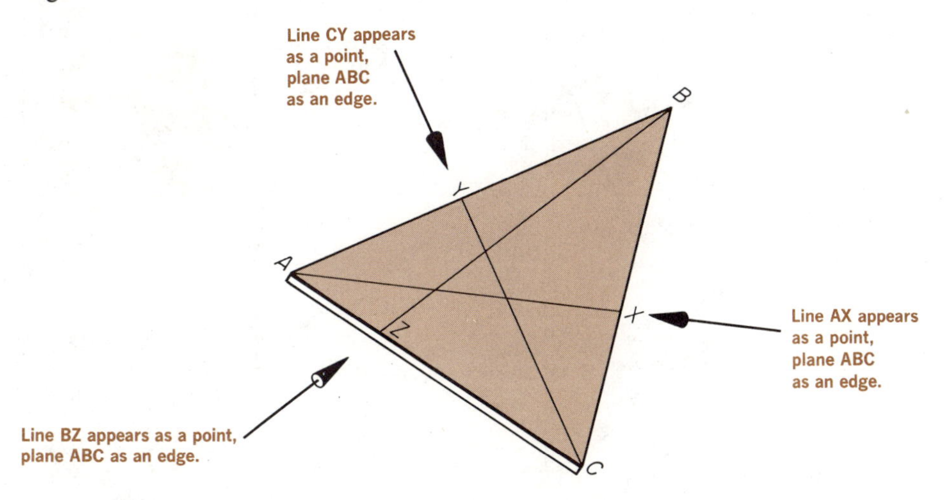

FIGURE 59

When you solve problems dealing with the edge view (EV) of a plane you must use a *direct approach* so that you can construct a view showing the plane's edge. Figure 60 pictorially presents an oblique plane ABC, the location of which in space is relative to the horizontal and frontal projection planes. In this figure you see that a horizontal line AX has been placed in the plane and projected onto an auxiliary elevation plane #3, in which AX appears as a point and the plane ABC as an edge.

Orthographic Example: Edge view of an oblique plane surface (Figs. 61 and 62).

Given Data: Horizontal and frontal projections of plane ABC.

Required: Edge view of plane ABC.

Construction Program #I: Utilizing a horizontal line—Fig. 61:

a. Draw a horizontal line a_2x_2 in plane ABC (view #2).

b. Project the horizontal line a_2x_2 into view #1 to get its true length projection a_1x_1.

FIGURE 60

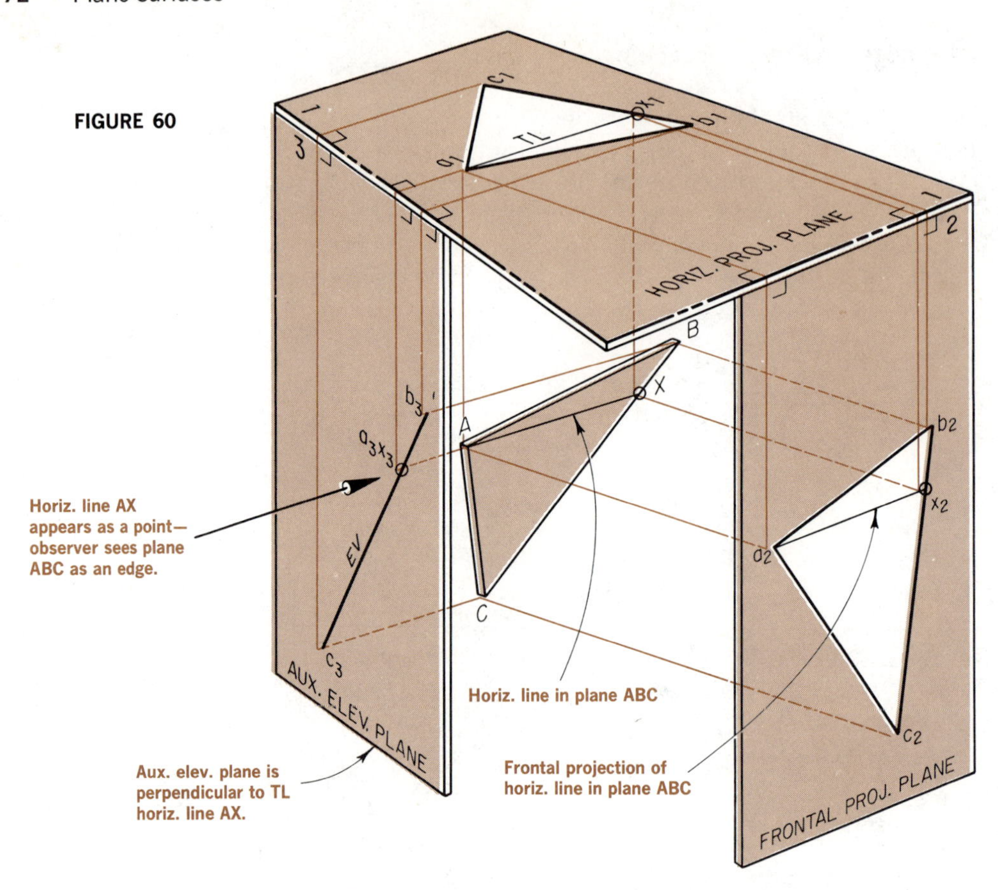

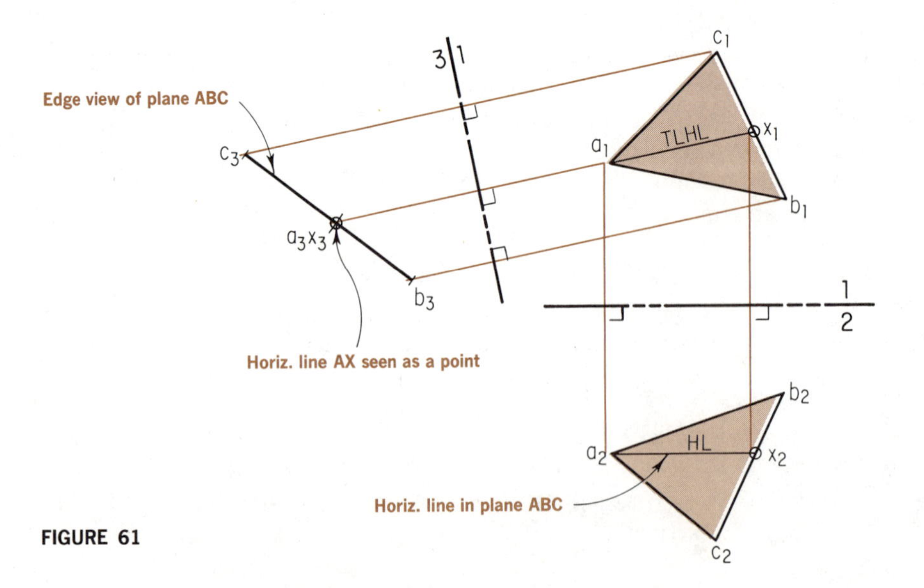

FIGURE 61

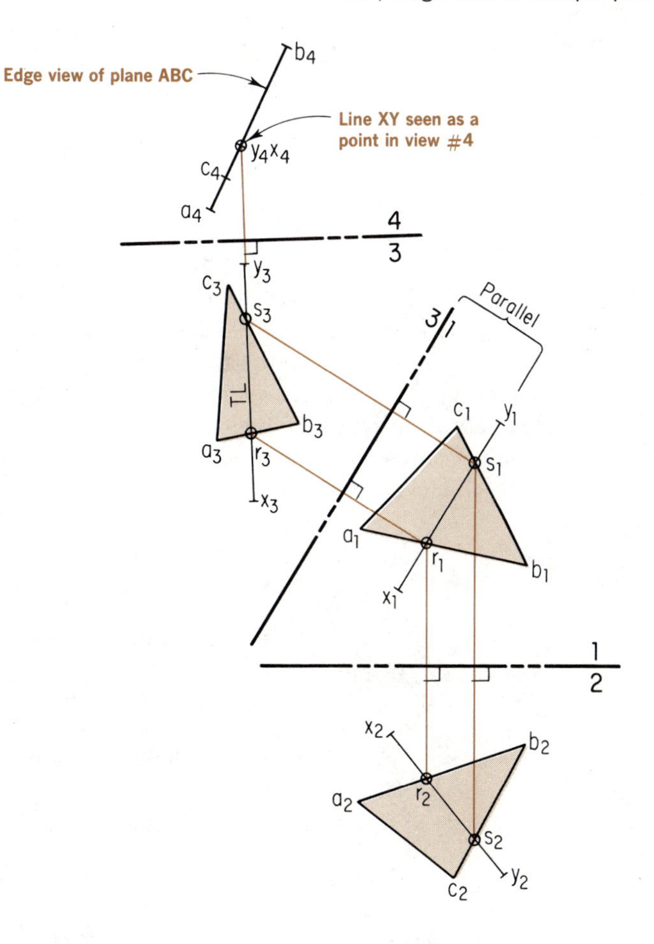

FIGURE 62

c. Pass a projection plane #3 *perpendicular* to the true length horizontal line a_1x_1.

d. Taking measurements below the horizontal projection plane in view #2 and transferring them with dividers to view #3, determine the horizontal line AX seen as a point a_3x_3. In view #3 plane ABC appears as an edge $a_3b_3c_3$.

Construction Program #II: Edge view of a plane by constructing a general line as a point—Fig. 62:

a. Construct any line XY on plane ABC in frontal view #2.

b. Determine the horizontal projection x_1y_1 of line XY.

c. Pass a projection plane #3 *parallel* to x_1y_1 in order to determine the true length projection of XY, which is noted as x_3y_3 in view #3.

d. Pass a projection plane (view #4) *perpendicular* to the true length projection x_3y_3 in order to see line XY as a point x_4y_4 in view #4. View #4 shows the edge view of plane ABC (measurements for view #4 were obtained from view #1).

25 True slope angle of a plane

The true slope angle (also referred to as "dip") is defined as the angle a plane makes with the *horizontal projection plane* (see Fig. 63).

In order for an observer to see the true slope angle of a given plane, he must obtain a view in which he can see *simultaneously* the *edge of the given plane* and the *edge of the horizontal projection plane.* This means that he can see the slope of a plane in an *elevation* view, since the horizontal projection plane can appear as an edge *only* in elevation views.

FIGURE 63

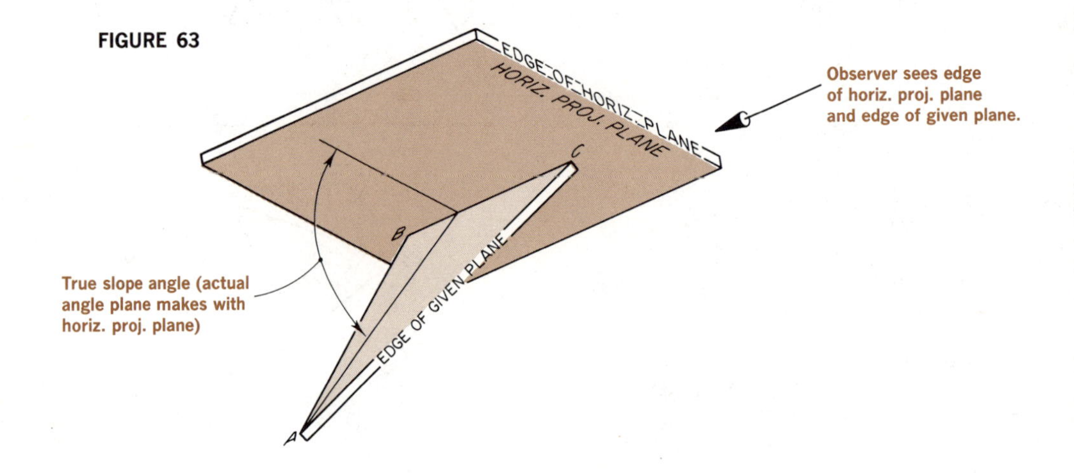

Orthographic Example: True slope angle of an oblique plane (Fig. 64).

Given Data: Horizontal and frontal projections of an oblique plane ABC.

Required: True slope angle of plane ABC.

Construction Program:

a. Draw a horizontal line a_2x_2 on plane ABC (see view #2).

b. Determine the horizontal projection a_1x_1 of horizontal line AX. (The horizontal projection a_1x_1 appears as a true length line.)

c. Determine a point view of horizontal line AX by passing a projection plane #3 *perpendicular* to the true length projection a_1x_1. In view #3, plane ABC appears as an edge, and since view #3 is an elevation view in which the horizontal projection plane is also seen as an edge, the true slope angle can be measured in this view. This is the angle the plane ABC makes with the horizontal plane. (Note that the construction is similar to that followed in Fig. 61 to find the edge view of a plane.)

FIGURE 64

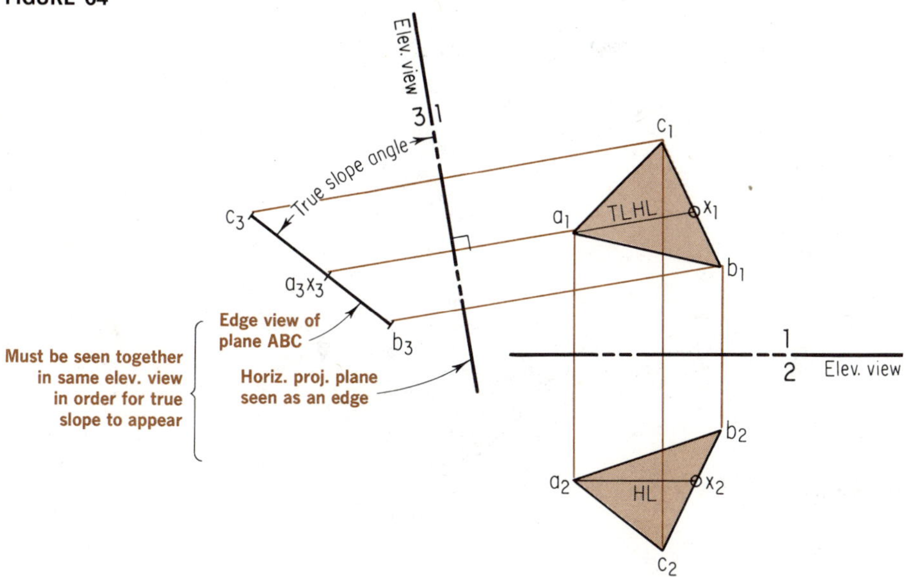

26 Normal, or true shape, view of a plane

The *normal view* of a given plane is a view in which the observer's lines of sight are *perpendicular* to the given plane (see Fig. 65).

A normal view of a plane results in a *true shape* view and appears on a projection plane that is perpendicular to the observer's lines of sight and, therefore, *parallel* to the given plane (refer to Fig. 65).

Before the true shape projection of the given plane can be determined, the observer must first construct a view in which the given plane appears as an edge. The edge view of the given plane can then be projected onto a parallel projection plane on which the true shape projection of the given plane will appear.

Orthographic Example: Normal, or true shape, view of a plane (Figs. 66 and 67).

Given Data: Horizontal and frontal projections of an oblique plane ABC.

Required: The normal or true shape view of plane ABC.

Construction Program:

a. Draw a frontal line b_1x_1 in the plane ABC (view #1).

b. Project the frontal line into the frontal projection (view #2) in order to determine its true length b_2x_2.

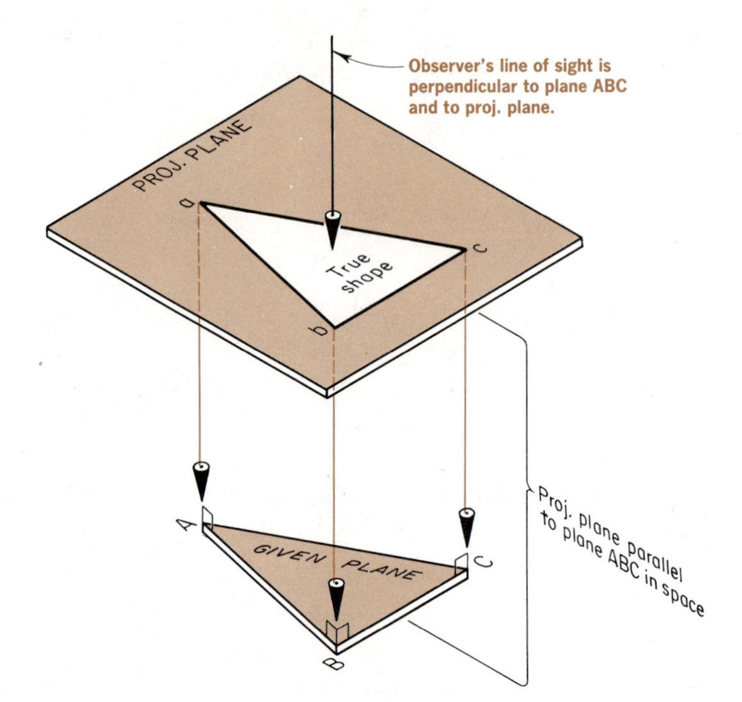

FIGURE 65

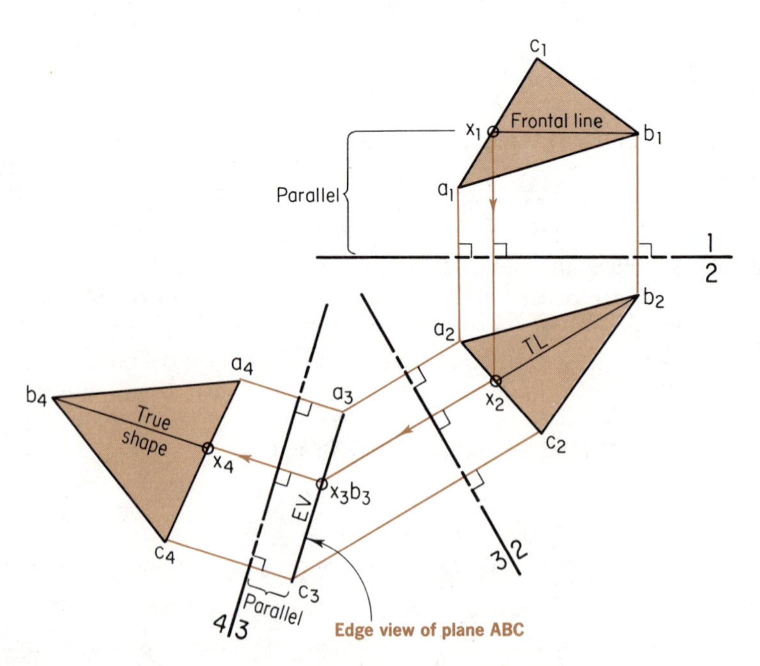

FIGURE 66

FIGURE 67

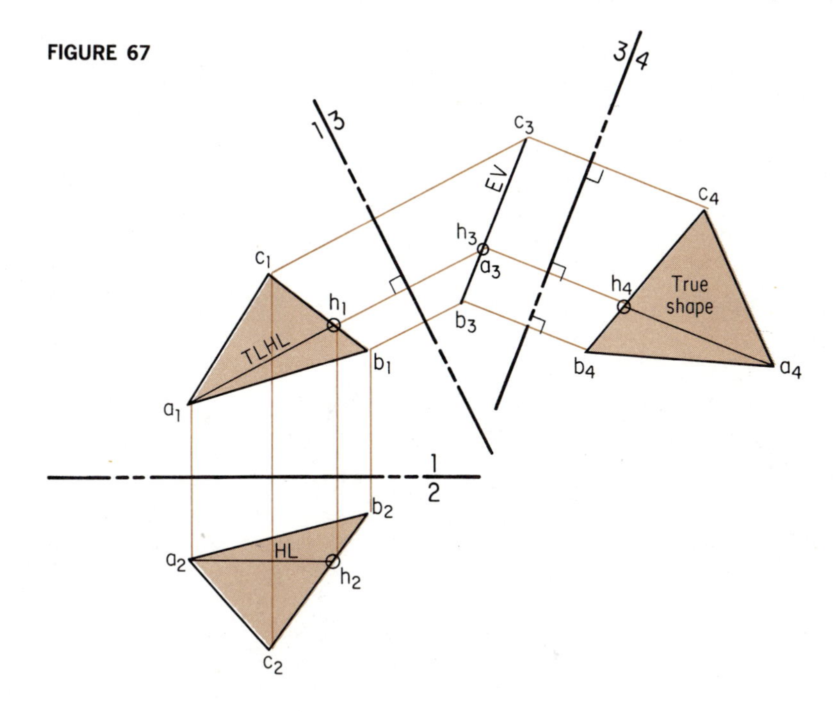

c. Draw projection plane #3 (represented by RL 2–3) *perpendicular* to the true length frontal line b_2x_2. In view #3, frontal line BX appears as a point x_3b_3 and, therefore, plane ABC appears as an edge $a_3b_3c_3$.

d. Construct a plane #4 (represented by RL 3–4) *parallel* to the edge view $a_3b_3c_3$ of plane ABC.

e. Project plane ABC into view #4 in order to determine the true shape $a_4b_4c_4$ of plane ABC (measurements for view #4 were obtained from view #2).

You can also determine the true shape of the plane ABC by first constructing a horizontal line on the plane ABC, thereby determining the plane as an edge (see Fig. 67).

SAMPLE QUIZZES

1. State three ways in which a plane may be created in space.

2. What conditions must be satisfied in order to see the true slope angle of an oblique plane?

3. What determines the position and direction of the observer's line of sight for a view showing a plane in its true shape?

4. What is generally the minimum number of views necessary, in addition to the given horizontal and frontal projection views, to obtain the following:

a. The edge view of a plane?
b. The true size of a plane?
c. The true slope angle of a plane?

5. Under what conditions can a line in a plane have the same slope as the plane itself?

PRACTICE PROBLEMS

1. Given Data: Horizontal and frontal projections of two parallel lines *AB* and *CD*, and the frontal projection of a point *P* in the plane formed by lines *AB* and *CD*.

Required: The horizontal projection p_1 of point *P* in the plane formed by lines *AB* and *CD*.

PROBLEM 1

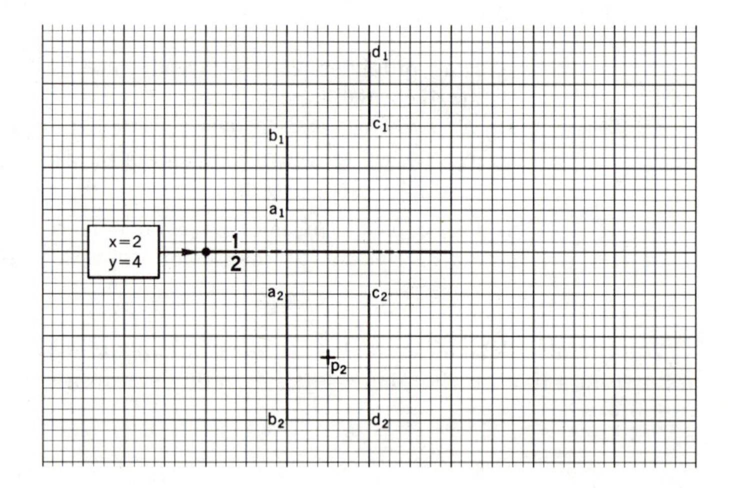

2. Given Data: A. RL 1–2 and RL 2–3.

B. Horizontal and frontal projections of a circular plane having a center *C*, and the horizontal projection p_1 of a point *P* in this plane.

Required: A. Auxiliary inclined view #3 of the given circular plane. (HINT: Assume points on the circle in view #1 and project them to views #2 and #3.)

B. The frontal projection p_2 and auxiliary projection p_3 of point *P* in the given plane.

PROBLEM 2

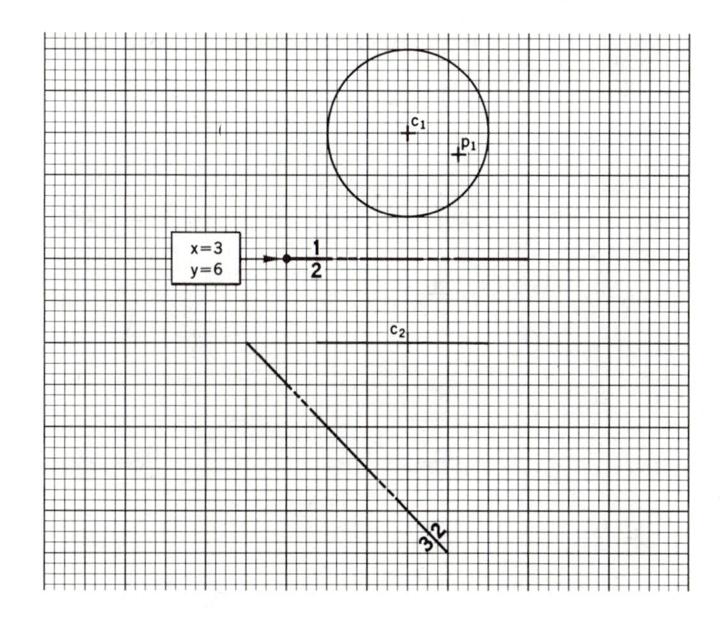

3. Given Data: Horizontal and frontal projections of two intersecting lines AB and CD, and the horizontal projection of a point P in the plane formed by lines AB and CD.

Required:

A. The frontal projection p_2 of point P in the plane formed by lines AB and CD.

B. The edge view of the plane (formed by lines AB and CD) in which line CD appears as a point.

C. The edge view of the plane (formed by lines AB and CD) in which line AB appears as a point. (SUGGESTION: Project from view #2 when working with line AB.)

D. Label and show all construction.

PROBLEM 3

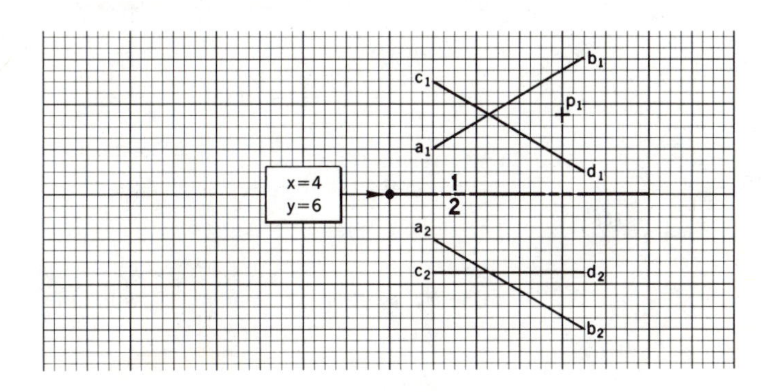

4. Given Data: Horizontal and frontal projections of plane *ABC*.

Required: A. The slope of plane *ABC*. (*Ans.:* 36°)

B. The true shape of plane *ABC*.

PROBLEM 4

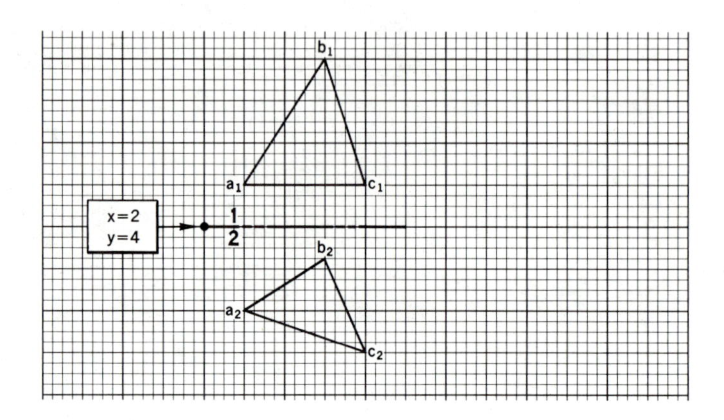

5. Given Data: Horizontal and frontal projections of plane *WXY*.

Required: A. The edge view of plane *WXY* in which a frontal line appears on the planc as a point.

B. The true shape or plane *WXY*.

PROBLEM 5

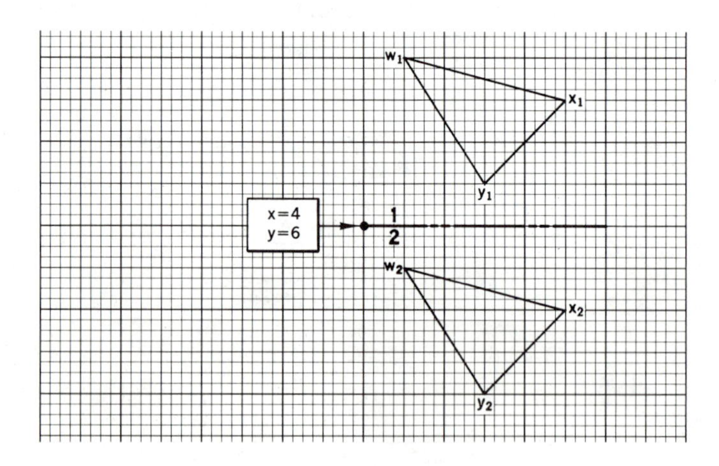

6. Given Data: Horizontal and frontal projections or plane *LMN*.

Required: A. The slope of plane *LMN*. (*Ans.:* 33°)

B. The horizontal and frontal projections, from point *N* to side *LM*, of a line *NP* that has an angle of 70° with side *LM* of plane *LMN*.

C. The true length, bearing, and slope of line *NP*. (*Ans.:* TL = $1\frac{11}{16}$ in., bearing = N31°E, slope = 31.5°)

PROBLEM 6

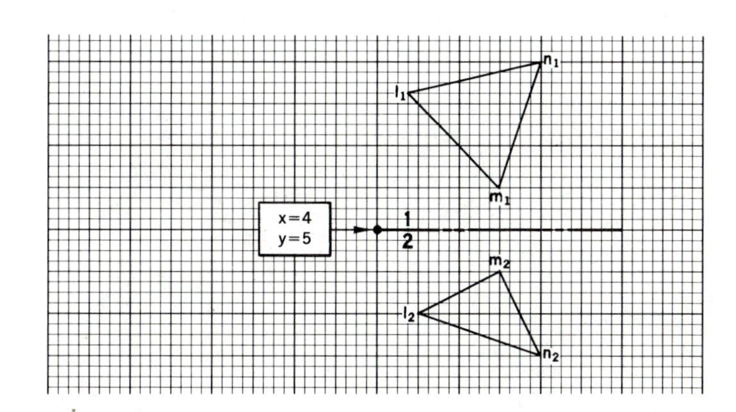

7. Given Data: Horizontal and frontal projections of planes *ABD* and *ACE*, which have a common point *A*.

Required: A. The slope and true shape of plane *ABD*. (*Ans.:* slope = 48°)

B. The slope and true shape of plane *ACE*. (*Ans.:* slope = 45°)

PROBLEM 7

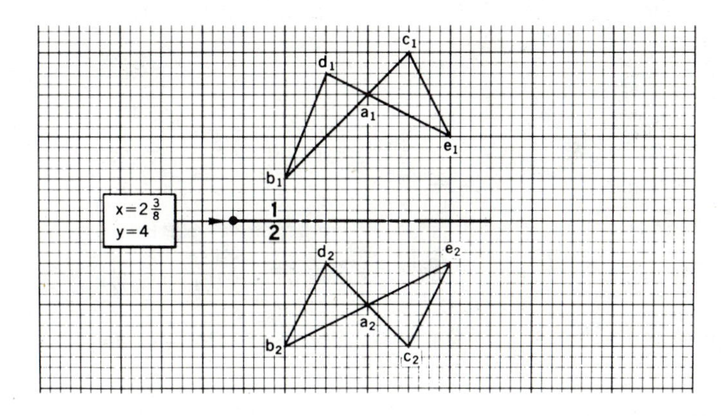

8. Given Data: Horizontal and frontal projections of a plane *ABD–ACE* formed by two intersecting lines *BC* and *DE*.

Required: The slope and true shape of plane *ABD–ACE*. (*Ans.:* slope = 48°)

PROBLEM 8

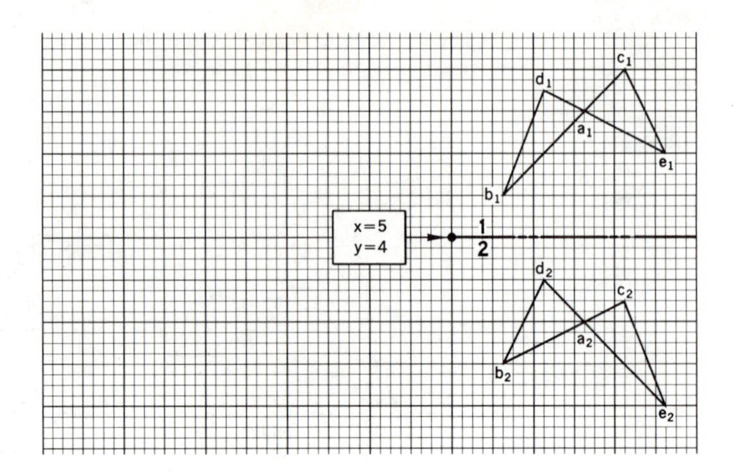

9. Given Data: A. Horizontal and frontal projections of point *A*.

B. The following spatial description of point *A*.

a. Point *A* is a vertex of an equilateral triangle *ABC* with $1\frac{1}{2}$ in. sides. Point *A* lies on a horizontal line in the triangle. The horizontal line has a bearing of N45°E from point *A*. The triangle has a slope of 45° downward and forward from point *A*.

b. Vertex *C* of the equilateral triangle is also located on the horizontal line containing point *A*.

Required: The horizontal and frontal projections of triangle *ABC*.

PROBLEM 9

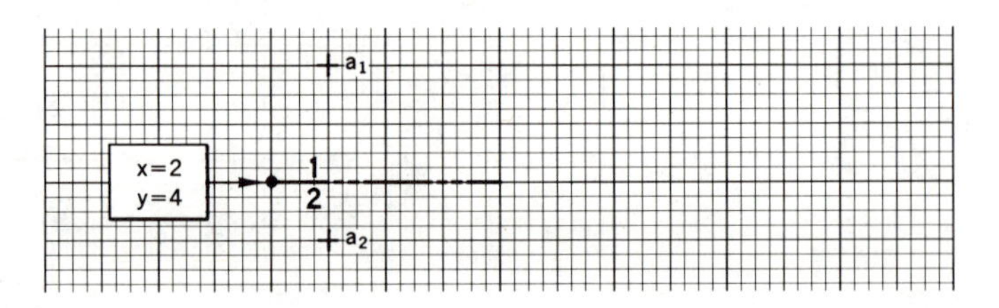

10. Given Data: Horizontal and frontal projections of plane *ABCD*.

Required: A. The slope and true shape of plane *ABCD*. (*Ans.:* slope = 47.5°)

B. The horizontal and frontal projections of a circle that has a $\frac{1}{2}$ in. radius and a center *O* located at the point of intersection of the two diagonals *AC* and *BD* in the given plane. The circle is located in plane *ABCD*.

PROBLEM 10

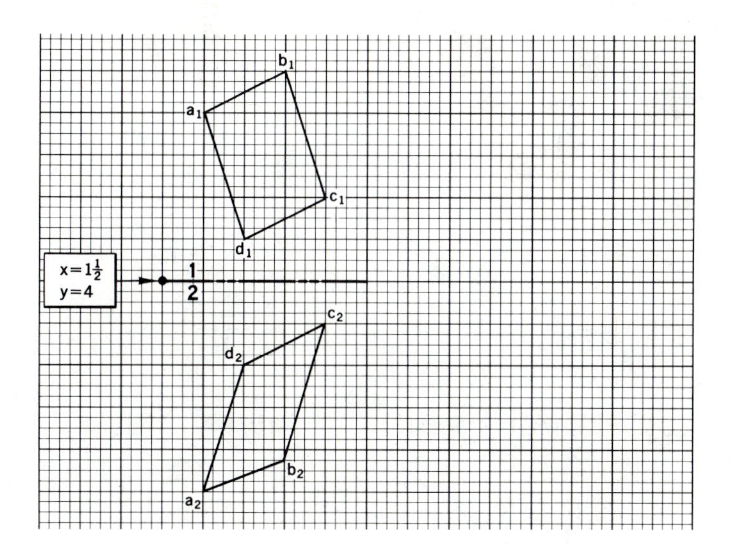

IV

Three-Dimensional Spatial Relationships of Lines and Planes

27 Basic relationships between lines and planes

You have now reached a point in the development of three-dimensional descriptive geometry where you will be able to apply the basic concepts involving true lengths and point views of lines, and true shapes and edge views of planes. You can now combine these concepts and apply them in such a way as to make it possible to solve spatial problems concerned with the actual measurement of distances, including straight line distances and angles. The basic construction programs discussed heretofore will play an important part in the solution of problems involving these combined relationships of lines and planes. The following topics will be presented in this chapter:

1. A line parallel to a plane surface.
2. A line perpendicular to a plane surface.
3. Shortest perpendicular distance from a point to a line.
4. Shortest perpendicular distance between two skew lines.
5. Shortest horizontal distance between two skew lines.
6. Shortest grade (slope) distance between two skew lines.
7. Shortest perpendicular distance from a point to a plane surface.
8. Relative positions of two skew lines as applied to the determination of visibility of lines on solid objects.
9. Point of intersection (piercing point) of a line and a plane surface.

10. Determination of the line of intersection between two limited planes.
11. Determination of the line of intersection between two unlimited planes.
12. Determination of the dihedral (acute) angle between two plane surfaces.
13. Determination of the angle (acute) between a line and a plane surface.

28 Parallel lines and parallel planes

You have seen that if two lines are parallel they are equidistant from each other in all points. This means, of course, that under these circumstances parallel lines never meet (see Fig. 68).

FIGURE 68

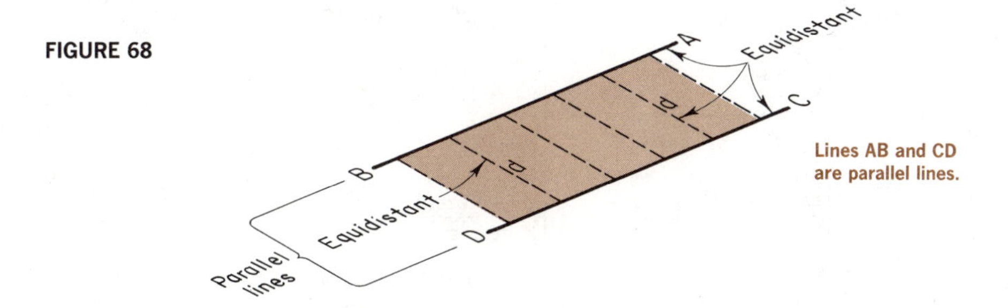

Lines AB and CD are parallel lines.

Parallel planes are planes that are equidistant from each other in all points. Figure 69 shows two parallel planes *ABC* and *DEF*, respectively. All of the points in one plane are equidistant from their respective points in the other plane.

FIGURE 69

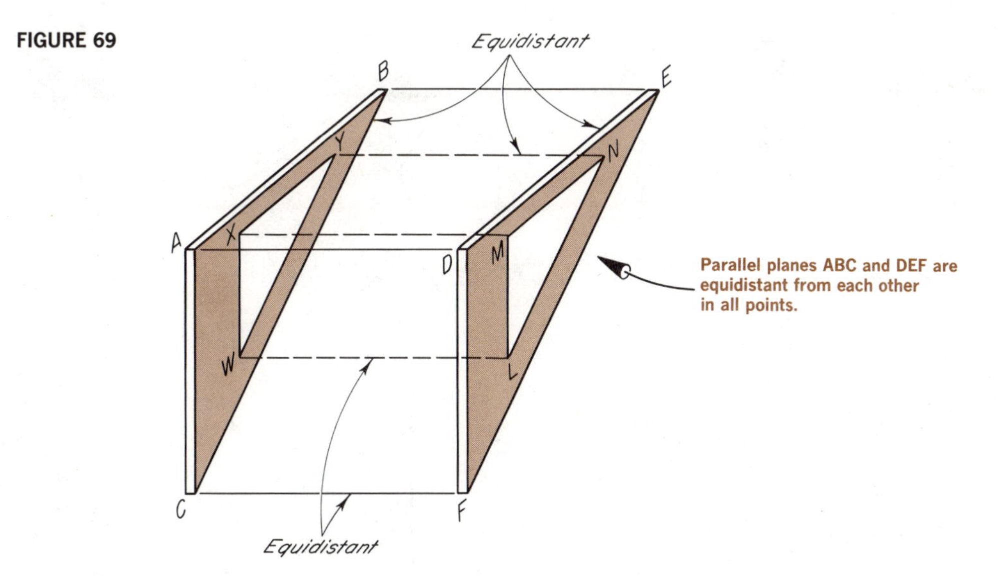

Parallel planes ABC and DEF are equidistant from each other in all points.

If we have two parallel planes *ABC* and *DEF* (as shown in Fig. 70), a line *LM* in the plane *ABC*, and a line *RS* in the plane *DEF*, this does not automatically mean that lines *LM* and *RS* are parallel. From Fig. 70 we can see that lines *LM* and *RS* are *not* parallel lines, since they are not equidistant from each other in all points. Lines *LM* and *RS* under this condition are lines that do not intersect and are not parallel. Therefore, they are called *skew lines.*

FIGURE 70

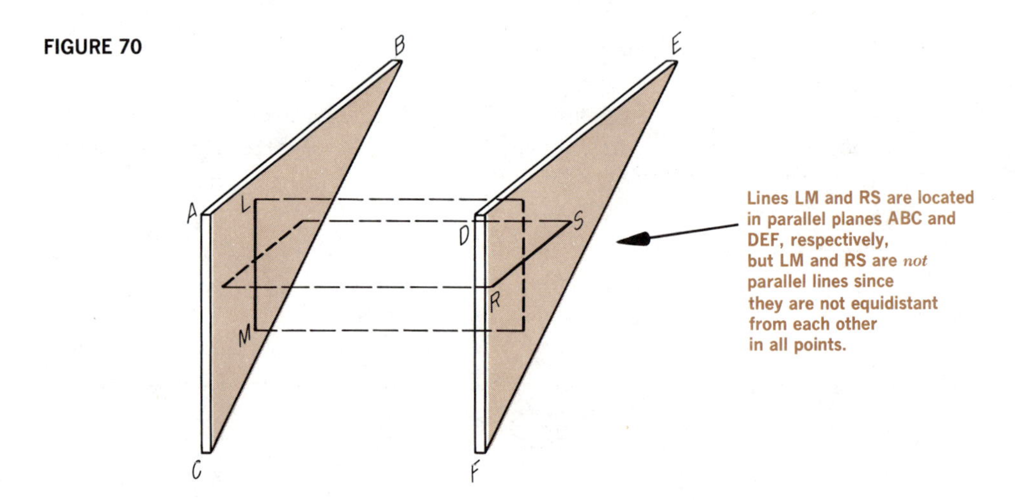

29 A line parallel to a plane surface

If a line outside a given plane is parallel to *any* line on the given plane, it is also parallel to the plane itself. In Fig. 71 you see a line *XY* (which is *not* on the plane

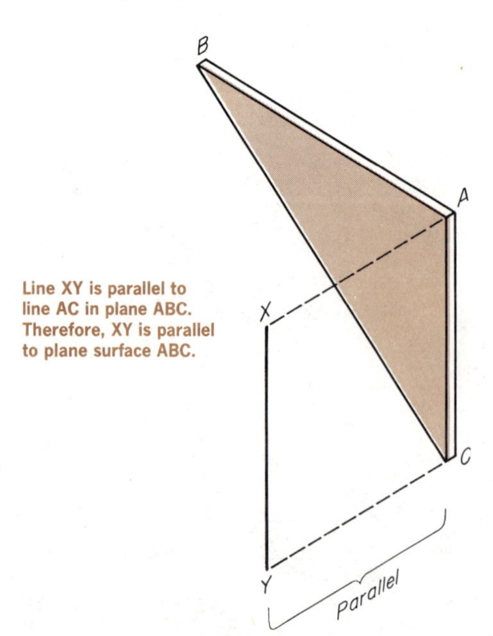

FIGURE 71

FIGURE 72

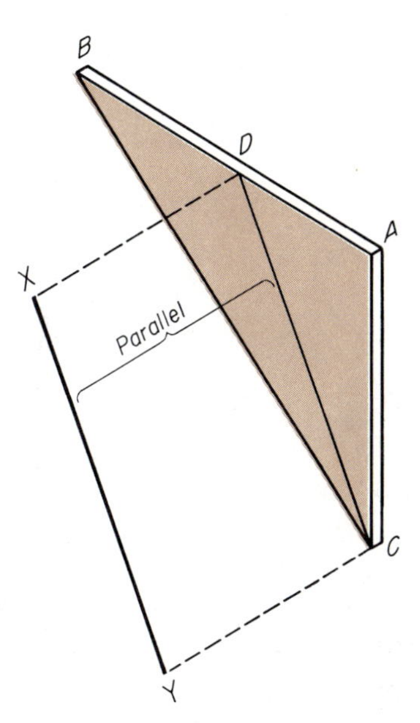

Line XY is parallel to line CD in plane ABC. Therefore, XY is parallel to plane surface ABC.

ABC) parallel to line *AC* of plane surface *ABC*. This makes the line *XY* equidistant at all points from line *AC*. Therefore, line *XY* is parallel to the plane surface *ABC*.

In Fig. 72 you see a line *XY* (which is *not* on the plane *ABC*) parallel to any line *CD* in the plane surface *ABC*. Using the same reasoning as you did in Fig. 71, you see that line *XY* must be parallel to the plane surface *ABC*.

When dealing with two skew lines in space, it is possible to create a plane that contains one of the skew lines and is parallel to the other.

Orthographic Example: Creation of a plane containing one given skew line and parallel to another (Fig. 73).

Given Data: Horizontal and frontal projections of two skew lines *AB* and *XY*.

Required:

a. A plane that contains the line *AB* and is parallel to the line *XY*.

b. Draw an edge view of the constructed plane to prove that the plane is parallel to line *XY*.

Construction Program:

a. In the horizontal projection view #1 draw a line a_1c_1 through point a_1, parallel to the horizontal projection x_1y_1 of line *XY*.

b. In the frontal view construct a line a_2c_2 through point a_2, parallel to the frontal projection x_2y_2 of line *XY*.

(NOTE: Intersecting lines *AB* and *AC*, as represented by their horizontal and frontal projections, form a plane. Since one line in this plane—line *AC*—is parallel to the line *XY* outside the plane, the plane *ABC* formed by the intersecting lines *AB* and *AC* is parallel to the line *XY* outside of this plane, and vice versa.)

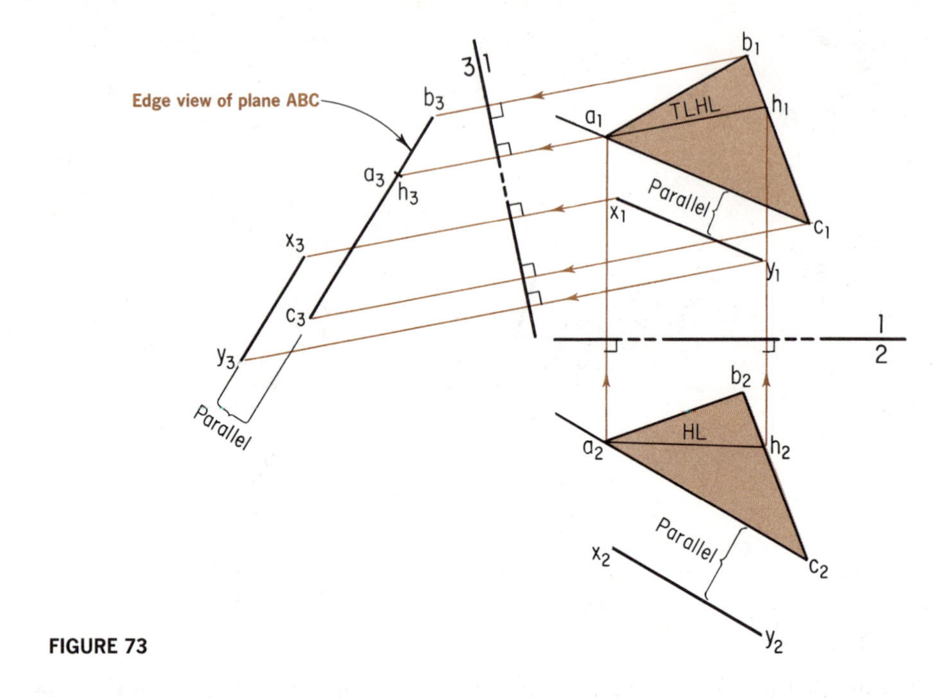

FIGURE 73

c. Limit the newly created plane by connecting points B and C with a straight line in both the horizontal and frontal views.

d. Draw a horizontal line a_2h_2 on the plane ABC in view #2.

e. Project the horizontal line a_2h_2 to view #1 in order to see its true length projection a_1h_1.

f. Construct a projection plane (represented by RL 1–3) that is perpendicular to the true length horizontal line projection a_1h_1. The edge view $a_3b_3c_3$ of plane ABC appears in view #3.

g. Project line XY into view #3. The projection x_3y_3 of line XY will appear parallel to the edge view projection $a_3b_3c_3$ of the plane ABC. This proves that the plane containing line AB is parallel to the line XY.

30 A line perpendicular to a plane surface

A line is perpendicular to a plane when every line in the plane that passes through the point of intersection of the given line and plane makes 90° with the given line. Figure 74 shows a line XY that is perpendicular to a given plane $ABCD$. Every line in the given plane passing through point Y makes 90° with the line XY.

Figure 75 shows the horizontal, frontal, and profile projections of a given horizontal plane $ABCD$ and a line XY perpendicular to this plane. When the given plane *ABCD* appears as an edge (in the frontal and profile projections in this case), you see the *actual* 90° that the line *XY* makes with this plane.

FIGURE 74

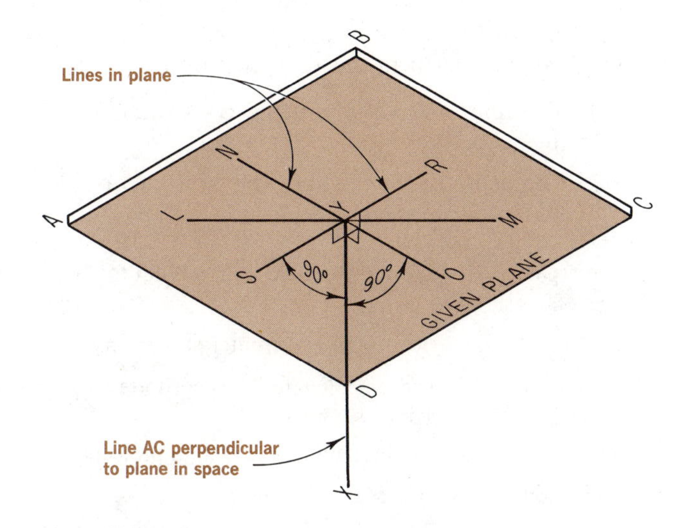

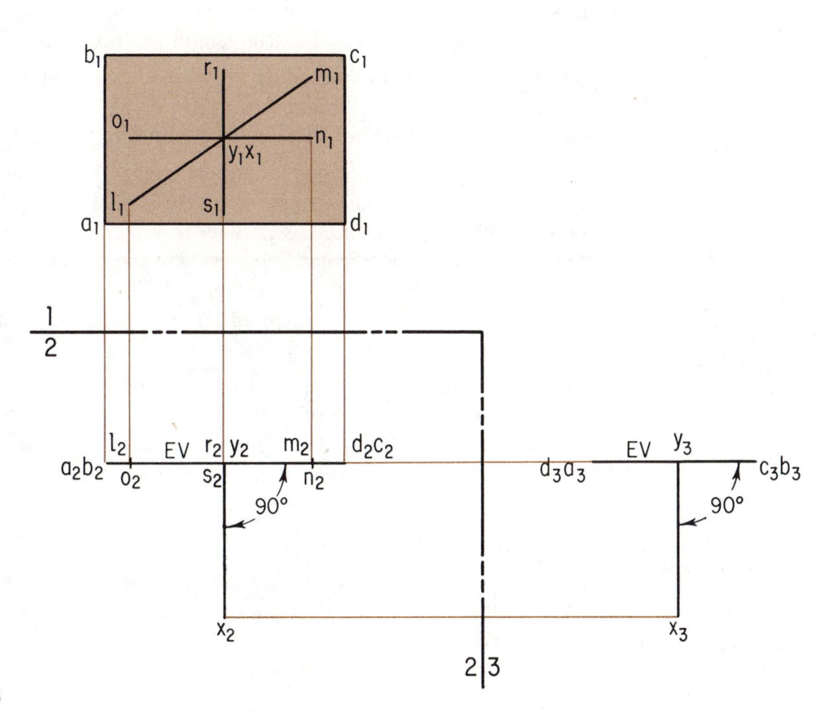

FIGURE 75

31 Shortest perpendicular distance from a point to a line

To determine the shortest distance from a point to a line, construct a line from the given point in space *perpendicular* to the given line. Two basic approaches are available to perform this construction:

1. The *line method,* where the construction deals directly with the given line.
2. The *plane method,* where the construction deals with the plane formed by the given point and line.

Orthographic Example: Shortest distance from a point to a line (Figs. 76 and 77).

Given Data: Horizontal and frontal projections of a line AB and a point X.

Required:

a. The true length of the shortest perpendicular distance XY from point X to line AB.

b. The horizontal and frontal projections of XY.

Construction Program #I: Line Method—Fig. 76:

a. Determine the true length projection of line AB by passing a projection plane #3 (represented by RL 1–3) parallel to the horizontal projection a_1b_1 of line AB.

b. In view #3, where the true length projection a_3b_3 appears, construct a *perpendicular* from x_3 to the true length projection a_3b_3.

(NOTE: This fulfills the reasoning presented in Section 20 where it is stated that a 90° angle between two intersecting lines will be seen in any view which shows at least one of the lines in true length.)

If you project the foot of the perpendicular y_3 into the horizontal projection view #1 and frontal projection view #2, you see that neither of these projections is parallel to either RL 1–3 or RL 1–2. Therefore, the perpendicular line x_3y_3 you see in view #3 does not appear true length in views #1, #2, or #3. This means that you must construct an additional view in order to determine its actual true length.

c. Construct a projection plane #4 (represented by RL 3–4) perpendicular to the true length of projection a_3b_3, and, therefore, parallel to x_3y_3. In view #4, line AB appears as a point a_4b_4 and the perpendicular distance from point X to line AB appears as true length x_4y_4.

Construction Program #II: Plane Method—Fig. 77:

a. Create a plane ABX using the line AB and the point X. (You see this plane in the horizontal and frontal projections as $a_1b_1x_1$ and $a_2b_2x_2$, respectively.)

b. Find the edge view of plane ABX by constructing a horizontal line (a_2h_2) on the plane in which this line appears as a point. You see the edge view $a_3b_3x_3$ of the plane ABX in view #3.

c. Construct a projection plane #4 (represented by RL 3–4) parallel to the edge view $a_3b_3x_3$ of ABX in order to determine the true shape of plane ABX, which you see in view #4 as $a_4b_4x_4$. Since all lines in a true shape view of a plane appear as true length, you can construct a perpendicular from x_4 to line a_4b_4. This perpendicular (x_4y_4) appears as true length and is the actual length of the shortest perpendicular distance from point X to line AB.

(NOTE: Point Y has not been projected back to the horizontal and frontal projection views in order to maintain clarity of presentation.)

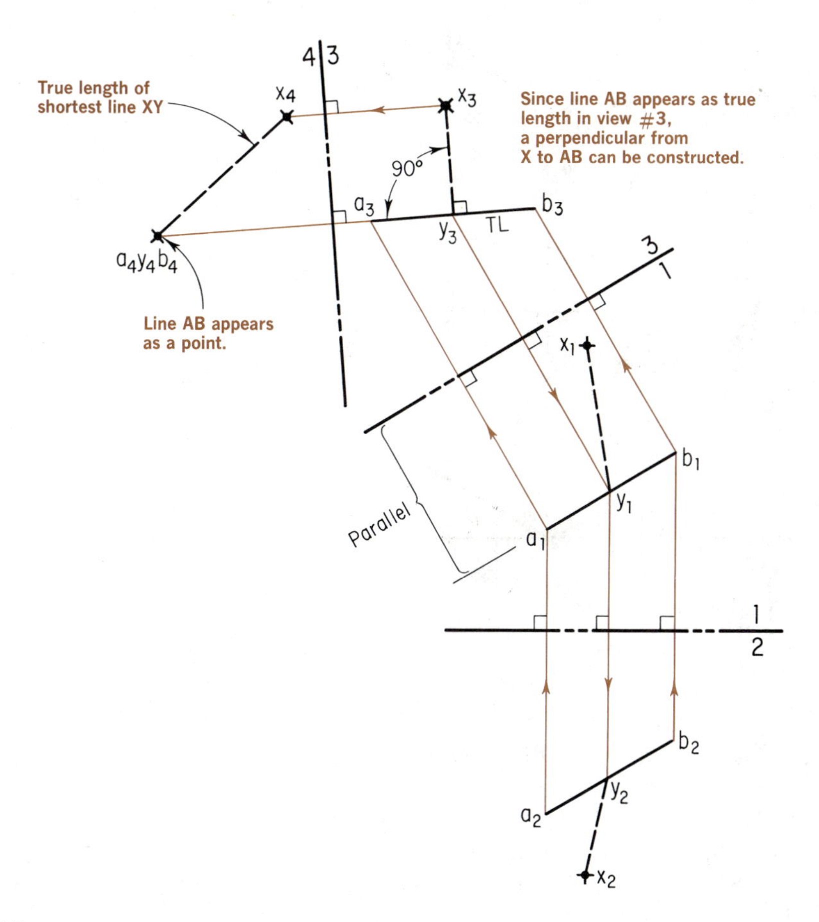

FIGURE 76

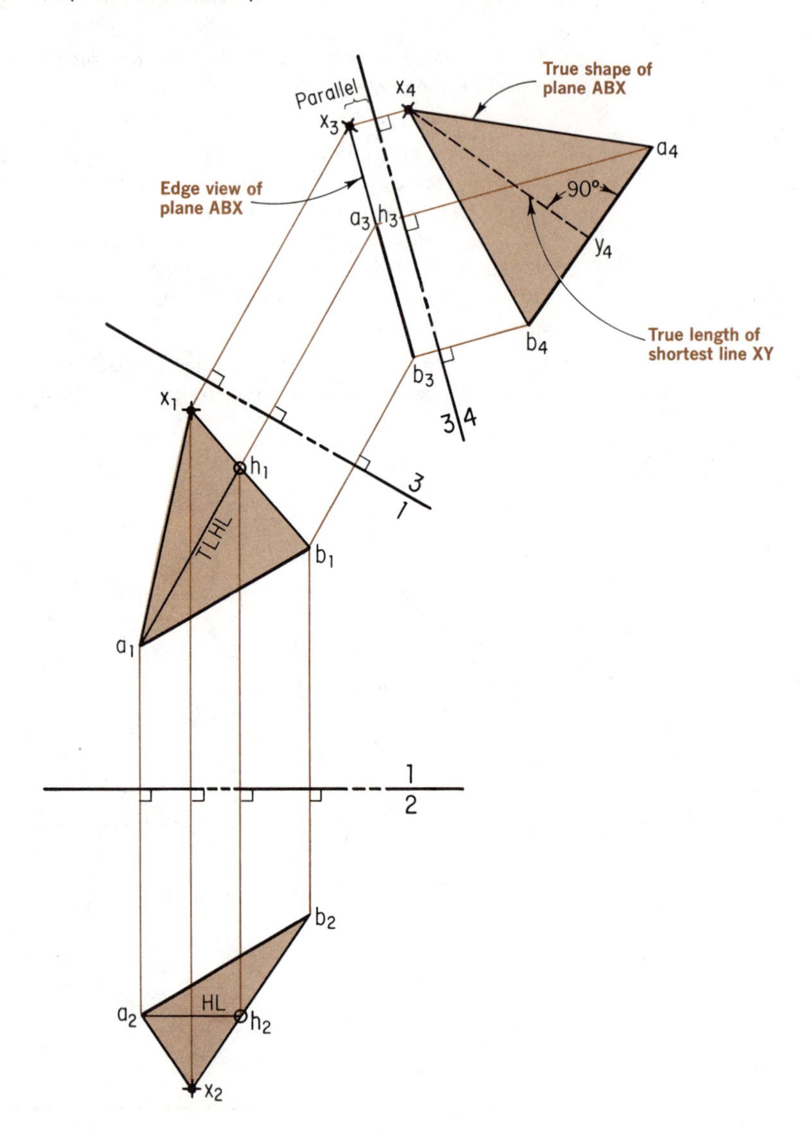

FIGURE 77

32 Shortest perpendicular distance between two skew lines

The shortest distance between two given skew lines (lines that are not parallel and do not intersect) is measured on a line that makes 90° with each given line. Figure 78 pictorially presents two skew lines AB and CD in space, with a perpendicular XY common to both lines.

There is only *one* common perpendicular possible between any two given skew lines. Because of this, the perpendicular has a unique location in space.

FIGURE 78

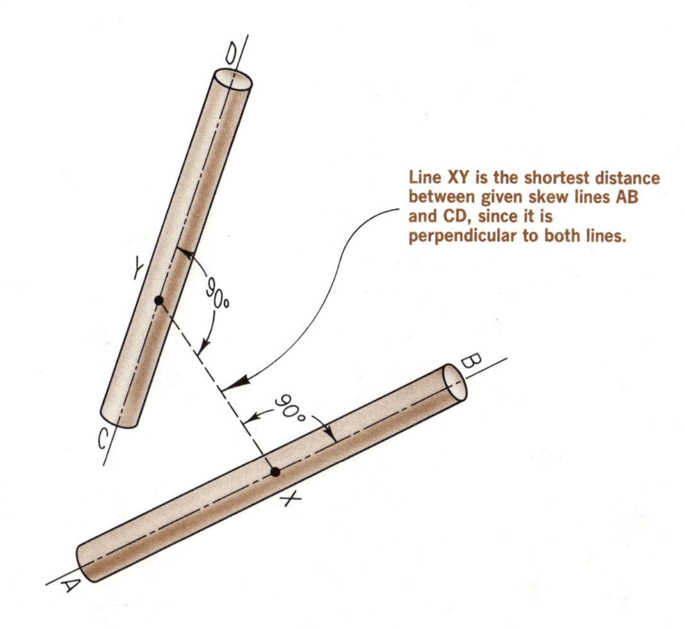

In determining the shortest perpendicular distance between two skew lines, two methods of approach are available:

1. The *line method*, by which we deal directly with one of the given skew lines in order to determine it as a point.
2. The *plane method*, by which we create a plane that contains one of the given skew lines and is parallel to the other line.

Orthographic Example: Determination of the shortest distance between two given skew lines (Figs. 79 and 80).

Given Data: Horizontal and frontal projections of two skew lines *AB* and *CD*.

Required:

a. The true length of the shortest distance *XY* from *AB* to *CD*.

b. The horizontal and frontal projections of *XY*.

Construction Program #I: Using the Line Method—Fig. 79:

a. Determine the true length of line *AB* by passing a projection plane #3 parallel to the horizontal projection a_1b_1 of line *AB*. The true length a_3b_3 of line *AB* appears in view #3.

b. Pass a projection plane #4 (represented by RL 3–4) perpendicular to the true length projection a_3b_3. This will show the line *AB* as a point (a_4b_4) in view #4.

c. From the point view a_4b_4 of line *AB* in view #4, construct a perpendicular line to the projection c_4d_4 of line *CD* in view #4.

[NOTE: You can definitely say that in this view the shortest distance from line *AB* (which appears as a point) to line *CD* (which appears as

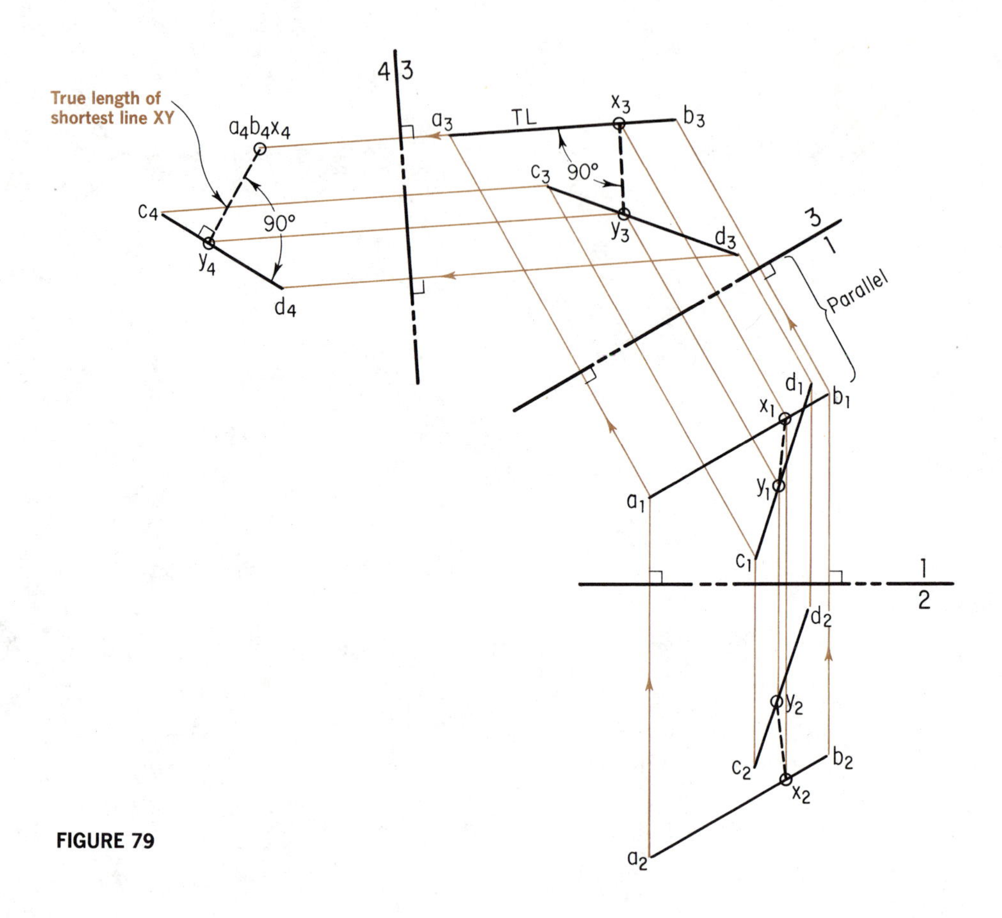

FIGURE 79

a line) is a perpendicular from the point view of *AB* to *CD*, since no other line is shorter.]

d. Note the foot of the perpendicular on c_4d_4 as y_4 and project it back to view #3.

e. From y_3 in view #3 draw a perpendicular line to (TL) line a_3b_3 to locate x_3.

f. Project x_3y_3 to views #1 and #2 in order to complete the problem. In view #4 you see the true length *XY*, which you can measure to any given or assumed scale.

Construction Program #II:

Using the Plane Method—Fig. 80:

a. Create a plane that contains the line *CD* parallel to the line *AB*. Do this by constructing a line *XY* through any *assumed* point *W* on *CD*.

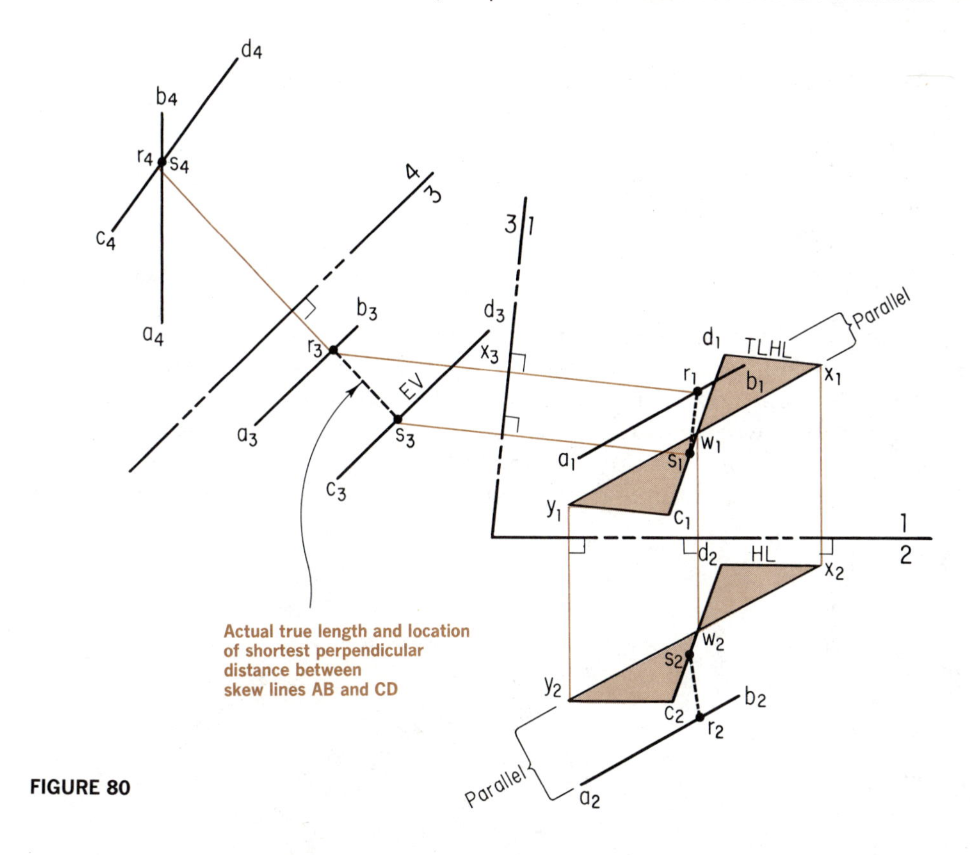

FIGURE 80

(x_2y_2 goes through assumed point w_2 and is parallel to a_2b_2 in frontal view #2; x_1y_1 goes through the projected point w_1 on c_1d_1 and is parallel to a_1b_1 in horizontal view #1.)

b. Determine the edge view of the newly created plane *CYDX* by finding a horizontal line d_2x_2 as true length. You see this line as a point d_3x_3 in view #3, which is perpendicular to the true length horizontal line d_1x_1. In view #3 the edge view of the created plane appears parallel to the line *AB*. In this view you can measure the perpendicular *distance,* but you must first determine its actual *location* by constructing an additional view.

c. To determine the actual location of the shortest perpendicular distance between the skew lines *AB* and *CD*, you must construct a view that is *perpendicular* to the *direction* of the shortest perpendicular distance known in view #3. Projection plane #4 is perpendicular to the direction of this distance. (This makes projection plane #4 automatically parallel to both lines *AB* and *CD*.) In view #4 lines *AB* and *CD appear* to intersect. The *apparent* point of intersection is the *actual* location r_4s_4 of the common perpendicular.

d. Project r_4 and s_4 back to views #3, #1, and #2 in order to complete the problem.

33 Shortest horizontal distance between two skew lines

In addition to your being able to determine the shortest common perpendicular between two skew lines, you can also determine the shortest *horizontal* distance between two skew lines. In order to do this, you *must* use the *plane method* of construction, by which you can obtain a view in which each skew line *appears parallel* to the other. The initial approach is similar to that presented in section 32.

Figure 81 pictorially presents two skew lines *AB* and *CD* in space related to a horizontal projection plane. Line *XY* is the shortest horizontal line between these two skew lines. You must remember that you can define line *XY* as horizontal only in *elevation* views. The reason why you *must* use the plane method in solving problems involving horizontal lines is that the plane method makes it possible to directly develop an elevation view in which two skew lines *appear* parallel.

FIGURE 81

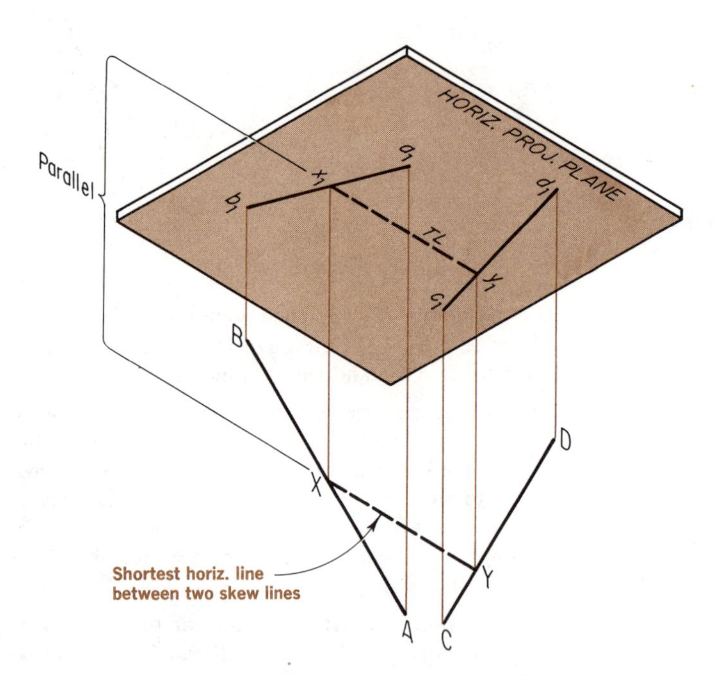

Orthographic Example: Determination of the shortest horizontal distance between two skew lines (Fig. 82).

Given Data: Horizontal and frontal projections of two skew lines *AB* and *CD*.

Required:

a. The true length of the shortest horizontal line *XY* between skew lines *AB* and *CD*.

b. The horizontal and frontal projections of *XY*.

Construction Program:

a. Create a plane that contains line AB and is parallel to line CD. In Fig. 82 you assumed a point M on line AB and constructed the line LM through this point parallel to line CD.

b. Find the edge view of the plane LBM that contains AB and is parallel to CD. In view #3 plane LBM appears as an edge. It is in this view that the projections a_3b_3 and c_3d_3 of the skew lines appear parallel. View #3 is an elevation view and, therefore, you can define a horizontal position here. Knowing the direction of a horizontal line in this view, construct a projection plane #4 *perpendicular* to this direction in order to determine a view in which lines a_4b_4 and c_4d_4 appear to intersect at point x_4y_4, which is the actual location of the shortest horizontal line and you see it as a point in view #4.

c. Project point x_4y_4 back to views #3, #1, and #2. You see the true length of the shortest horizontal line in views #3 and #1.

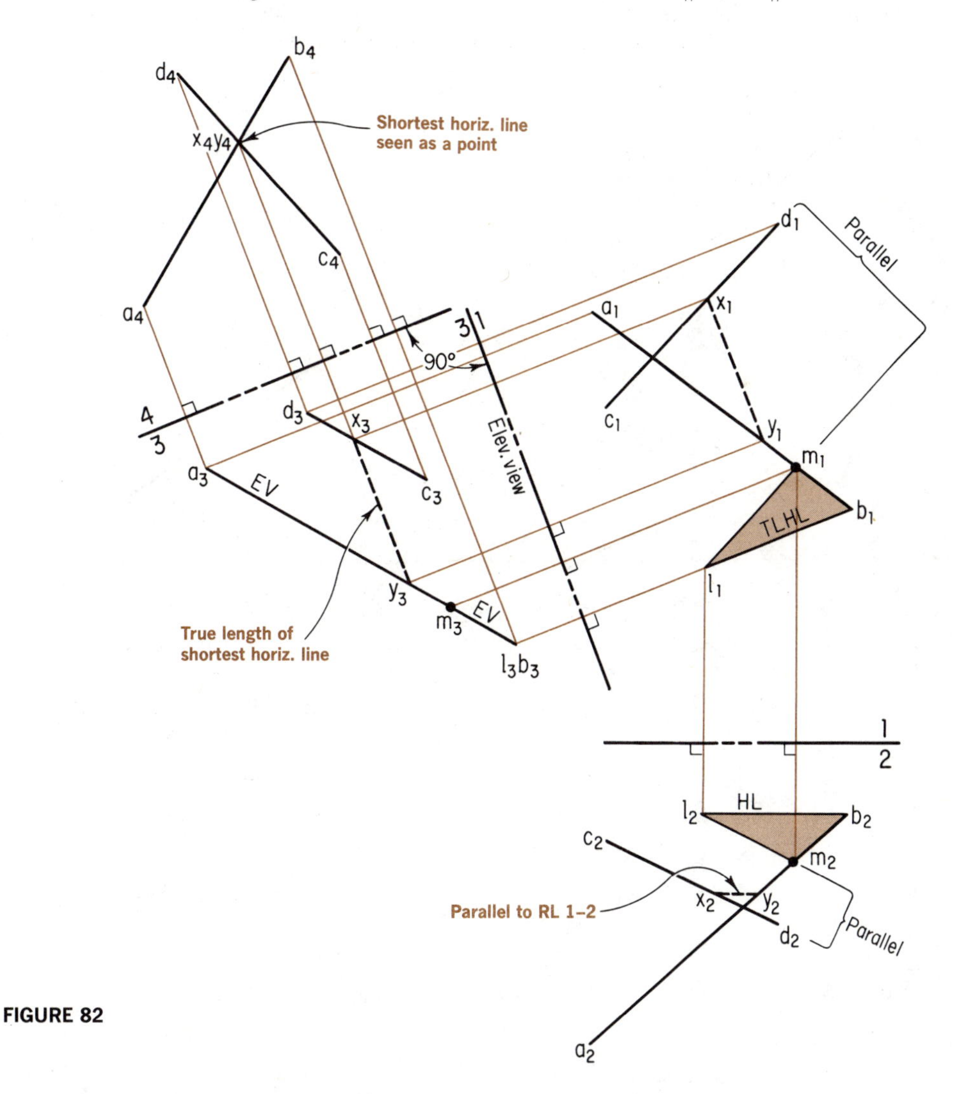

FIGURE 82

34 Shortest grade distance between two skew lines

Another type of shortest distance measurement involves the shortest distance having a specific grade (slope) between two given skew lines. For a given grade there is a unique position for this shortest distance.

Since we want to determine the *slope* of a line, we must use the *plane method* of construction to obtain an elevation view in which the given skew lines appear parallel to each other.

Orthographic Example: Determination of the shortest grade distance between two skew lines (Fig. 83).

Given Data: Horizontal and frontal projections of two skew lines AB and CD.

Required: a. The true length of the shortest grade distance XY, which has a slope of 20% downward from line CD to line AB.

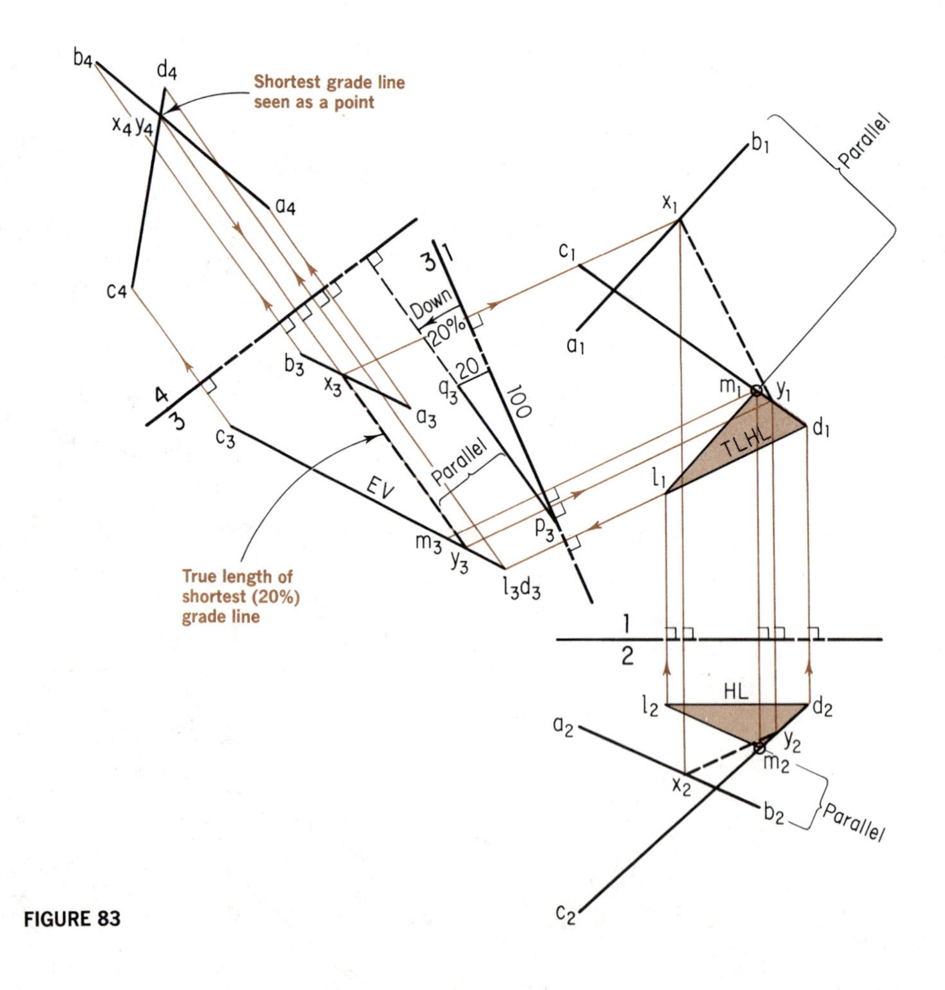

FIGURE 83

b. The horizontal and frontal projections of the shortest grade distance XY.

Construction Program:

a. Create a plane that contains line CD and is parallel to line AB. In Fig. 83 you assumed point M on line CD and constructed line LM through this point parallel to AB (see views #1 and #2).

b. Find the edge view of the plane LDM that contains line CD and is parallel to line AB. In view #3 plane LDM appears as an edge, and in this view the projections a_3b_3 and c_3d_3 of the skew lines appear parallel. Since view #3 is an elevation view, the slope can be determined.

c. Choosing any assumed point p_3 on RL 1–3, construct a line p_3q_3 that has a downward slope of 20%. The line p_3q_3 now determines the slope of the required shortest grade distance XY.

d. Construct a projection plane #4 (represented by RL 3–4) *perpendicular* to the *direction* of the 20% slope as represented by line p_3q_3. In view #4, the projections a_4b_4 and c_4d_4 of the given skew lines appear to intersect. The apparent point of intersection (noted as x_4y_4) is the *actual* location of the shortest grade distance XY, which has a slope of 20% downward from line CD to line AB.

e. Project points x_4 and y_4 back to views #3, #1, and #2. You see the true length (x_3y_3) of the shortest 20% grade line in view #3.

35 Shortest perpendicular distance from a point to a plane surface

By applying the principle of a line perpendicular to a plane surface presented in section 30, you can determine the shortest distance from a point in space to a given plane surface. You can measure the shortest *perpendicular* distance from the point in space to the plane surface.

Figure 84 shows a point M in space, from which a line has been constructed perpendicular to a given plane ABC. In addition, this figure shows one possible

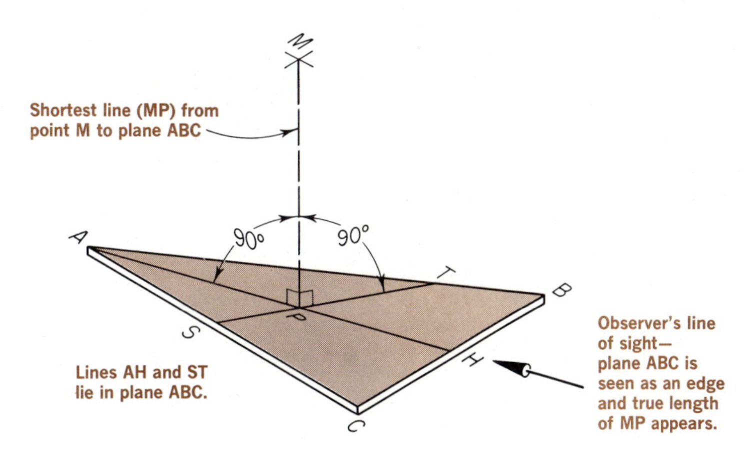

FIGURE 84

position you can take in order to see the true length of the shortest distance from point *M* to the plane *ABC*. Note that the view in which you see the true length of the shortest distance *MP* is also the view in which the plane *ABC* appears as an edge.

FIGURE 85

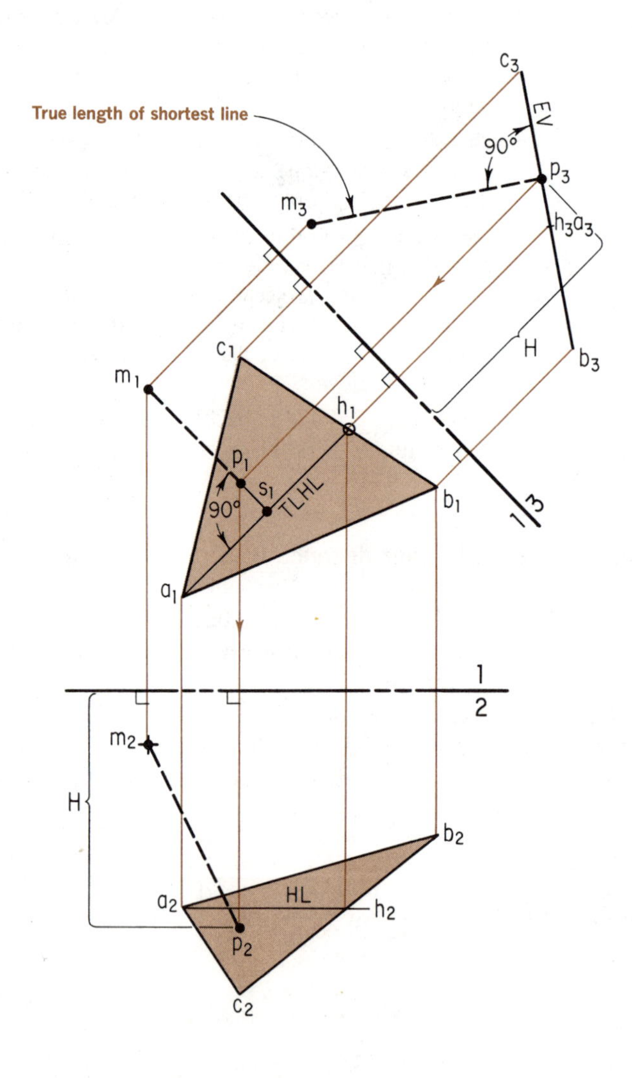

Orthographic Example: Determination of the shortest distance from a point in space to a given plane surface (Fig. 85).

Given Data: The horizontal and frontal projections of a plane *ABC* and a point *M* outside the plane.

Required:

a. The true length of the shortest distance *MP* from the point in space to the plane surface *ABC*.

b. The horizontal and frontal projections of the shortest distance *MP*.

Construction Program:

a. Construct an edge view of the given plane *ABC* by drawing a horizontal line a_2h_2 on the plane in which this line appears as a point in view #3.

b. From m_3 in view #3 construct a perpendicular line to the edge view $a_3b_3c_3$ of plane *ABC*. The foot of the perpendicular is p_3.

c. Project p_3 back to view #1. In order to locate p_1 in view #1, you must determine the *direction* of the perpendicular line from *M* to the plane *ABC*. This direction is determined by drawing a line from m_1 *perpendicular* to a *true length* line in the plane *ABC*. (Since a_1h_1 is a true length horizontal line, the 90° direction is established relative to this line.) Note that the point s_1, at which the perpendicular line from m_1 intersects the true length horizontal line a_1h_1, is *not* the foot of the perpendicular line on the plane surface *ABC*. The horizontal line a_2h_2 is a *random* horizontal line and, therefore, only by *coincidence* could s_1 correspond to p_1.

d. To locate the frontal projection of the shortest distance from *M* to the plane *ABC*, you must project p_1 into view #2. To locate the frontal projection p_2, measure the distance *H* (in view #3) of point *P below* the horizontal projection plane and transfer this distance to view #2 with dividers.

36 Relative positions of two skew lines as applied to the determination of visibility of lines on solid objects

By considering the relative positions of two skew lines as seen by an observer from any given location, you can determine which of two skew lines is nearer to the observer. In the same way, you can also determine the *visibility* of lines in solid objects.

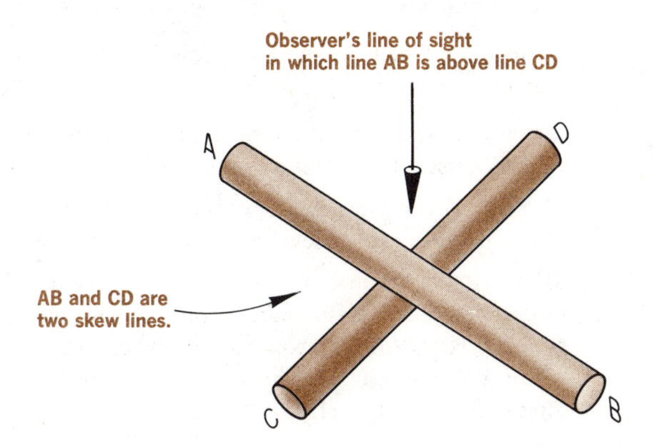

FIGURE 86

Figure 86 shows two skew lines AB and CD in space. The indicated observer's line of sight is such that the observer would see the line AB above line CD at the point where the two lines *appear* to intersect. An orthographic analysis of "visibility" is presented in Fig. 87.

FIGURE 87

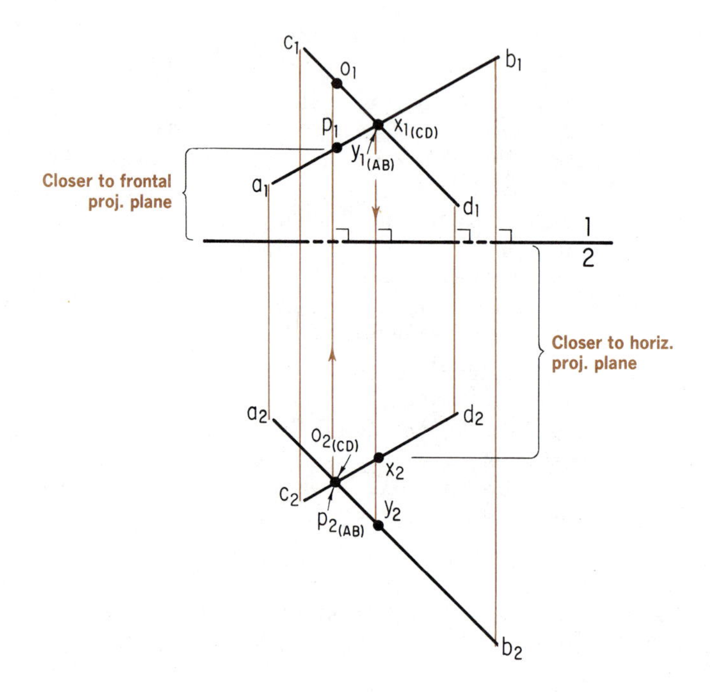

Orthographic Example: Visibility: relative positions of two skew lines (Fig. 87).

Given Data: Horizontal and frontal projections of two skew lines AB and CD.

Required: Determine which of the two skew lines is *above* and *in front of* the other.

Construction Program:

a. In view #1, at the *apparent* point of intersection of skew lines AB and CD, assume a point x_1 on line CD and a point y_1 on line AB.

b. Project x_1 into view #2 in order to determine x_2 on line c_2d_2. Likewise, project y_1 into view #2 in order to locate y_2 on line a_2b_2.

c. From view #2 you can now see that the point x_2 is closer to the horizontal projection plane than is point y_2. This means that x_2 is *above* y_2. Since x_2 is on line CD, you can say (when you return to view #1) that point X (on line CD) is above point Y (on line AB), and, therefore, that line CD is above line AB.

d. In view #2, assume a point p_2 on line a_2b_2 and a point o_2 on line c_2d_2.

e. Project points o_2 and p_2 to lines c_1d_1 and a_1b_1, respectively, in view #1 in order to locate points p_1 and o_1. From view #1 you see that p_1 is closer to the frontal projection plane than is o_1. Returning to view #2, then, you can say that since point P (on line AB) is closer to the frontal projection plane than is point O (on line CD), line AB is in front of line CD.

You can apply the analysis in the previous example to determine whether the lines on a solid object that is orthographically represented are visible or hidden. Figure 88(a) shows the horizontal and frontal projections of a cube, the visibility of which has not been indicated. In frontal view #2, the apparent point of intersection between the lines d_2h_2 and f_2g_2 is indicated as x_2y_2. Project this apparent point of intersection to view #1, and you will see that the projector strikes the line d_1h_1 (with the point y_1 on it) before it strikes line f_1g_1 (with the point x_1 on it). This means that the line d_2h_2 is seen before the line f_2g_2. We have represented d_2h_2 as a solid line, and f_2g_2 as a dotted line to indicate the fact that f_2g_2 is hidden in view #1. [See Fig. 88(b).] By determining the visibility between the lines d_2h_2 and f_2g_2, you have determined the entire visibility of the cube in the frontal projection view #2. Since f_2 is hidden, the line from b_2 to f_2 and the line from e_2 to f_2 must also be hidden.

Return to view #1 in Fig. 88(a). The apparent point of intersection between the lines b_1c_1 and e_1f_1 is indicated as w_1z_1. If you project w_1 and z_1 downward to the frontal projection view #1, you will see that the projector strikes the line b_2c_2 (with the point w_2 on it) before it strikes the line b_2f_2 (with the point z_2 on it). This means that W on the line BC is above Z on the line BF. When you return to the horizontal view #1, line b_1c_1 is above line e_1f_1. Thus we have represented b_1c_1 as a solid line and e_1f_1 a dotted (hidden) line. [See Fig. 88(b).]

FIGURE 88

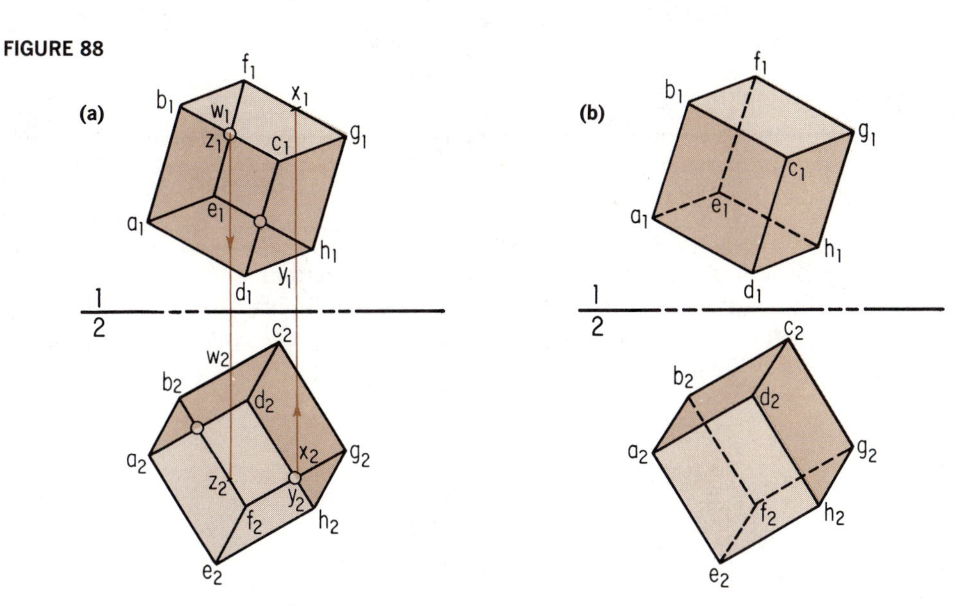

When you determine the visibility of any solid object, you should remember that the *outer* boundary lines of the object are visible in all views. These are the first lines that you can draw with complete confidence. After that, you need only determine the visibility of the *interior* lines.

37 Point of intersection (piercing point) of a line and a plane surface

A straight line intersects a flat plane surface only once. This point of intersection is referred to as a *piercing point.* Figure 89 pictorially presents a line *XY* intersecting a plane *ABC* at a piercing point *P.*

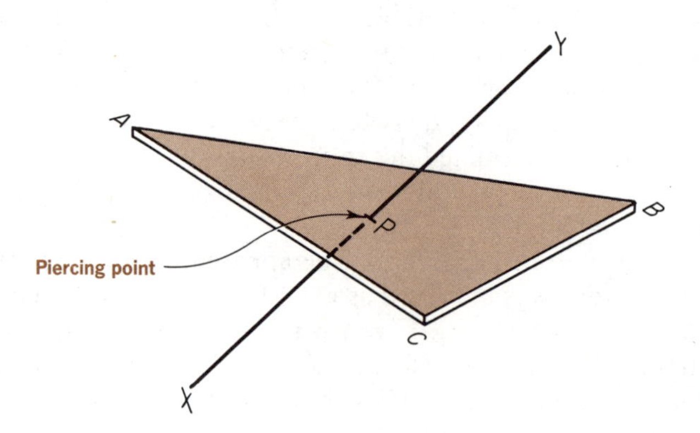

FIGURE 89

To determine the piercing point between a line and a plane, in orthographic projection, two basic approaches are available:

A. THE EDGE VIEW METHOD

In this method for determining the piercing point between a line and a plane, you must determine the edge view of the given plane, thereby locating the piercing point.

B. THE AUXILIARY CUTTING PLANE METHOD

In this method, a cutting plane that appears as an edge either in the horizontal *or* frontal projection views contains the given line. The piercing point is located at the intersection of the cutting plane with the given plane.

Orthographic Example: Piercing point of a line and a plane (Fig. 90).

Given Data: Horizontal and frontal projections of a plane surface *ABC* and a line *XY*.

FIGURE 90

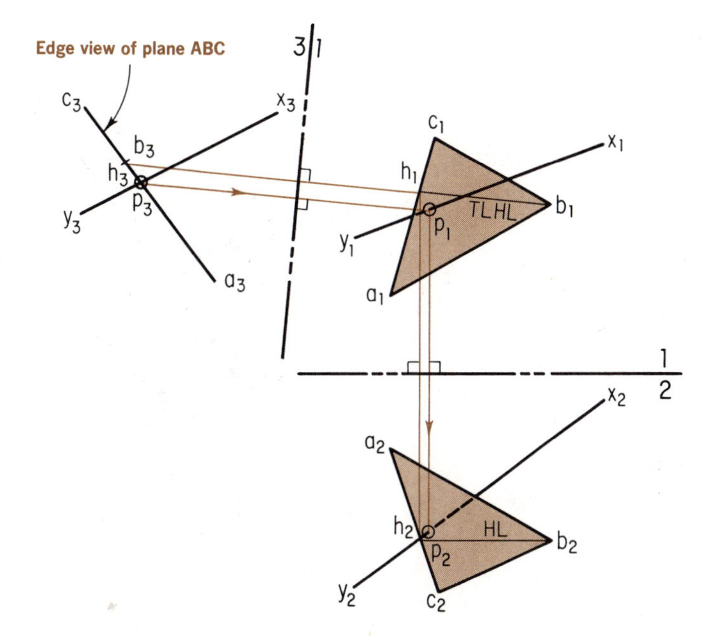

Required: The horizontal and frontal projections of the piercing point P of line XY on plane ABC.

Construction Program #I: Using the Edge View Method—Fig. 90:

a. Determine the edge view of plane ABC by constructing a horizontal line b_2h_2 on the plane in which the line will appear as a point (h_3b_3). (See views #2 and #3.)

b. In view #3 you can note that the line x_3y_3 intersects the plane $a_3b_3c_3$ at the piercing point p_3.

c. Project point p_3 back to the horizontal and frontal projection views (p_1 is on line x_1y_1 in view #1, and p_2 is on line x_2y_2 in view #2).

d. Using the method of analysis for determining visibility presented in section 36, establish the visibility of line XY relative to plane ABC in the horizontal and frontal projection views.

Construction Program #II: Using the Auxiliary Cutting Plane Method—Figs. 91 and 92:

(NOTE: Figure 91 is a pictorial representation of how the piercing point P of a line XY and a plane ABC is determined by using an *auxiliary vertical cutting plane XMN* that contains the line XY and intersects the given plane ABC on the line WZ. The line WZ also intersects the given line XY at point P. Point P is the piercing point of the line XY and the given plane ABC because it is common to the given plane ABC, the auxiliary vertical cutting plane XMN, and the line of intersection WZ between the given plane and the cutting plane.)

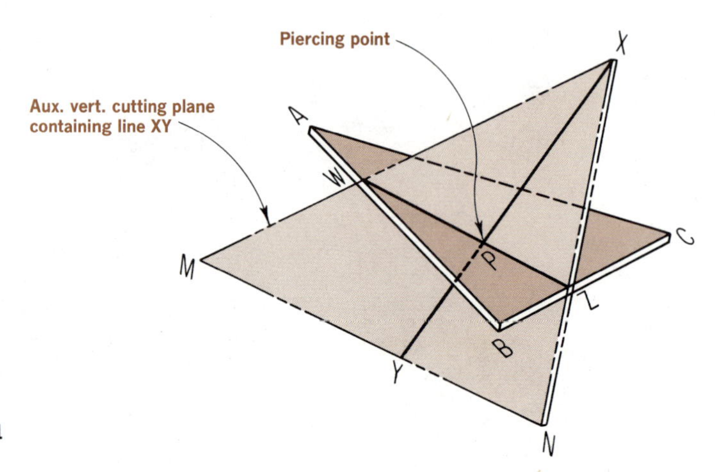

FIGURE 91

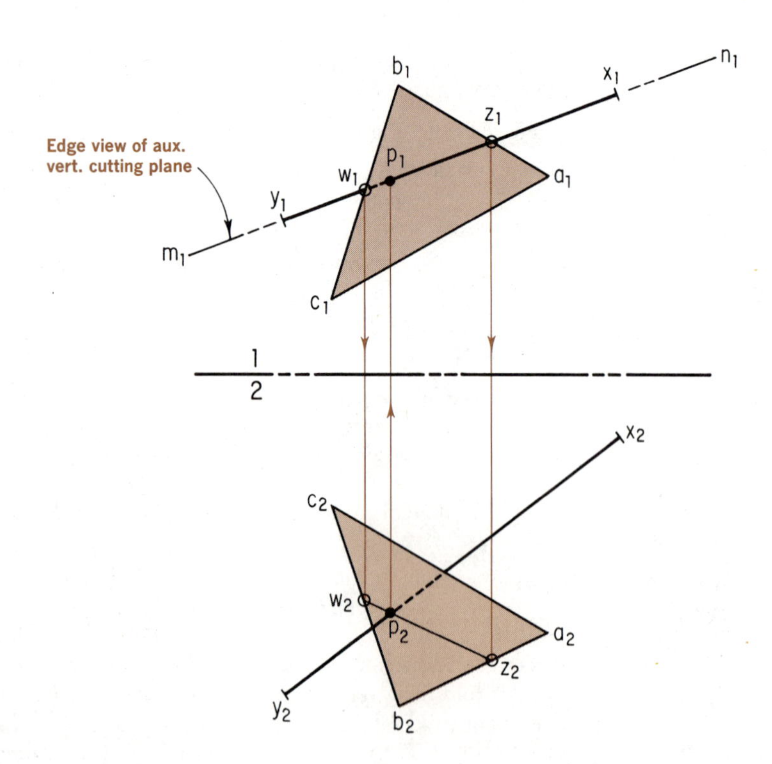

FIGURE 92

a. In the horizontal projection view #1, draw an edge view of an auxiliary vertical cutting plane that contains the line x_1y_1. (The cutting plane is represented as m_1n_1 in this view.)

b. The cutting plane m_1n_1 intersects lines b_1c_1 and b_1a_1 on plane *ABC* at points w_1 and z_1, respectively.

c. Project w_1 and z_1 into the frontal projection view #2 in order to locate w_2 and z_2 on lines b_2c_2 and b_2a_2, respectively.

d. Connect w_2 and z_2 with a straight line. Piercing point p_2 is located at the intersection of lines w_2z_2 and x_2y_2. This is the frontal projection view of the piercing point of the given line and plane.

e. Project p_2 into the horizontal projection view #1 in order to determine p_1 on line x_1y_1.

f. Determine the visibility of the line *XY* relative to the given plane *ABC* in both the horizontal and frontal projection views.

38 Determination of the line of intersection of two limited planes

By applying the methods for determining the piercing point of a line on a plane surface, described in section 37, you can determine the line of intersection between two limited planes. Figure 93 pictorially presents a line *XY*, which is the line of intersection between the limited planes *ABC* and *MNOP*.

From the figure you can see that point *X* is the piercing point of line *MP* on the plane surface *ABC*, and point *Y* is the piercing point of line *NO* on the plane surface *ABC*. Since you are still dealing with piercing points of lines and planes, you may use either the edge view method or the cutting plane method to determine the line of intersection between the two limited planes. The edge view and cutting plane methods are similar, in principle, to those methods presented in section 37.

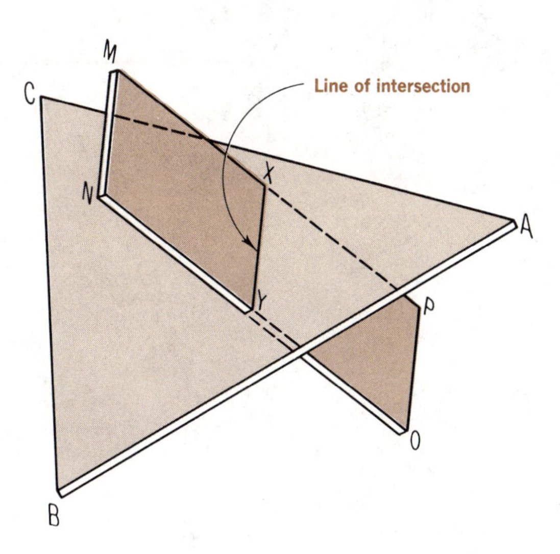

FIGURE 93

Orthographic Example: Determination of the line of intersection of two limited planes (Fig. 94).

Given Data: Horizontal and frontal projections of two intersecting planes *ABC* and *MNOP*.

Required:

a. The horizontal and frontal projections of the line of intersection *XY* of the two given planes.

b. The relative visibility of the given planes in the horizontal and frontal projection views.

Construction Program #I: Using the Edge View Method—Fig. 94:

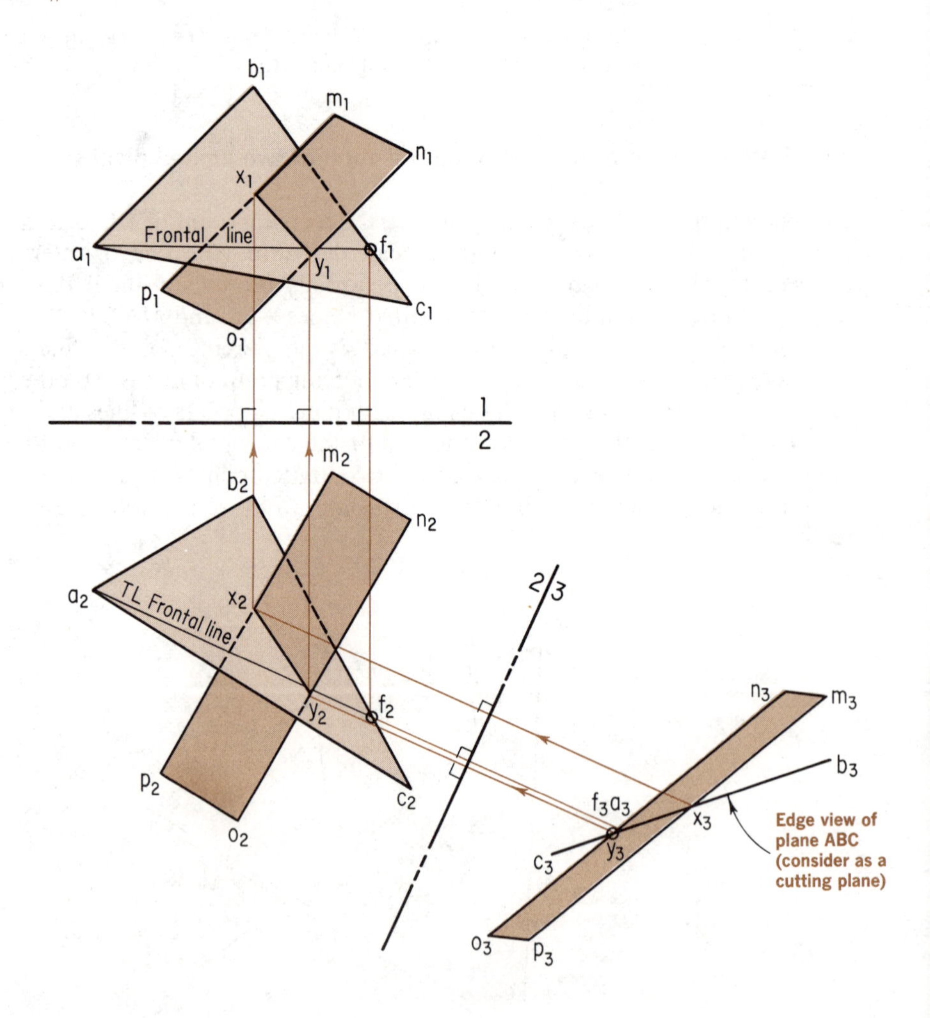

FIGURE 94

a. Determine the edge view of plane ABC. (In Fig. 94, a frontal line a_1f_1 was constructed in the plane ABC. It appears as a point in view #3, which is perpendicular to the true length frontal line a_2f_2.)

b. In view #3, in which plane ABC appears as an edge $a_3b_3c_3$, you can see where plane ABC *cuts* line m_3p_3 at point x_3 and line n_3o_3 at point y_3.

c. Project points x_3y_3 into view #2 in order to locate x_2 on m_2p_2 and y_2 on line n_2o_2. Connect x_2 and y_2 with a straight line. This is the required frontal view of the line of intersection.

d. Project x_2 and y_2 into view #1 in order to locate x_1 on line m_1p_1 and y_1 on n_1o_1. Connect x_1 and y_1 with a straight line. This is the required horizontal view of the line of intersection.

e. Determine the relative visibility of the given planes in views #1 and #2.

Construction Program #II:

Using the Cutting Plane Method—Figs. 95, 96, and 97:

a. Determine the piercing point X of line MN with the plane ABC by passing a cutting plane (#I) through m_2p_2 in the frontal view #2. This cutting plane locates point t_2 on line b_2c_2 and point u_2 on line a_2c_2 of plane ABC.

b. Project points t_2 and u_2 to lines b_1c_1 and a_1c_1, respectively, in order to determine t_1 and u_1 in view #1. Connect t_1 and u_1 with a straight line. Piercing point x_1 is located at the intersection of lines u_1t_1 and m_1p_1 on the plane surface $a_1b_1c_1$.

c. Project x_1 downward to the frontal view to determine point x_2 on line m_2p_2.

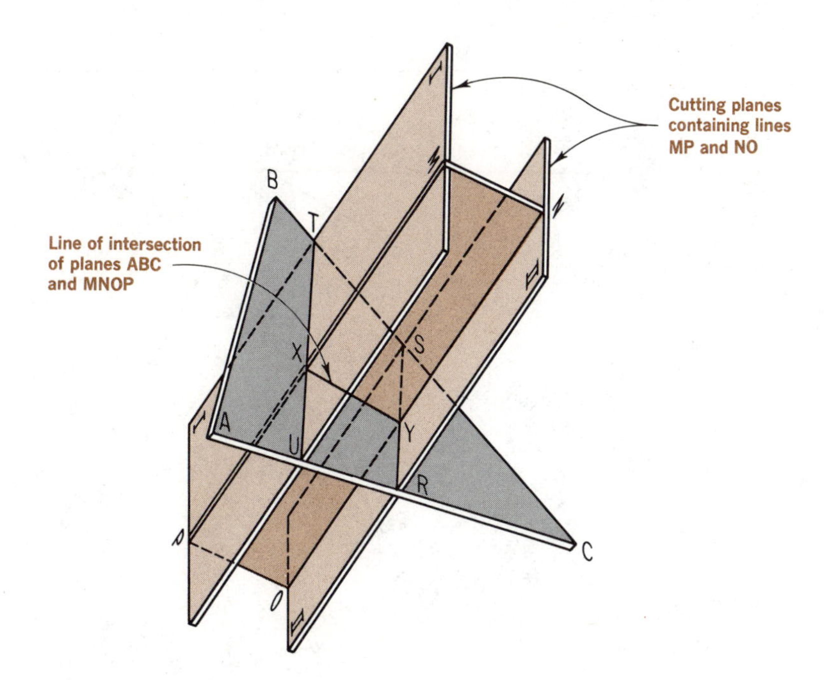

FIGURE 95

d. Pass a cutting plane (#II) through the line n_2o_2 in order to determine points r_2 and s_2 on lines a_2c_2 and b_2c_2, respectively.

e. Project r_2 and s_2 upward to horizontal view #1 in order to locate point r_1 on line a_1c_1 and point s_1 on line b_1c_1.

f. Connect r_1 and s_1 with a straight line. Piercing point y_1 is located at the intersection of lines r_1s_1 and n_1o_1 on the plane surface $a_1b_1c_1$. Project y_1 downward to frontal view #2 in order to locate point y_2 on line n_2o_2.

g. Connect x_1 and y_1 with a straight line, and x_2 and y_2 with a straight line. This is the required line of intersection between planes *ABC* and *MNOP*.

h. Determine the relative visibility of the planes in each view.

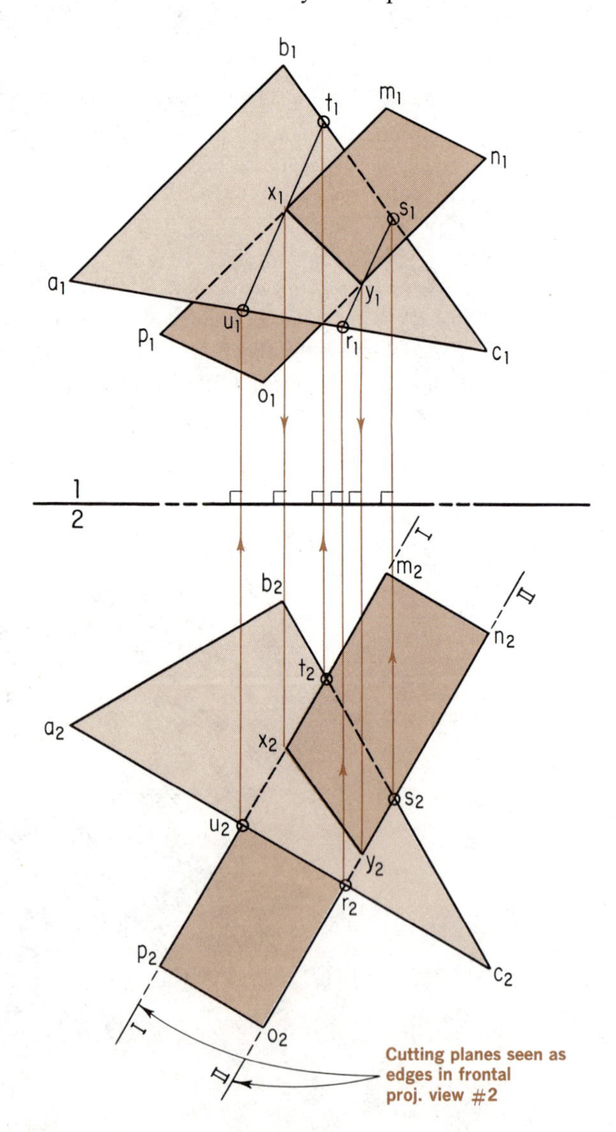

FIGURE 96

[NOTE: Cutting planes could be passed through lines on plane ABC. For example, in Fig. 97 cutting planes #I and #II were passed through lines BC and AC, respectively, of plane ABC (see view #2). Cutting plane #I determines points e_2 and f_2. The extended line drawn through e_1 and f_1 in view #1 intersects line b_1c_1 at point j_1, which is on the line of intersection between the two extended planes. Cutting plane #II determines points g_2 and h_2. The extended line drawn through g_1 and h_1 in view #1 intersects line a_1c_1 at point k_1, which is also on the line of intersection between the two extended planes. Connecting j_1 and k_1 with a straight line and then extending this line, you see that the line j_1k_1 intersects lines n_1o_1 and m_1p_1 at points y_1 and x_1, respectively. The line segment x_1y_1 is the required horizontal view #1 of the line of intersection. You can determine the frontal view of this line by projection from view #1.]

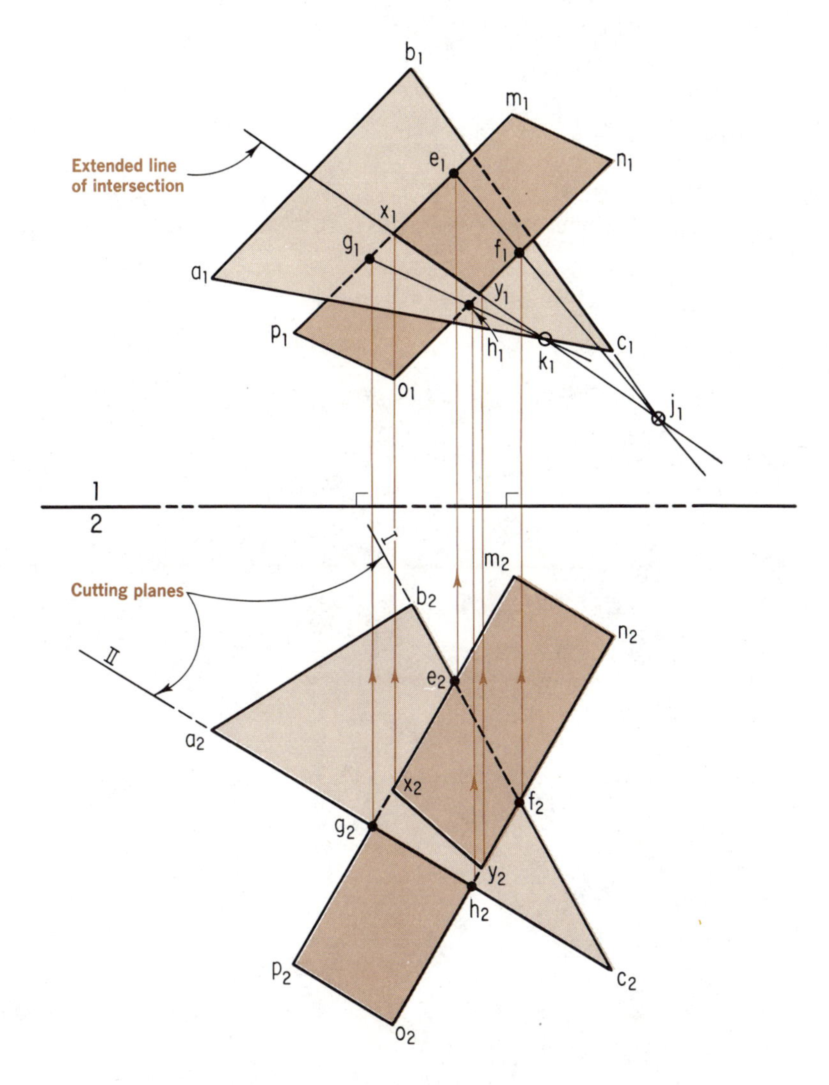

FIGURE 97

39 Determination of the line of intersection of two unlimited planes

To determine the line of intersection of two planes that are indefinite in extent (unlimited in size), you usually must apply the cutting plane method. (The edge view method is also a valid approach, except that normally the available space on a standard drawing sheet cannot accommodate the solution of this type of problem.)

Figure 98 pictorially presents "pieces" of two unlimited planes *ABC* and *MNOP*. Auxiliary cutting planes #I and #II cut lines on each given unlimited plane. The intersection of the lines determine points (*X* and *Y*) on the line of intersection of the two unlimited planes since these points are common to the auxiliary cutting planes, to the lines of intersection between the cutting planes and the given planes, and to the given unlimited planes.

FIGURE 98

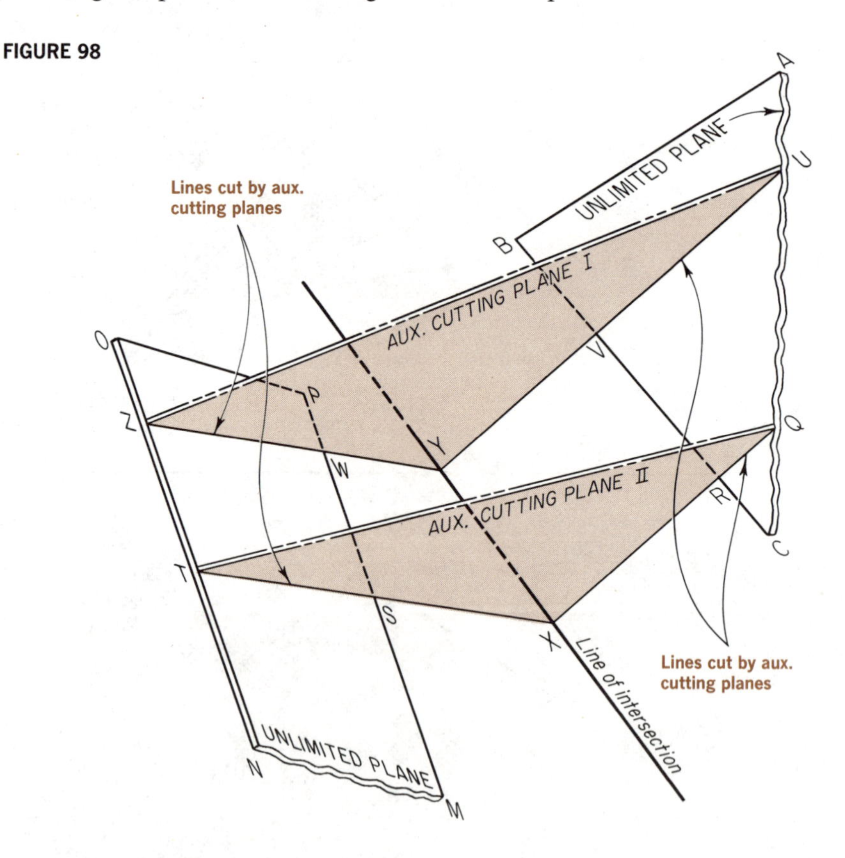

Orthographic Example: Line of intersection between two unlimited planes (Fig. 99).

Given Data: Horizontal and frontal projections of unlimited planes *ABC* and *MNOP*.

FIGURE 99

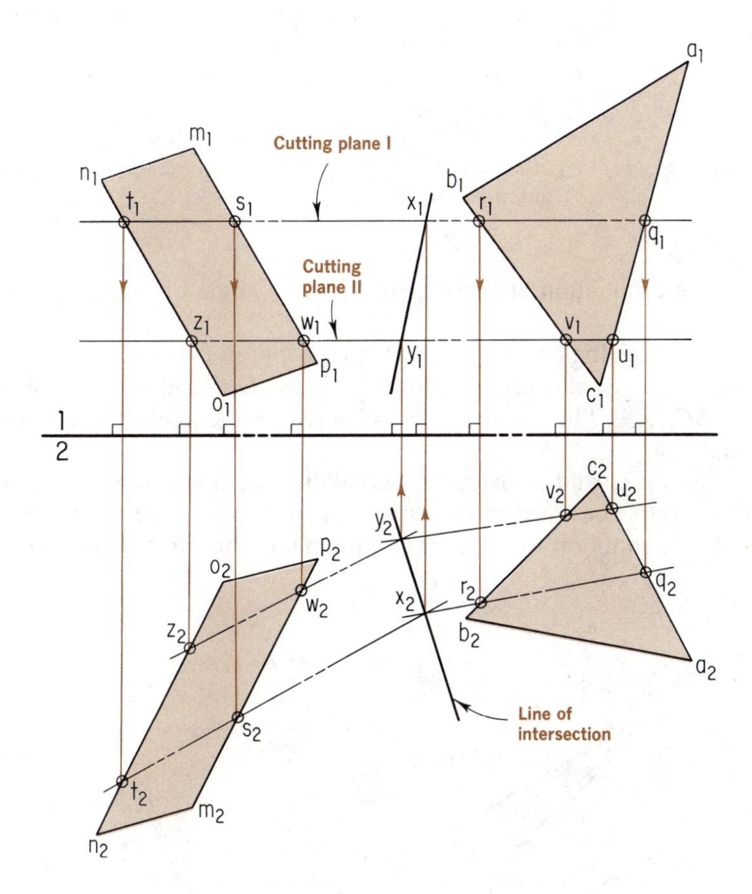

Required: A portion of the line of intersection XY between the two given planes.

Construction Program:

a. Draw a cutting plane (#I) that appears as an edge in view #1. This cutting plane determines points t_1 and s_1 on lines n_1o_1 and m_1p_1, respectively, and points r_1 and q_1 on lines b_1c_1 and a_1c_1, respectively.

b. Project points t_1, s_1, r_1, and q_1 to their respective lines n_2o_2, m_2p_2, b_2c_2, and a_2c_2 in order to determine points t_2, s_2, r_2, and q_2 in view #2.

c. Connect points t_2 and s_2 and points r_2 and q_2 with extended lines. These extended lines intersect at point x_2. This is one point on the line of intersection between two given planes.

d. Project x_2 upward to horizontal projection view #1 until it intersects the edge view of cutting plane #I. This will locate point x_1.

e. In view #1, pass another cutting plane (#II) that cuts plane $m_1n_1o_1p_1$ at points z_1 and w_1, and plane $a_1b_1c_1$ at points v_1 and u_1.

f. Project points z_1, w_1, v_1, and u_1 downward to their respective lines in view #2 in order to determine points z_2, w_2, and v_2, u_2.

g. Connect z_2 and w_2 with an extended straight line. Do the same for v_2 and u_2. At point y_2, where the extended lines intersect, is a second point on the line of intersection.

h. Project y_2 upward to view #1 in order to determine y_1, which is located on the edge view of cutting plane #II. Connect points x_1 and y_1 in the top view and points x_2 and y_2 in the front view #2 in order to obtain the required line of intersection between the two given unlimited planes.

40 Determination of the dihedral (acute) angle between two plane surfaces

The dihedral angle is the angle formed between two intersecting planes. Figure 100 is a pictorial representation of planes *ABC* and *BCD*, which intersect on line *BC*. The dihedral angle is measured as the acute angle between the two planes.

You can see and measure the actual dihedral angle in a view in which the *line of intersection* between two given planes appears as a *point*. Since the line of intersection is common to both planes, the planes in this view will appear as edges.

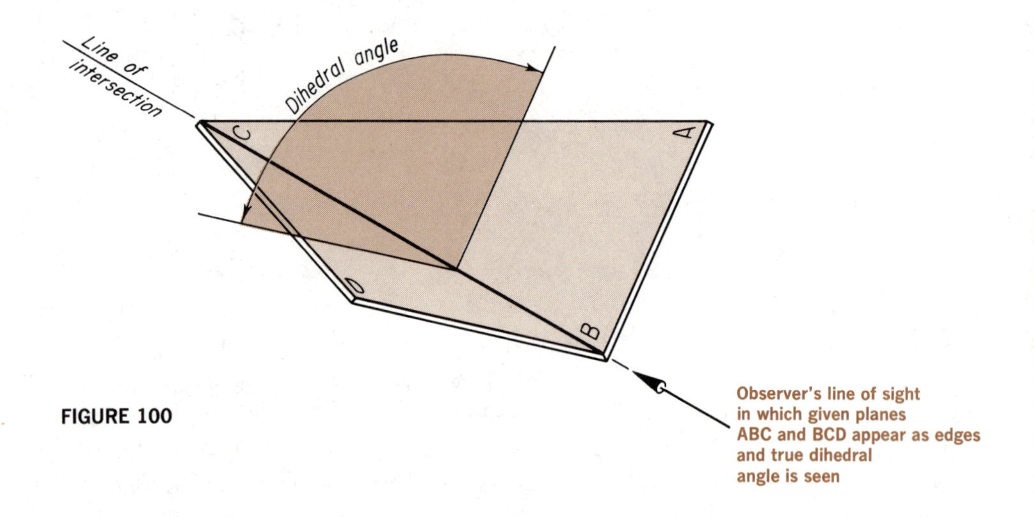

FIGURE 100

Orthographic Example: Determination of the dihedral angle between two intersecting planes (Fig. 101).

Given Data: Horizontal and frontal projections of two intersecting planes *ABC* and *BCD*.

Required: The true dihedral angle between planes *ABC* and *BCD*.

Construction Program:

a. Determine the true length of the line of intersection *BC* by passing a plane #3 (represented by RL 1–3) parallel to b_1c_1 in view #1.

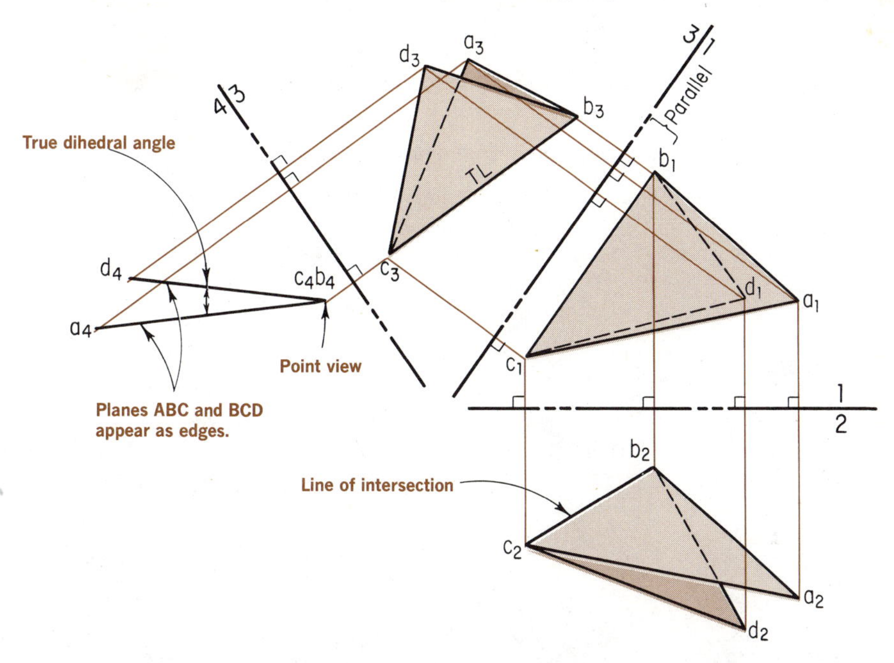

b. Construct a view (#4) *perpendicular* to the true length line of intersection b_3c_3, and in which the line of intersection appears as a point b_4c_4. (In view #4 the line of intersection appears as a point and, therefore, the planes *ABC* and *BCD* appear as edges.) It is in this view that you can see and measure the true dihedral angle.

41 Determination of the angle (acute) between a line and a plane surface

You can see the angle between a line and a plane surface *only* in a view in which (1) the line appears as true *length* and (2) the plane appears as an *edge* (see Fig. 102).

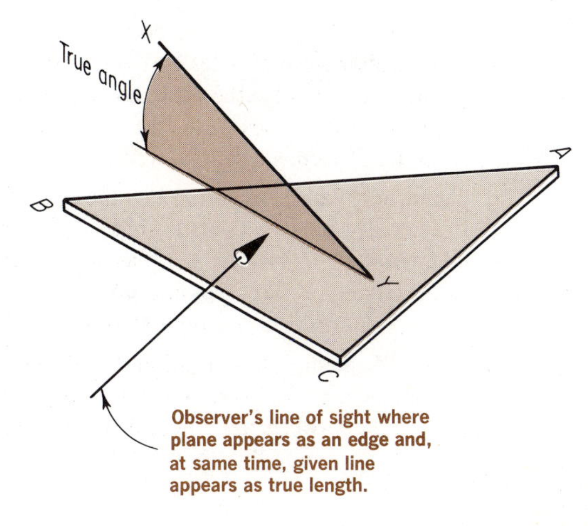

FIGURE 102

Orthographic Example: Determination of the angle between a line and a plane surface (Fig. 103).

Given Data: Horizontal and frontal projections of plane ABC and line XY.

Required: The true acute angle line XY makes with plane ABC.

FIGURE 103

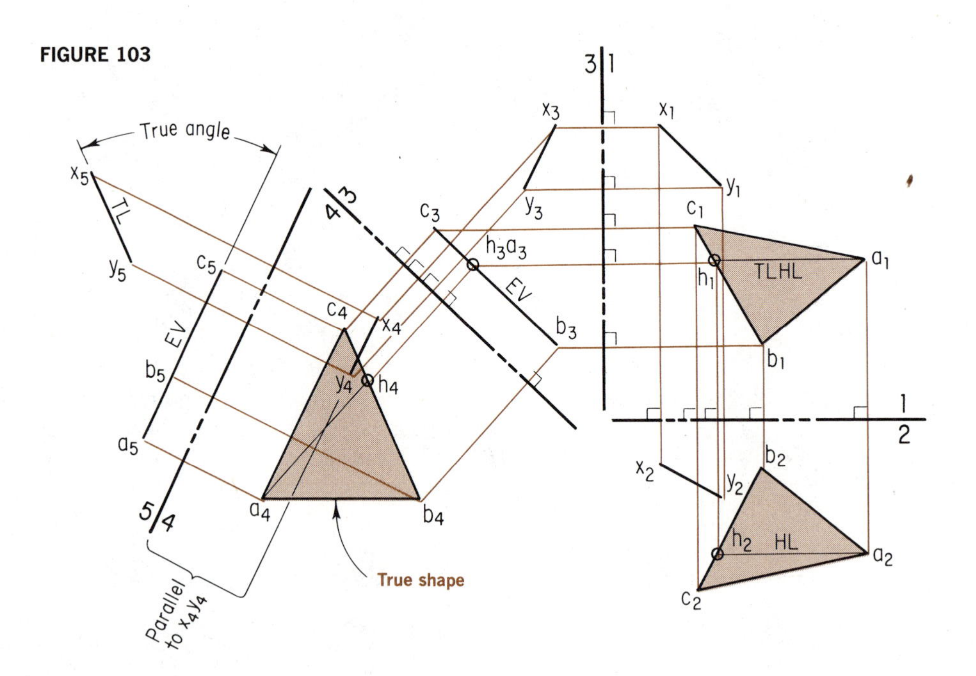

Construction Program:

a. A general analysis dictates that you must find a true shape of the plane before you can solve the problem. Therefore, determine the edge view of plane ABC by locating a line a_1h_1 as true length and constructing a plane (#3) *perpendicular* to it.

b. Construct plane #4 *parallel* to the edge view $a_3b_3c_3$ of plane ABC. In view #4 you see the true shape $a_4b_4c_4$ of plane ABC. In any view taken *directly* from view #4, the given plane ABC will *always* appear as an edge. It is now possible to construct plane #5 *parallel* to the line x_4y_4 in order to locate it as true length in view #5.

c. View #5 satisfies the condition that the given plane ABC must appear as an edge and the given line XY as true length. You can measure the true angle between the line and the plane in this view.

SAMPLE QUIZZES

1. Describe the view in which you can measure the true dihedral angle formed by two given intersecting planes.

2. Having determined the plane that contains one given line and is parallel to another given line, what new view can you draw to check the solution? What is the check?

3. Referring to question 1, assume that the line of intersection of the two planes is not given and describe the procedure by which you could obtain a view in which both planes would appear as edges.

4. How must you measure the shortest distance between any two nonintersecting, nonparallel lines, relative to the given lines?

5. Describe the key view in the line method, by which you can determine the shortest distance between any two nonintersecting, nonparallel lines.

6. Referring to the plane method of determining the shortest grade distance between the two nonintersecting, nonparallel lines, name the view in which the required distance is exactly located.

7. A. Describe two views in which two parallel lines will not appear parallel.

B. In what view is the true distance between two parallel lines seen?

8. Referring to the plane method of determining the shortest distance between any two nonintersecting, nonparallel lines, describe the view in which the true length and true slope of the required perpendicular appear.

9. Under what condition can two lines that lie in two different intersecting planes be parallel to each other? Show by freehand sketch the actual construction necessary to locate the shortest line connecting two skew lines, one of which is vertical and the other inclined. Label fully and indicate each true length view.

10. The line of intersection of two limited planes is determined by two points. What are they?

11. If the line of intersection is horizontal, in which view will the dihedral angle appear in its true size?

12. If the line of intersection is parallel to a frontal projection plane, in which view will the true dihedral angle appear?

13. To determine the piercing point of a line and an oblique plane, we use the edge view of a cutting plane. Assuming that the horizontal and frontal projections of the given line and oblique plane are given, in which view can the cutting plane be indicated?

14. In solving for the line of intersection of two oblique planes, we find that the lines on which the auxiliary plane intersects the two given planes are parallel instead of intersecting. We use another cutting plane, but the result is the same. What does this indicate?

15. To draw a line perpendicular to a plane from a point that is not on the plane, we construct a third view in order to locate the perpendicular. If this third view is an auxiliary elevation view, how must we draw the horizontal projection view of the perpendicular?

16. If a line is perpendicular to an oblique plane in space, its horizontal projection view will appear at right angles to ________________?

17. In drawing a line perpendicular to a plane from a point that is not on the plane, and using two views only, what two things are essential in order to fully determine the direction of this perpendicular?

18. In what view does the true size of the angle that a line makes with a plane appear?

19. If the horizontal and frontal projections of two lines are given, and if their points of intersection are also in projection, what kind of lines are the given lines?

PRACTICE PROBLEMS

1. Given Data: Horizontal and frontal projections of two skew lines *AB* and *CD*.

Required:

A. The horizontal and frontal projections of a plane that contains line *AB* and is parallel to line *CD*.

B. A view that actually shows line *CD* parallel to the edge view of the plane containing line *AB*.

PROBLEM 1

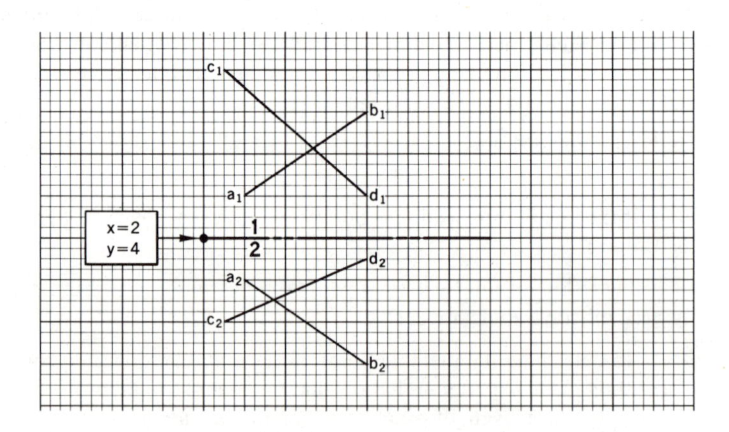

2. Given Data: Horizontal and frontal projections of two skew lines *AB* and *CD*.

Required:

A. The horizontal and frontal projections of a plane that contains line *CD* and is parallel to line *AB*.

B. A view that shows the edge view of the plane containing line *CD* and line *AB* in a position that proves parallelism.

PROBLEM 2

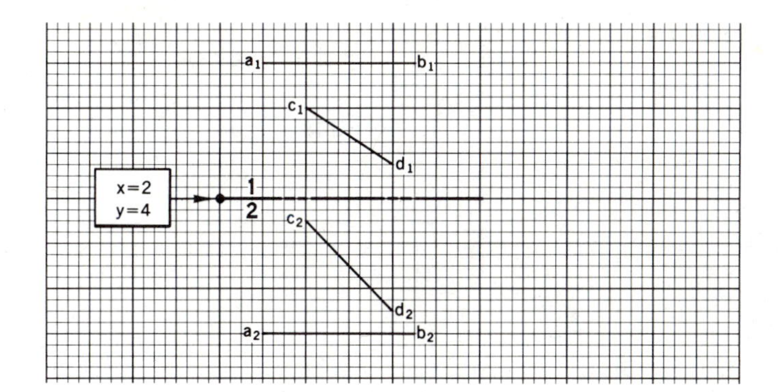

3. Given Data: Horizontal and frontal projections of two skew lines *WX* and *YZ*.

Required: A. The horizontal and frontal projections of a plane that contains line *WX* and is parallel to line *YZ*.

B. A view in which line *YZ* appears parallel to the edge view of the plane containing *WX*.

PROBLEM 3

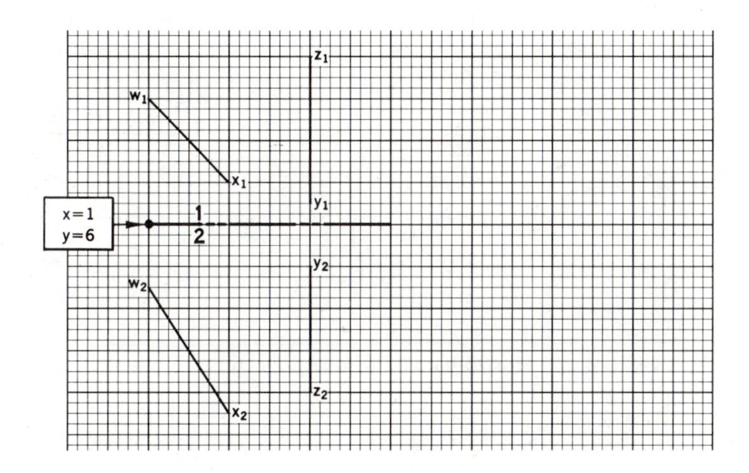

4. Given Data: Horizontal and frontal projections of line *AB* and a point *P* outside the line.

Required: A. Using the line method, determine the true length, slope, and bearing of the shortest line *PQ* from point *P perpendicular* to line *AB*. (*Ans.:* TL = $1\frac{3}{16}$ in., slope = 46°, bearing = N81°E)

B. Draw the horizontal and frontal projections of line *PQ*.

C. Reproduce the given data on another sheet of cross-section paper and solve using the plane method.

PROBLEM 4

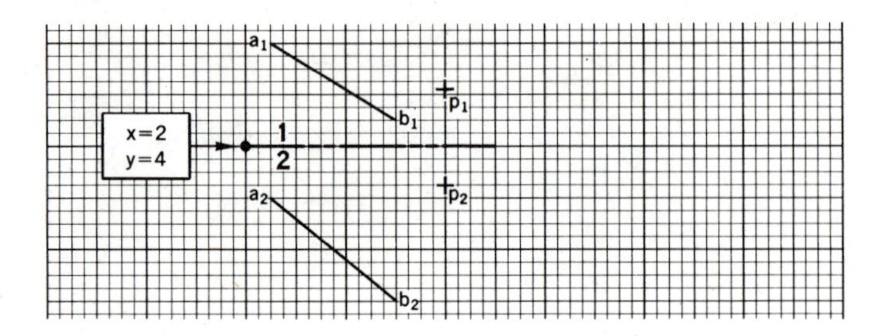

5. Given Data: Horizontal and frontal projections of two skew lines *AB* and *CD*.

Required:

A. Using the plane method, determine the true length, bearing, and slope of the shortest line *XY* between skew lines *AB* and *CD*. (*Ans.:* TL = $\frac{11}{32}$ in., bearing = N7°E, slope = 45°)

B. The horizontal and frontal projections of *XY*.

C. Label and show all construction in all views.

D. Reproduce the given data on another sheet of cross-section paper and solve using the line method.

PROBLEM 5

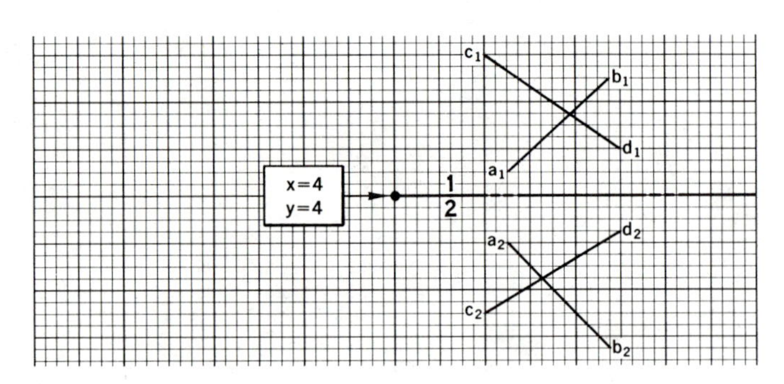

6. Given Data: Horizontal and frontal projections of two skew lines *CD* and *EF*.

PROBLEM 6

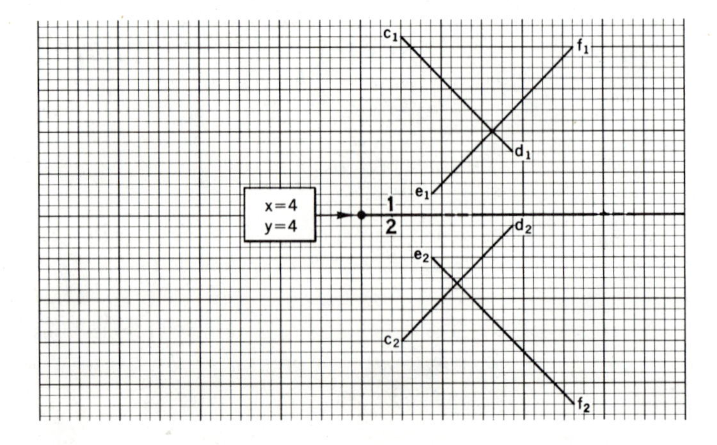

Required:

A. The true length and bearing of the shortest horizontal line *LM* between skew lines *CD* and *EF*. (*Ans.:* TL = $\frac{7}{8}$ in., bearing = due north)

B. The horizontal and frontal projections of *LM*.

7. Given Data: Horizontal and frontal projections of two skew lines *AB* and *CD*.

Required:

A. The true length, bearing, and slope of the shortest line *XY* between skew lines *AB* and *CD*. (*Ans.:* TL = $\frac{3}{4}$ in., bearing = S75°E, slope = 22°)

B. The true length and bearing of the shortest line *RS* from *AB* to *CD* that has a grade of 60% downward from *AB*. (*Ans.:* TL = $1\frac{9}{32}$ in., bearing = S75°E)

C. The frontal projections of *XY* and *RS*.

D. Label and show all construction.

E. What characteristics are common to lines *XY* and *RS*?

PROBLEM 7

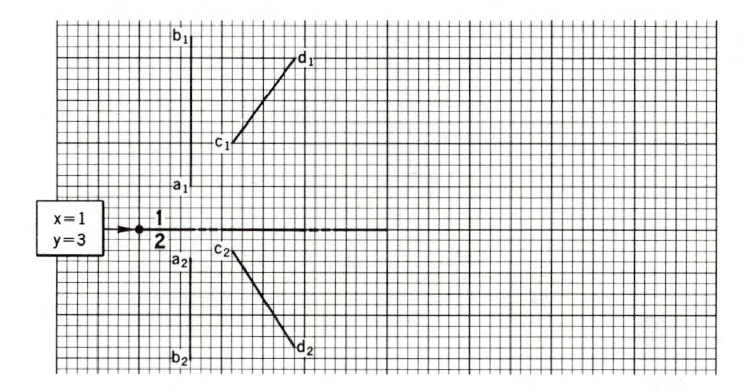

8. Given Data: Horizontal and frontal projections of two skew lines *AB* and *CD*. (Lines *AB* and *CD* are segments of lines of indefinite length and may be extended if necessary.)

PROBLEM 8

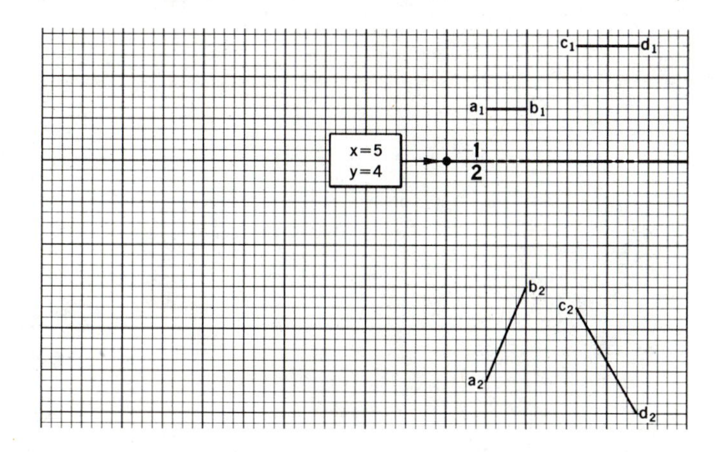

Required: A. The true length and bearing of the shortest line XY between AB and CD having a slope of 30° downward from AB. (*Ans.:* TL = $\frac{7}{8}$ in., bearing = due north)

B. The frontal projection of XY.

C. Label and show all construction.

9. Given Data: Horizontal and frontal projections of two skew lines LM and NO. (Lines LM and NO are segments of lines of indefinite length and may be extended if necessary.)

Required: A. The true length and bearing of the shortest line RS from LM to NO that has a grade of 50% upward from LM. (*Ans.:* TL = $\frac{1}{2}$ in., bearing = N16°W)

B. The frontal projection of RS.

C. Label and show all construction.

PROBLEM 9

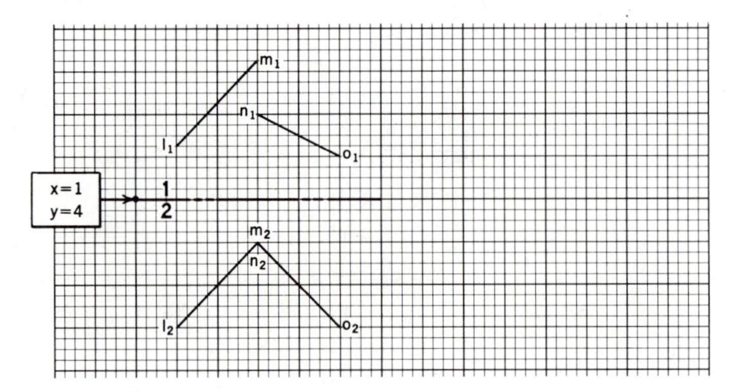

10. Given Data: Horizontal and frontal projections of a plane ABC and a point P outside the plane.

PROBLEM 10

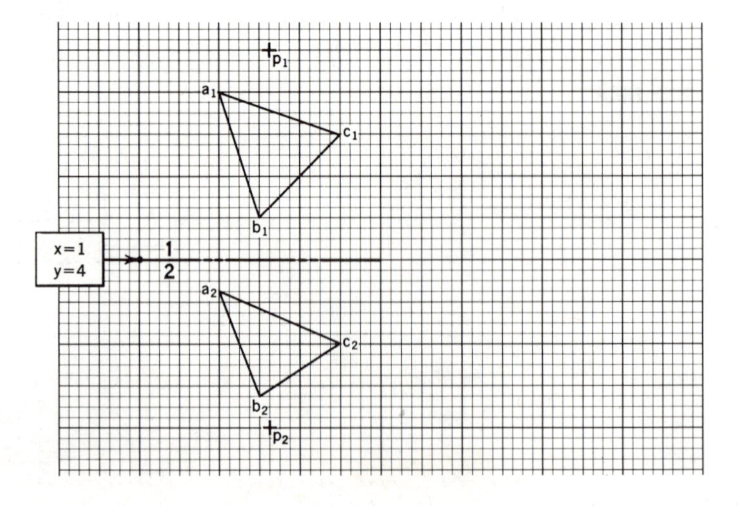

Required: A. Using the edge view method, draw the horizontal and frontal projections of a line PQ that is *perpendicular* to the plane ABC from the point P.

B. The true length, bearing, and slope of perpendicular line PQ. (*Ans.:* TL = $1\frac{1}{2}$ in., bearing = S11°E, slope = 52°)

C. Label and show all construction.

11. Given Data: Horizontal and frontal projections of two intersecting lines AB and CD and a point P outside the planes formed by these lines. (Lines AB and CD are segments of lines of indefinite length lines and may be extended if necessary.)

Required: A. Horizontal and frontal projections of the shortest line PQ from point P to the plane formed by lines AB and CD.

B. The true length, bearing, and slope of PQ. (*Ans.:* TL = $1\frac{1}{2}$ in., bearing = N10°W, slope = 62°)

C. Label and show all construction.

PROBLEM 11

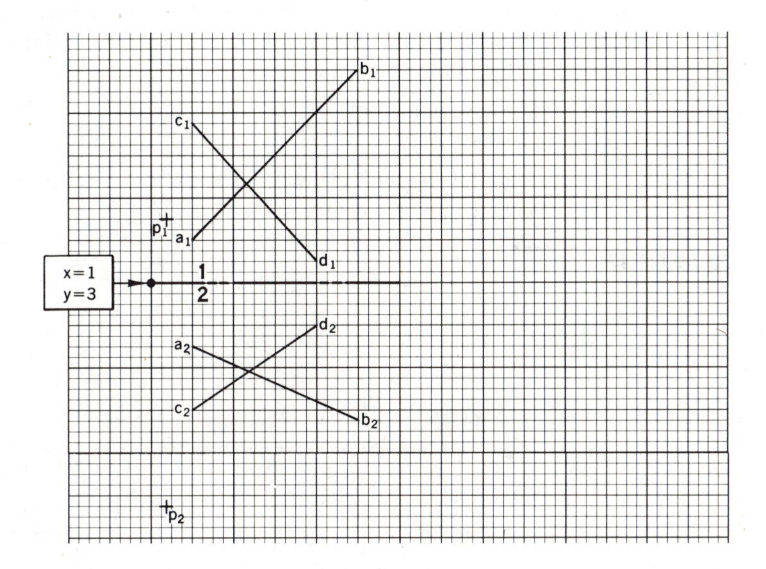

12. Given Data: Horizontal and frontal projections of a plane ABC and an intersecting line LM.

Required: A. Using the cutting plane method, draw the horizontal and frontal projections of the piercing point P of line LM and plane ABC.

B. Show the complete visibility of line LM in the horizontal and frontal views.

C. Use the edge view method to check the results.

D. Label and show all construction.

PROBLEM 12

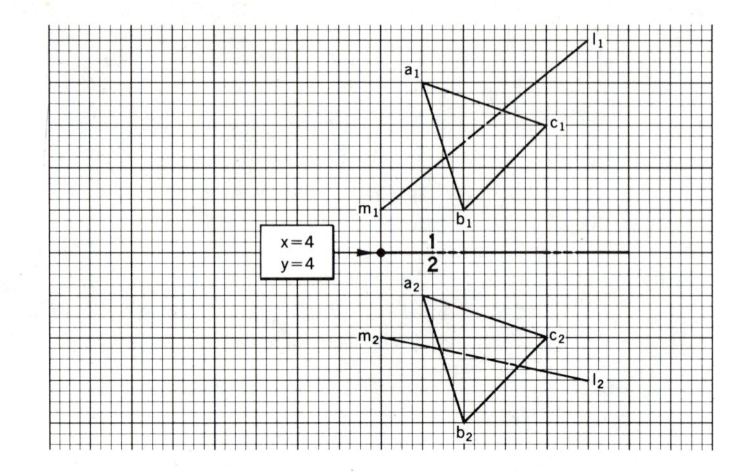

13. Given Data: Horizontal and frontal projections of three noncollinear points *E*, *F*, and *G* and a line *LM*.

Required:

A. Using the edge view method, draw the horizontal and frontal projections of the piercing point *P* of the line *LM* and the plane formed by points *E*, *F*, and *G*.

B. Show the complete visibility of line *LM* and the plane *EFG* in all views.

C. Use the cutting plane method to check the results.

D. Label and show all construction.

PROBLEM 13

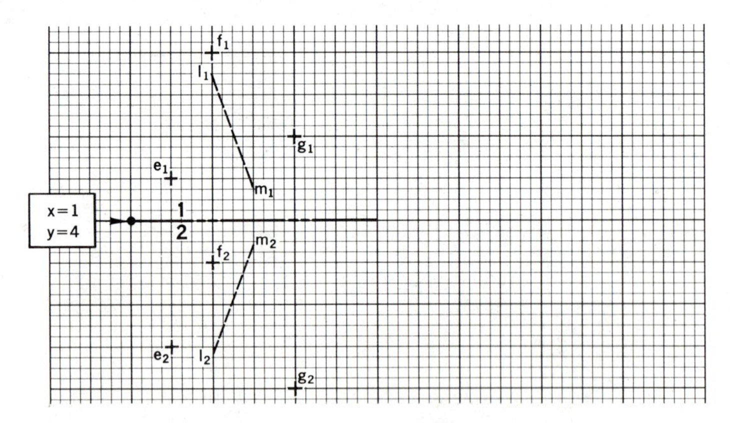

14. Given Data: Horizontal and frontal projections of two intersecting planes *ABC* and *DEF*.

Required:

A. Using the edge view method, draw the horizontal and frontal projections of the line of intersection between planes *ABC* and *DEF*.

B. Show the complete visibility of planes *ABC* and *DEF* in all views.

C. Label and show all construction.

PROBLEM 14

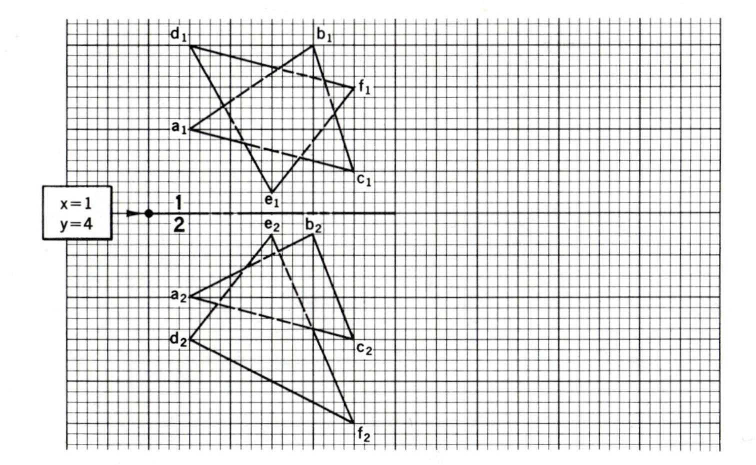

15. Given Data: Horizontal and frontal projections of plane *ABC* and intersecting plane *YXWZ*, formed by intersecting lines *XY* and *WX*.

Required: A. Using the cutting plane method, determine the line of intersection between planes *ABC* and *YXWZ* in the horizontal and frontal views.

B. Show the complete visibility of the given planes in all views.

PROBLEM 15

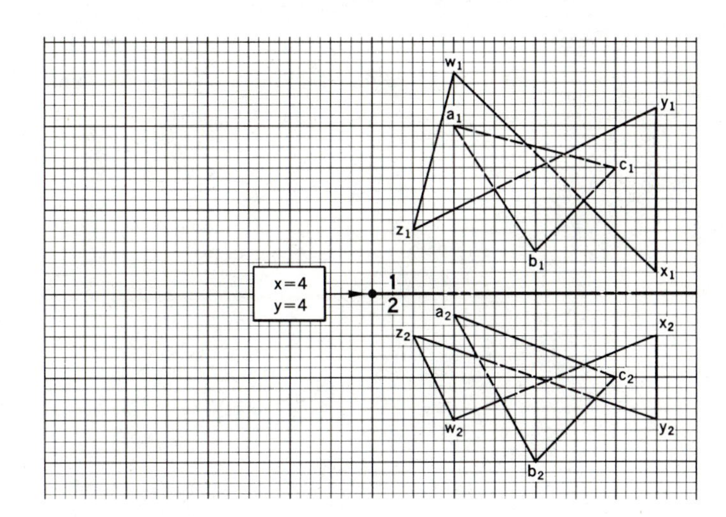

16. Given Data: Horizontal and frontal projections of two unlimited intersecting planes *LMN* and *RST*.

Required: A. Using only the two given views, draw the horizontal and frontal projections of the line of intersection between planes *LMN* and *RST*.

B. Label and show all construction.

PROBLEM 16

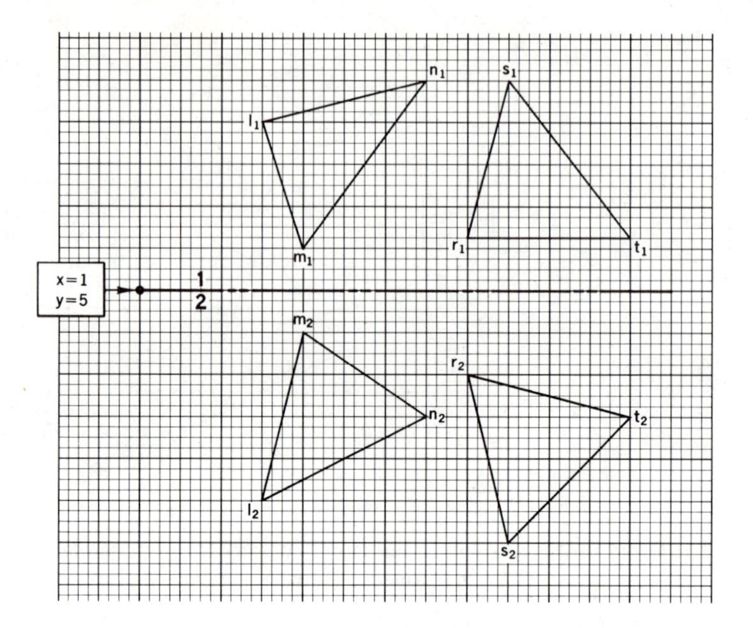

17. Given Data: Horizontal and frontal projections of two intersecting planes *ABC* and *ABD* (visibility incomplete).

Required: A. The true dihedral angle between planes *ABC* and *ABD*. (*Ans.:* 14°)

B. The complete visibility of planes *ABC* and *ABD* in all views.

C. Label and show all construction.

PROBLEM 17

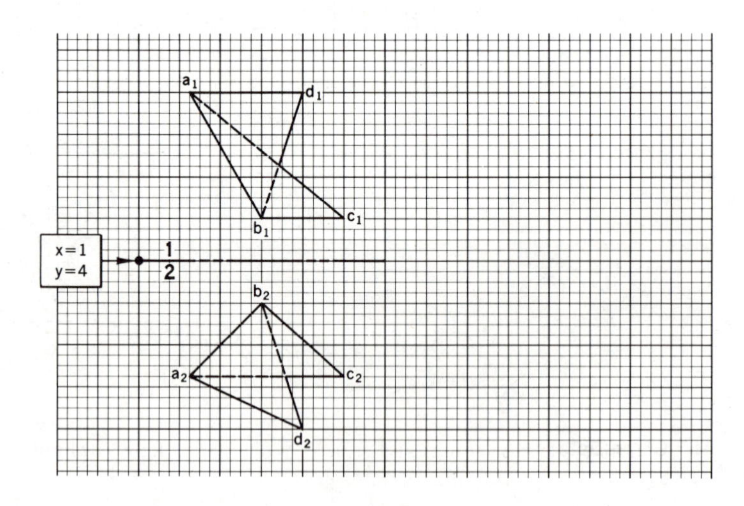

18. Given Data: Horizontal and frontal projections of two unlimited intersecting planes *ABC* and *DEF*.

Required: A. The horizontal and frontal projections of the line of intersection between planes *ABC* and *DEF*. (SUGGESTION: Use the cutting plane method.)

B. The true dihedral angle between planes *ABC* and *DEF*. (*Ans.*: 122°)

C. Label and show all construction.

PROBLEM 18

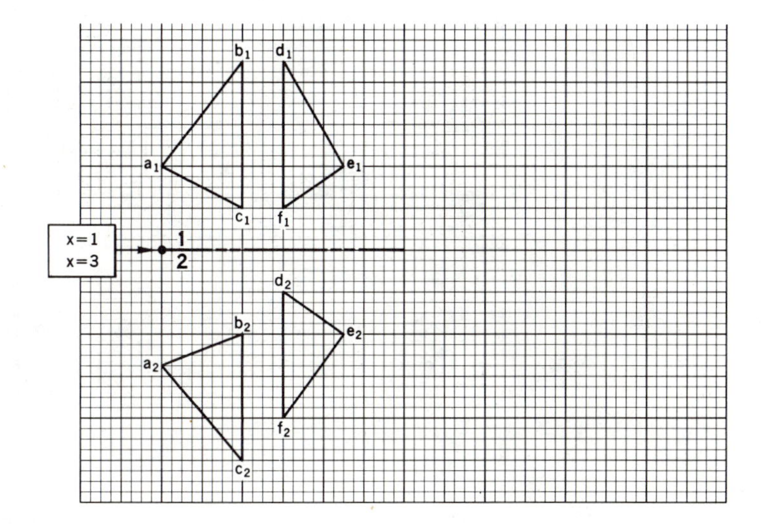

19. Given Data: Horizontal and frontal projections of plane *ABC* and line *LM*.

Required: A. The true acute angle line *LM* makes with plane *ABC*. (*Ans.*: 35°)

B. The piercing point and visibility of line *LM* and plane *ABC* in all views.

C. Label and show all construction.

PROBLEM 19

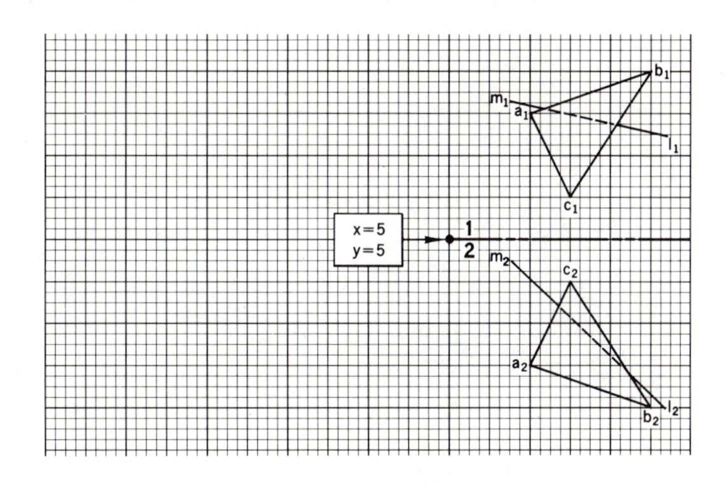

20. Given Data: Horizontal and frontal projections of plane *ABCD* and line *LM*.

Required: A. The true acute angle line *LM* makes with plane *ABCD*. (*Ans.:* 43°)

B. Label and show all construction.

PROBLEM 20

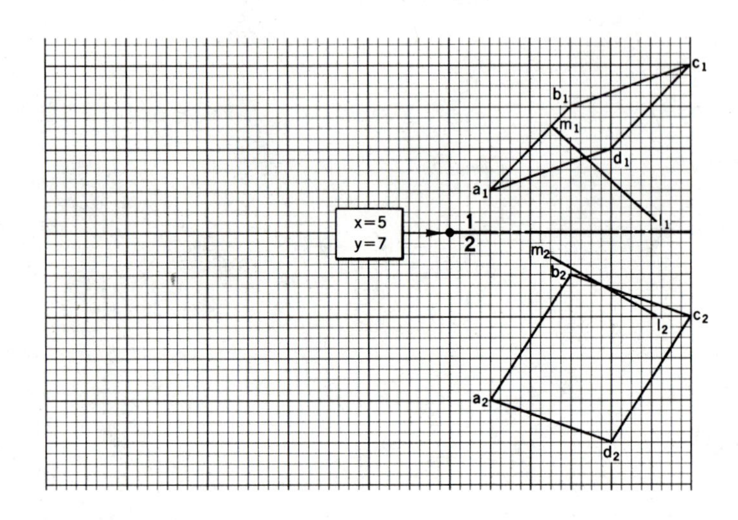

Rotation of Geometric Elements in a Three-Dimensional Space

42 The notion of rotation

In the system of orthographic projection, we have assumed that an observer *changes his position* in space in order to see the different possible projections of a geometric element. The geometric element (whether it be a point, line, plane, or solid) is *fixed in position* relative to a frame of reference made up of the horizontal and frontal projection planes.

Problems involving true lengths and point views of lines, edge views and true shapes of planes, dihedral angles between two planes and the angle a line makes with a given plane can be solved faster if certain geometric elements in each type of problem are permitted to *change their position* relative to the given horizontal and frontal projection planes. This change is accomplished by rotating, or revolving, the geometric elements around specific axes and in specific directions. Involved in the process of rotation are (1) an axis of rotation, (2) a center of rotation, and (3) a path of rotation. These elements define a circle, or part of a circular plane, of rotation that has a specific radius.

43 Rotation of a point about a line axis

When a given point is revolved about a given line axis, the point is located at a specific distance from the axis. Figure 104 shows a point P in an initial position, located at a perpendicular distance R from a line axis AX. The perpendicular

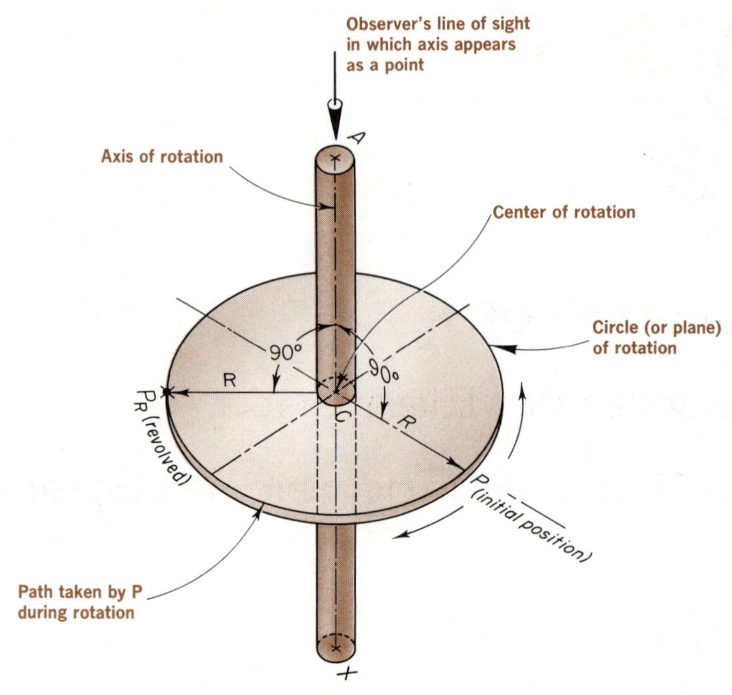

FIGURE 104

distance R is the *radius* of rotation, and it describes a complete circle when it travels from the initial position of P through an angle of 360°. The *path* of rotation taken by P is a *circle,* and together with the radius of rotation R, a *plane* of rotation is generated. The plane of rotation is *perpendicular* to the axis of rotation AX.

The two key points to remember when applying the rotation procedure are:

1. When a point is revolved about a line axis, the plane of rotation (which contains the path of rotation) is *perpendicular* to the axis of rotation.
2. The path of rotation is a circle in which the axis of rotation is seen as a point.

This is true of *any* geometric element revolved since, in the final analysis, *all geometric elements consist of points.*

Orthographic Example: Rotation of a point about a line axis (Fig. 105).

Given Data: Horizontal and frontal projections of a line axis AX and a point P *not* on the axis.

Required: Revolve point P through an angle of 240°, *counterclockwise* from its initial position.

FIGURE 105

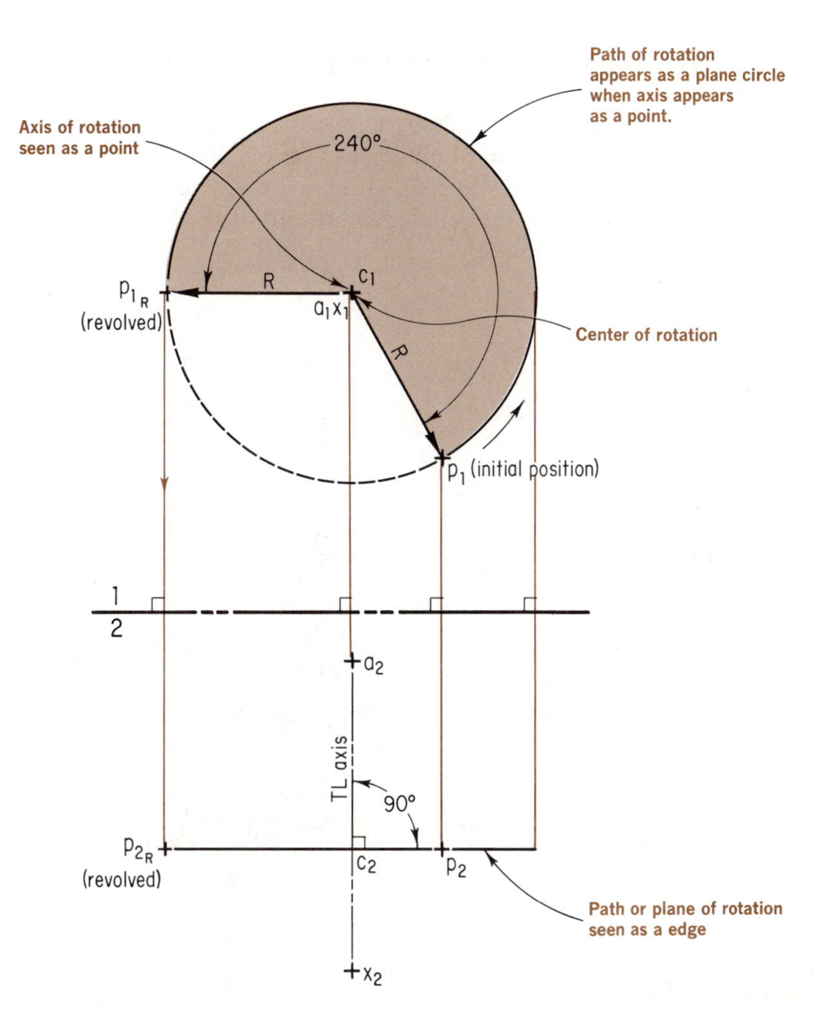

Construction Program:

a. The axis of rotation AX appears as a point a_1x_1 in view #1. The center of rotation c_1 also appears on the point view of the axis.

   The true length a_2x_2 of the axis AX appears in view #2.

b. In view #1, using c_1 as the center of rotation, strike an arc of 240°, having a radius R (from c_1 to p_1). Measure in a counterclockwise direction from the initial position of point p_1.

c. In view #2, construct a line through p_2 *perpendicular* to the *true length axis of rotation.* This line represents the edge view of the plane of rotation.

d. Project p_{1_R} (final position of revolved point p_1) downward to view #2 in order to locate its position p_{2_R} on the edge view of the plane of rotation. This completes the requirements of the problem.

44 Rotation of a line about a line axis

As you rotate a given line about a line axis, if one end of the given line is located *on* the line axis and the rotation goes through 360°, you will generate a cone of revolution. In Fig. 106 you see a given line *DB* with a point *D* on a line axis *AX*. (We show only a few intermediate positions of the line *DB* in order to maintain clarity of presentation.)

FIGURE 106

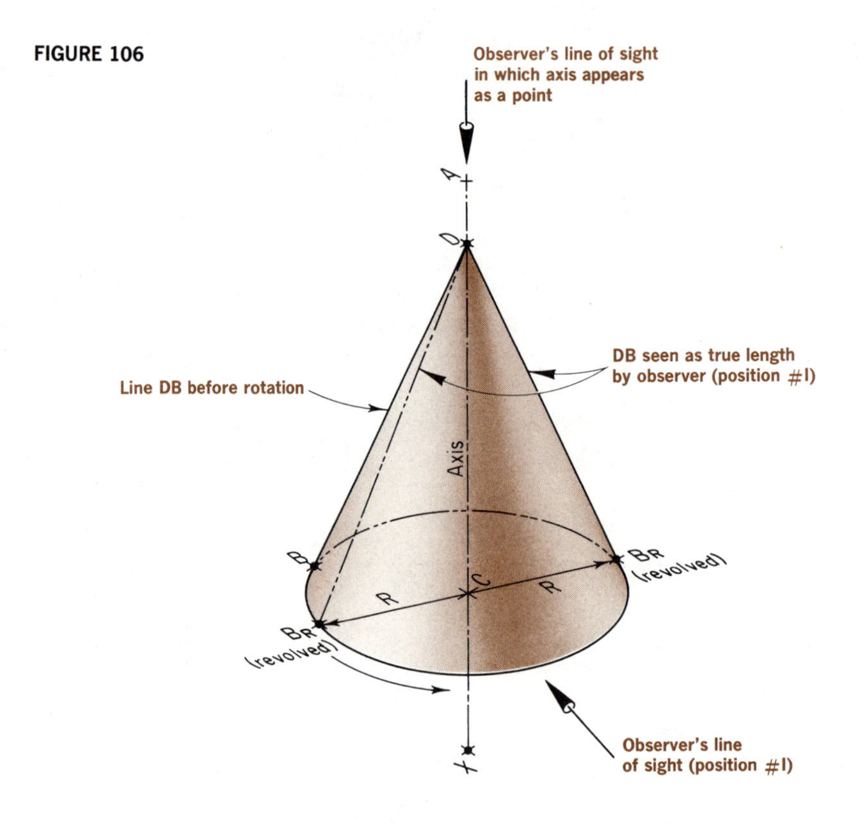

If an observer in position #I wishes to see the true length of the given line *DB*, *DB* can be revolved into such a position that the observer's line of sight will be perpendicular to it. (Note that in Fig. 106 there are two such positions available, since the line *DB* can be revolved either clockwise or counterclockwise.)

Utilizing the rotation procedure, we can easily determine the true length and true slope of a line.

Orthographic Example: True length and true slope of a line determined by the rotation procedure (Fig. 107).

Given Data: Horizontal and frontal projections of a line *DB*.

FIGURE 107

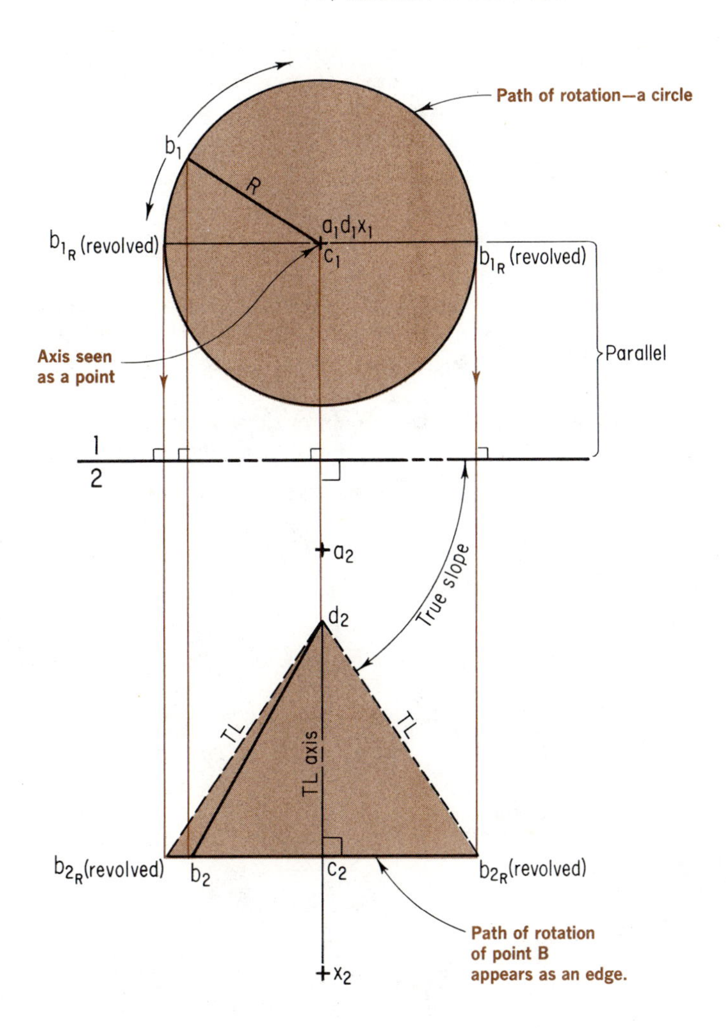

Required: The true length and true slope of line *DB*.

Construction Program:

a. Choose a line axis of rotation that passes through point *D* on line *DB*. In view #1, the axis appears as a point a_1x_1.

b. In view #2, the line axis appears as true length a_2x_2. (This must be so, because a line can appear as a point only if its adjacent view is true length and perpendicular to a reference line.)

c. Using c_1 in view #1 as a center and the horizontal view b_1d_1 of line *DB* as a radius *R*, revolve the point b_1 about the center of rotation c_1. Do this until b_1d_1 is parallel to RL 1–2. (The rotation in this case could be either clockwise or counterclockwise, giving two possible locations for the final position b_{1_R}.)

d. Project b_{1_R} (either one) downward to view #2.

e. The path of rotation of b_2 appears in view #2 as an edge that is perpendicular to the true length axis of rotation. Therefore, from b_2 construct a line representing the path of rotation *perpendicular* to the true length axis. Point b_{2_R} is located on this path and can be determined by projection from view #1.

f. Connect d_2 and b_{2_R} with a straight line. You see the true length of line DB as $b_2d_{2_R}$ and, since view #2 is a frontal *elevation* view, you also see the true slope.

(NOTE: *The line axis of revolution may be located anywhere in space relative to a given line.* The following orthographic example demonstrates how the true length and true slope of a line are determined by using an axis location *outside* the given line.)

Orthographic Example: True length and true slope of a line by the rotation procedure when an axis is located outside the line (Fig. 108).

Given Data: Horizontal and frontal projections of a line DB and an axis AX.

Required: The true length and true slope of the line DB.

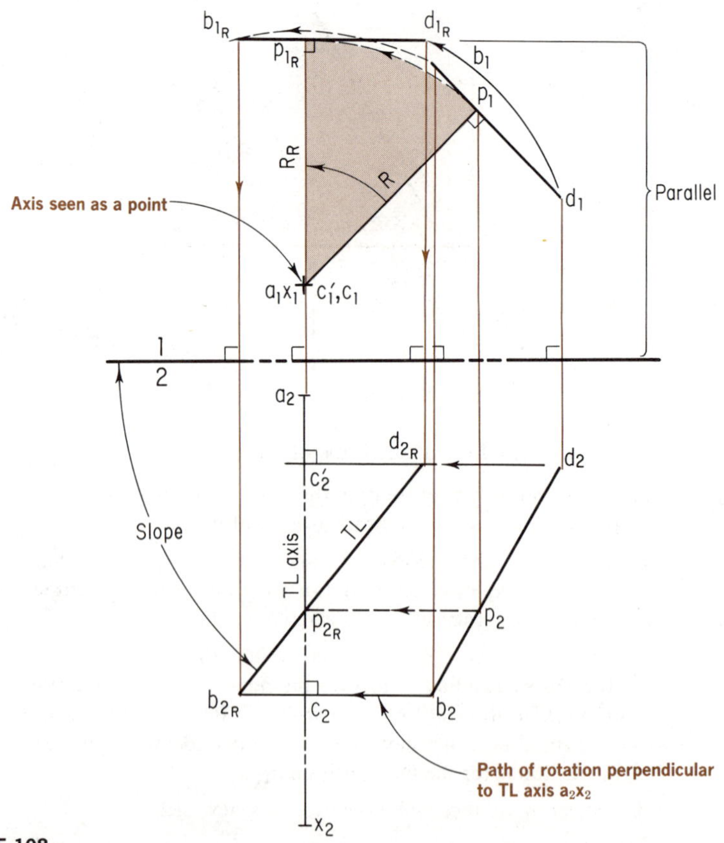

FIGURE 108

Construction Program:

a. Use the point view a_1x_1 of the axis in view #1 as a center (c_1), and a radius R constructed from the point view of the axis *perpendicular* to the horizontal view b_1d_1 of line DB (the foot of this perpendicular radius is point p_1). Revolve p_1 into a position in the top view so that R is perpendicular to RL 1–2. (Perpendicular position is noted as R_R.)

b. In view #1, at p_{1_R}, construct a line that is *parallel* to RL 1–2.

c. From the center c_1, revolve point b_1 until the arc of rotation intersects the parallel line at point b_{1_R}.

d. Likewise, from center c_1' (which is also located on the point view of axis a_1x_1), revolve point d_1 until its arc of rotation intersects the parallel line at d_{1_R}. Connect b_{1_R} and d_{1_R} with a straight line. This is the new position $b_{1_R}d_{1_R}$ of the line DB that is parallel to the frontal projection plane.

e. Project the newly revolved points b_{1_R} and d_{1_R} downward to frontal view #2.

f. The paths of revolution for points b_2 and d_2 in view #2 are perpendicular to the true length axis of rotation a_2x_2. Construct lines from b_2 and d_2 perpendicular to the true length axis a_2x_2. The locations of points b_{2_R} and d_{2_R} are at the intersection of the projectors from b_{1_R} and d_{1_R}, respectively (view #1), with the perpendiculars from b_2 and d_2 (view #2). Connect b_{2_R} and d_{2_R} with a straight line. This is the required true length $b_{2_R}d_{2_R}$ of line DB. Since view #2 is a frontal *elevation* view, the true slope of line DB is also seen.

45 Rotation of a plane about a line axis

You can also apply rotation procedures to plane surfaces. A given plane surface BCD, such as that shown in Fig. 109, can be revolved about a line axis so that it

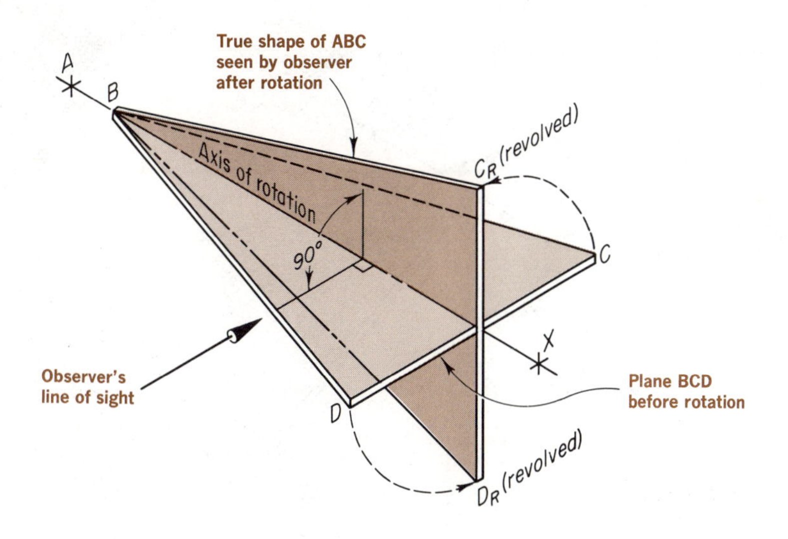

FIGURE 109

is perpendicular to the observer's line of sight. In this position the observer can see the true shape of the given plane surface.

The axis of rotation in Fig. 109 is a line axis AX, which lies in the plane BCD. (The axis of rotation could also lie outside the given plane—see Fig. 110.)

Orthographic Example: True shape of a plane using the rotation procedure (Fig. 111).

Given Data: Horizontal and frontal projections of plane BCD.

Required: The true shape view of plane BCD.

Construction Program:

a. In view #2, draw a horizontal line b_2d_2, extended, in plane $b_2c_2d_2$. This line will be used as the axis of revolution a_2x_2.

b. The true length b_1h_1 of the horizontal line can be seen in view #1 by direct projection. (This is part of the true length axis of revolution a_1x_1.)

c. Draw a projection plane #3 (represented by RL 1–3) *perpendicular* to the true length axis a_1x_1. The axis appears as a point (a_3x_3) in view #3, and the plane appears as an edge ($b_3c_3d_3$) in the same view.

d. Using the point view a_3x_3 of the axis AX as a center of rotation, revolve the edge view $b_3c_3d_3$ of plane BCD so that it assumes a position *parallel* to RL 1–3.

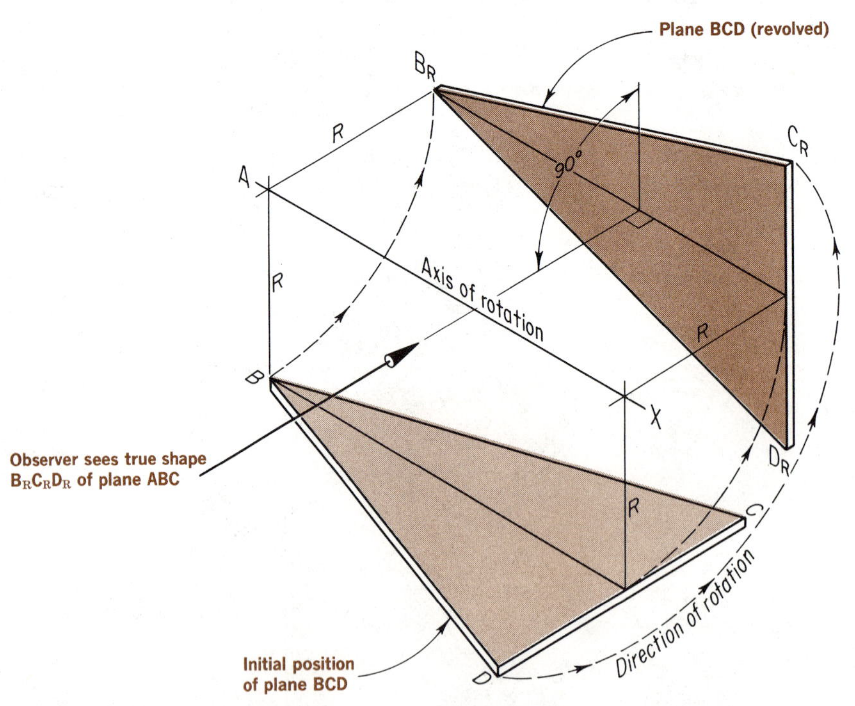

FIGURE 110

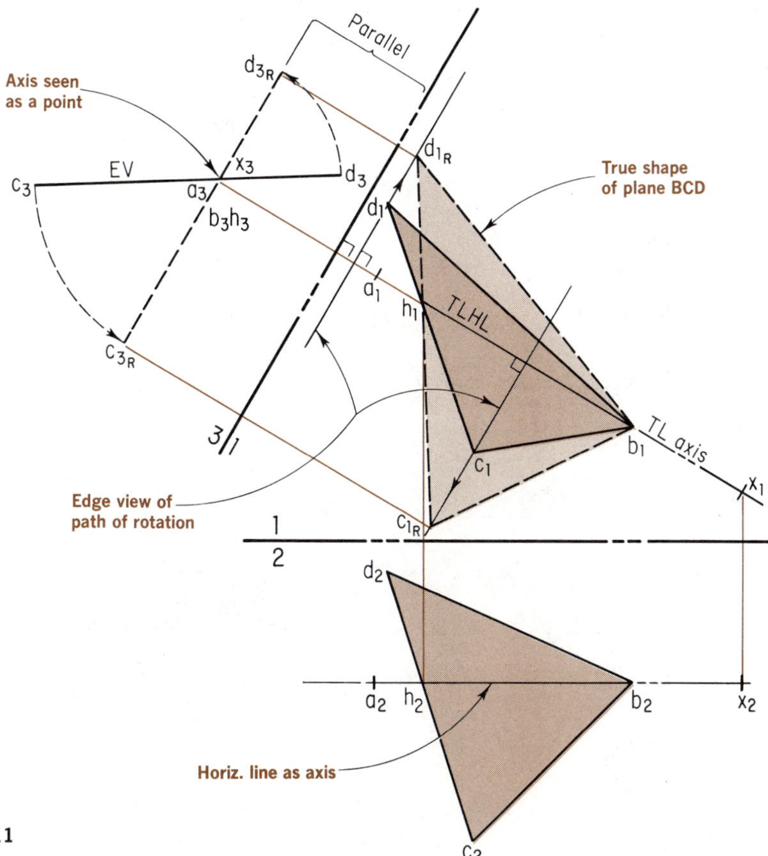

FIGURE 111

e. Determine the new position of points c_1 and d_1 in view #1. The paths of revolution for these points are perpendicular to the true length axis a_1x_1. Therefore, by projection from view #3, you can locate c_{1_R} and d_{1_R}. (Since point b_1 is on the axis of revolution it does not move.)

f. Connect points $b_1d_{1_R}$, $b_{1_R}c_{1_R}$, and $c_{1_R}d_{1_R}$ with straight lines. This is the required true shape of plane *BCD*.

Figure 112 shows plane *BCD* being revolved about an axis *AX*, which is located *outside* the plane. Note that the point view of the axis appears in the view in which the plane is seen as an edge. In this view the axis may be located anywhere. It is recommended that you write out the construction program for this figure.

The rotation procedure also makes it possible to determine the true shape of a given plane using *only two* given views.

Figure 113 shows how this is done. Study this figure carefully and write out the construction program for the example. (Note that two axes #I and #II were used in Fig. 113. Axis #I was used to locate the edge view of plane *BCD* in view #2, and axis #II was used to rotate the edge view into a position parallel to the horizontal projection plane, thus causing the true shape plane *BCD* to appear in the horizontal projection view #1.)

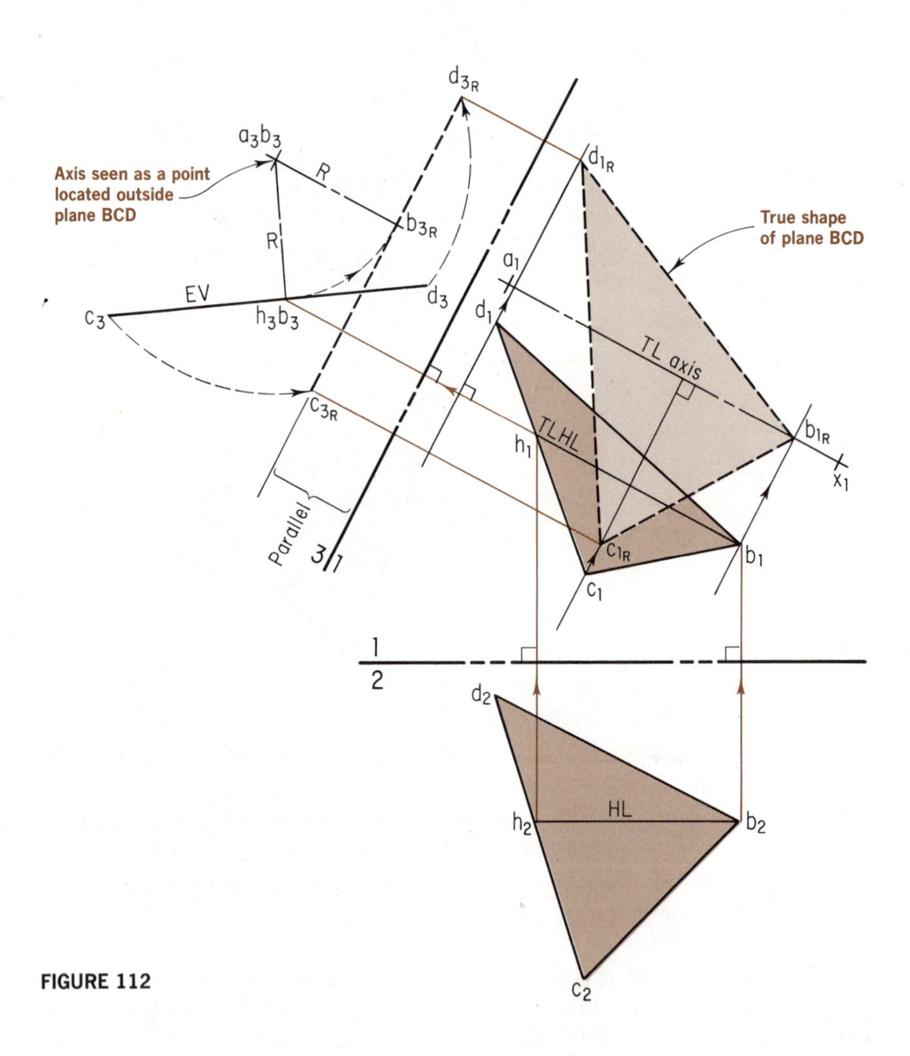

FIGURE 112

FIGURE 113

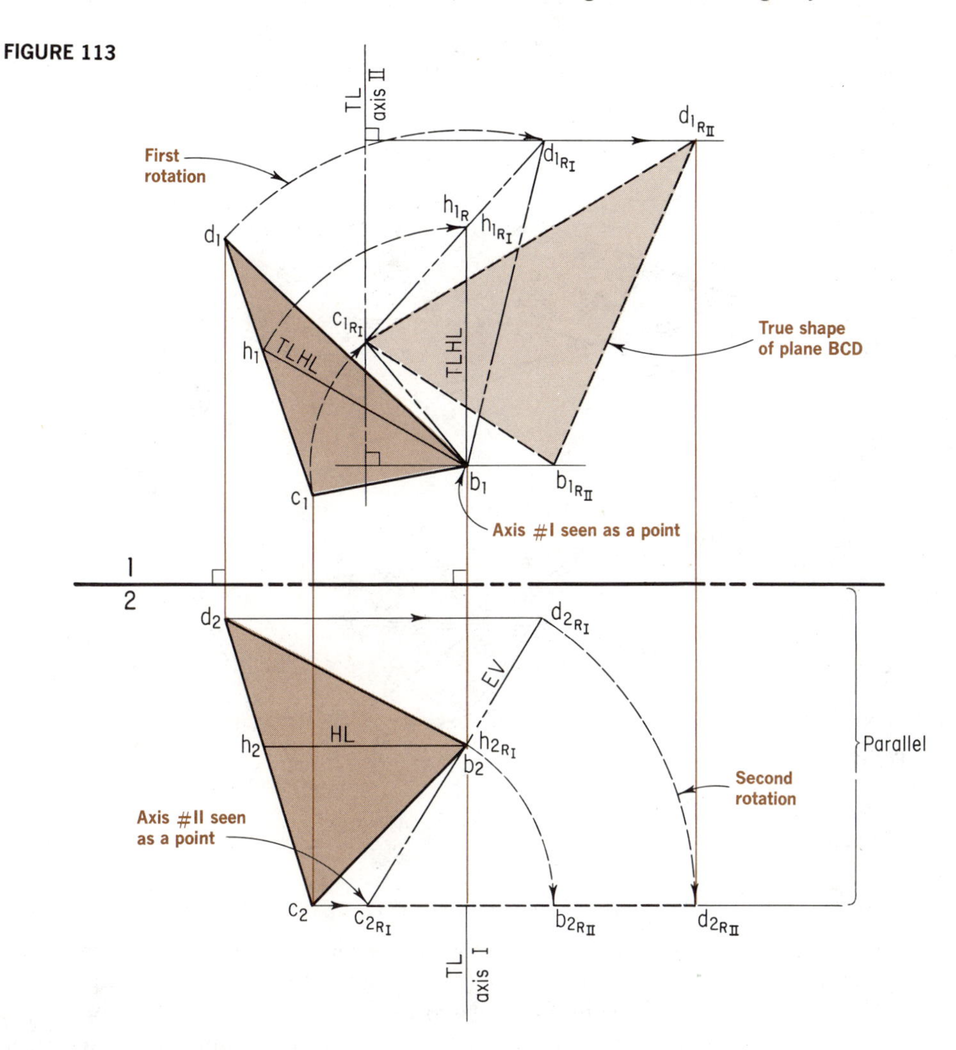

46 Determination of the dihedral angle by rotation

To determine the dihedral angle between two given planes by rotation, you must pass a cutting plane *perpendicular* to the line of intersection between the given planes. This cutting plane cuts a "plane section," the *true shape* of which contains the dihedral angle. By revolving the plane section so that it is perpendicular to the observer's line of sight, you can see the true dihedral angle (see the pictorial representation in Fig. 114).

Orthographic Example: Determination of the dihedral angle between two planes by rotation.

Given Data: Horizontal and frontal projections of two intersecting planes *ABC* and *ABD*, the line of intersection of which is *AB*.

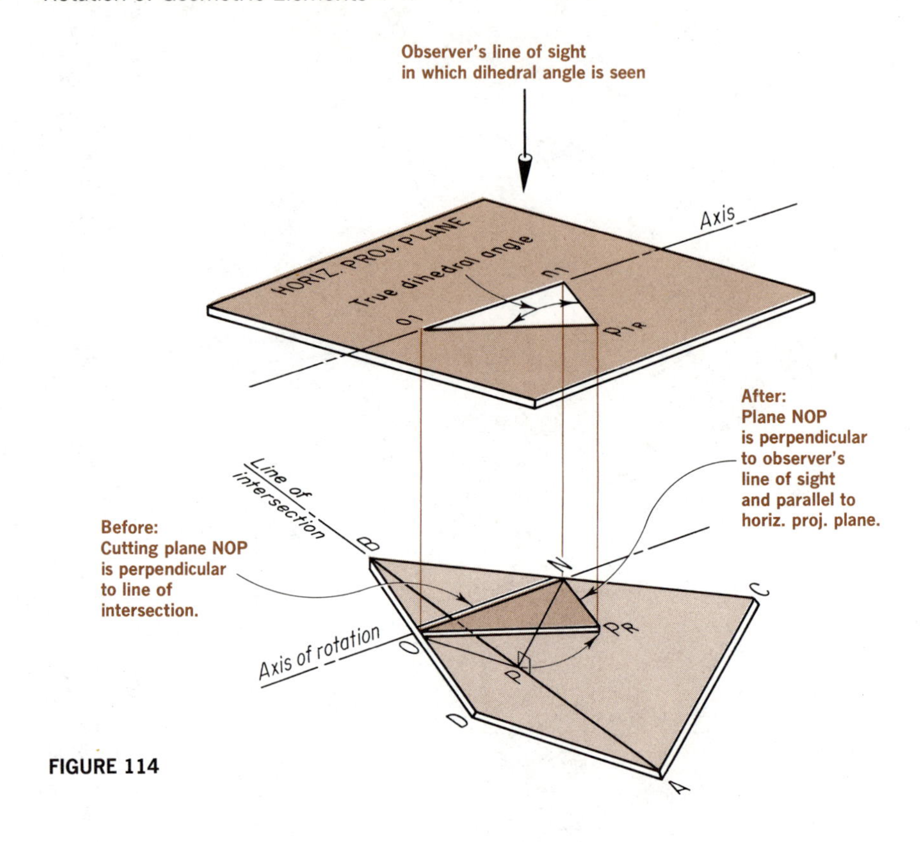

FIGURE 114

Required: The dihedral angle between planes ABC and ABD.

Construction Program:

a. Determine the true length of the line of intersection AB between the two given planes by constructing a projection plane #3 (represented by RL 1–3) *parallel* to a_1b_1.

b. In view #3, construct the edge view of the cutting plane (indicated as x_3y_3) *perpendicular* to the true length line of intersection a_3b_3. This cutting plane cuts a plane section indicated as $o_3p_3n_3$. The line o_3p_3 is on plane $a_3b_3c_3$, and the line p_3n_3 is on plane $a_3b_3d_3$.

c. Project points o_3, p_3, and n_3 to view #1 in order to determine points o_1, p_1, and n_1.

d. Choose for an axis a point at which the edge view of the cutting plane appears in view #3 (in view #3 the axis is to be located on a line connecting o_3n_3). Revolve the plane section $o_3p_3n_3$ so that it is parallel to RL 1–3. This determines the location of p_{3_R}.

e. Project p_{3_R} into view #1 in order to locate p_{1_R}, which lies on a path that is perpendicular to the true length axis of revolution.

f. Connect p_{1_R} and o_1, and p_{1_R} and n_1 with extended straight lines. The dihedral angle is measured between these lines.

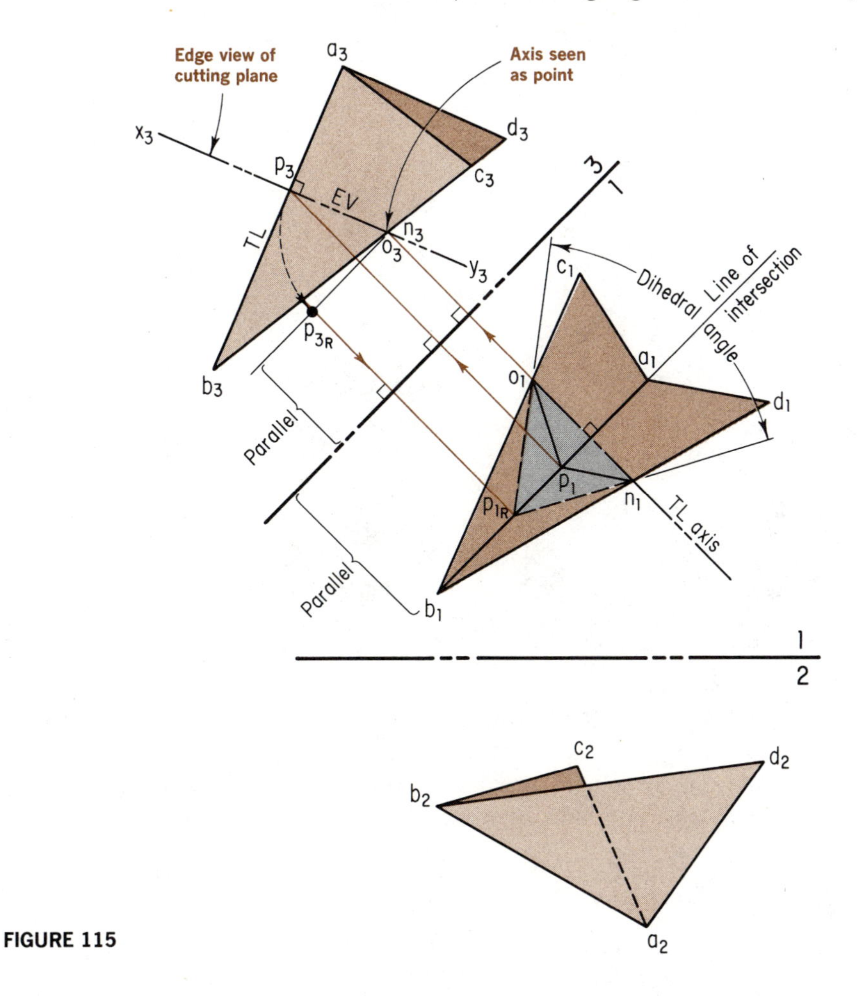

FIGURE 115

47 Determination of the angle between a line and a plane

In order to measure the angle that a line makes with a given plane, you must determine a view in which the *line* and the plane simultaneously appear as true length and an edge, respectively. To obtain such a view by rotation, first determine the *true shape* of the plane by direct orthographic projection. The true shape view is necessary because the line, which makes an angle with the plane, *must* be determined in its *true length* by revolving it about an axis that is *perpendicular* to the plane. This means that such an axis will appear as a *point* in the *true shape* view of the plane in which the path of rotation appears as a circle. (This is important to remember, for the rotation of the line about an axis *not* perpendicular to the given plane will *void* the original angular relationship of the line and plane—see Fig. 116.)

FIGURE 116

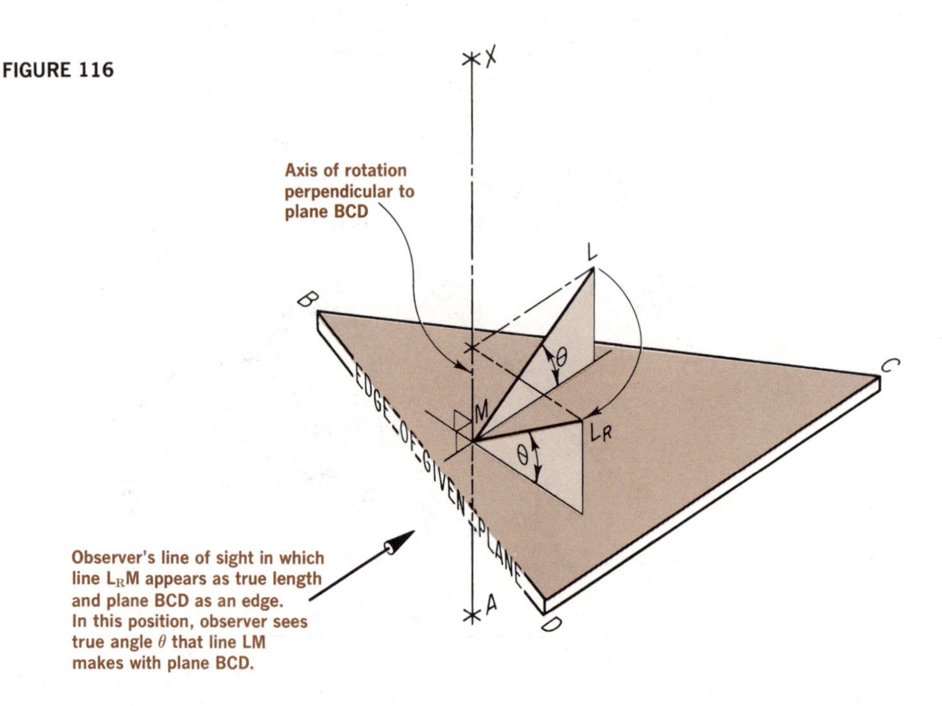

The *true shape* view of the plane is also a key view, since in any view *adjacent* to it the plane *always* appears as an edge.

Orthographic Example: Angle between a line and a plane using the rotation procedure (Fig. 117).

Given Data: Horizontal and frontal projections of a plane *BCD* and a line *LM* *outside* the plane. (*LM* is *not* parallel to *BCD*.)

Required: A view in which we can measure the true angle θ that line *LM* makes with plane *ABC*.

Construction Program:

a. Draw the true shape view ($b_4c_4d_4$) of plane *BCD* by using direct orthographic projection (view the horizontal line *CH* in plane *BCD* as a point in order to determine the edge view and true shape view of the plane in Fig. 117).

b. In true shape view #4, select a point view of the axis of rotation on line l_4m_4 (m_4 is used as an example). In order to determine the true length ($l_{3_R}m_3$) of line *LM* in view #3, revolve line l_4m_4 until $l_{4_R}m_4$ is parallel to RL 3–4. (In view #3 the axis of rotation appears as true length; therefore, point l_3 must follow a path perpendicular to this true length axis in order to locate l_{3_R}.)

c. Since $l_{3_R}m_3$ is the true length of line *LM* and is seen in the same view (#3) in which plane *BCD* appears as an edge ($b_3c_3d_3$), the *true angle* θ that the line makes with the plane can be measured.

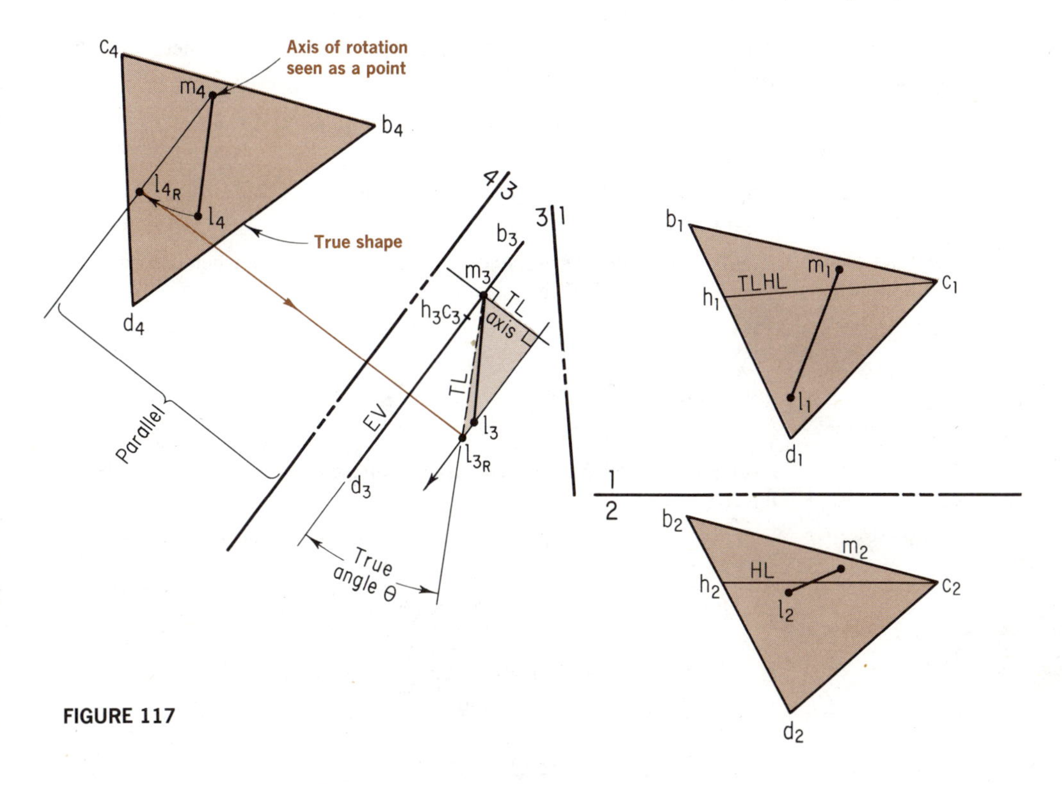

FIGURE 117

SAMPLE QUIZZES

1. In order to find the true length of any line by the rotation procedure, where must the axis be located? In which view do you see the axis as a point?

2. In order to find the true size of the dihedral angle between two intersecting planes having a level line of intersection, where must you locate the axis of rotation? In which view will the dihedral angle appear in its true size?

3. In order to find the true size of the dihedral angle between two intersecting planes having a line of intersection parallel to a frontal plane, where must the axis of rotation be located? In which view will the dihedral angle appear in its true size?

4. In order to find the true size of the dihedral angle between two intersecting planes having a line of intersection that is a profile line, where must the axis of rotation be located? In which view will the dihedral angle appear in its true size?

5. Give a brief analysis of the rotation procedure used to find the true size of any dihedral angle.

6. In revolving a point about any straight line axis, the path of rotation is a circle.

a. In what view will this circle appear as an edge?
b. In what view will it always appear as a circle?

7. In order to find the true size of the angle that a line makes with a plane, where must the axis of rotation be located?

8. Develop the complete construction program for determining the true shape of a plane when the plane is revolved about a line axis outside it.

9. Develop the complete construction program for determining the true shape of a plane by using only two given views of the plane.

10. Why must the axis of rotation be perpendicular to a given plane when determining by rotation the angle between a line and the given plane?

PRACTICE PROBLEMS

1. Given Data: Horizontal and frontal projections of a line *AX* and a point *P* outside the line.

Required:

A. Using line *AX* as an axis, revolve point *P* clockwise through an angle of 120°.

B. Draw the horizontal and frontal projections of the new position of point *P*.

C. Label and show all construction, including the circular arc and the straight line path of rotation.

PROBLEM 1

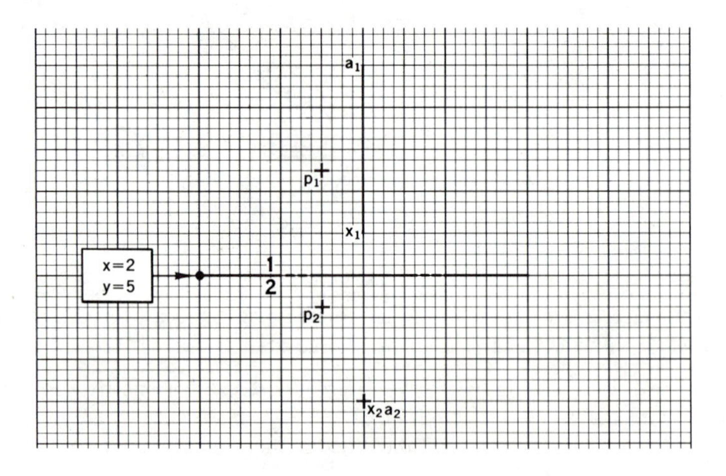

2. Given Data: Horizontal and frontal projections of a line *AX* and a point *P* outside the line.

Required:

A. Using line *AX* as an axis, revolve point *P* so that it assumes a position $\frac{1}{2}$ in. below the horizontal projection plane. Show this new position in all views.

B. What is the smallest angle through which point *P* must be revolved in order to assume its new position? (*Ans.:* 33°)

C. Label and show all construction, including the circular arc and the straight line path of rotation.

PROBLEM 2

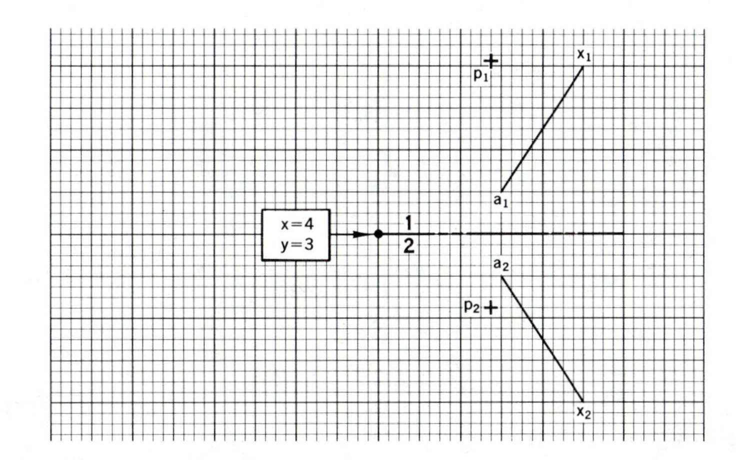

3. Given Data: Horizontal and frontal projections of a line *LM*.

Required:

A. Using *only* the two given views, determine the true length of line *LM* in view #1. (*Ans.:* TL = $2\frac{19}{32}$ in.)

B. Through what angle of rotation did line *LM* travel to reach its true length position? (*Ans.:* 44°)

C. Label and show all construction, including the circular arc and the straight line path of rotation.

PROBLEM 3

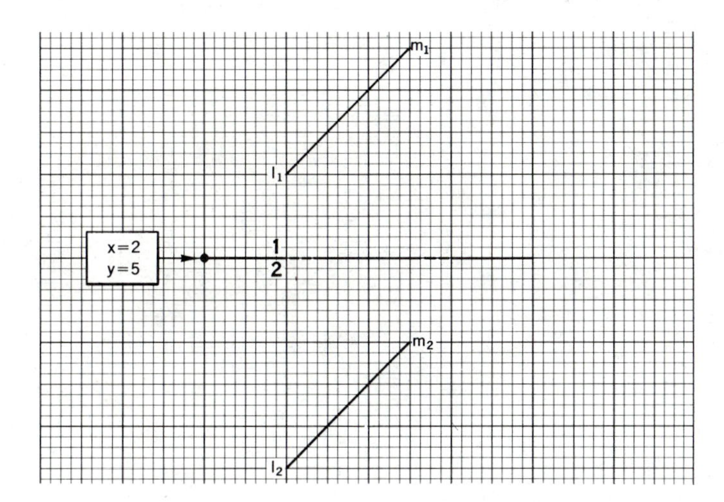

4. Given Data: Horizontal and frontal projections of two skew lines *RS* and *AX*.

Required:

A. Using line *AX* as an axis of rotation, determine the true length of line *RS*. (*Ans.:* TL = $2\frac{5}{16}$ in.)

B. Show the new position of line *RS* in all views.

C. Label and show all construction, including the circular arcs and the straight line paths of rotation.

PROBLEM 4

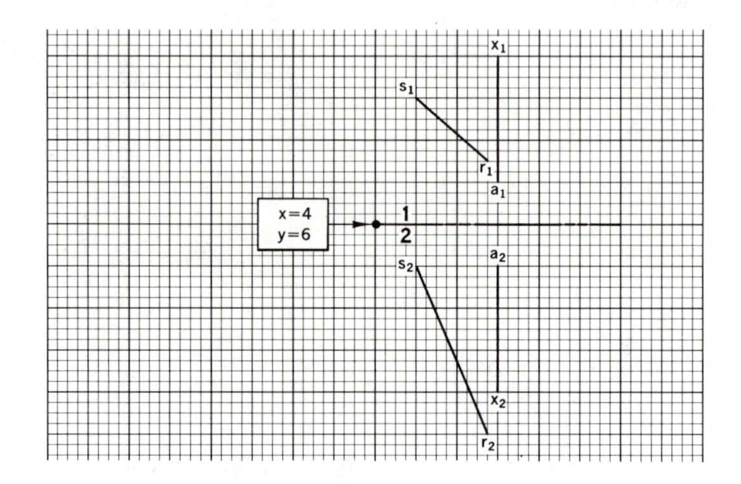

5. Given Data: Horizontal and frontal projections of plane *ABC*.

Required: A. Using line *AB* of the plane *ABC* as an axis, revolve the plane in order to determine its true shape view.

B. Through what angle was the plane revolved to obtain the true shape view? (*Ans.:* 120° or 60°)

C. Label and show all construction, including the circular arc and the straight line path of rotation.

PROBLEM 5

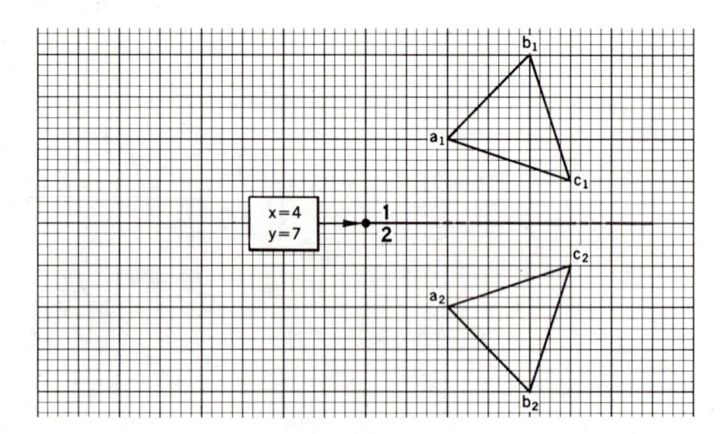

6. Given Data: Horizontal and frontal projections of plane *ABCD*.

Required: A. Using a frontal line in the plane *ABCD* as an axis, revolve the plane so that its true shape view appears in frontal view #2.

B. What is the smallest angle through which the plane could be revolved to obtain its true shape view? (*Ans.:* 46°)

C. Label and show all construction, including circular arc and straight line paths of rotation.

PROBLEM 6

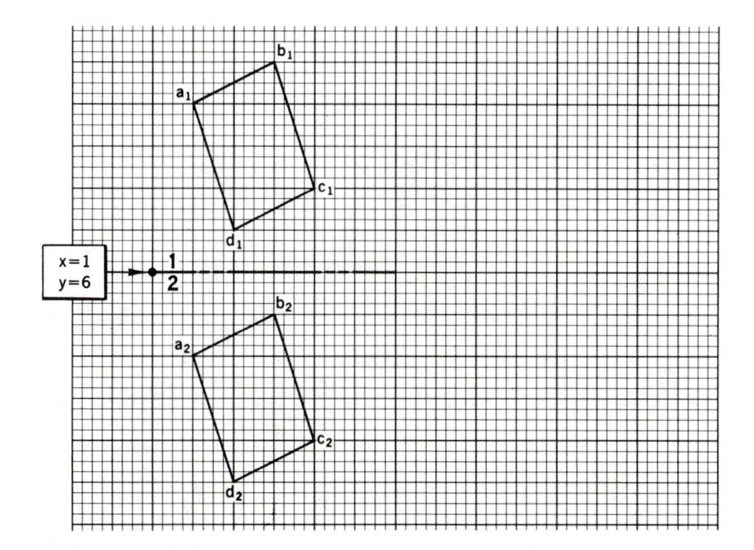

7. Given Data: Horizontal and frontal projections of intersecting planes *ABCD* and *ABE*.

Required: A. Determine by rotation the dihedral angle between planes *ABCD* and *ABE*. (*Ans.:* 12°)

B. Label and show all construction, including circular arcs and straight line paths of rotation.

PROBLEM 7

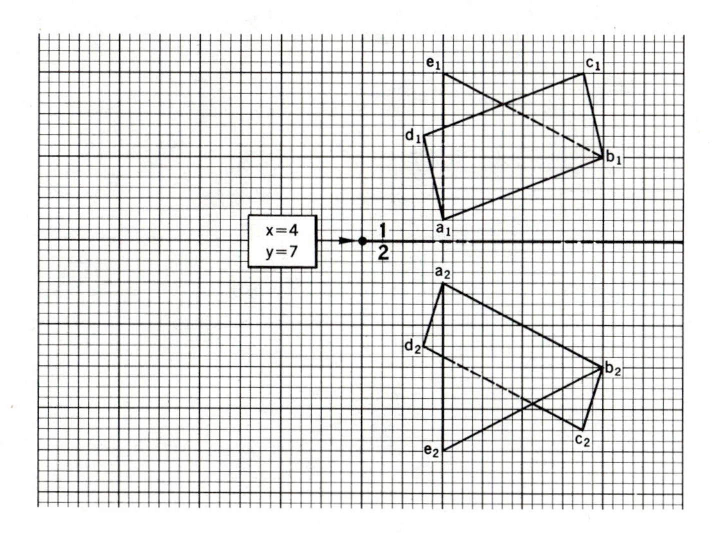

8. Given Data: Horizontal and frontal projections of two unlimited intersecting planes *ABC* and *WXYZ*.

Required:

A. Determine by rotation the dihedral angle between planes *ABC* and *WXYZ*. (*Ans.:* 92°±)
(SUGGESTION: First determine the line of intersection between the given planes. If necessary, extend the lines of the planes.)

B. Label and show all construction, including circular arcs and straight line paths of rotation.

PROBLEM 8

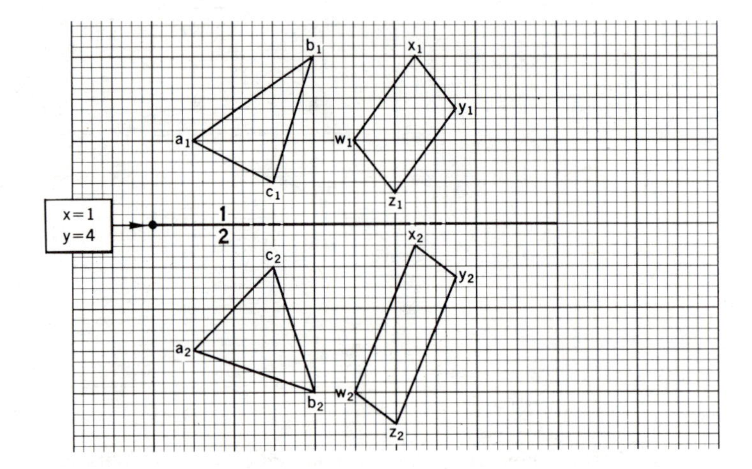

9. Given Data: Horizontal and frontal projections of plane *ABC* and a line *LM*.

Required:

A. Determine by rotation the true acute angle between line *LM* and plane *ABC*. (*Ans.:* 7°)

B. Label and show all construction, including the circular arc and straight line paths of rotation.

PROBLEM 9

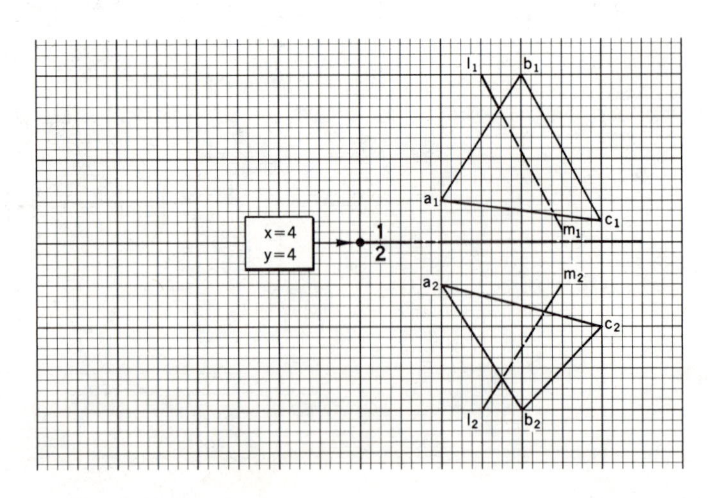

10. Given Data: Horizontal and frontal projections of plane *ABC* and a line *XY*.

Required: A. Determine by rotation the true acute angle line *XY* makes with plane *ABC*. (*Ans.:* 39°)

B. Label and show all construction, including circular arc and straight line path of rotation.

PROBLEM 10

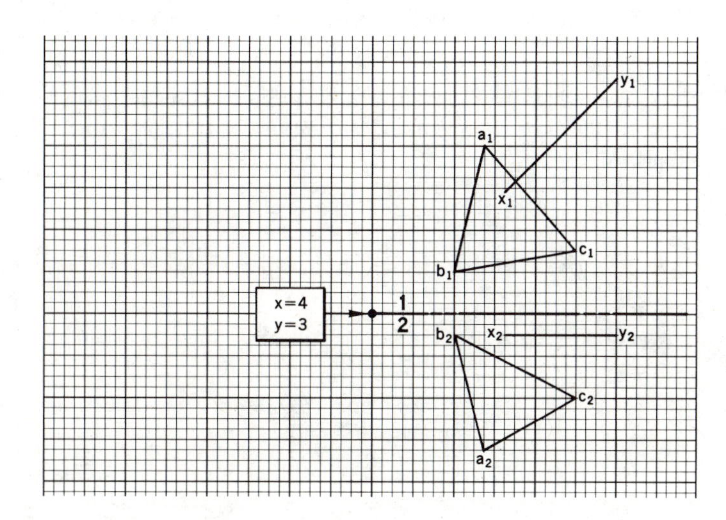

Sample Descriptive Geometry Test #2 (Covering Chapters III–V)

COURSE NAME: ________ COURSE NO.: ________ STUDENT NAME: ____________ DATE: ________

TIME LIMIT: 90 MINUTES

NOTE: Reproduce problems on 8 in. × 10 in. cross-section paper having 8 in. × 8 in. divisions to the inch, horizontally and vertically (see the Appendix for solutions to test problems).

1. GIVEN DATA: Horizontal and frontal projections of two intersecting planes *ABC* and *ABD*.

REQUIRED: A. Determine the dihedral angle between the given planes by direct projection.

B. Revolve plane *ABD* through an angle so that it lies in the same plane as plane *ABC*. Indicate the new position of plane *ABD* in all views.

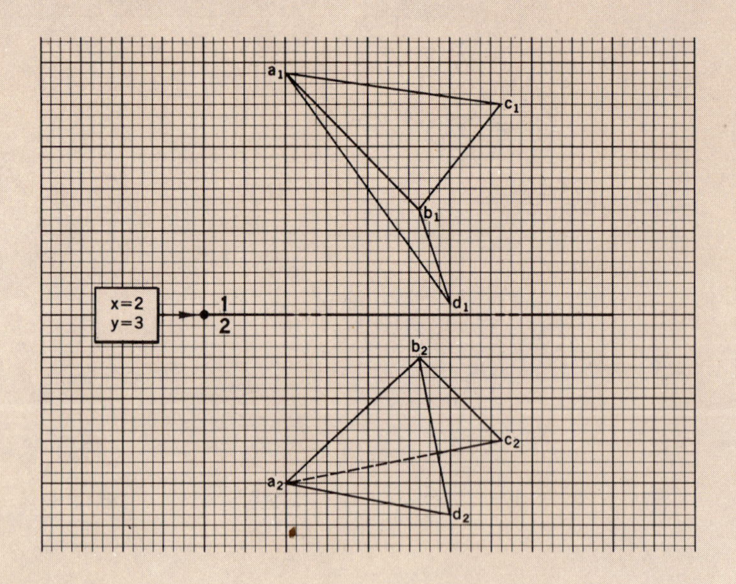

2. GIVEN DATA: Horizontal and frontal projections of two skew lines *AB* and *CD* and a point *P* outside the lines.

REQUIRED: A. The horizontal and frontal projections of the shortest line from *P* to the common perpendicular line between *AB* and *CD*.

B. The bearing, slope, and true length of the shortest line from *P* to the common perpendicular line between *AB* and *CD*.

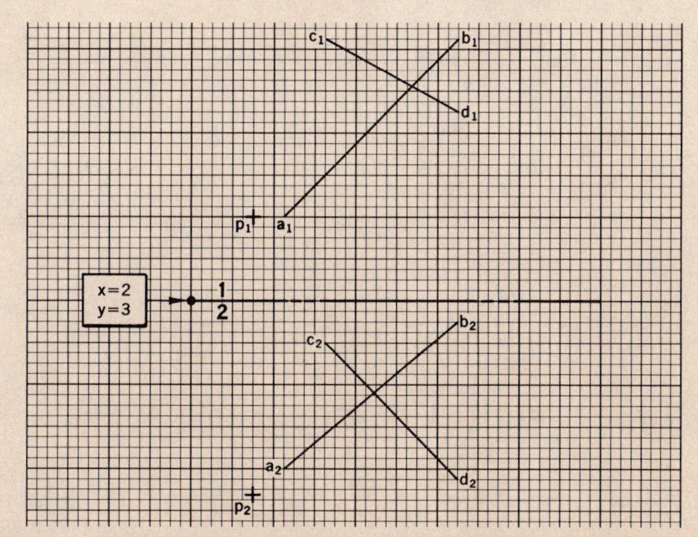

3. GIVEN DATA: Horizontal and frontal projections of a parallelepiped (prism) and an intersecting plane WXY.

REQUIRED: Using *only* the given views, draw the horizontal and frontal projections of the line of intersection between the given parallelepiped and the plane. Indicate the visibility of the line of intersection and of the given solid and plane in both views.

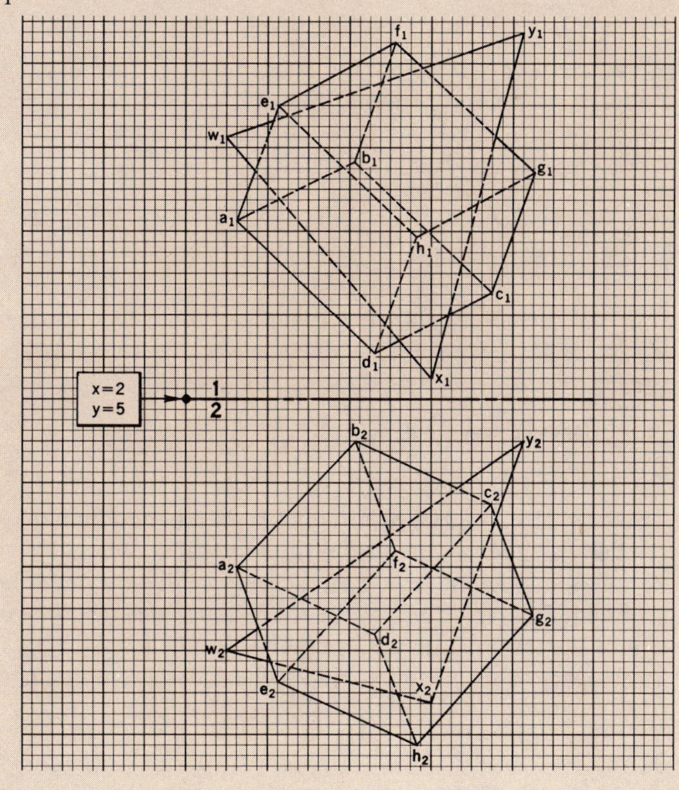

VI

Location of Points and Tangent Planes on Geometric Solids and Surfaces

48 Introduction

In this chapter, we review the concepts of certain basic geometric solids and surfaces, and present methods of locating points on and planes tangent to them. This material is fundamental to determining lines of intersection between common geometric solids and surfaces.

49 Basic polyhedrons

Common forms of geometric solids are those having sides, or "faces," which consist of flat intersecting plane surfaces. These solids generally fall into two categories:

1. Regular polyhedrons.
2. Irregular polyhedrons.

Figures 118 and 119 pictorially present some familiar types of polyhedrons. Regular polyhedrons are those with congruent faces. (This means that the faces are alike in size and shape.) Figure 118(a) illustrates a *tetrahedron* with four congruent faces and four vertices. The four faces intersect on lines called *lateral edges.* (The tetrahedron is the simplest geometric solid that can enclose a three-dimensional space.) Figure 118(b) illustrates the familiar cube, or *hexahedron.*

FIGURE 118

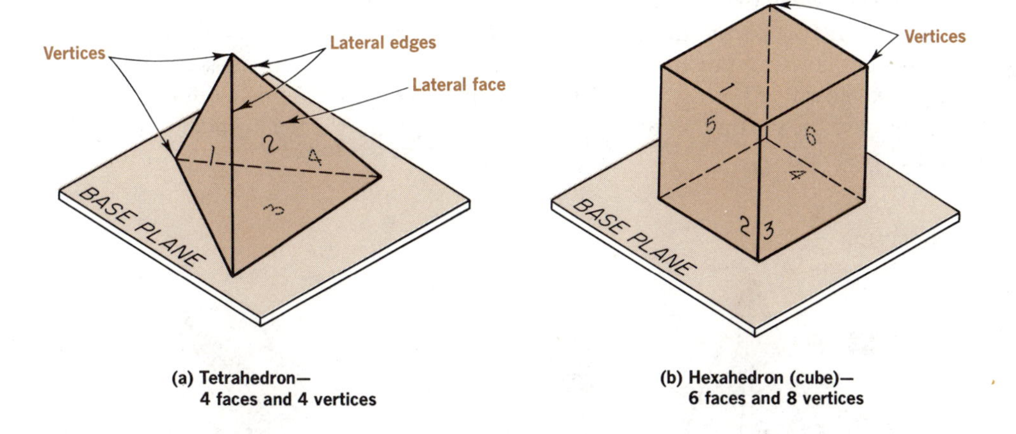

(a) Tetrahedron—
4 faces and 4 vertices

(b) Hexahedron (cube)—
6 faces and 8 vertices

FIGURE 119

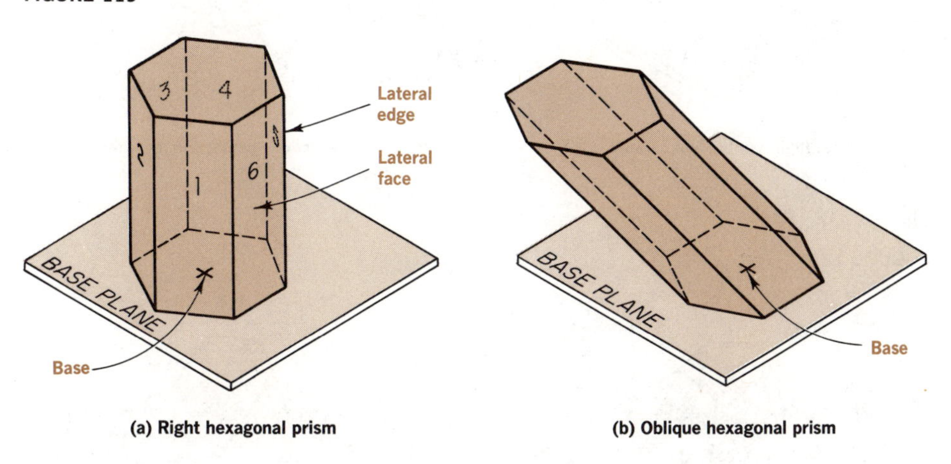

(a) Right hexagonal prism

(b) Oblique hexagonal prism

The cube has six congruent faces and eight vertices. In Fig. 119(a) you see a right hexagonal prism. The base of the prism is a *hexagon,* and the lateral faces and edges of the prism are *perpendicular* to the base. This characteristic classifies the prism as a *right* prism. Figure 119(b) illustrates an *oblique* hexagonal prism in which the lateral faces and edges of the prism are *not* perpendicular to the base.

Figure 120(a) depicts a *right* pentagonal pyramid. The base of the pyramid is a *pentagon,* and the vertex of the pyramid is directly above the *center* of the base. Figure 120(b) pictorially presents an *oblique* pentagonal pyramid. The base of this pyramid is a pentagon, but because the vertex is *not* directly over the center of the base, the pyramid leans to one side. This characteristic classifies it as an oblique pyramid. Note that the lateral faces of the right pyramid in Fig. 120(a) are congruent, but the bottom *base* is not. In the oblique pyramid shown in Fig. 120(b), *none* of the faces are congruent—including the base.

Figure 121(a) pictorially presents a *truncated* right pyramid. This means that a cutting plane has cut the pyramid *parallel* to its *base,* leaving a section, or top face, as shown in the figure. Figure 121(b) shows a *frustum* of a right pyramid. In this case the cutting plane has cut a section so that it is *not* parallel to the base of the pyramid.

FIGURE 120

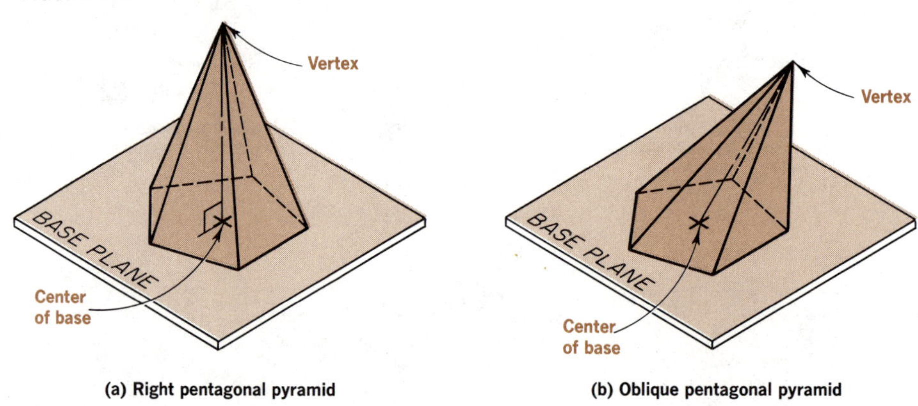

(a) Right pentagonal pyramid

(b) Oblique pentagonal pyramid

FIGURE 121

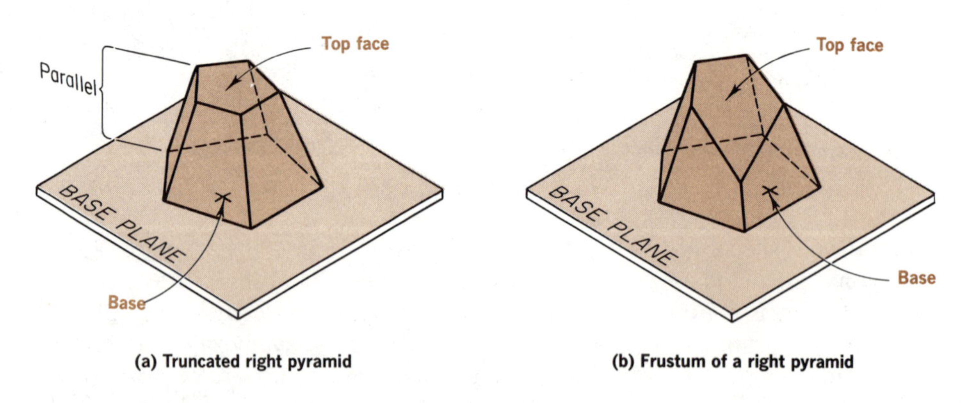

(a) Truncated right pyramid

(b) Frustum of a right pyramid

50 Common geometric surfaces

In Chapter III you saw how flat plane surfaces were generated: Two parallel lines, for example, form a plane, as do two intersecting lines. [See Figs. 122(a) and 122(b).]

FIGURE 122

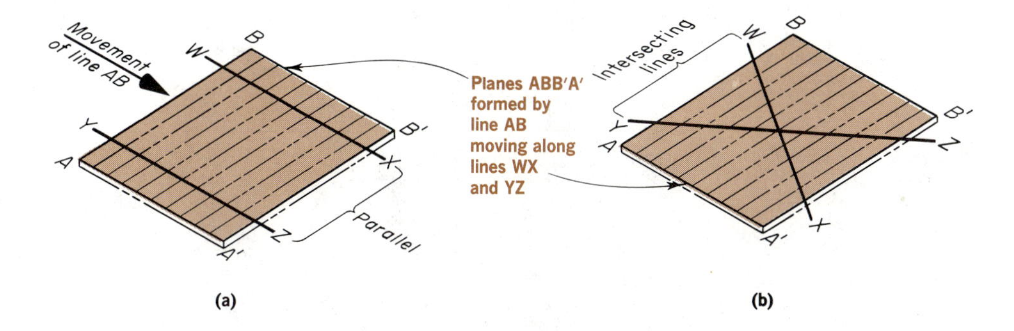

Note that in Fig. 122(a) the surface $ABB'A'$ was generated by a line AB, which moved parallel to itself and yet remained in contact with two other parallel lines WX and YZ. In Fig. 122(b) the plane $ABB'A'$ was generated by a line AB moving parallel to itself and, at the same time, remaining in contact with intersecting lines WX and YZ.

In Fig. 123 you see a line AB in contact with a curved line XY. When line AB moves parallel to itself and continues to remain in contact with the curved line, the surface it generates is a *general cylindrical surface*.

In Fig. 124 you see a line AB that moves parallel to itself and at the same time, remains in contact with a circular curved line. As the line AB travels parallel to itself through 360° it generates a right *circular* cylinder (cylindrical surface).

Figure 125 pictorially presents a line AB, that is held fixed at point A, as the free end of the line (at point B) moves along a circular curve through 360°. The surface generated by line AB is a *cone of revolution.*

Figures 122–24 depict surfaces that are generated by a line moving parallel to itself, following either a straight or a curved path. Such surfaces are classified as *ruled surfaces.* This means that a straight line *on* the surface may be drawn through *any point* on the surface.

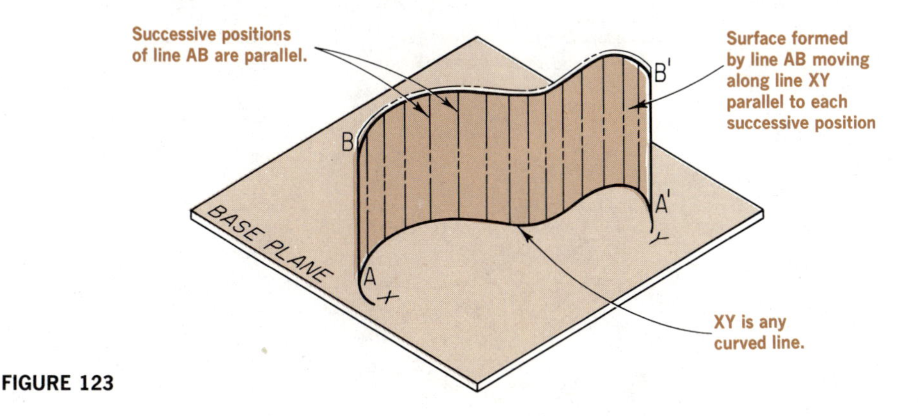

FIGURE 123

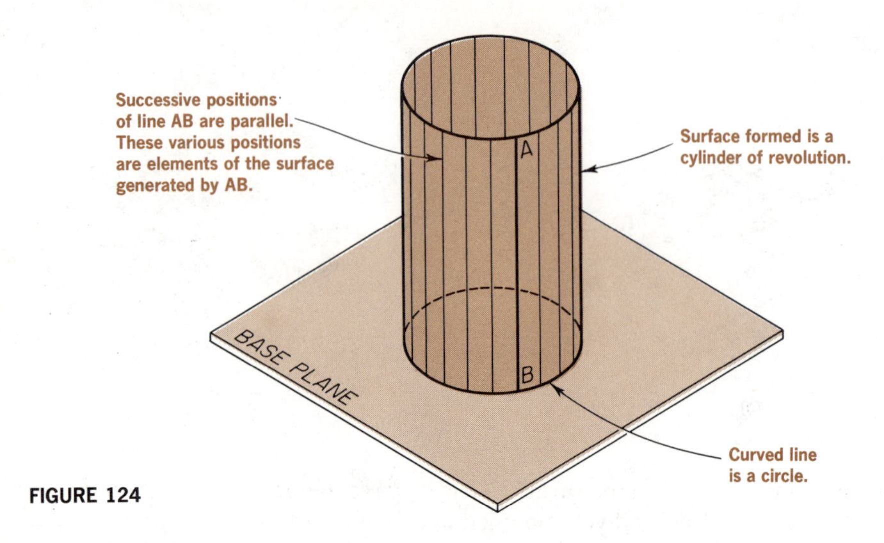

FIGURE 124

Figure 125 also illustrates a ruled surface generated by a straight line that is fixed at one end, while the other end remains in contact with a curved line.

Other classifications of surfaces involve *warped* and *double-curved surfaces.* Figure 126 illustrates the generation of a hyperbolic paraboloid, created by a line AB, and which moves parallel to a given plane while remaining in contact with two skew lines XX and YY. This warped surface cannot be developed accurately: that is, no two consecutive lines on the surface may lie in the same plane if the surface is laid out onto a flat plane. (The surface would be distorted.)

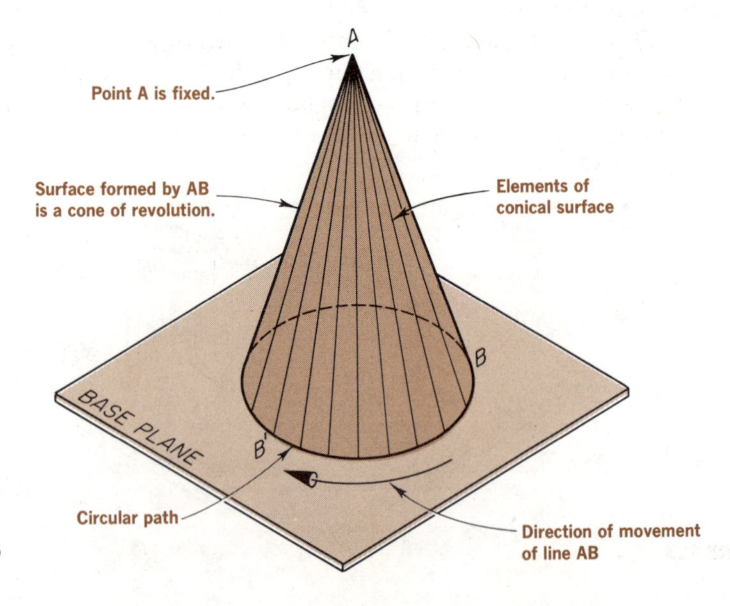

FIGURE 125

FIGURE 126

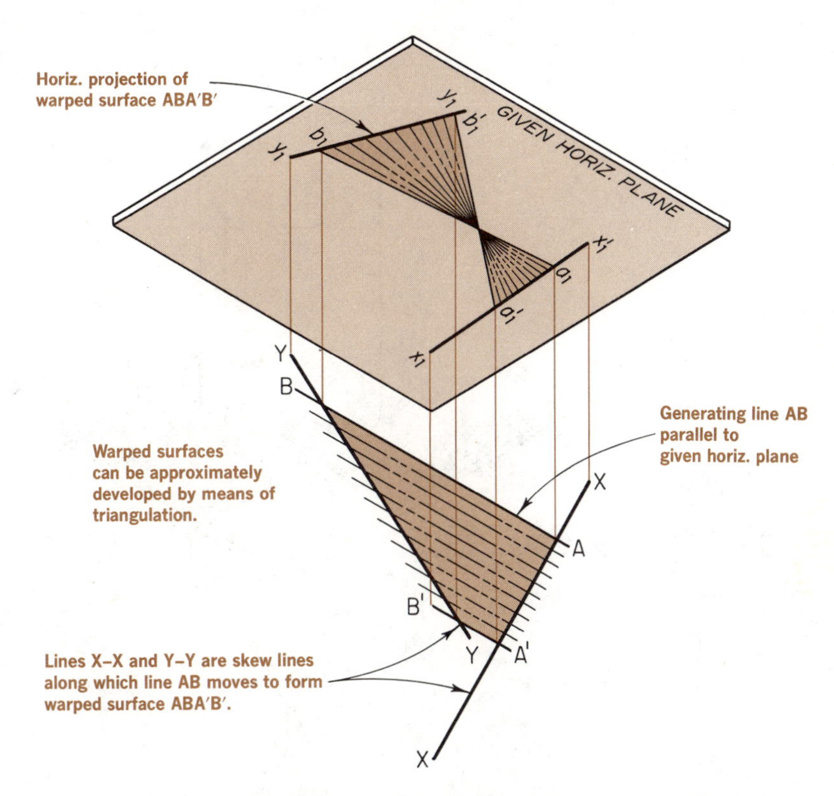

In Fig. 127 you see a sphere that is a double-curved surface, generated by a semicircle rotating about an axis of revolution. The double-curved surface also cannot be accurately developed.

Surfaces may be formed by revolving either curved or straight lines. Figures 124, 125, and 127–29 show examples of *surfaces of revolution.* The *successive positions* of a line that generates a surface are referred to as *elements* of the surface.

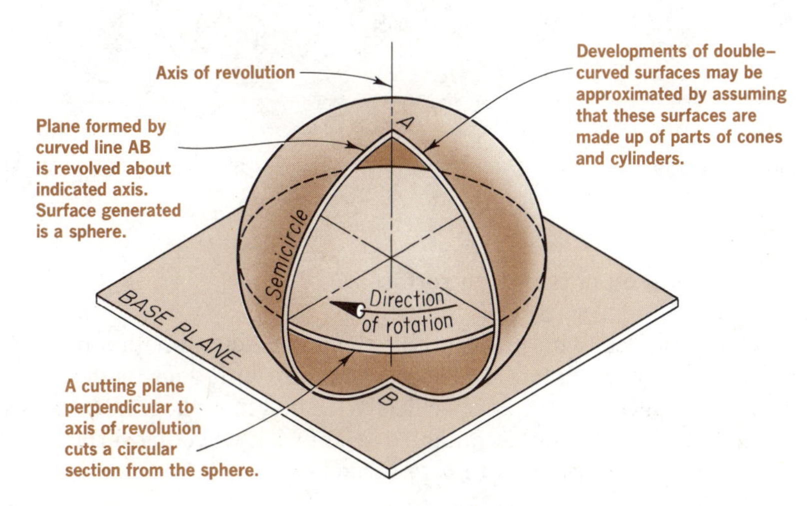

FIGURE 127

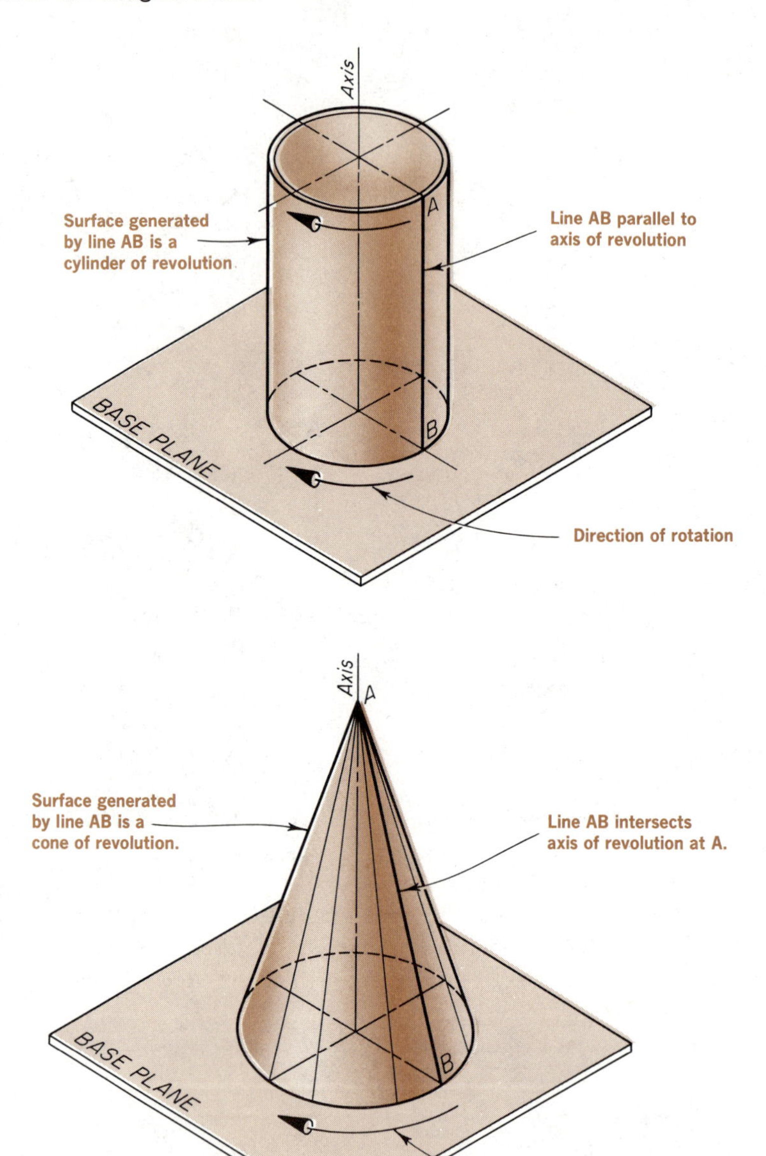

FIGURE 128

FIGURE 129

51 Location of points on surfaces

When you deal with problems involving the determination of the lines of intersection between different surfaces, you must introduce methods by which *points* on these surfaces can be specifically located in orthographic projection. Determining the line of intersection between two intersecting surfaces actually means locating *points* common to *both* surfaces.

52 Location of points on the surface of a prism

In order to locate a point on a prism in orthographic projection, you must apply the cutting plane method: through the given point, pass a cutting plane that cuts the prism face containing the point. (See Fig. 130, in which a vertical cutting plane is used to illustrate this point.)

FIGURE 130

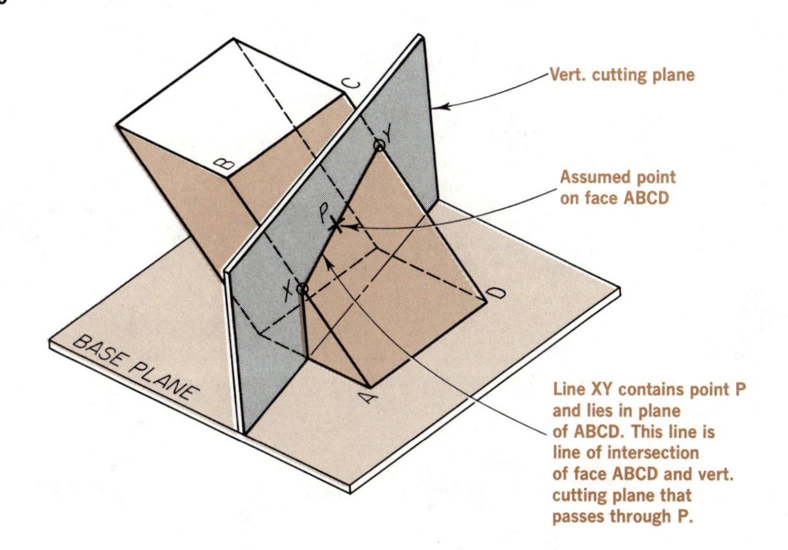

Orthographic Example: Location of a point on the surface of a prism (Fig. 131).

Given Data: Horizontal and frontal projections of an oblique prism and the horizontal projection p_1 of a point P on one face ($ABCD$) of the given prism.

Required: The frontal projection p_2 of point P.

Construction

a. Construct a vertical cutting plane—seen as an edge in view #1—through point p_1. (In this view, the position of the vertical cutting plane is arbitrary as long as it contains p_1.) The cutting plane determines the line x_1y_1 on the prism face $a_1b_1c_1d_1$.

b. Project x_1y_1 downward to view #2 in order to determine its frontal view x_2y_2.

c. Since p_1 is on the line x_1y_1, you must locate the frontal projection p_2 on the frontal projection x_2y_2. Therefore, project p_1 downward to view #2 and onto line x_2y_2 in order to determine the required frontal view p_2 of point P.

FIGURE 131

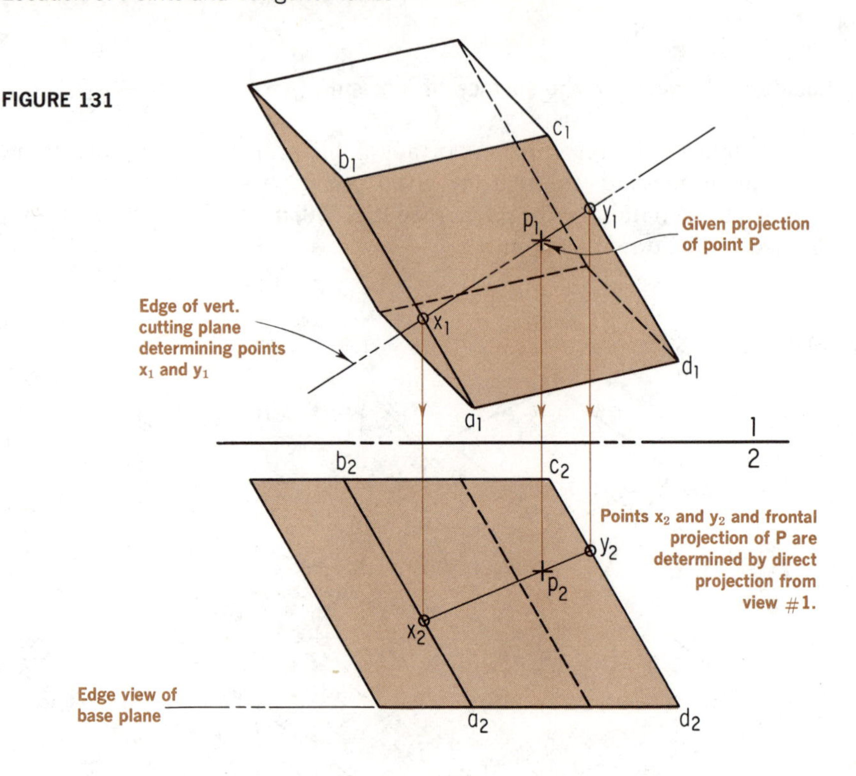

53 Points of intersection (piercing points) of a line and a prism

Figure 132 pictorially presents a line XY intersecting two faces of a given prism at points P and M. You can locate P and M on the respective prism faces by applying the cutting plane method presented in section 52. The cutting plane contains the intersecting line (XY in this case) and intersects two faces of the prism.

Orthographic Example: Determination of the piercing points of a line and a given prism (Fig. 133).

Given Data: Horizontal and frontal projections of an oblique rectangular prism and a line XY that intersects two lateral faces of the prism.

Required: The points of intersection (piercing points) of line XY and the prism.

Construction Program:

a. Construct a vertical cutting plane—seen as an edge in view #1—through the line x_1y_1. The plane intersects the faces $a_1b_1c_1d_1$ and $a_1b_1f_1e_1$ of the given prism.

This cutting plane determines line o_1l_1 on face $a_1b_1c_1d_1$, and line o_1n_1 on face $a_1b_1f_1e_1$.

b. Project lines o_1l_1 and o_1n_1 downward to view #2 in order to determine their frontal projections o_2l_2 and o_2n_2, respectively.

FIGURE 132

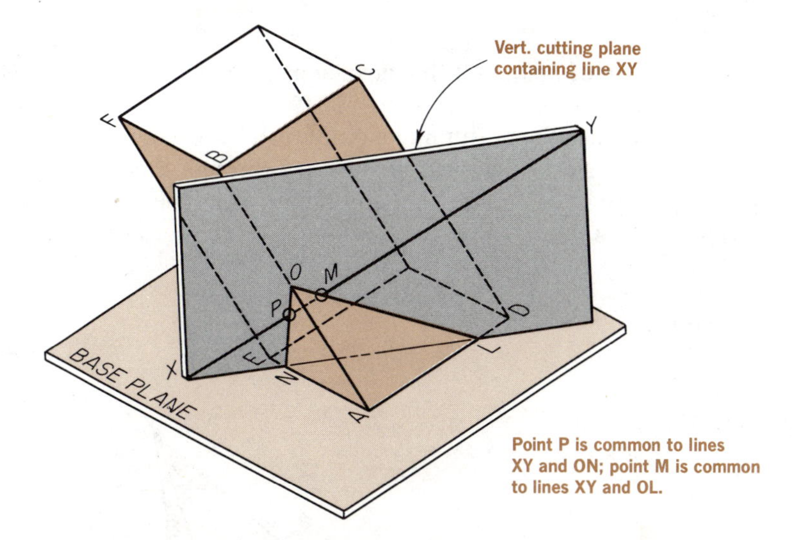

c. The piercing point m_2 of line XY with lateral face $a_2b_2c_2d_2$ is determined by the intersection of o_2l_2 and the frontal view x_2y_2 of XY.

d. The piercing point p_2 of line XY with face $a_2b_2f_2e_2$ is determined by the intersection of o_2n_2 and x_2y_2.

e. Project m_2 and p_2 upward to view #1 in order to determine m_1 and p_1, which are both on line x_1y_1. Points M and P are the required piercing points of the given line and prism.

f. Determine the visibility of the line relative to the prism by the usual method.

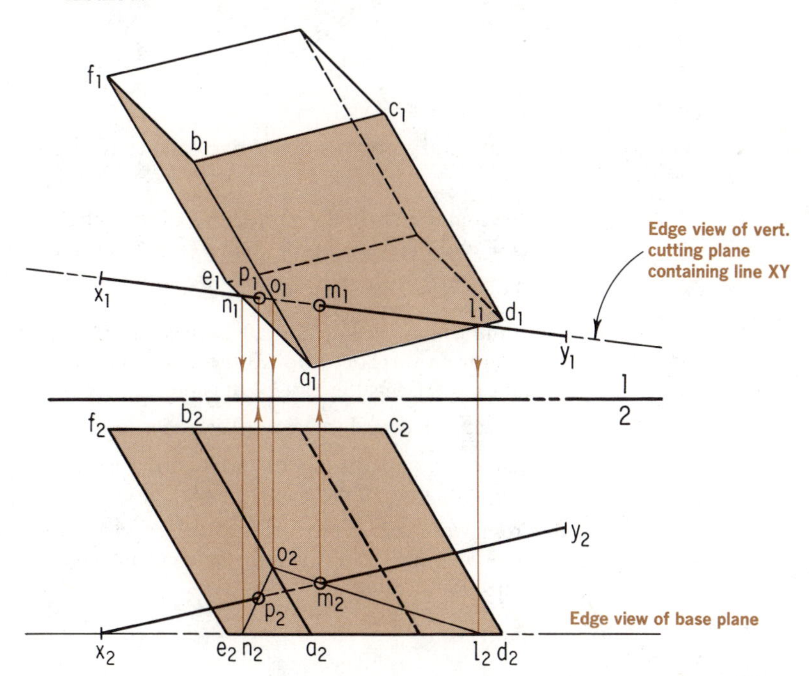

FIGURE 133

54 Location of points on the surface of a cone

You can establish the position of a point on a cone in orthographic projection by passing a *straight line element* of the cone through the given point. Figure 134 pictorially presents a point P on an oblique cone, the vertex of which is V. The element VM contains the point P, which is on the surface of the cone.

FIGURE 134

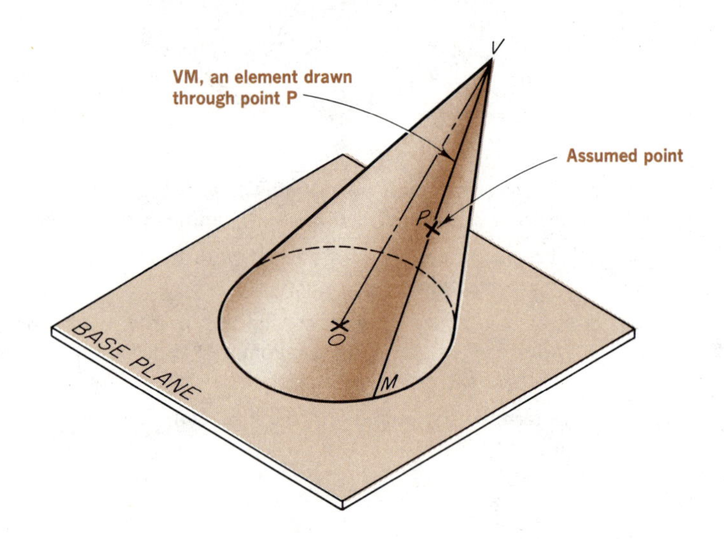

Orthographic Example: Location of a point on the surface of a cone (Fig. 135).

Given Data: Horizontal and frontal projections of an oblique cone having a vertex V and the frontal projection p_2 of a point P on its surface.

Required: The horizontal projection p_1 of P.

Construction Program:

a. In the frontal view #2, draw an element (line on the surface of the cone) from vertex v_2 through the given point p_2. Extend this element until it intersects the base of the cone at point m_2.

b. Project m_2 upward to view #1 in order to determine m_1.

c. Connect v_1 with m_1 in view #1 in order to determine the horizontal projection v_1m_1 of the element VM.

d. From view #2, project p_2 upward to view #1 in order to determine p_1, which is located on the element v_1m_1. This is the required projection of point P.

FIGURE 135

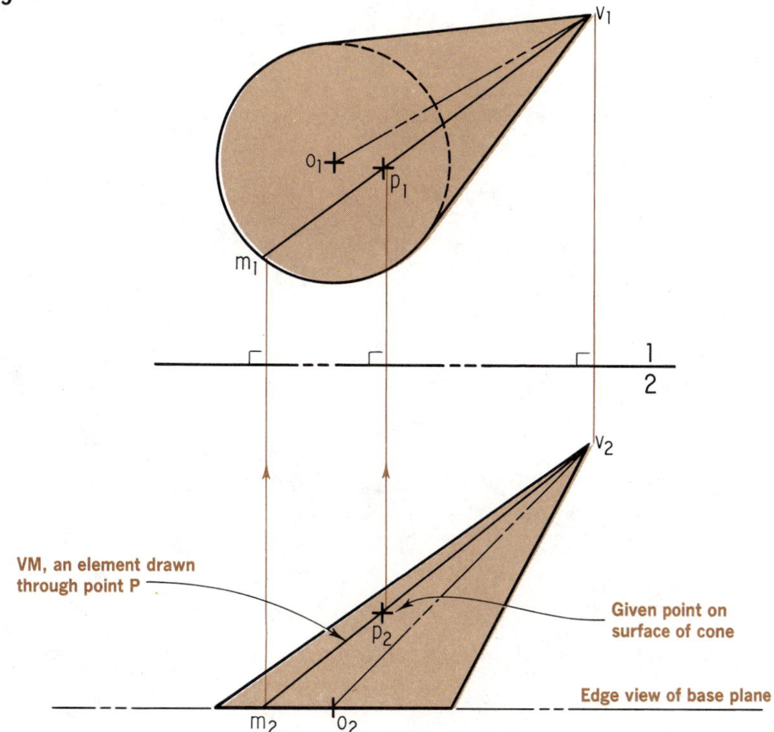

55 Points of intersection (piercing points) of a line and a cone

A straight line that intersects the surface of a cone will intersect it at two points. To determine these piercing points you must utilize a cutting plane containing the *vertex* of the cone and the given *intersecting line.* Such a plane will intersect the given cone and locate *straight line elements* on the surface of the cone, thus making it possible to determine the piercing points. Figure 136 is a pictorial representation of this principle.

Orthographic Example: Determination of the piercing points of a line and a cone (Fig. 137).

Given Data: Horizontal and frontal projections of an oblique cone having a vertex V and a line AB that intersects the cone.

Required: The piercing points of line AB on the given cone.

Construction Program:

a. In view #2, extend line a_2b_2 until it intersects the edge view of the *base plane* at d_2. (The base plane is the plane upon which the base of the cone rests.)

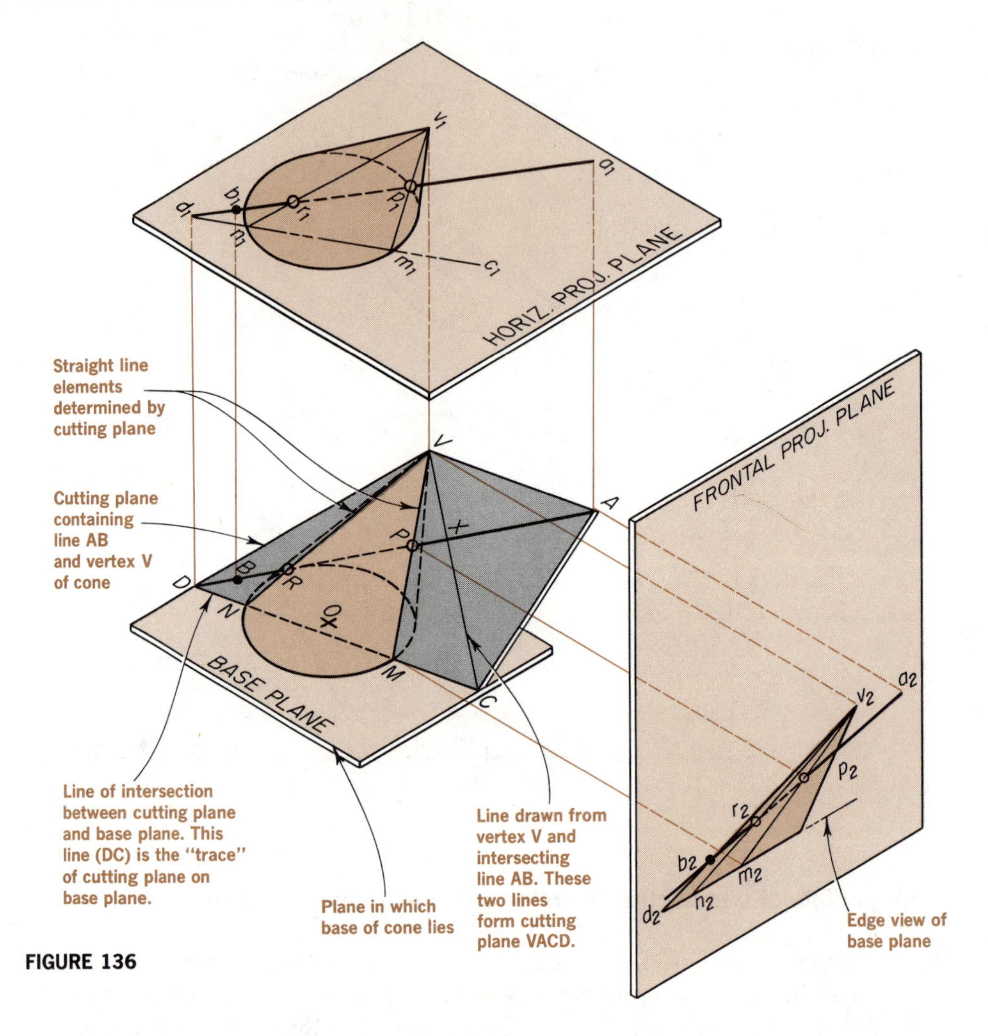

FIGURE 136

b. In view #2, draw a line from vertex v_2 intersecting a_2b_2 at *any* point x_2; extend it until it intersects the base plane at point c_2. (Intersecting lines a_2b_2 and v_2c_2 determine a plane containing line a_2b_2 and vertex v_2.)

c. Project d_2, x_2, and c_2 upward to view #1 and locate the positions of line v_1c_1 and point d_1.

d. In view #1, connect d_1 and c_1 with a straight line. This is the line of intersection (referred to as a *trace*) between the newly created cutting plane (formed by lines VC and AB) and the base plane. Note in view #1 that this trace determines two points, m_1 and n_1, on the base of the given cone. In view #1, connect vertex v_1 with points m_1 and n_1, respectively (v_1m_1 and v_1n_1 are straight line elements that have been determined by the cutting plane formed by intersecting lines VC and AB).

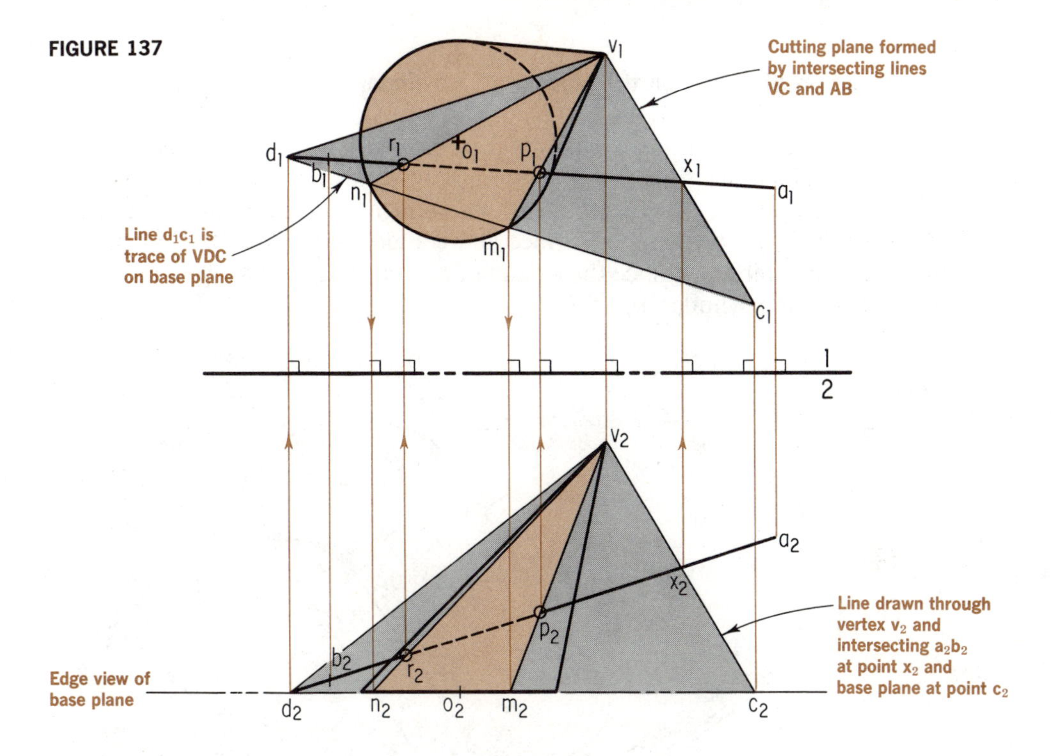

FIGURE 137

e. Determine the frontal projections v_2m_2 and v_2n_2 of these elements by projecting m_1 and n_1 downward to view #2 to m_2 and n_2, respectively.

f. The straight line element v_2m_2 intersects a_2b_2 at p_2, and the straight line element v_2n_2 intersects a_2b_2 at r_2. These are the frontal projections of the piercing points of the given line and cone.

g. Project p_2 and r_2 upward to view #1 in order to determine the horizontal projections of piercing points p_1 and r_1, respectively. These are the required piercing points of the line AB on the given cone.

h. Determine the visibility in view #2 of the line AB by noting whether the elements on which the points p_2 and r_2 are located are visible or hidden in the frontal projection view. You can do this by referring to view #1 and determining whether points m_1 and n_1 are on the frontal "side" of the cone. You can then see that both elements v_2m_2 and v_2r_2 are visible in view #2 and, therefore, p_2 and r_2 are also visible. Line AB is visible from a_2 to p_2, and r_2 to b_2, but invisible from p_2 to r_2, since it is inside the cone.

i. Establish the visibility of p_1 and r_1 by determining whether elements v_1m_1 and v_1n_1 are visible or hidden in view #1. Do this by determining whether points m_1 and n_1 occur on the visible portion of the base of the cone in view #1. You can tell that both of these occur on the visible portion of the base of the cone and, therefore, elements v_1m_1 and v_1n_1 are both visible, thus making points p_1 and r_1 also visible.

56 Location of points on the surface of a cylinder

In order to locate a point on a cylindrical surface, utilize *straight line elements* of the cylindrical surface. Figure 138 pictorially presents a point R on the surface of a cylinder that has an axis OP. Through point R, construct an element MN. This element is on the cylinder surface and contains the point R, which is also on the cylinder surface. (Since the element MN is a straight line element it must be parallel to the cylinder axis OP.)

FIGURE 138

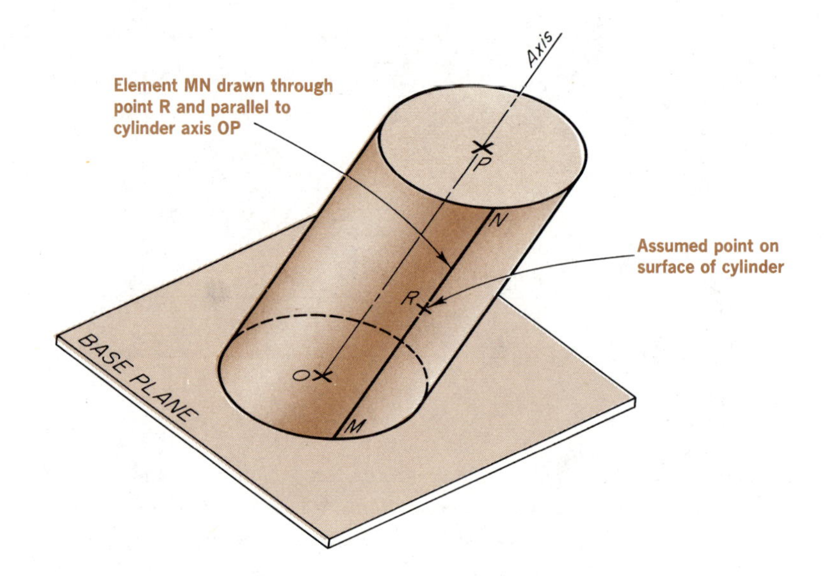

Orthographic Example: Location of a point on the surface of a cylinder (Fig. 139).

Given Data: Horizontal and frontal projections of an oblique cylinder containing the frontal projection r_2 of a point R on its surface.

Required: The horizontal projection r_1 of R on the cylinder.

Construction Program:

a. Draw a straight line element m_2n_2 through r_2 in frontal view #2. This straight line element will be parallel to the frontal projection o_2p_2 of the given cylinder's axis.

b. Project m_2 and n_2 to their respective bases in view #1 in order to locate m_1 and n_1 (m_1 is on the lower base of the cylinder, while n_1 is on the upper base). Element m_1n_1 is the horizontal projection of the straight line element passing through point R.

c. Project r_2 upward to view #1 in order to locate r_1 on the straight line element m_1n_1. This satisfies the requirements of the problem, since r_1 is the desired horizontal view of the given point R on the cylinder's surface.

FIGURE 139

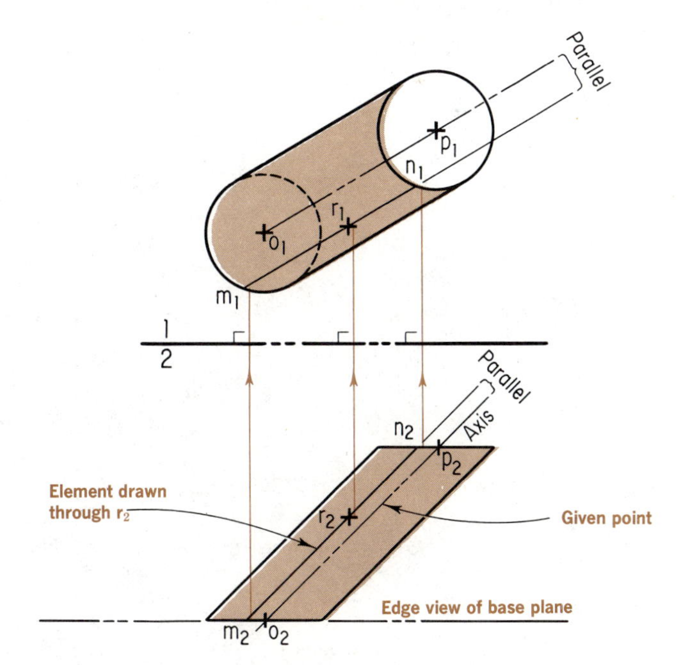

57 Piercing points of a line and a cylinder

In order to determine the piercing points of a line and the surface of a cylinder, you must utilize a cutting plane that contains the intersecting line and is *parallel* to the cylinder *axis.* Since the cutting plane is parallel to the cylinder axis, it will cut straight line elements on the cylinder surface. When you extend the cutting plane so that it intersects the base plane of the cylinder, it will have a trace on the base plane that will cut the cylinder base at two points. These two points are the points through which the straight line elements, determined by the cutting plane, pass. Since the straight line elements are on the surface of the cylinder, and have been determined by a cutting plane containing the line that intersects the cylinder, the points of intersection of the straight line elements with the line intersecting the cylinder are the piercing points of the given line and cylinder (see Fig. 140).

Orthographic Example: Determination of the piercing points of a line and a cylinder (Fig. 141).

Given Data: Horizontal and frontal projections of an oblique cylinder having an axis OP and a line AB that intersects the cylinder.

Required: The piercing points of the given line with the given cylinder.

Construction Program:

a. Assume a point X on the line AB, and find its horizontal and vertical projections. (You assumed point x_2 in view #2 and projected it upward to view #1 in order to determine x_1.)

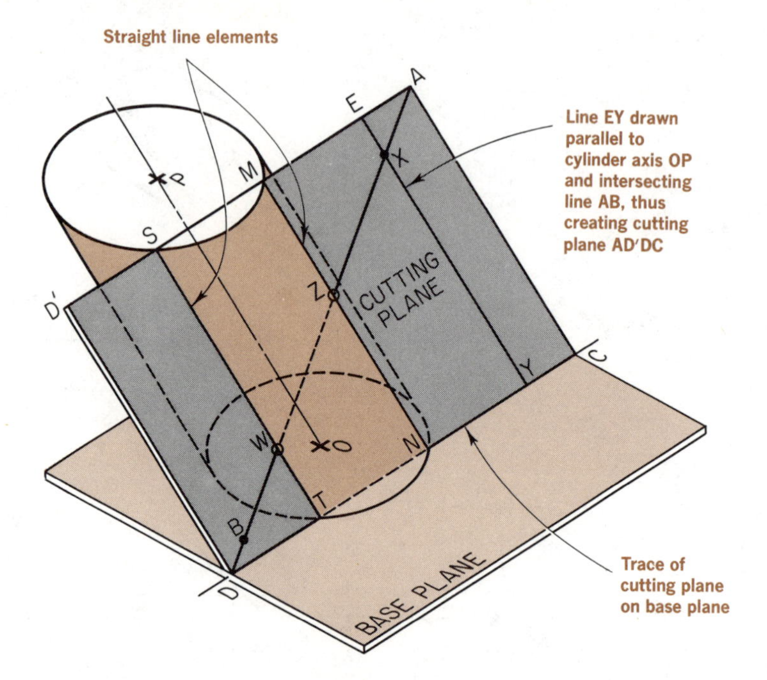

FIGURE 140

b. Through x_2 construct a line *parallel* to the frontal view o_2p_2 of the cylinder axis. Extend this line until it intersects the base plane at point y_2.

c. From x_1 in view #1, construct a line *parallel* to the horizontal view o_1p_1 of the cylinder axis.

d. Project y_2 upward to view #1 in order to determine y_1, which is on the line that you have constructed through x_1.

e. Extend line a_2b_2 until it intersects the base plane at point d_2.

f. Project d_2 upward to view #1 in order to determine d_1. (Lines AB and XY intersect to form a plane. One line of this plane, XY, is parallel to cylinder axis OP; therefore, the plane is parallel to the cylinder axis.)

g. In view #1, connect d_1 and y_1 with a straight line. Line d_1y_1 is the trace of the plane that contains the line AB and is parallel to the cylinder axis OP on the base plane of the cylinder. In view #1, the trace intersects the base of the *cylinder* at points n_1 and t_1.

h. From n_1 and t_1 construct straight line elements that are parallel to the cylinder axis o_1p_1. These elements are n_1m_1 and t_1s_1.

i. Project n_1 and t_1 downward to view #2 in order to locate n_2 and t_2 on the base of the given cylinder.

j. From points n_2 and t_2 construct straight line elements that are parallel to the cylinder axis o_2p_2. These elements are n_2m_2 and t_2s_2. The points z_2 and w_2, respectively, where these elements intersect line

FIGURE 141

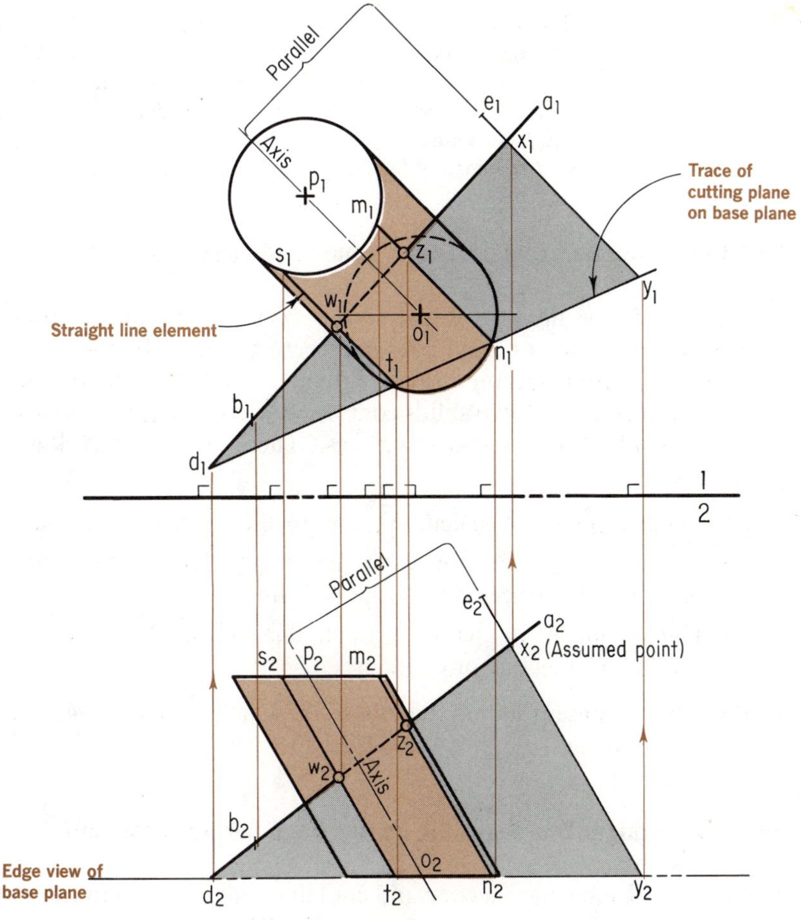

a_2b_2, are the frontal views of the piercing points of line AB with the given cylinder.

k. Project z_2 and w_2 upward to view #1 in order to determine z_1 and w_1 on line a_1b_1. The horizontal projections of the piercing points of line AB and the given cylinder are z_1 and w_1.

l. In view #1, establish the visibility of the line a_1b_1 relative to the cylinder by determining whether elements n_1m_1 and t_1s_1 can be seen. Since elements n_1m_1 and t_1s_1 each go through points n_1 and t_1, respectively (which occur on the *visible* portion of the *cylinder base* in view #1), they are visible elements in this view. Therefore, piercing points z_1 and w_1 are also visible.

m. You can establish the visibility of the piercing points in view #2 by determining whether elements n_2m_2 and t_2s_2 are visible. This is done by referring to view #1 and determining whether these straight line elements pass through points on the cylinder base that are nearest the frontal projection plane and are still on the visible side of the cylinder. (As long as points n_1 and t_1 remain on the front half of the cylinder base, the elements passing through these points will be visible in view #2.) From this analysis you can determine that in view #2, since elements n_2m_2 and t_2s_2 are visible, points z_2 and w_2 are also visible. This fulfills the requirements of the problem.

58 Planes tangent to surfaces of cones and cylinders

You can determine the line of intersection between two intersecting cones, two intersecting cylinders, or a cylinder intersecting a cone, by using cutting planes in order to locate common straight line elements that intersect *on the line of intersection* between the two solids concerned. Certain cutting planes are tangent to the individual intersecting surfaces. There are various kinds of plane tangency:

1. Planes tangent to specific points on the surface of a cone or cylinder.
2. Planes that are tangent to a cone or cylinder and that contain a point outside the surface of the cone or cylinder.
3. Planes tangent to a cone or a cylinder and parallel to a line outside the surface of the cone or cylinder.

Sections 59–64 present in detail methods by which tangent planes satisfying the above conditions may be determined.

59 Plane tangent to a specific point on the surface of a cone

A plane that is tangent to a cone and contains a given point on the surface of the cone is tangent along a straight line element on the cone's surface containing the given point *P*. The tangent plane has a *trace* on the base plane of the cone, *tangent* to the *base* at the point where this element intersects the base (see Fig. 142).

Orthographic Example: Determination of a plane tangent to a cone at a specific point on the cone's surface (Fig. 143).

Given Data: Horizontal and frontal projections of an oblique cone having an axis *VO* and a frontal projection p_2 of a point *P* on its surface.

Required: The horizontal and frontal projections of a plane tangent to cone *VO* at point *P*.

Construction Program:

a. In view #2, construct an element from v_2 through point p_2 intersecting the base of the cone at m_2.
b. Project m_2 upward to view #1 in order to locate m_1 on the base of the cone.
c. Draw element v_1m_1 in view #1.
d. Project p_2 upward to view #1 on element v_1m_1 in order to locate point p_1.

FIGURE 142

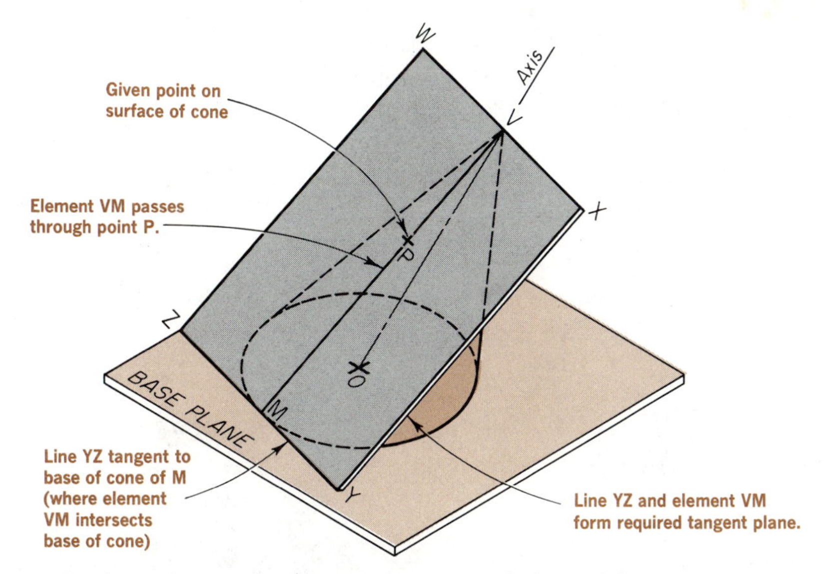

e. In view #1, construct a line y_1z_1 tangent to the base of the cone at m_1. The element v_1m_1 of the cone and the tangent line y_1z_1 are intersecting lines and therefore determine a plane. This plane is tangent to the cone along the element v_1m_1, and has a trace on the base plane of y_1z_1. This is the required tangent plane containing p_1.

(NOTE: In view #2, the tangent plane is determined by the frontal projection view of the element v_2m_2 and the trace y_2z_2. The tangent plane has been *arbitrarily* limited in size as $WXYZ$. Actually, it can be *limitless* in size.)

FIGURE 143

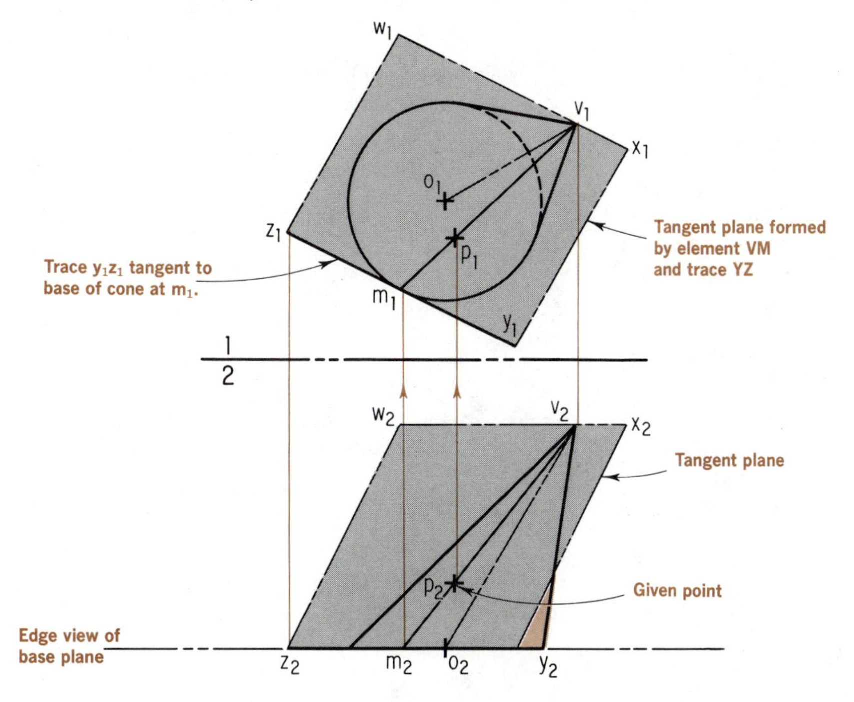

60 Planes tangent to a cone and containing a point outside the cone's surface

In order to construct a plane that is tangent to a given cone and that also contains a point outside the cone's surface, the tangent plane must contain the *vertex* of the cone. Figure 144 pictorially presents a plane that is tangent to an oblique cone and that contains a point *P* outside the cone's surface. Notice that the tangent plane is determined by a line from the vertex *V* that passes through *P* (and intersects the base plane at *R*) and the element *VM*.

FIGURE 144

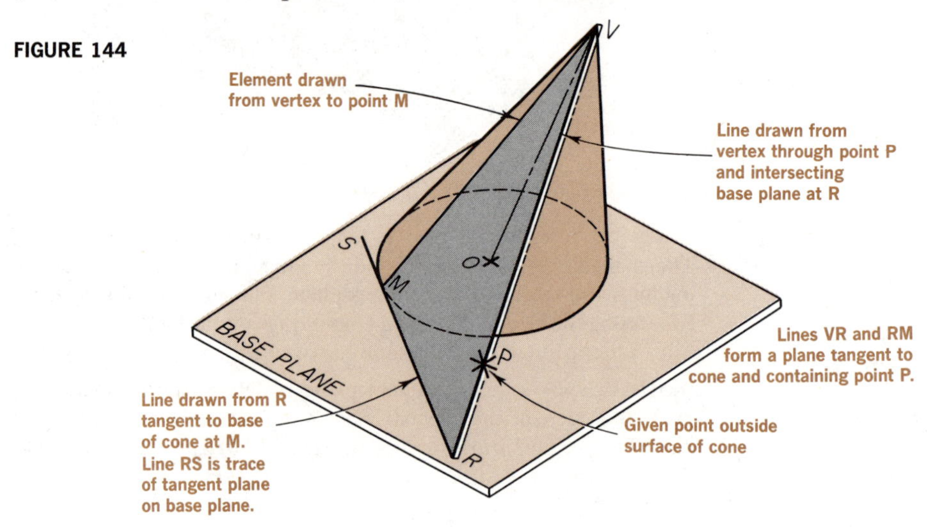

Orthographic Example: Determination of a plane that is tangent to a cone and that contains a point outside the cone's surface (Fig. 145).

Given Data: Horizontal and frontal projections of an oblique cone, the axis of which is *VO*, and of a point *P* outside its surface.

Required: A plane that contains point *P* and is tangent to the surface of the given cone.

Construction Program:

a. Starting with view #2, connect the vertex v_2 and point p_2 with a straight line. Extend this line until it intersects the base plane at point r_2.

b. In view #1, connect vertex v_1 with point p_1 and extend this line.

c. Project r_2 upward to view #1 until it intersects the extended line between v_1 and p_1 at point r_1.

d. From r_1 draw a line r_1s_1 tangent to the base of the cone at point m_1.

e. In view #1, draw element v_1m_1 (point m_1 is the tangent point of the line r_1s_1). The tangent plane is now defined by the intersecting line v_1r_1 and the element v_1m_1. The trace of the tangent plane $v_1m_1r_1$ on the base plane of the cone is the line r_1s_1.

FIGURE 145

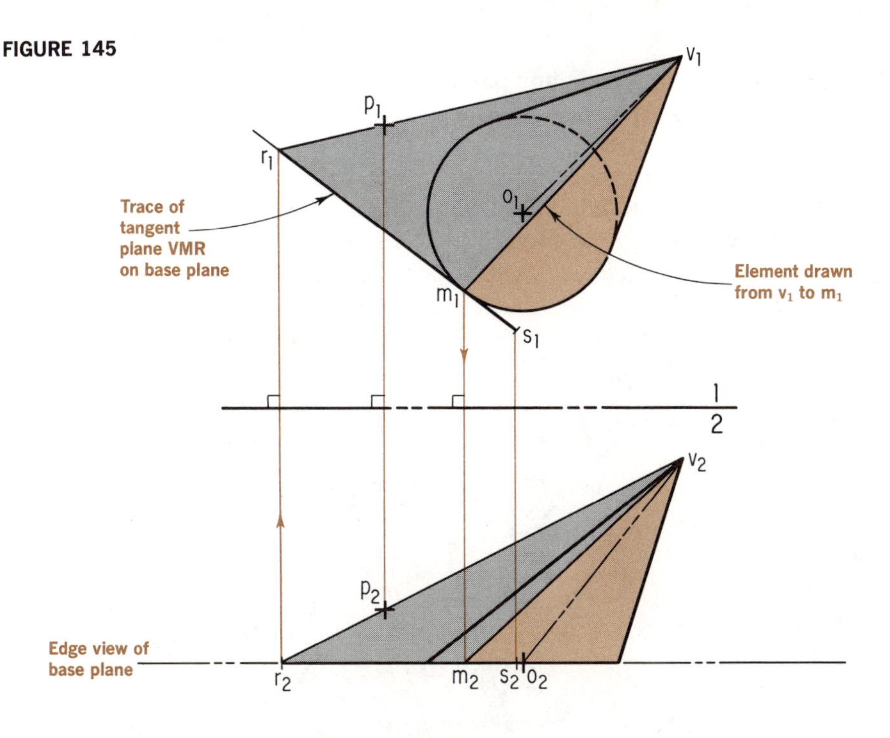

f. Project m_1 downward to view #2 in order to determine m_2. The tangent plane in view #2 is represented by $v_2m_2r_2$.

[NOTE: A second plane containing point P may be constructed tangent to the given cone: draw a trace from point R tangent to the trace of the cone on the *opposite side* of the first point of tangency (Fig. 146).]

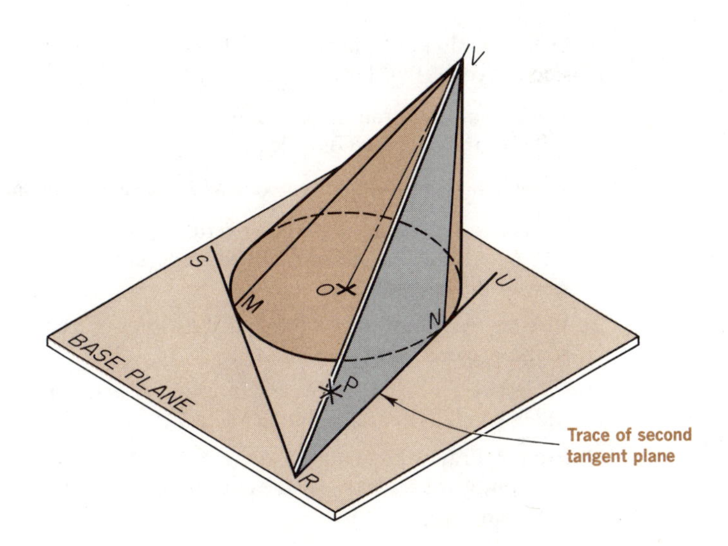

FIGURE 146

61 Plane tangent to a cone and parallel to a line outside the cone's surface

A plane that is tangent to the surface of a given cone and parallel to a line outside the cone may be determined by constructing a plane that contains the *vertex* of the cone and that is *parallel* to the *line* outside the cone. Such a plane determines the *straight line element,* along which the plane is tangent, on the cone's surface (see Fig. 147).

FIGURE 147

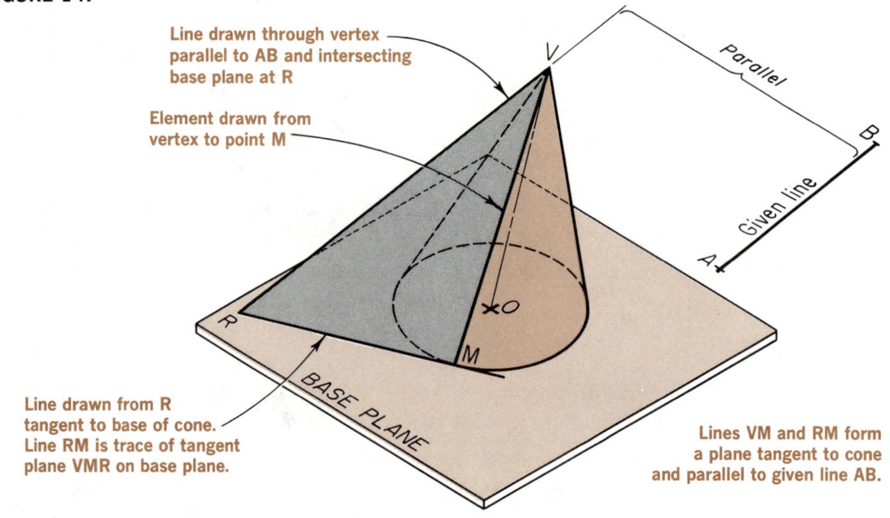

Orthographic Example: Determination of a plane tangent to a cone and parallel to a line outside the cone (Fig. 148).

Given Data: Horizontal and frontal projections of an oblique cone having an axis VO and a line AB outside it.

Required: A plane tangent to the given cone and parallel to the line AB.

Construction Program:

a. In view #2, draw a line through vertex v_2 parallel to the frontal projection a_2b_2 of line AB. Extend this line until it intersects the edge view of the base plane at point r_2.

b. In view #1, draw a line through the vertex v_1 parallel to the horizontal projection a_1b_1 of line AB.

c. Project r_2 upward to view #1 in order to determine r_1 on the line which contains v_1 and is parallel to a_1b_1.

d. From r_1 draw a line tangent to the base of the cone at point m_1. The line r_1m_1 is the trace of a plane tangent to the given cone and parallel to the line a_1b_1.

FIGURE 148

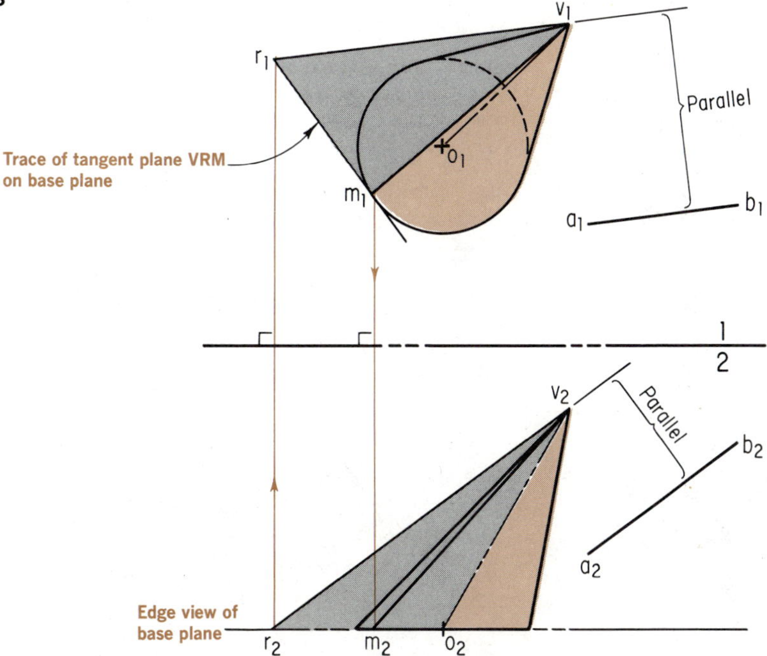

e. Project m_1 downward to view #2 in order to determine m_2 on the base plane. The tangent plane appears as $v_1r_1m_1$ in view #1, and as $v_2r_2m_2$ in view #2.

(NOTE: A second plane, tangent to the given cone and parallel to the line AB outside the cone, may be drawn as illustrated in Fig. 149.)

FIGURE 149

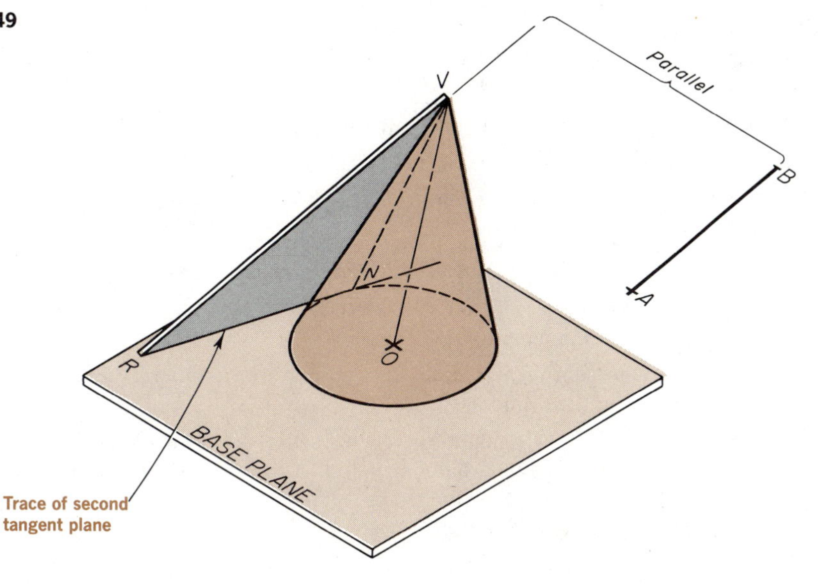

62 Plane tangent to a specific point on the surface of a cylinder

A plane is tangent to a cylinder along a *straight line element* only when the plane is *parallel* to the cylinder *axis*. Figure 150 pictorially presents a plane $WXYZ$ that is tangent to a cylinder at a specific point P. The line of tangency of the cylinder and the plane is a straight line element MN on the cylinder surface on which point P is located.

FIGURE 150

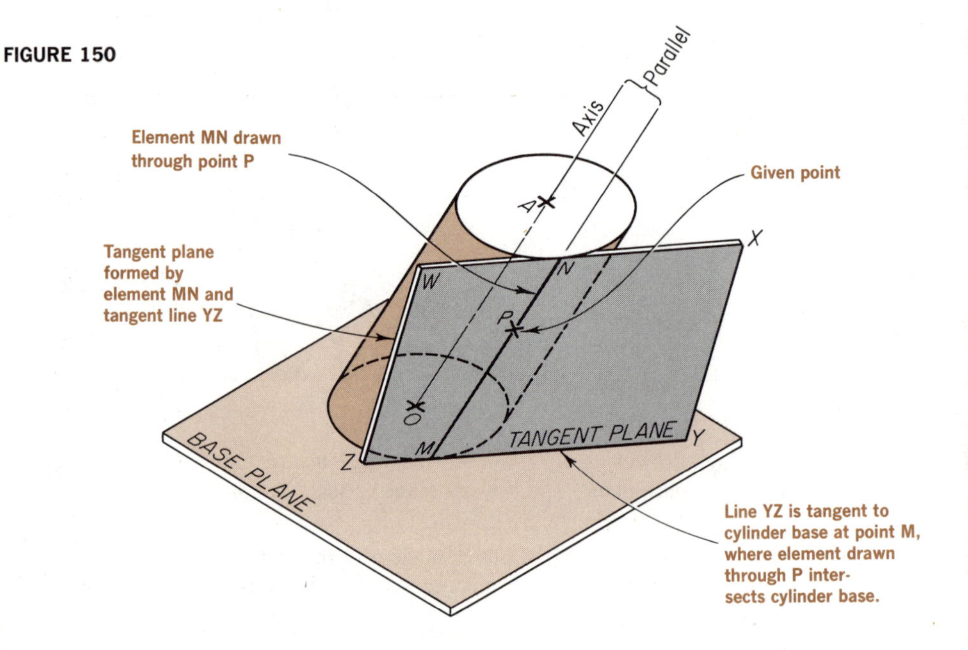

Orthographic Example: Determination of a plane tangent to a specific point on the surface of a cylinder (Fig. 151).

Given Data: Horizontal and frontal projections of an oblique cylinder, the axis of which is AO, and of a point P on its surface.

Required: A plane tangent to the given cylinder at point P.

Construction Program:

a. In view #2, draw a straight line element n_2m_2 parallel to the cylinder axis a_2o_2 and passing through point p_2.

b. Project m_2 on the base plane upward to view #1 in order to determine point m_1.

c. Through point m_1 in view #1, draw a line y_1z_1 tangent to the base of the cylinder.

d. Draw straight line element n_1m_1 from m_1 parallel to axis a_1o_1. Elements m_1n_1 and line y_1z_1 intersect to form the required tangent plane. Line y_1z_1 is the trace of this tangent plane on the base plane of the given cylinder.

FIGURE 151

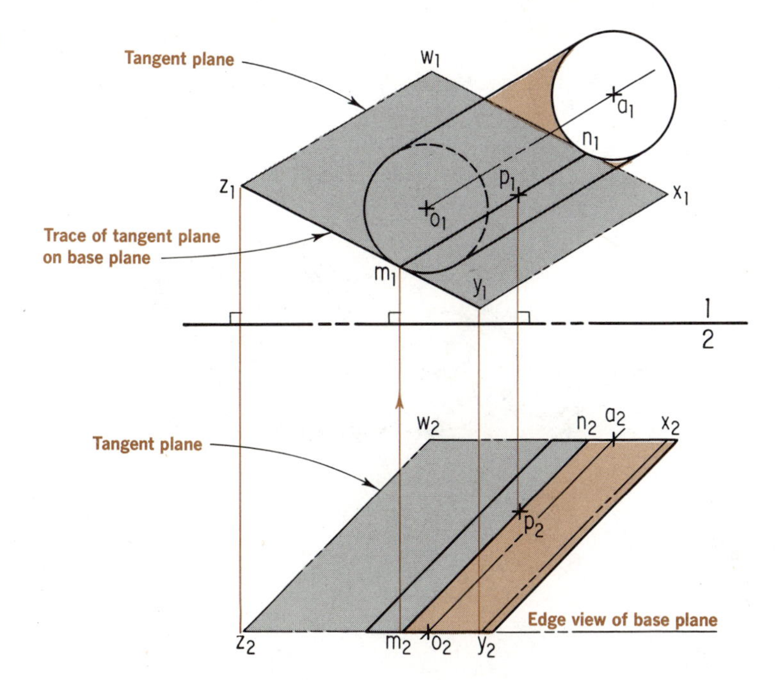

e. The tangent plane is represented in view #2 by the intersecting straight line element n_2m_2 and the trace y_2z_2. (The tangent plane is shown as a plane limited by $WXYZ$.)

63 Plane tangent to a cylinder and containing a point outside the cylinder surface

A plane that is tangent to a given cylinder and that contains a point outside the cylinder is parallel to the cylinder axis. Figure 152 pictorially presents a plane $WXMN$ containing a point P outside a cylinder having an axis AO. (The line WX of the tangent plane is parallel to the cylinder axis; therefore, the plane $WXMN$ is also parallel to the cylinder axis.) The line of tangency between the tangent plane and the given cylinder is the straight line element MN.

Orthographic Example: Determination of a plane that is tangent to a cylinder and that contains a point outside the cylinder surface (Fig. 153).

Given Data: Horizontal and frontal projections of an oblique cylinder having an axis AO and a point P outside it.

Required: A plane tangent to the cylinder and containing the point P.

Construction Program:

a. Starting with view #2, draw a line through point p_2 parallel to the cylinder axis a_2o_2. Extend this line until it intersects the base plane at point x_2.

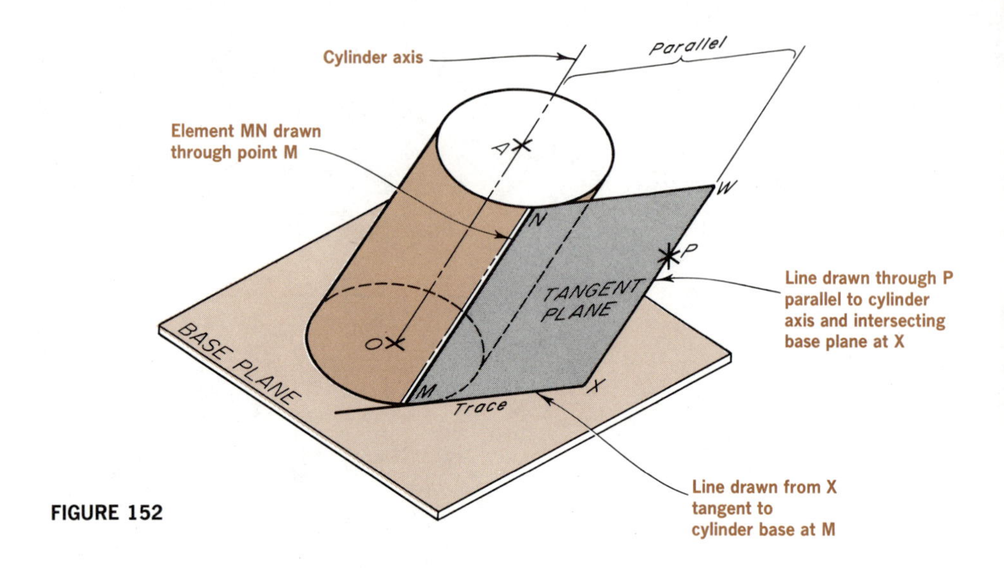

FIGURE 152

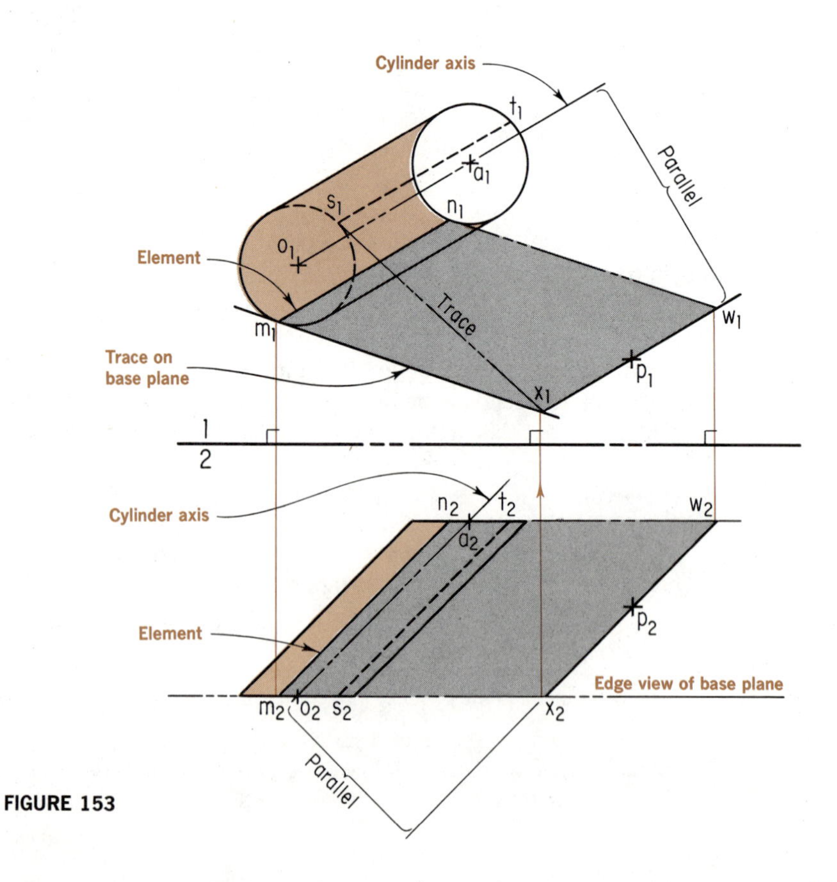

FIGURE 153

b. In view #1, construct a line through p_1 parallel to the cylinder axis a_1o_1.

c. Project x_2 upward to view #1 in order to determine point x_1 on the line that is parallel to the cylinder axis a_1o_1 and that contains point p_1.

d. From x_1 draw a line tangent to the cylinder base at point m_1. The lines w_1x_1 and x_1m_1 intersect to form a plane that contains the point p_1 and that is parallel to the cylinder axis a_1o_1. This plane is tangent to the cylinder along the straight line element m_1n_1 (view #1). The trace of this plane on the base plane of the cylinder is x_1m_1.

e. Project m_1 downward to view #2 and determine m_2. The tangent plane appears as $w_2n_2m_2x_2$. The line of tangency in this view (#2) is the straight line element m_2n_2.

[NOTE: In view #1, a second trace (x_1s_1) has been indicated. It represents a second plane that contains the point *P*, that is parallel to the cylinder axis, and that is therefore tangent to the given cylinder. This second plane determines a straight line element s_1t_1 on the surface of the cylinder. (The frontal view of this element is s_2t_2.) Also note that the visibility of the element *ST* is hidden in both views. Why?]

64 Plane tangent to a cylinder and parallel to a line outside the cylinder surface

In order to determine a plane that is tangent to a given cylinder and, at the same time, parallel to a *line* located *outside* the cylinder, you must create a plane that contains the line and that is parallel to the cylinder axis (and, therefore, parallel to all elements of the cylinder). Any other plane traces parallel to the *trace* of the *constructed plane* on the cylinder base plane represent traces of planes parallel to the cylinder axis. For example, in Fig. 154 you see a tangent plane *CDEF*. Its trace, *CD*, is parallel to the trace *XY* of the plane formed by intersecting lines *RS* and *WX* (*WX* is parallel to the cylinder axis *AO*). The tangent plane *CDEF* is therefore parallel to the given line *RS*, which is located outside the cylinder.

Orthographic Example: Determination of a plane tangent to a given cylinder and parallel to a line outside the cylinder (Fig. 155).

Given Data: Horizontal and frontal projections of an oblique cylinder having an axis *AO* and a line *RS* located outside the cylinder.

Required: A plane that is tangent to the given cylinder and parallel to line *RS*.

Construction Program:

a. Starting in view #2, assume point p_2 on line r_2s_2. Construct a line w_2x_2 that passes through p_2, that is parallel to the cylinder axis a_2o_2, and that intersects the base plane at point x_2.

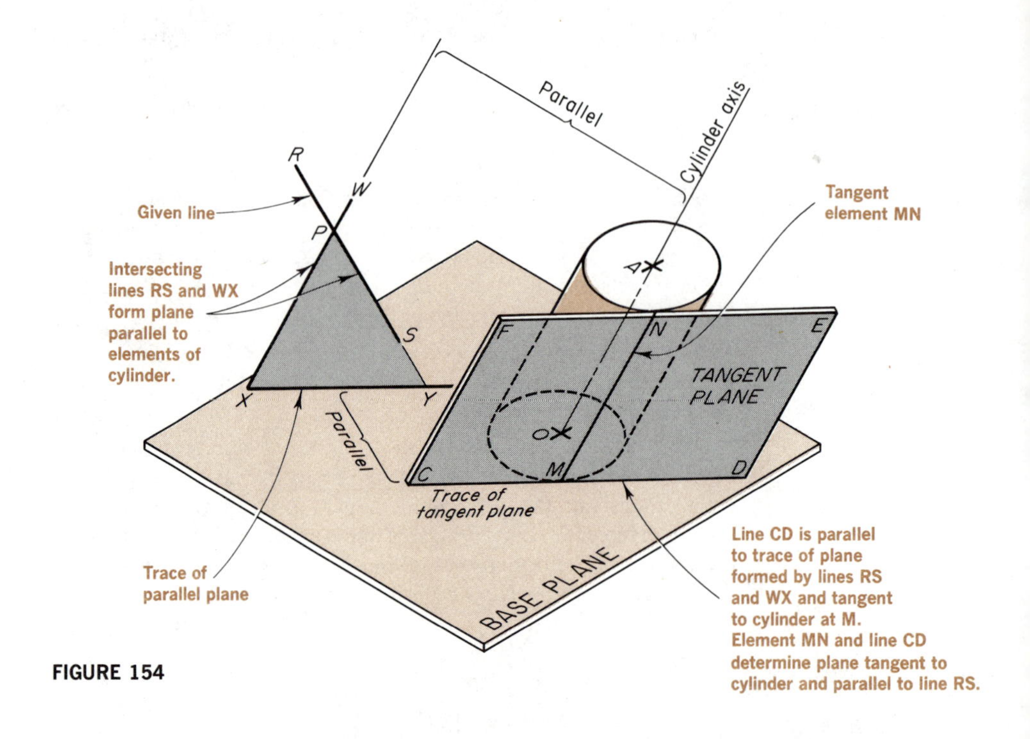

FIGURE 154

b. Extend r_2s_2 until it intersects the base plane at point y_2. Line x_2y_2 is the frontal projection of the trace of the plane formed by the intersecting lines r_2s_2 (extended) and w_2x_2 (parallel to the cylinder axis a_2o_2).

c. Project p_2 upward to view #1 in order to determine p_1 on r_1s_1. Through p_1 construct a line w_1x_1 parallel to the horizontal projection of the cylinder axis a_1o_1.

d. Project x_2 and y_2 upward to view #1 in order to determine x_1 and y_1. Line x_1y_1 is the trace of the plane, formed by intersecting lines w_1x_1 and r_1s_1, which is parallel to the cylinder axis.

e. Draw a line parallel to trace x_1y_1 and *tangent* to the base of the cylinder at point m_1.

f. In view #1, construct at m_1 the element m_1n_1 parallel to the cylinder axis. Element m_1n_1 and the parallel trace c_1d_1 (which determines m_1) determine a plane that is tangent to the cylinder and parallel to the line r_1s_1.

g. Project m_1 downward to view #2 and locate m_2. Draw element m_2n_2 and completely determine the tangent plane with line c_2d_2. (The tangent plane has been limited as $m_2n_2d_2e_2$.)

(NOTE: In view #1, a tangent trace has been constructed at point t_1 parallel to the trace x_1y_1. In other words, there are two planes tangent to the cylinder and parallel to the given line *RS* outside the cylinder.)

FIGURE 155

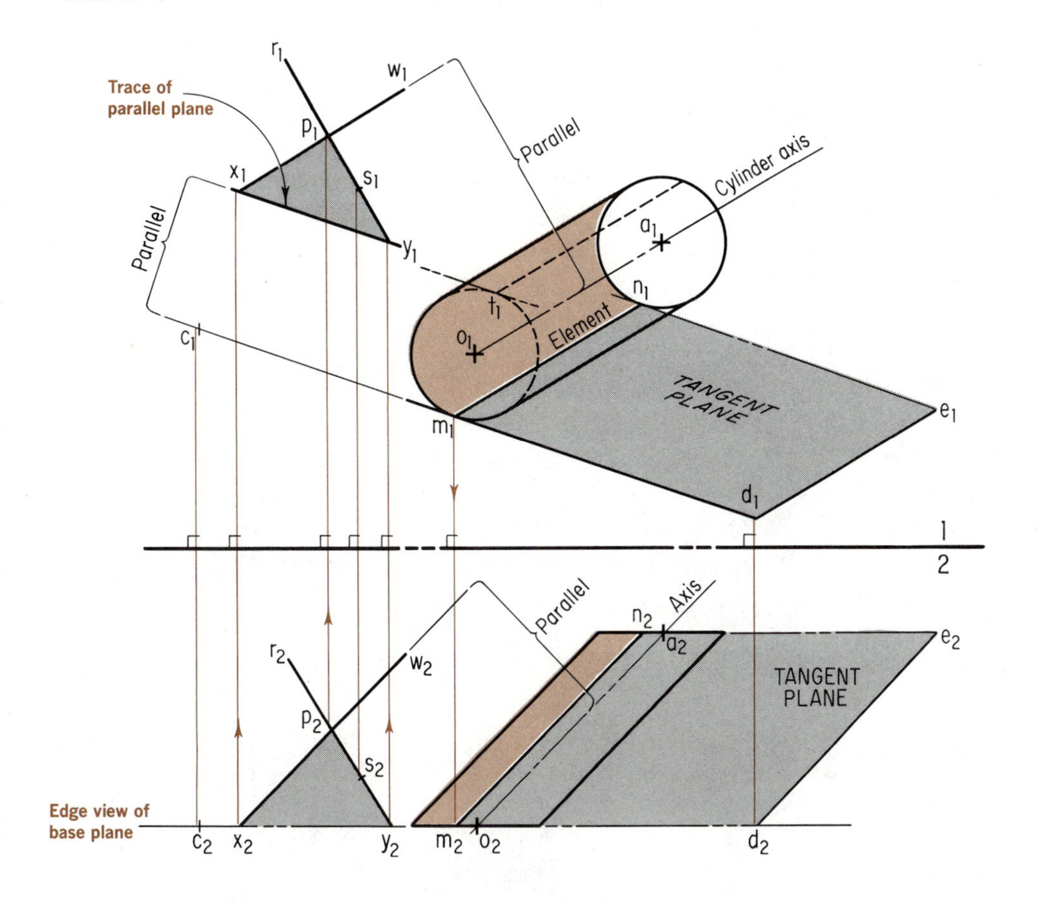

SAMPLE QUIZZES

1. Give the general analysis (not the detailed procedure) necessary for determining the piercing points of a line with a given prism.

2. Consider a plane tangent to a cone. How is the tangent plane determined?

3. Consider a plane tangent to a cylinder. How is the tangent plane determined?

4. Consider a point in space and the front and top views of a cone the base of which lies in a horizontal plane. What procedure must you use to establish a plane that contains the point and is tangent to the cone?

5. Consider a cylinder and a line in space. Describe how a plane may be constructed tangent to the cylinder and parallel to the line.

6. An oblique cone of revolution has a circular base. True or false?

7. Explain briefly—without drawing a sketch—how to set up a plane that will indicate the direction of a plane tangent to a given cylinder and parallel to a line outside the cylinder.

8. Define an oblique cone of revolution.

9. Give a brief general analysis (not the detailed procedure) of the method for determining the point at which a line pierces a cylinder by using *only* the horizontal and frontal views.

10. Give a brief general analysis (not the detailed procedure) of the method for determining the piercing points of a line and a cone.

PRACTICE PROBLEMS

1. Given Data: A. Horizontal and frontal projections of a rectangular prism *ABCD–EFGH*.

B. Frontal projection of a point *P* on face *BCGF* of the given prism.

C. Horizontal projection of a point *Q* on face *ABFE* of the given prism.

Required: A. The horizontal projection p_1 of point *P*.

B. The frontal projection q_2 of point *Q*.

C. Label and show all construction.

PROBLEM 1

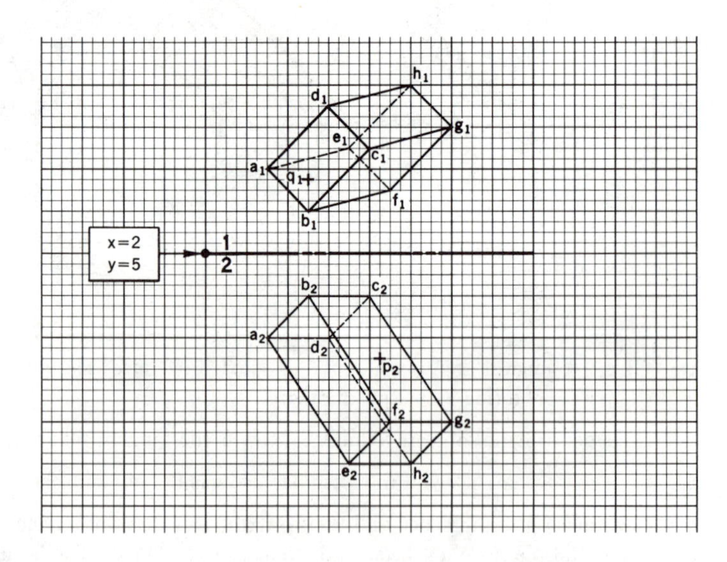

2. Given Data: A. Horizontal, frontal, and right profile projections of an oblique cone having a vertex *V*; the base of the cone is parallel to the right profile projection plane.

B. The frontal projection p_2 of point *P* on the surface of the given cone

(p_2 is visible in frontal view #2).

C. The right profile projection q_3 of point Q on the surface of the given cone (q_3 is hidden in view #3).

Required:

A. The horizontal and right profile projections of point P.

B. The horizontal and frontal projections of point Q.

C. Determine the visibility of points P and Q in all views by indicating whether they occur on visible or hidden elements.

D. Label and show all construction.

PROBLEM 2

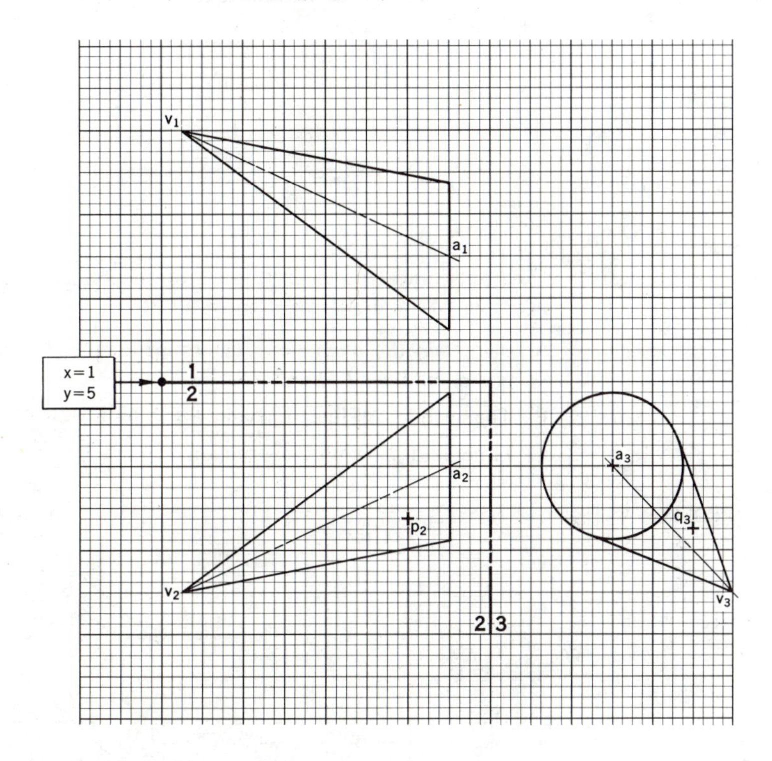

3. Given Data:

A. Horizontal, frontal, and right profile projections of an oblique cylinder.

B. The right profile projection p_3 of a point P on the surface of the cylinder (p_3 is hidden in view #3).

C. The frontal projection q_2 of a point Q on the surface of the cylinder (q_2 is hidden in view #2).

Required:

A. The horizontal and frontal projections of point P.

B. The horizontal and right profile projections of point Q.

C. Determine the visibility of points P and Q in all views by indicating whether they occur on visible or hidden elements.

D. Label and show all construction.

PROBLEM 3

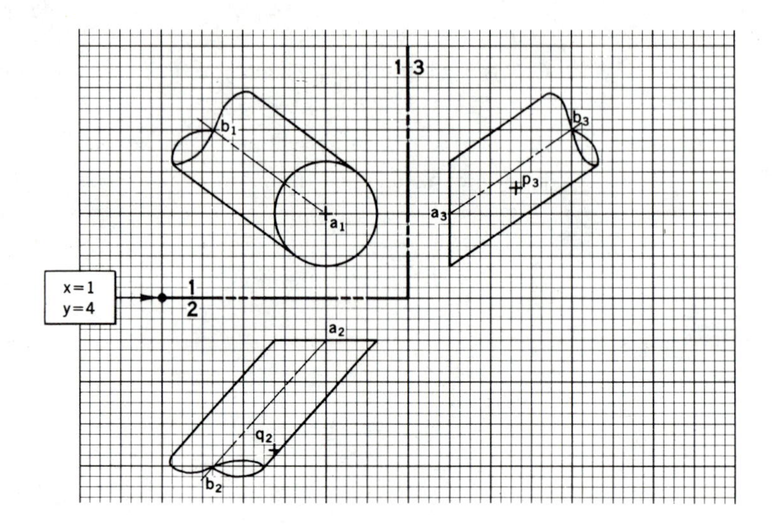

4. Given Data: Horizontal and frontal projections of an oblique triangular prism and two lines *LM* and *RS*, which intersect it.

Required: A. Using *only* the given views, determine the piercing points of the given prism and lines *LM* and *RS*.

B. The visibility of lines *LM* and *RS* in both views.

C. Label and show all construction.

PROBLEM 4

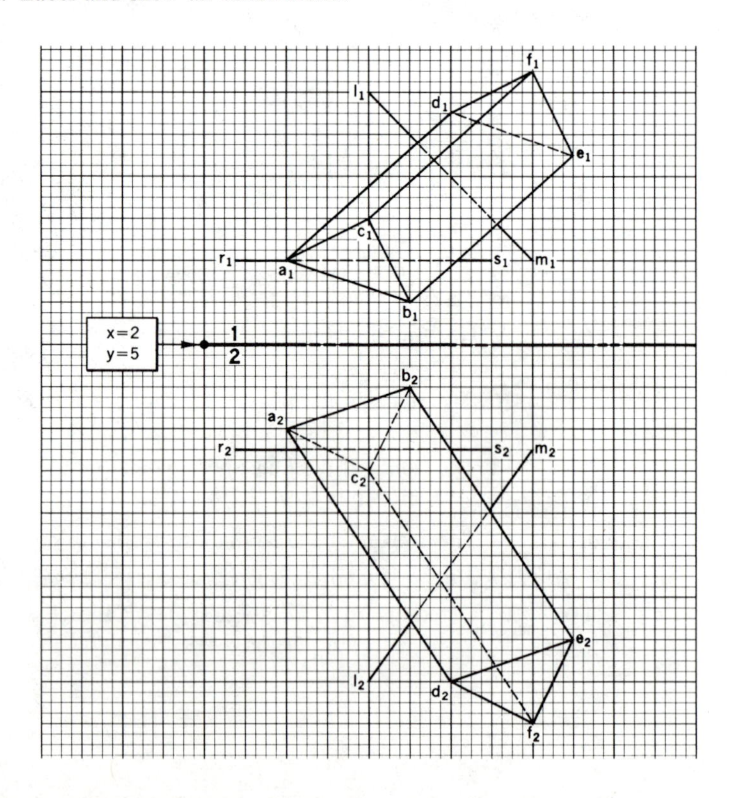

5. Given Data: Horizontal and frontal projections of an oblique cone with a vertex *V*; the cone is intersected by a line *LM*.

Required:

A. The piercing points of line *LM* and the given cone.

B. The complete visibility of line *LM*.

C. Label and show all construction.

PROBLEM 5

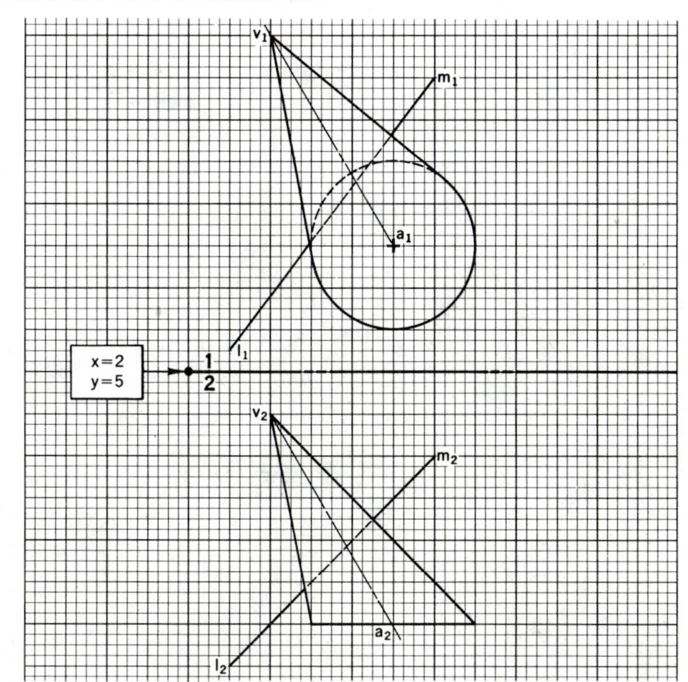

6. Given Data: Horizontal and frontal projections of an oblique cylinder, the axis of which is *AB*. Its bases are parallel to the frontal projection plane, and a line *RS* intersects it.

PROBLEM 6

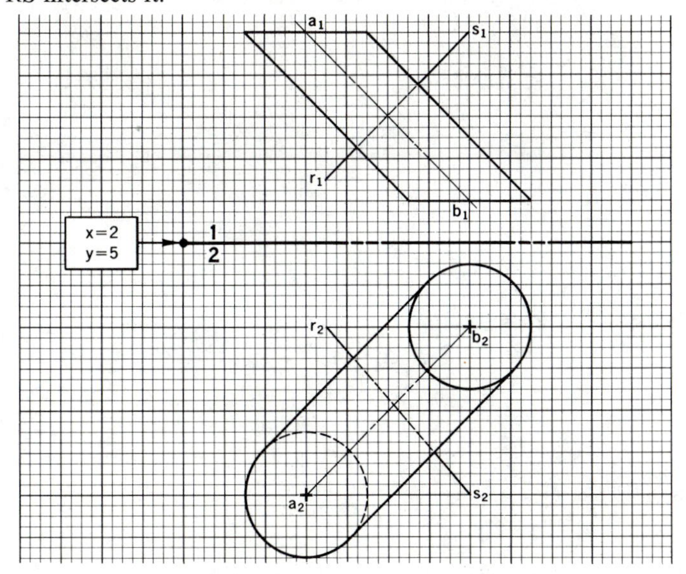

Required:

A. The piercing points of line RS and the given cylinder.

B. The complete visibility of line RS.

C. Label and show all construction.

7. Given Data:

A. Horizontal and frontal projections of an oblique cone, the vertex of which is V.

B. The horizontal projection p_1 of point P on the surface of the given cone (p_1 is visible in view #1).

C. The frontal projection q_2 of a point Q on the surface of the given cone (q_2 is hidden in view #2).

Required:

A. Horizontal and frontal projections of a plane tangent to the given cone at point P.

B. Horizontal and frontal projections of a plane tangent to the given cone at point Q.

C. Indicate the visibility of the tangent elements.

D. Label and show all construction.

PROBLEM 7

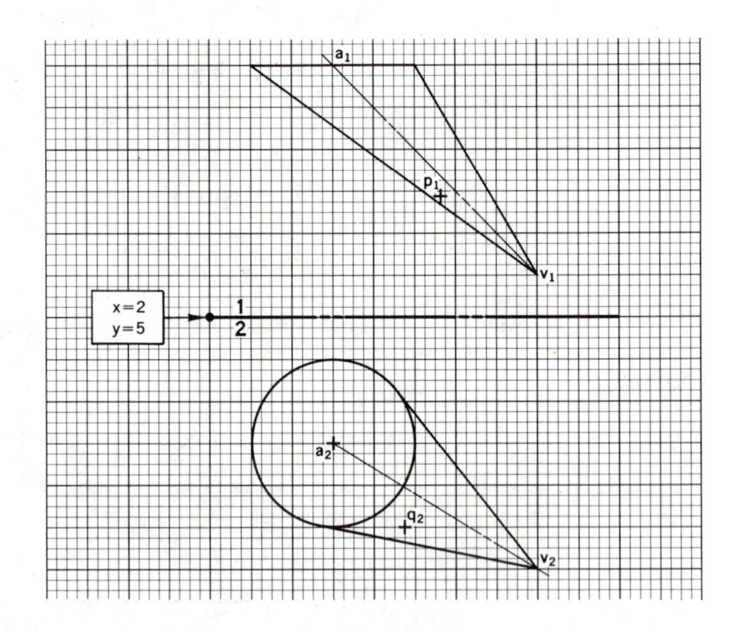

8. Given Data:

A. Horizontal and frontal projections of an oblique cylinder, the axis of which is AB.

B. The horizontal projection p_1 of a point P on the surface of the cylinder (p_1 is visible in view #1).

C. The frontal projection q_2 of a point Q on the surface of the cylinder (q_2 is visible in view #2).

Required:

A. Horizontal and frontal projections of a plane tangent to the given cylinder at point P.

B. Horizontal and frontal projections of a plane tangent to the given cylinder at point Q.

C. Indicate the visibility of the tangent elements.

D. Label and show all construction.

PROBLEM 8

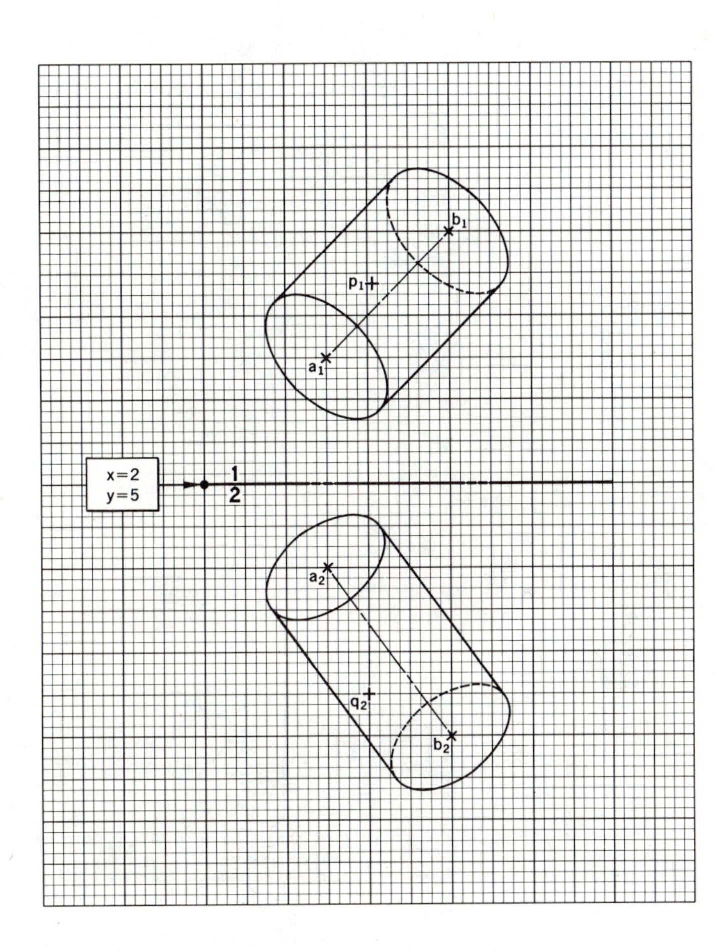

9. Given Data: Horizontal and frontal projections of an oblique cone having a vertex V; a point P is located outside the cone.

Required:

A. The horizontal and frontal projections of two planes that contain point P and are tangent to the given cone.

B. Indicate the visibility of the tangent elements.

C. Label and show all construction.

PROBLEM 9

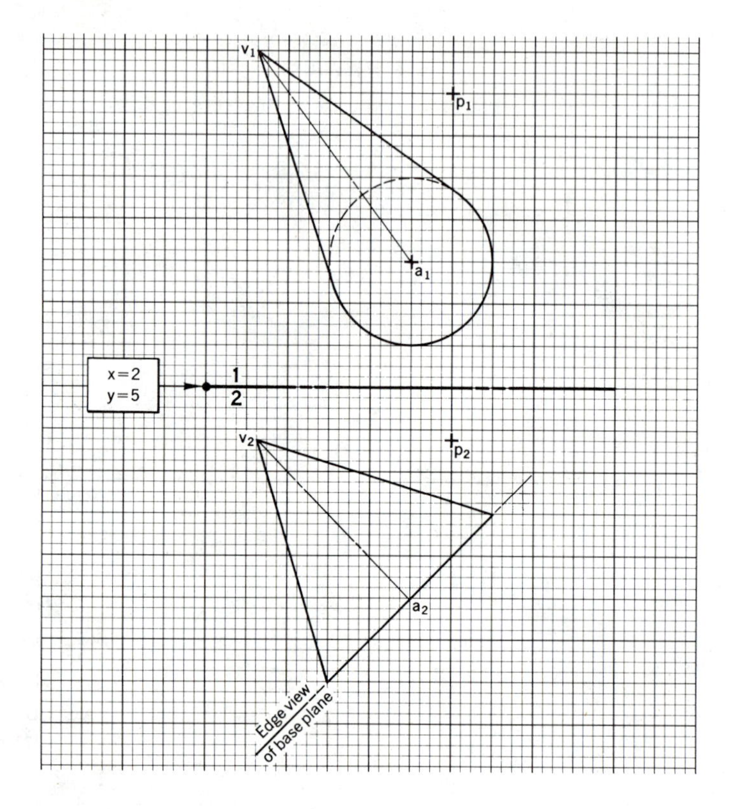

10. Given Data: Horizontal and frontal projections of an oblique cylinder having an axis AB; a point S is located outside the cylinder.

PROBLEM 10

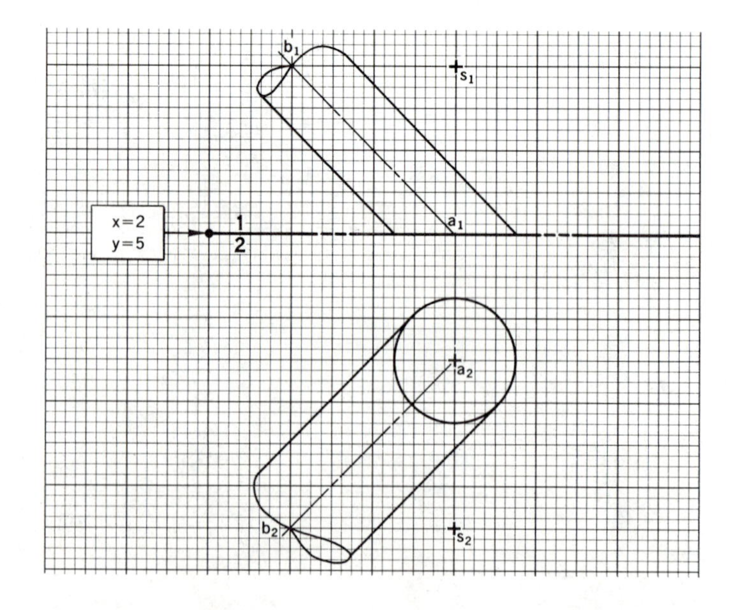

Required: A. The horizontal and frontal projections of two planes that contain point *S* and are tangent to the given cylinder.

B. Indicate the visibility of the tangent elements.

C. Label and show all construction.

11. Given Data: Horizontal and frontal projections of an oblique cone, the vertex of which is *V*; a line *LM* is located outside the cone.

Required: A. Horizontal and frontal projections of two planes that are tangent to the given cone and parallel to line *LM*.

B. Indicate the visibility of the tangent elements.

C. Label and show all construction.

PROBLEM 11

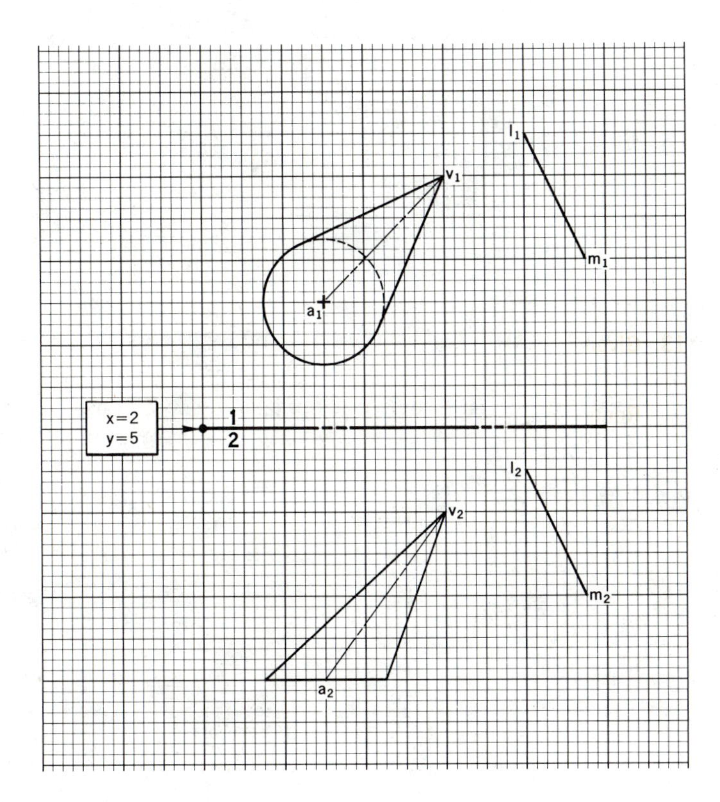

12. Given Data: Horizontal and frontal projections of an oblique cylinder, the axis of which is *AB*; a line *RS* is located outside the cylinder.

Required: A. Horizontal and frontal projections of two planes that are parallel to line *RS* and tangent to the given cylinder.

B. Indicate the visibility of the tangent elements.

C. Label and show all construction.

PROBLEM 12

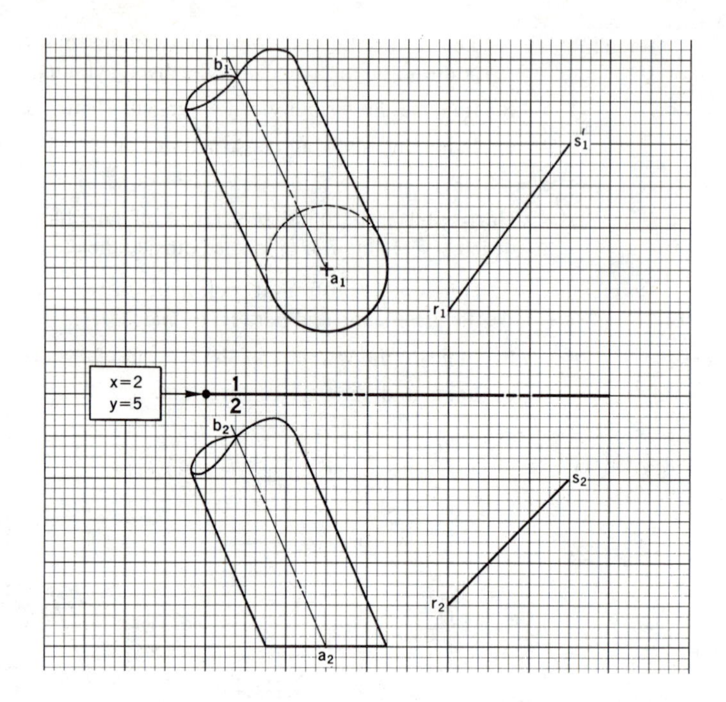

13. Given Data: Horizontal and frontal projections of an oblique cone, the vertex of which is V, and an oblique cylinder, the axis of which is AB.

Required: A. Horizontal and frontal projections of two planes tangent to the given cone and parallel to the axis of the given cylinder.

PROBLEM 13

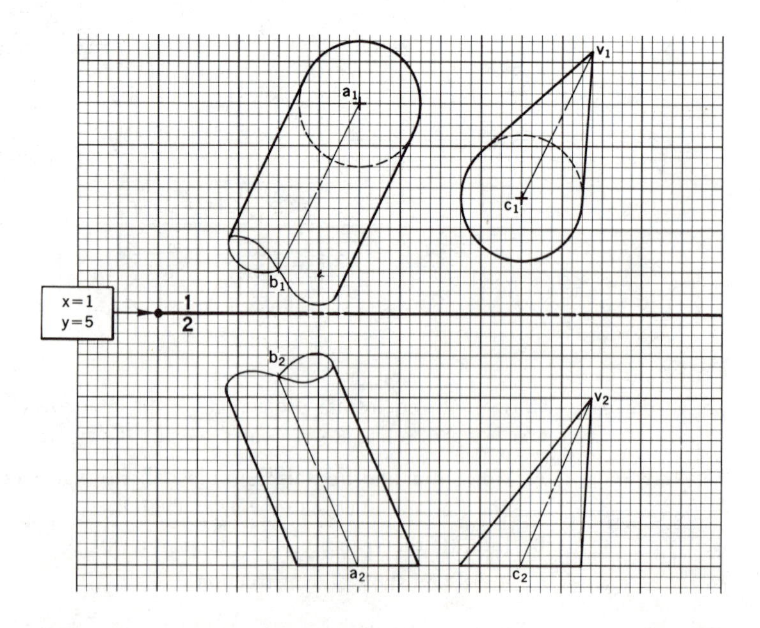

B. Indicate the visibility of the tangent elements.

C. Label and show all construction.

14. Given Data: Horizontal and frontal projections of two oblique cylinders having axes *AB* and *CD*, respectively, and containing bases in the same horizontal plane.

Required: A. Horizontal and frontal projections of two planes tangent to cylinder *AB* and parallel to cylinder *CD*.

B. Horizontal and frontal projections of two planes tangent to cylinder *CD* and parallel to cylinder *AB*.

C. Indicate the visibility of all the tangent elements.

D. Label and show all construction.

PROBLEM 14

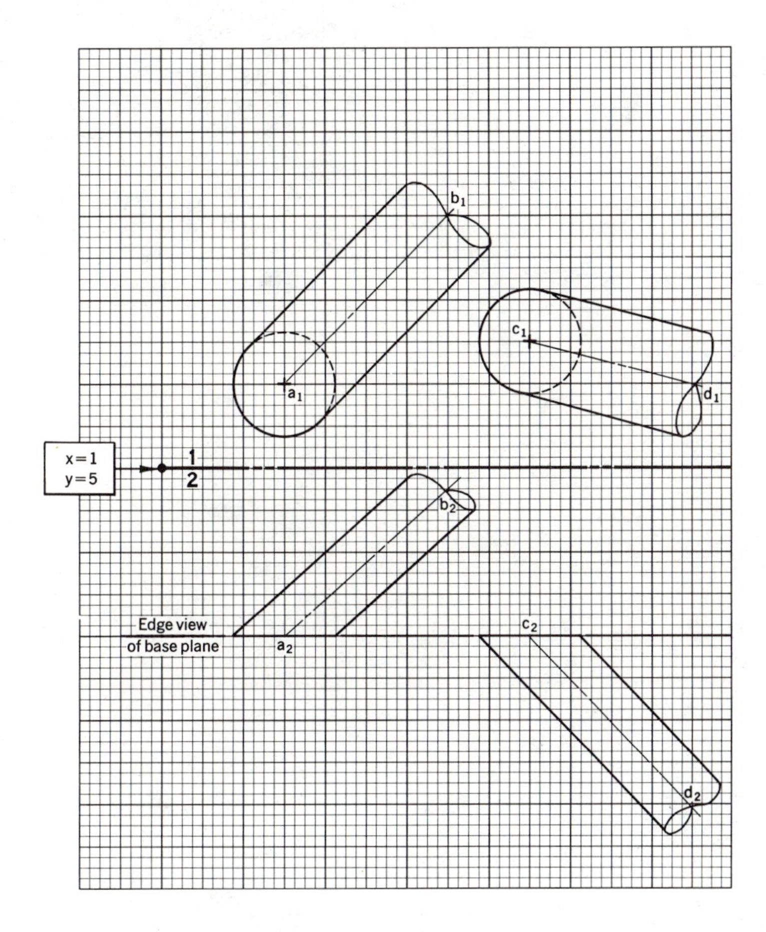

15. Given Data: Frontal and right profile projections of an oblique cone, the vertex of which is *V*, and an oblique cylinder, the axis of which is *BC*.

Required:

A. Horizontal, frontal, and right profile projections of two planes that are tangent to the given cone and parallel to cylinder axis *BC*.

B. Indicate the visibility of the tangent elements.

C. Label and show all construction.

PROBLEM 15

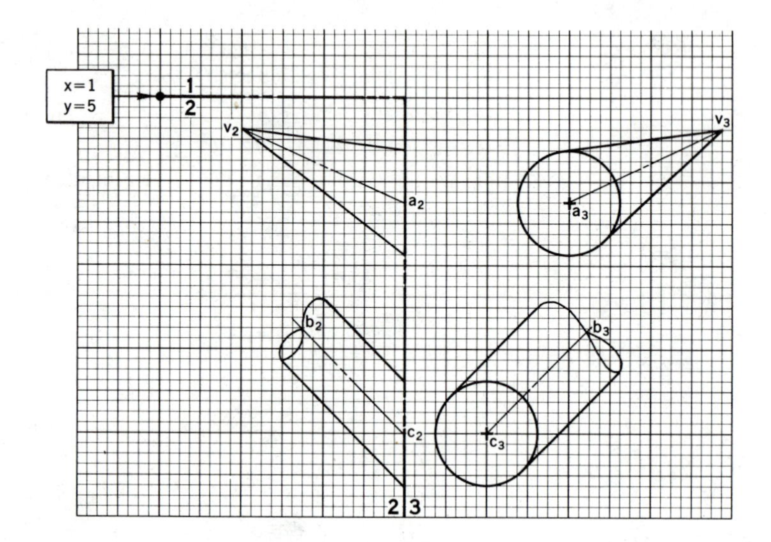

VII

Intersections of Common Geometric Solids and Surfaces

65 Principles of descriptive geometry involved in the determination of lines of intersections between solids and between solids and surfaces

Because many principles of descriptive geometry are necessary to the solution of problems involving the intersections of solids and surfaces, the study of their lines of intersection promotes visualization of three-dimensional space and the relations, in this space, between various geometric elements. To solve these problems you need (1) an understanding of planes and of their properties, (2) the application of relationships existing between lines and planes, and (3) a knowledge of the basic characteristics of solids and surfaces, including the location of points on surfaces and the construction of tangent planes to various curved surfaces.

66 Determination of the line of intersection between a plane surface and the faces of a prism

The intersection of two flat plane surfaces is a line. Therefore, when a given plane surface intersects the faces of a prism, it intersects each face along a definite line. The *individual* lines of intersection between the intersecting plane and the faces of the prism form the *complete* line of intersection between the given

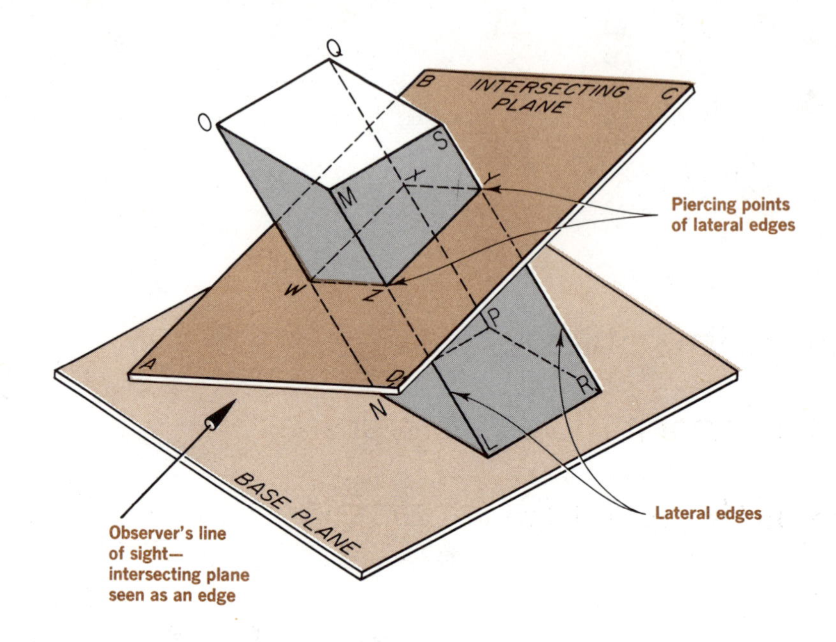

FIGURE 156

plane surface and the prism. Figure 156 illustrates a plane surface *ABCD* intersecting an oblique prism along the line *WXYZ*.

You have noted that the line of intersection between two planes can be determined by (1) viewing one intersecting plane as an *edge*, or (2) using the *cutting plane* method to locate the piercing points of one plane on another. Both methods may be used to determine the line of intersection between a plane surface and a given prism (see Figs. 156 and 158).

Orthographic Example: Determination of the line of intersection between a plane surface and a prism (Fig. 157).

Given Data: Horizontal and frontal projections of plane *ABCD* and an oblique prism, the base of which is *LNPR*. Assume that the plane and the prism are opaque.

Required:

a. The *complete* line of intersection between the plane and prism in the given horizontal and frontal views.

b. The combined visibility of the line of intersection and the given plane and prism (that is, the visibility of the line of intersection and that of the plane relative to the prism).

Construction Program #I: Using the Edge View of the Intersecting Plane—Fig. 157:

a. Determine the edge view of the given plane *ABCD* by passing a projection plane perpendicular to a true length line in the plane *ABCD*. (Line b_2c_2 is a horizontal line and appears as true length b_1c_1 in view #1. RL 1–3 is perpendicular to true length b_1c_1.)

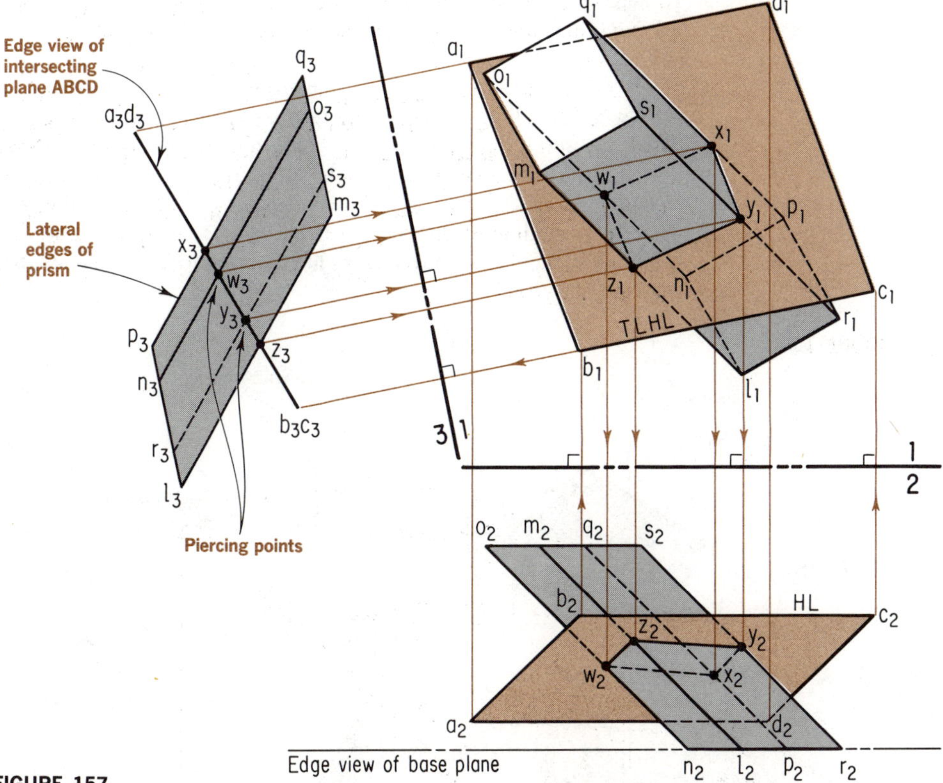

FIGURE 157

b. In view #3 the intersecting plane appears as an edge ($a_3d_3b_3c_3$). You can see where each *lateral edge* of the prism pierces the intersecting plane *ABCD* at points x_3, w_3, y_3, and z_3.

c. Project x_3, w_3, y_3, and z_3 to their respective lateral edges p_1q_1, o_1n_1, s_1r_1, and m_1l_1 in view #1. Connect points x_1, y_1, z_1, and w_1 with straight lines. This is the horizontal projection of the required line of intersection (visibility not applied).

d. Project points w_1, x_1, y_1, and z_1 to their respective lateral edges to determine w_2, x_2, y_2, and z_2 in view #2. Connect these points with straight lines. This is the frontal projection of the required line of intersection (visibility not applied).

e. Determine the visibility of the plane *ABCD* relative to the given prism in view #2. You can note from view #1 that line b_2c_2 is in front of lateral edge s_2r_2. You can now determine the complete visibility of the plane relative to the prism in view #2.

f. In view #1, determine the visibility of the plane relative to the prism by considering the positions of b_1c_1 and l_1m_1. From view #2 you can determine that b_1c_1 is above l_1m_1. Thus, in view #1, you can also determine the visibility of the plane *ABCD* relative to the given prism.

g. To determine the visibility of the line of intersection in views #1 and #2, you must consider whether the points located on it lie on visible lateral edges of the prism in these views. In view #1 you can see that points x_1, y_1, and z_1 are located on the visible lateral edges. Therefore, in view #1, the line of intersection from x_1 to y_1 to z_1 is visible (indicated by solid lines), while the lines connecting w_1 have been dotted to indicate that they are not visible. In view #2, you can see that the points w_2, z_2, and y_2 are located on visible lateral edges. Therefore, the line of intersection from w_2 to z_2 to y_2 is visible and solid, while the lines connecting x_2 have been dotted to indicate that they are not visible.

FIGURE 158

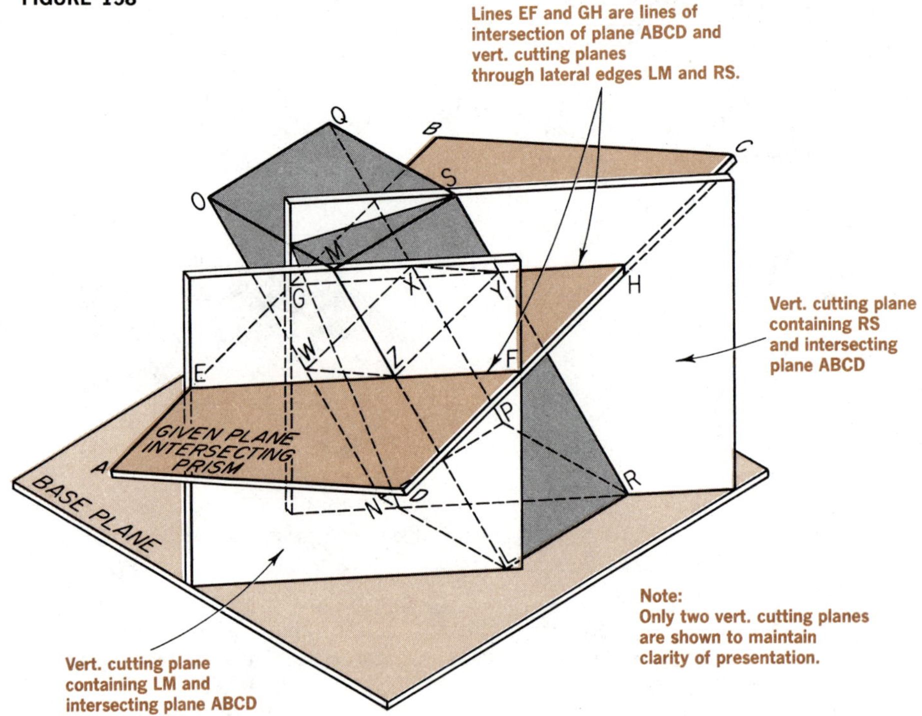

Construction Program #II:

Using the Cutting Plane Method—Figs. 158 and 159:

a. The piercing point of each lateral edge of the given prism can be determined by constructing vertical cutting planes (which appear as edges in view #1) through each lateral edge of the prism. In Fig. 159, vertical cutting plane #I passes through lateral edge q_1p_1. This cutting plane intersects the given plane $a_1b_1c_1d_1$ at point t_1 on line b_1c_1 and at point u_1 on line a_1d_1.

b. Project t_1 and u_1 downward to view #2 in order to determine t_2 and u_2 on lines b_2c_2 and a_2d_2, respectively. The piercing point x_2 of this

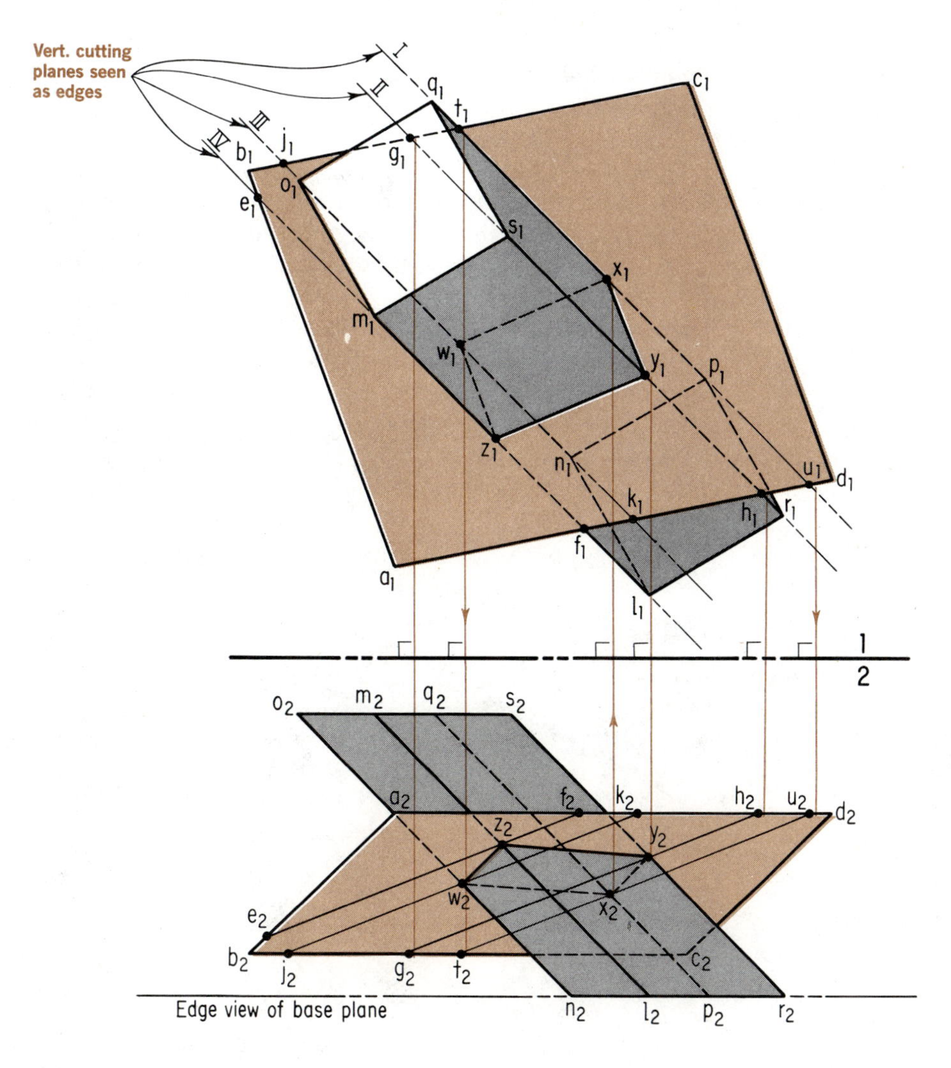

FIGURE 159

lateral edge on the plane $a_2b_2c_2d_2$ can be found where the line connecting t_2u_2 intersects lateral edge q_2p_2. Project x_2 upward to view #1 in order to determine x_1 on lateral edge q_1p_1.

c. Repeat this process by passing vertical cutting planes #II, #III, and #IV through lateral edges s_1r_1, o_1n_1, and m_1l_1, respectively. This will determine points y_1, w_1, and z_1, respectively. Determine y_2, z_2, and w_2 in view #2 by projection. Connect points X, Y, Z, and W in views #1 and #2, and determine the horizontal and frontal views of the line of intersection between the given plane and prism.

d. Determine the visibility as before.

67 Determination of the line of intersection between the lateral faces of a prism and a pyramid

By applying *both* the *edge view* and *cutting plane methods* of determining the piercing points of lines and planes, we can determine the line of intersection between a prism and a pyramid. In this case, the lines will be lateral edges of a prism and a pyramid, and the planes will be the lateral faces of a prism and a pyramid.

Figure 160 indicates a position in which the observer's line of sight is such that the *end view* of the prism is seen. In this end view, *all* lateral faces of the prism appear as *edges,* and each lateral edge of the pyramid can be seen to intersect the faces. To determine where the lateral edges of the prism intersect the faces of the pyramid, you must utilize *cutting planes* that contain the lateral edges of the prism and that intersect the faces of the pyramid.

FIGURE 160

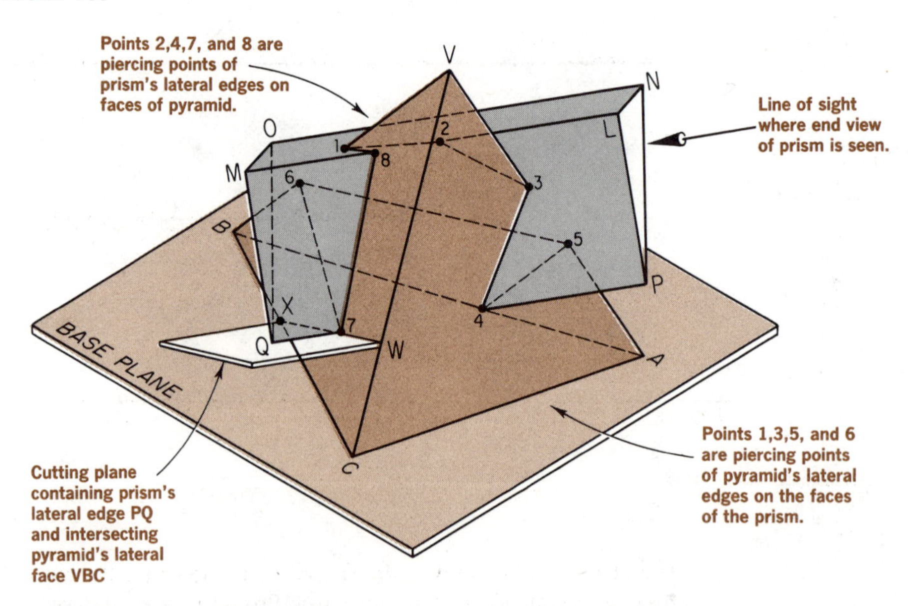

Orthographic Example: Determination of the line of intersection between a prism and a pyramid (Fig. 161).

Given Data: Horizontal and frontal projections of a triangular pyramid, the vertex of which is *V*. A triangular prism intersects the pyramid.

Required:

a. The line of intersection between the given prism and pyramid.

b. The combined visibility of the prism and pyramid and the line of intersection.

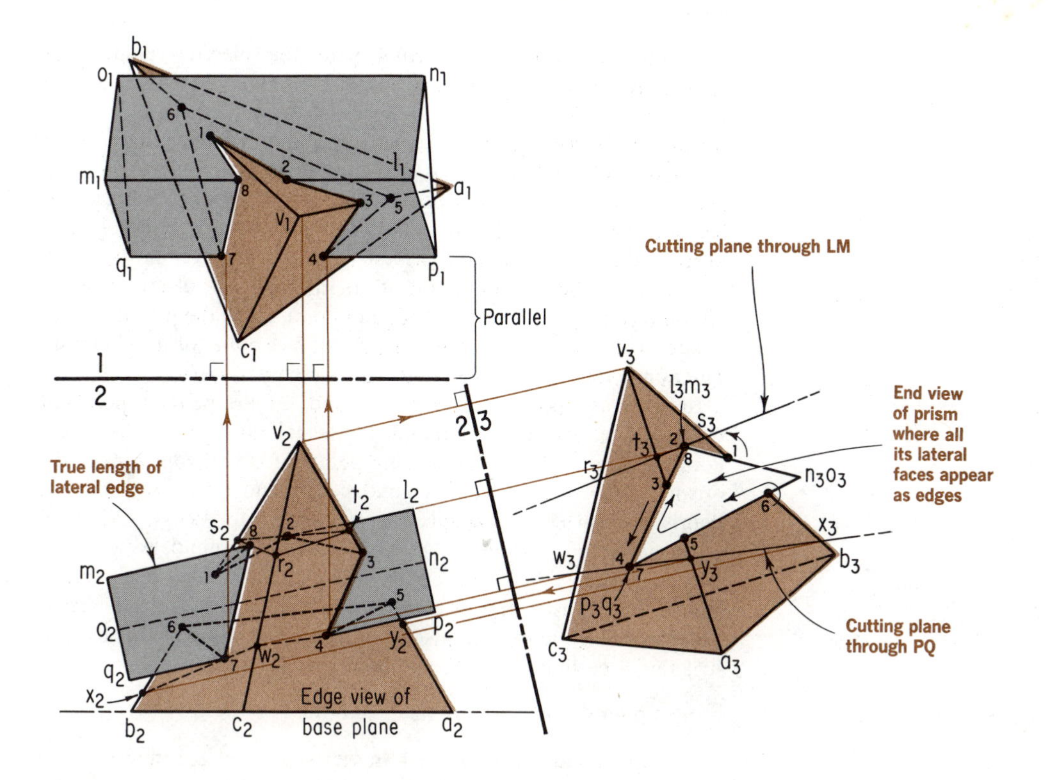

FIGURE 161

[NOTE: Consider the prism and the pyramid to be opaque and to be one *solid* integral casting where the pyramid meets the prism and the prism meets the pyramid (meaning that neither goes through the other).]

Construction Program:

a. Determine an *end view* of the prism by viewing its lateral edges as points. (Since the lateral edges of the prism are parallel to the frontal projection plane, they appear as true length in frontal projection view #2.) Construct projection plane #3 perpendicular to the true length lateral edges in order to develop the end view of the prism (view #3).

b. In view #3 (in which the end view of the prism appears) the faces of the prism appear as edges; therefore, we can determine the piercing points of the *lateral edges* of the *pyramid* on these faces. The lateral edges of the prism intersect the faces of the pyramid. Since the lateral edges appear as points, we must utilize the cutting plane method in order to locate these piercing points in the given horizontal and frontal projection views. In view #3 the line of intersection between the prism and the pyramid follows the outline (within the confines of the pyramid) of the edge view of the prism faces.

Number consecutively, starting with the piercing points that appear *nearest* the observer in view #3. Your first point is the piercing point of lateral edge v_3b_3 of the *pyramid* on one lateral face of the prism. Your second point is the piercing point of lateral edge l_3m_3 of the *prism* on lateral face $v_3a_3b_3$ of the pyramid. Continue in this manner to identify the next four points. The numbering of piercing points becomes *reversed* in direction (see curved arrow at point #6) to identify the piercing points *farthest* from the observer in view #3. (Point #7 is the piercing point of lateral edge p_3q_3 of the prism on lateral face $v_3b_3c_3$ of the pyramid, and point #8 is the piercing point of lateral edge l_3m_3 of the prism on lateral face $v_3b_3c_3$ of the pyramid.) The numbering ends at the point from which you started.

In summary, points #1, #3, #5, and #6 are piercing points of the lateral edges of the pyramid on the lateral faces of the prism. Points #2 and #8 are the piercing points of lateral edge l_3m_3 of the prism on the faces of the pyramid $v_3a_3b_3$ and $v_3b_3c_3$, respectively. Points #4 and #7 are the piercing points of lateral edge p_3q_3 of the prism on lateral faces $v_3a_3c_3$ and $v_3b_3c_3$ of the pyramid, respectively. You can locate the second and eighth piercing points in views #2 and #1 by constructing a cutting plane through lateral edge l_3m_3. You can locate piercing points #4 and #7 in the same views by constructing a cutting plane through lateral edge p_3q_3.

c. Project points #1–#8 to views #2 and #1.

d. Connect piercing points #1–#8 in views #2 and #1 in *consecutive* order, describing the line of intersection between the given prism and pyramid.

e. Determine the visibility of the line of intersection and of the given prism and pyramid in the frontal and horizontal projection views (#2 and #1, respectively) by considering the visibility of each piercing point. Observe that in order to be visible in a particular view, a point *must* appear on a visible face, or edge, of *both* solids *simultaneously*. In view #2, points #3, #4, #7, and #8 are visible since they fulfill the above requirement. Points #1, #2, #5, and #6 are hidden since they appear on hidden faces of the pyramid and prism.

In view #1, points #1, #2, #3, #4, #7, and #8 are visible since they appear on visible edges as well as visible faces of the pyramid and prism. Points #5 and #6 are hidden in this view since they appear on faces and edges that are hidden.

f. The line of intersection is a continuous series of straight lines running from points #1 to #2 to #3 to #4 to #5 to #6 to #7 to #8 and back to #1.

g. Determine the visibility of the prism and the pyramid in views #2 and #1 by the usual method. You can note from both the horizontal and frontal projection views that a line that connects with a hidden point must also be hidden.

68 Determination of the line of intersection of two oblique cylinders having bases in one plane

You can determine the line of intersection between two intersecting cylinders by establishing on both cylinders straight line elements that intersect at *points common* to their surfaces. These common points lie on the line of intersection between the two cylinders. Such straight line elements are determined by cutting planes that are *parallel* to both cylinder axes and that are represented by their *traces* on the base plane of the two cylinders. The *direction* of the traces is established by a "master" plane created from the intersection of two lines, each of which is parallel to one of the cylinder axes. Figure 162 illustrates two intersecting cylinders having one common base plane. Line *WX*, outside the cylinders, is parallel to cylinder axis *CD*, and line *YZ*, intersecting *WX* at point *O*, is parallel to cylinder axis *AB*. These intersecting lines form a "master" plane that is parallel to the axes of both cylinders. The trace *YW* of this plane establishes the direction of the traces of all planes *parallel* to the cylinder axes *AB* and *CD*. Figure 163 illustrates the "master" plane formed by intersecting lines *WX* and *YZ* and two cutting planes, the traces of which are parallel to the trace of the "master" plane. (Note that the cutting planes are *parallel* to the "master" plane.)

Orthographic Example: Determination of the line of intersection of two oblique cylinders, the bases of which lie in one plane (Figs. 164–167).

Given Data: Horizontal and frontal projections of two intersecting cylinders, the axes of which are *AB* and *CD*, respectively. Base *B* of cylinder *AB* and base *D* of cylinder *CD* are both located in the same base plane.

Required:

a. The complete line of intersection between the two given cylinders.

b. The complete visibility of the line of intersection and the combined cylinders.

(NOTE: Assume that both cylinders are opaque and of one integral piece—a casting, for example—so that one cylinder meets the other but does not pass through it.)

Construction Program:

a. Through an assumed point o_2 in view #2, construct a line y_2z_2 parallel to cylinder axis a_2b_2; through the same point construct a line x_2w_2 parallel to cylinder axis c_2d_2. The plane $o_2w_2y_2$, formed by these intersecting lines, creates a "master" plane that is parallel to both cylinder axes.

b. The horizontal view (#1) of the parallel "master" plane can be determined by constructing a line y_1z_1 parallel to cylinder axis a_1b_1 in view #1, and a line x_1w_1 parallel to cylinder axis c_1d_1 (of course, y_1z_1 and x_1w_1 intersect at point o_1).

c. In view #2, the trace of the "master" parallel plane on the base plane is y_2w_2. Project y_2w_2 upward to view #1 in order to determine the

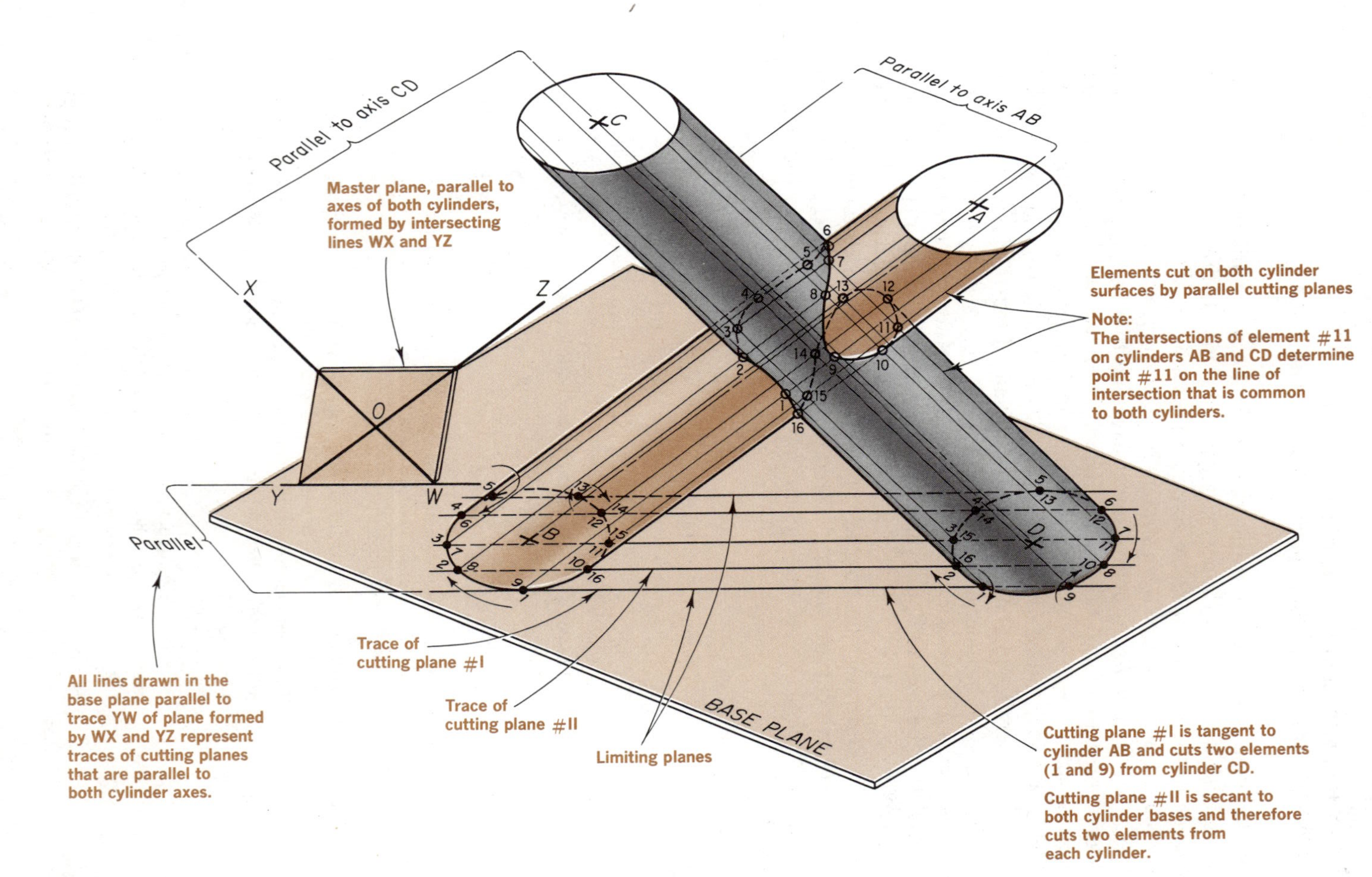

Parallel to axis CD
Parallel to axis AB
Master plane, parallel to axes of both cylinders, formed by intersecting lines WX and YZ
X
Z
O
Y
W
C
A
B
D
Elements cut on both cylinder surfaces by parallel cutting planes
Note:
The intersections of element #11 on cylinders AB and CD determine point #11 on the line of intersection that is common to both cylinders.
Parallel
Trace of cutting plane #I
Trace of cutting plane #II
Limiting planes
BASE PLANE
All lines drawn in the base plane parallel to trace YW of plane formed by WX and YZ represent traces of cutting planes that are parallel to both cylinder axes.
Cutting plane #I is tangent to cylinder AB and cuts two elements (1 and 9) from cylinder CD.
Cutting plane #II is secant to both cylinder bases and therefore cuts two elements from each cylinder.

FIGURE 162

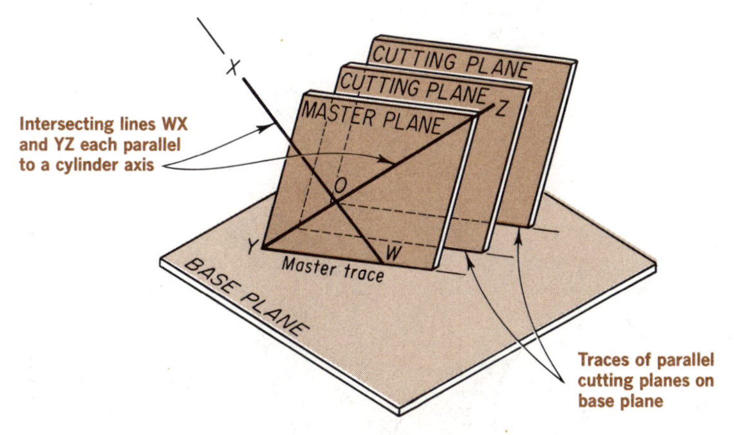

FIGURE 163

"master" trace y_1w_1. Any lines drawn parallel to y_1w_1 represent the traces of planes that are parallel to both cylinder axes. Such planes can therefore determine *straight line elements* on *both* cylinders.

d. Construct a convenient number of traces parallel to the trace y_1w_1. (In our example, only five traces were used.) Note that trace #I is tangent to the cylinder whose base center is b_1 and *secant* to the cylinder whose base center is d_1. Cutting plane #V, on the other hand, is secant to the cylinder with base center b_1 and tangent to the cylinder whose base center is d_1. The intermediate traces (#II, #III, and #IV) are secant to both bases. Cutting plane traces #I and #V are called *limiting cutting* plane *traces* because if trace #I were tangent to base d_1, it would *not* touch the base of the other cylinder. Likewise, if cutting plane trace #V were tangent to base b_1, it would *not* touch the cylinder whose base center is d_1.

(NOTE: At this point consider the effects of the cutting planes on each cylinder. Cutting planes that are *tangent* to the base of a cylinder determine only *one* straight line element on the cylinder surface, while cutting planes that are *secant* to the base of a cylinder determine *two* straight line elements on the cylinder surface. For example, in the case of cutting plane #I, *one* straight line element is determined on cylinder a_1b_1 and *two* straight line elements are determined on cylinder c_1d_1. Note that the single element on cylinder a_1b_1 is intersected twice—once by each element determined on cylinder c_1d_1. These points of intersection are common to both cylinders and, therefore, lie on the line of intersection between them.)

e. In order to maintain an orderly record of each element determined by the cutting planes traces, apply a numbering system as illustrated in Figs. 164 and 165. Starting with a trace that is *tangent* to one base of a given cylinder (for example, trace #I), number the tangent point #1, and proceed to the *next nearest* cutting plane trace #II. This

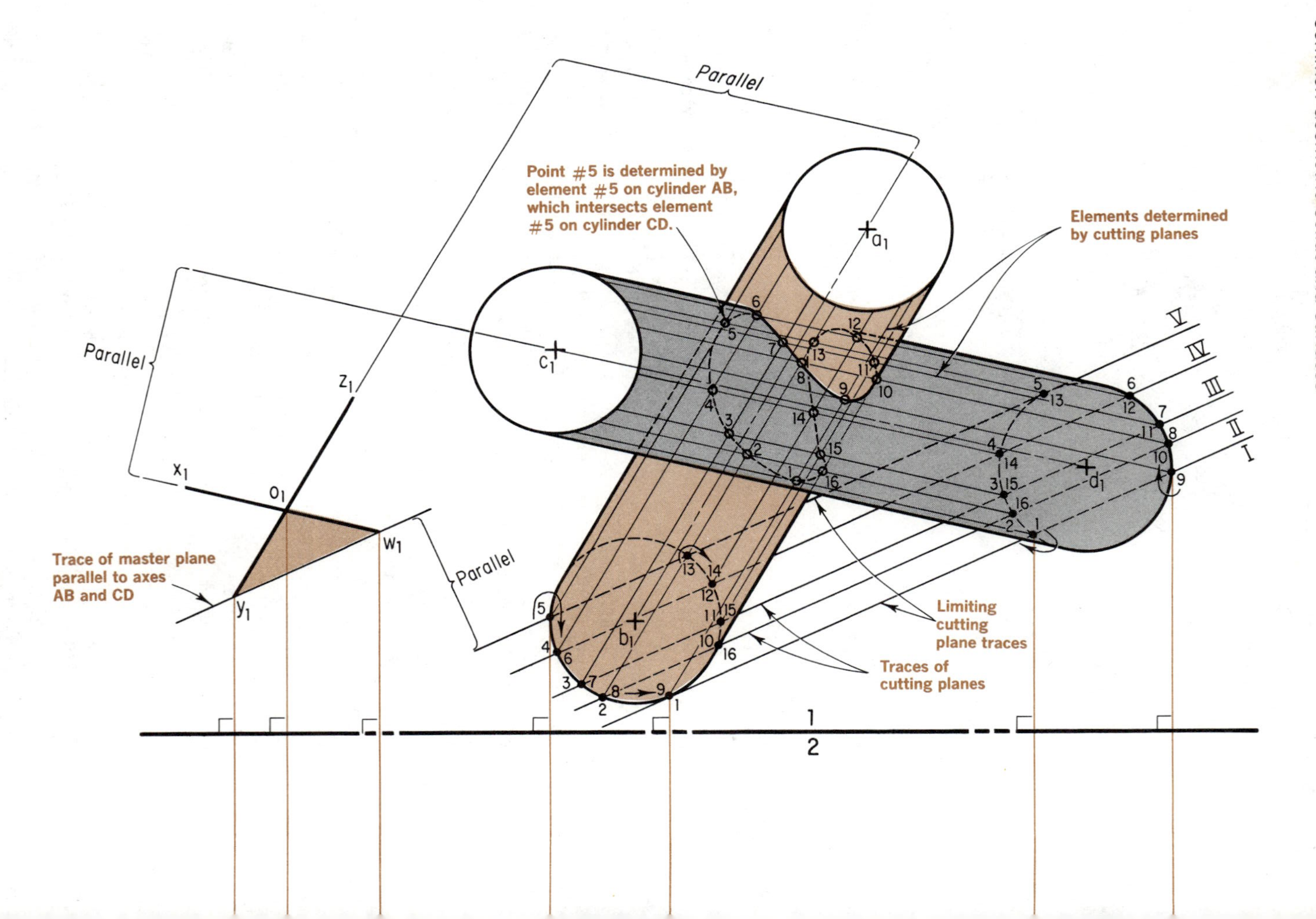
Parallel
Point #5 is determined by element #5 on cylinder AB, which intersects element #5 on cylinder CD.
Elements determined by cutting planes
Parallel
Trace of master plane parallel to axes AB and CD
Parallel
Limiting cutting plane traces
Traces of cutting planes

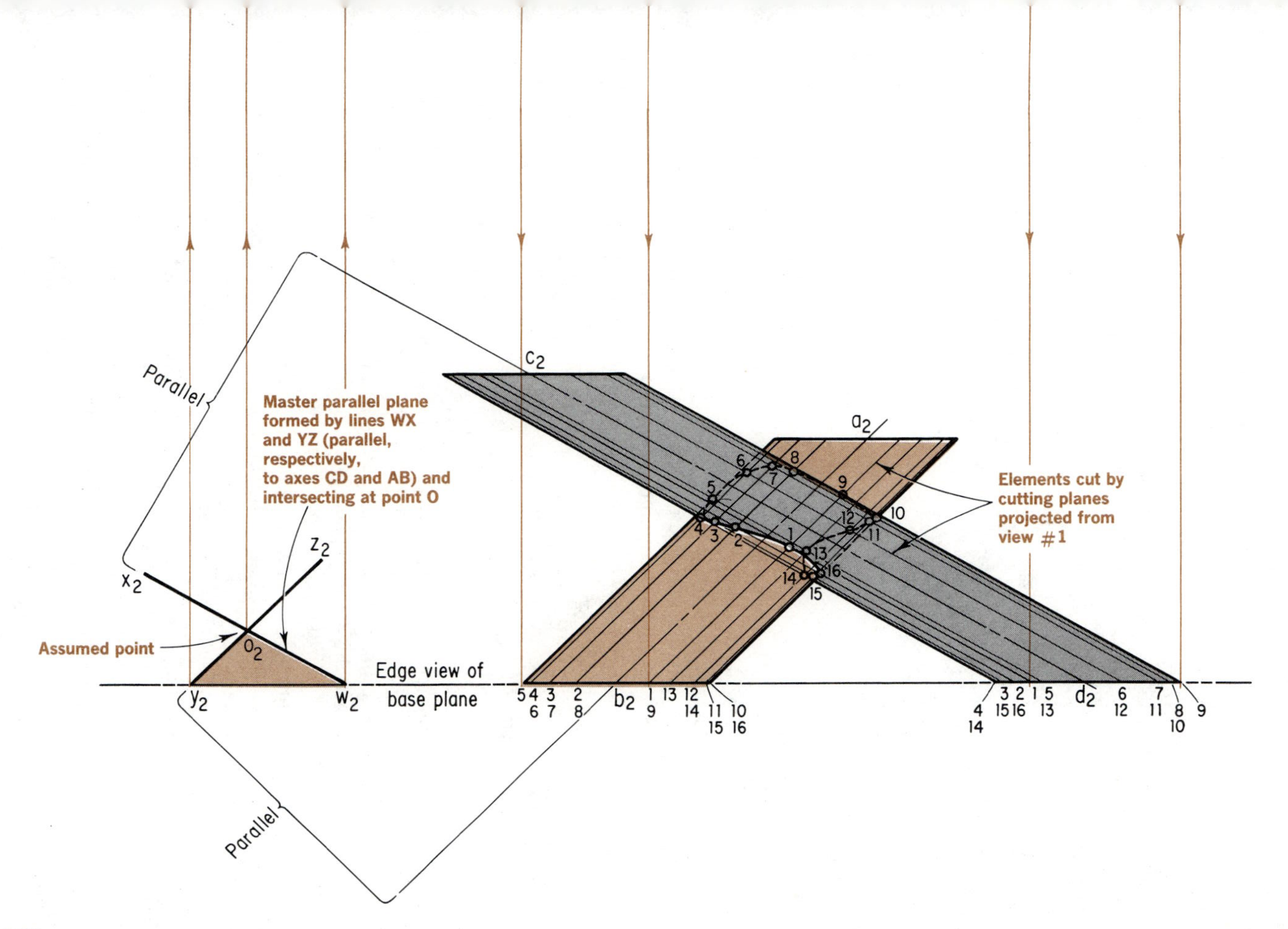
Master parallel plane formed by lines WX and YZ (parallel, respectively, to axes CD and AB) and intersecting at point O
Elements cut by cutting planes projected from view #1
Assumed point
Edge view of base plane
Parallel
Parallel
c2
a2
b2
d2
x2
z2
o2
y2
w2

FIGURE 164

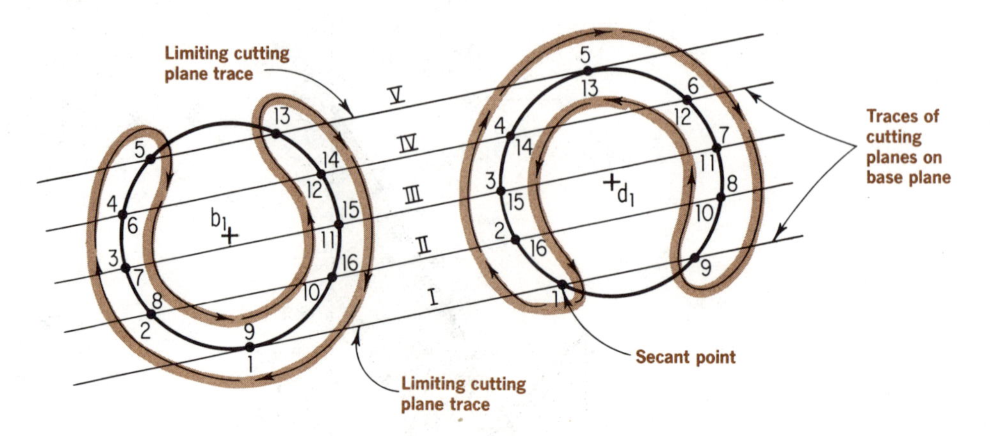

FIGURE 165

trace is secant to the same base at point #2. Proceed in numerical order to determine points #3–#5. At point #5, you can see from Fig. 165 that the direction of numbering must be reversed, since the next point (#6) is determined by cutting plane trace #IV. This means that two numbers may appear at one point (#4 and #6, for example). Continuing, you see that at point #13, which was determined by cutting plane #V, you must again reverse direction, since point #14 is determined by cutting plane trace #IV. Progressing to points #15, and #16, you return to the original point at which the numbering started.

Notice that a complete, continuous loop of numbers (#1–#16 and back to #1) has resulted. Also note that each of the *secant* cutting plane traces has determined *four numbers* (trace #III, for example, has determined points #3, #7 and #11, #15). Proceeding to the cylinder base whose center is d_1, note that point #1, determined by cutting plane #I, is a secant point. This point determines the starting point for our numbering system on this base. Continue in numerical order, as indicated in Fig. 165. (Note that cutting plane trace #III, which is secant to this cylinder base, has determined points #3, #15 and #11, #7.) Using as an example cutting plane trace #III, you can see that this trace has determined a total of four numbers—#3, #7, #11, and #15. In other words, this trace has determined *two* straight line elements on *each* cylinder, and since the two elements on *each* cylinder intersect each other *twice*, *four* points of intersection have been located.

(NOTE: The numbering system described above enables you to easily construct the actual line of intersection between two cylinders.)

f. In view #1 in Fig. 164 (from the points determined on the bases of each cylinder by the cutting plane traces), construct straight line elements on each cylinder.

g. Where *like-numbered* elements intersect, points *common* to both cylinders are located. For example, element #5 on cylinder a_1b_1 intersects element #5 on cylinder c_1d_1. The intersection of these two elements determines point #5 on the line of intersection between the two cylinders.

Construct a smooth curve to connect the newly determined points in numerical order.

h. Project the elements on each cylinder determined in view #1 to view #2. Like-numbered elements intersect in this view to determine points on the line of intersection (frontal view #2). Construct a smooth curve to connect these points in a *continuous, consecutive numerical* order.

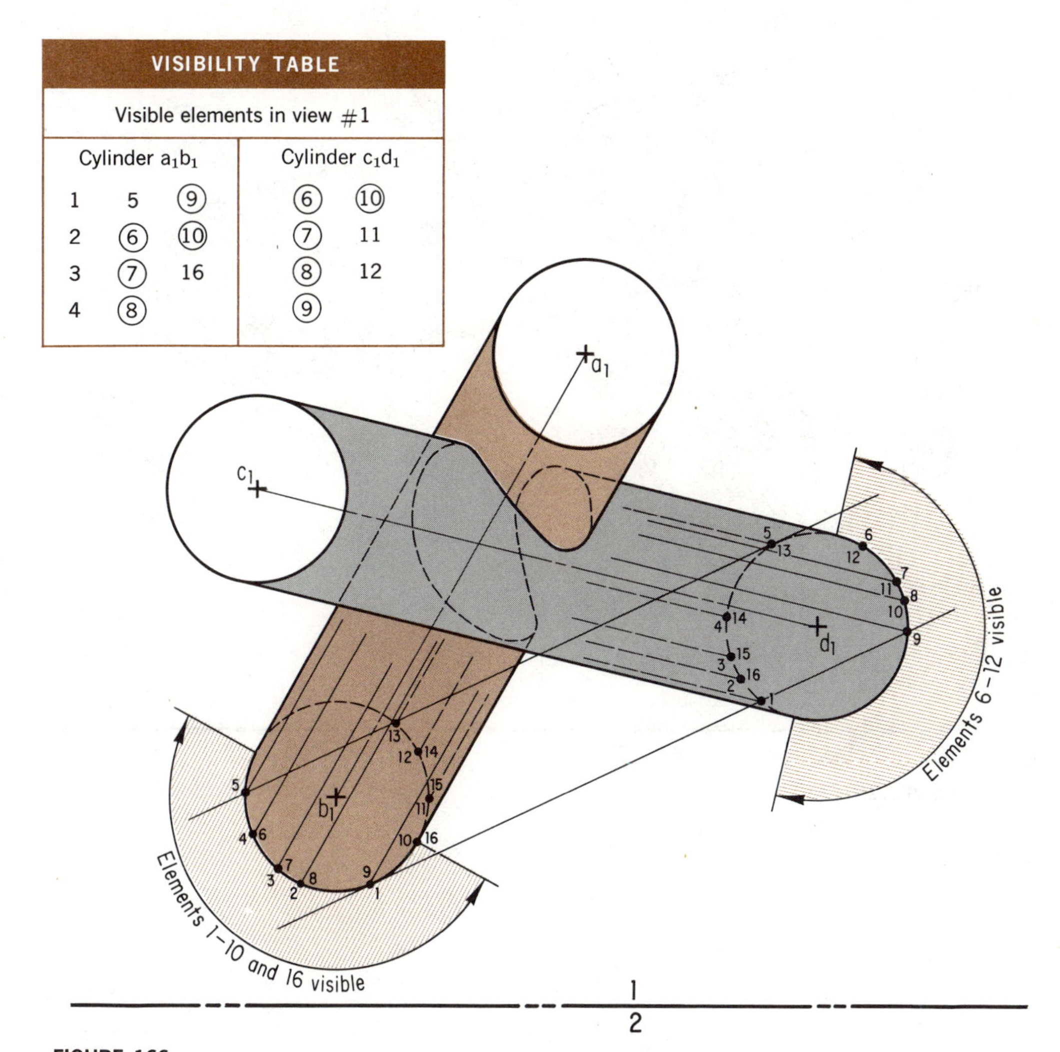

VISIBILITY TABLE					
Visible elements in view #1					
Cylinder a_1b_1			Cylinder c_1d_1		
1	5	⑨	⑥	⑩	
2	⑥	⑩	⑦	11	
3	⑦	16	⑧	12	
4	⑧		⑨		

FIGURE 166

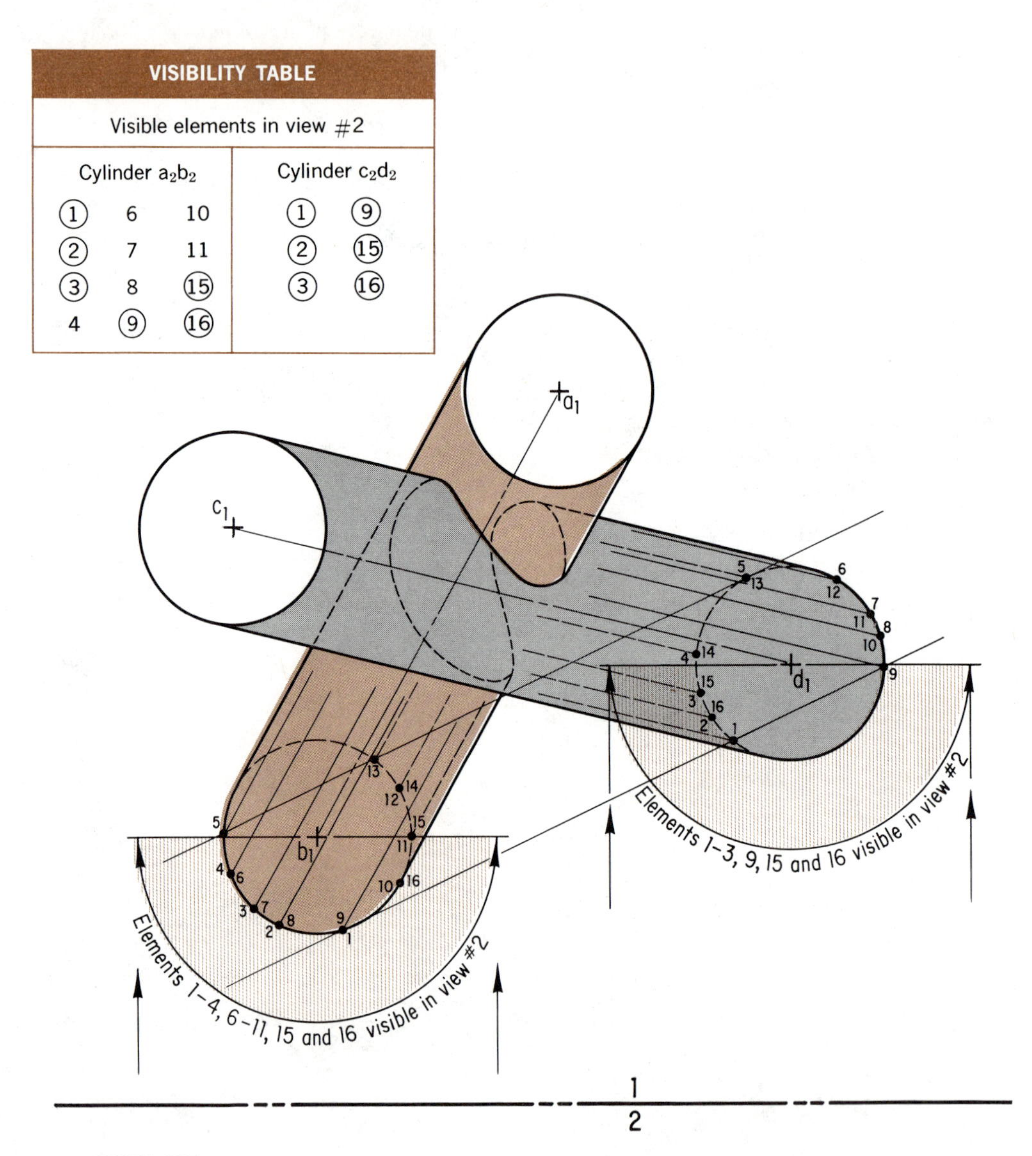

VISIBILITY TABLE

Visible elements in view #2				
Cylinder a_2b_2			Cylinder c_2d_2	
①	6	10	①	⑨
②	7	11	②	⑮
③	8	⑮	③	⑯
4	⑨	⑯		

FIGURE 167

i. In order to determine the visibility of the line of intersection between the cylinders, locate, in *each* view, *visible* elements on *both* cylinders (to be visible, a point on the line of intersection must be determined by *two visible elements:* one on cylinder a_1b_1 and one on cylinder c_1d_1). To facilitate determination of the visibility, Fig. 166 illustrates view #1 of the given cylinders. From the cylinder bases in this view you can see that for the cylinder with base center b_1, elements #1–#10 and element #16 are visible. For the cylinder with base center d_1, elements #6–#12 are visible. Figure 166 contains a tabular summary of this information. Elements #6–#10 are listed in both columns for the bases of each cylinder. The circled numbers represent points common to each cylinder and visible in view #1. Therefore, the portion of the line of intersection passing through points #6–#10 must also be visible in view #1, while the remaining portions are not.

j. In order to determine the visbility of the line of intersection in view #2, refer again to the horizontal projection (view #1) of the given cylinders. The elements located on the *front half* of each base (nearest to the frontal plane) are visible in view #2. The visible elements are summarized in tabular form in Fig. 167. Elements #1–#4, #6–#11, and #15 and #16 are visible on the cylinder with base center b_1. Elements #1–#3, #9, #15, and #16 are visible on the cylinder with base center d_1. The circled elements in each column represent points that, in view #2, are common to both cylinders and visible on the line of intersection.

69 Determination of the line of intersection of two oblique cones having bases in one plane

The *cutting planes* that pass through the *vertices* of two oblique cones will cut *straight line elements* on each cone. The intersection of these straight-line elements determines points common to both cones. The points thus located lie on the line of intersection between the two cones.

Figure 168 illustrates two intersecting cones, the vertices of which are O and V, respectively. The line connecting the vertices and intersecting the base plane of the two cones at point P is the *vertex line.* This is the line through which all cutting planes must pass in order to cut straight line elements on both cones. It also may be referred to as a hinge line, since the cutting planes "hinge" on the vertex line.

In Fig. 169, three cutting planes pass through a vertex line PV.

The cutting planes have traces on the base plane, and the straight line elements that they determine on each cone intersect at points on the line of intersection. For example, in Fig. 168 notice that element O–1 on the cone with vertex O intersects element V–1 on the cone with vertex V, thus determining point #1 on the line of intersection between these cones.

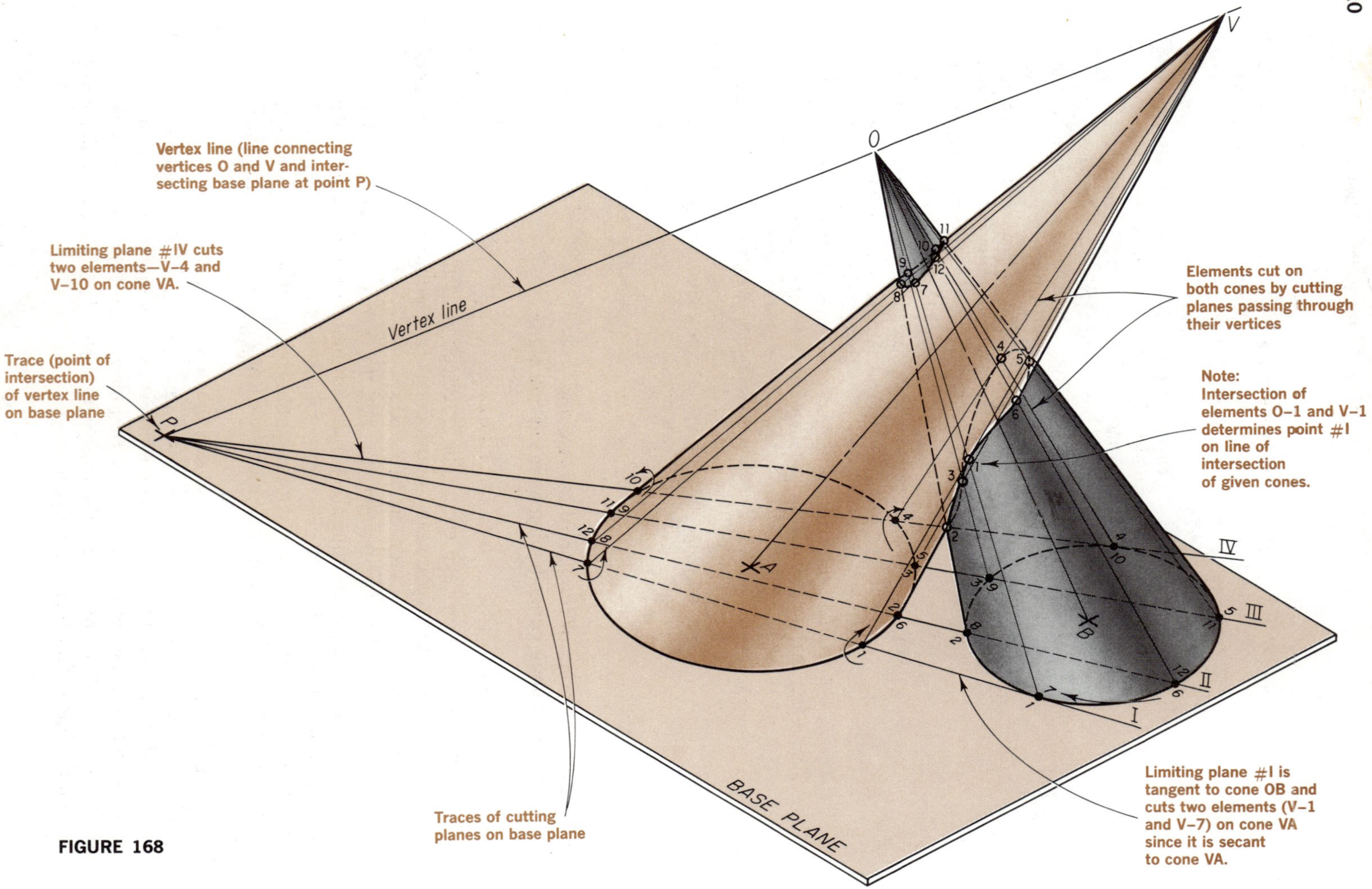
Vertex line (line connecting vertices O and V and intersecting base plane at point P)
Limiting plane #IV cuts two elements—V–4 and V–10 on cone VA.
Trace (point of intersection) of vertex line on base plane
Vertex line
P
O
V
A
B
Elements cut on both cones by cutting planes passing through their vertices
Note: Intersection of elements O–1 and V–1 determines point #I on line of intersection of given cones.
I
II
III
IV
Limiting plane #I is tangent to cone OB and cuts two elements (V–1 and V–7) on cone VA since it is secant to cone VA.
Traces of cutting planes on base plane
BASE PLANE

FIGURE 168

FIGURE 169

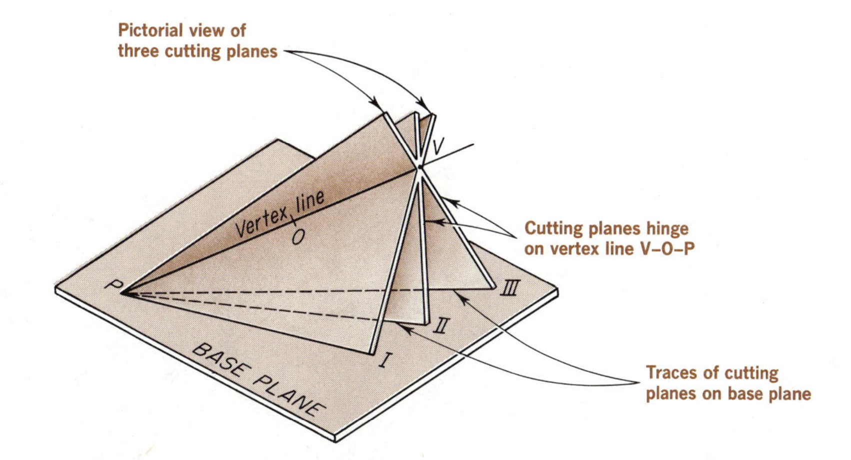

Orthographic Example: Determination of the line of intersection between two oblique cones, the bases of which lie in one plane (Figs. 170 and 171).

Given Data: Horizontal and frontal projections of two oblique cones with vertices O and V, respectively, and bases in the same plane.

Required:

a. The line of intersection between the two oblique cones.

b. The visibility of the line of intersection and the given cones. (Assume that the cones are opaque and of one solid piece.)

Construction Program:

a. Starting with view #2, connect vertices v_2 and o_2 with a straight line, extending this line until it intersects the edge view of the base plane at point p_2. (This line is the vertex line.)

b. In view #1 connect, with an extended line, vertices v_1 and o_1 of the given cones. (This is the horizontal projection of the vertex line.)

c. From view #2, project p_2 to the vertex line in view #1 in order to determine p_1. Since all cutting planes must pass through the vertex line, their traces will "radiate" from point p_1 in view #1, where the vertex line intersects the base plane. Therefore, from p_1 construct a series of radiating cutting plane traces (traces #I–#IV are used in our figure).

d. Apply a numbering system which is similar in principle to the system presented earlier, when we discussed the intersection of two cylinders (see Figs. 170 and 171).

Cutting plane traces for the two intersecting cones are shown in Fig. 171. The limiting cutting plane traces #I and #IV are tangent to the base with center b_1 of the cone with vertex O and secant to the base with center a_1 of the cone with vertex V. Starting at the tangent point of limiting cutting plane trace #I, and continuing in a clockwise direction (actually, either clockwise or counterclockwise direc-

VISIBILITY TABLE

Visible elements in view #1			
Cone VA		Cone OB	
①	10	①	⑦
⑦	⑪	2	⑧
⑧	⑫	5	⑪
9		6	⑫

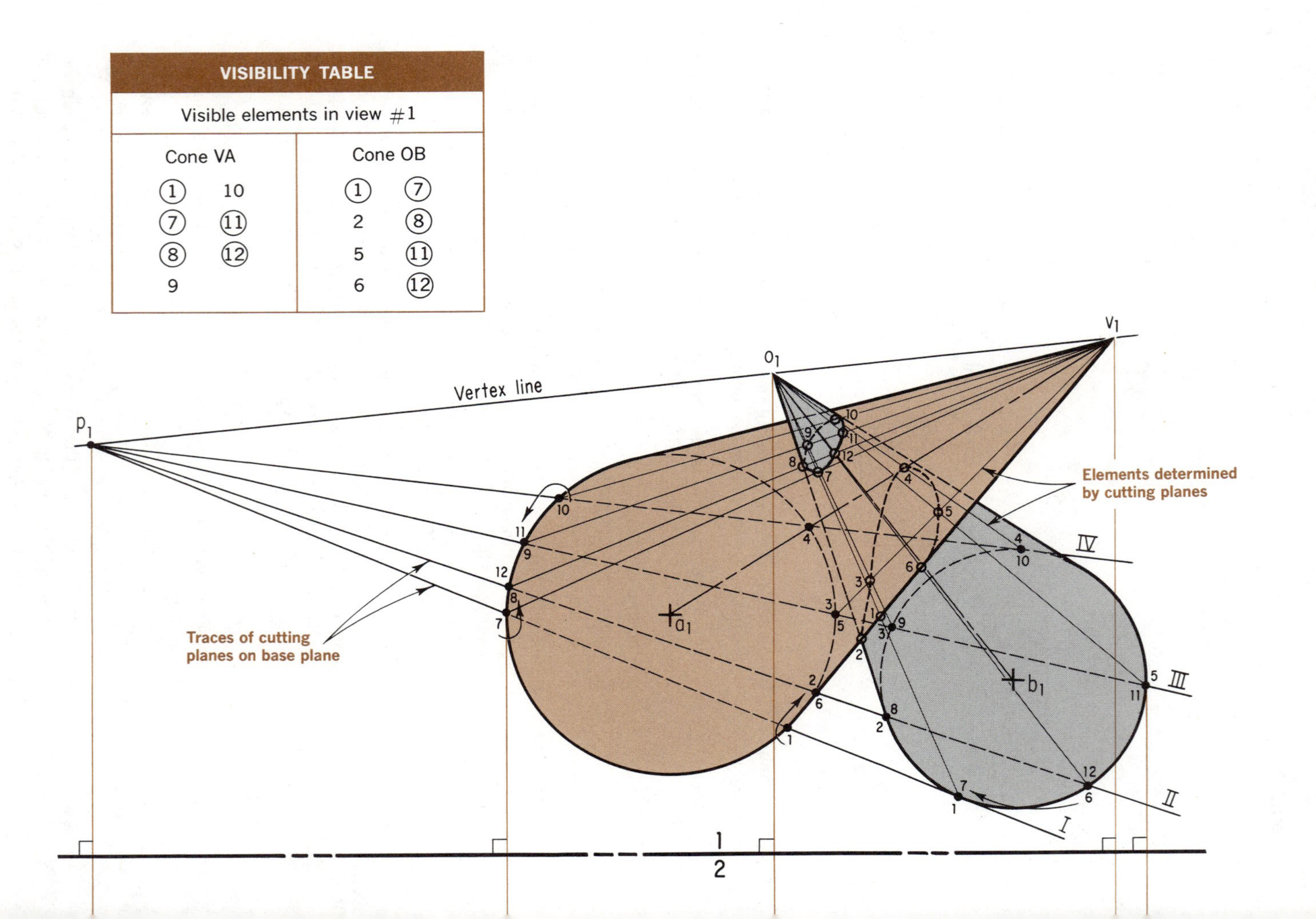

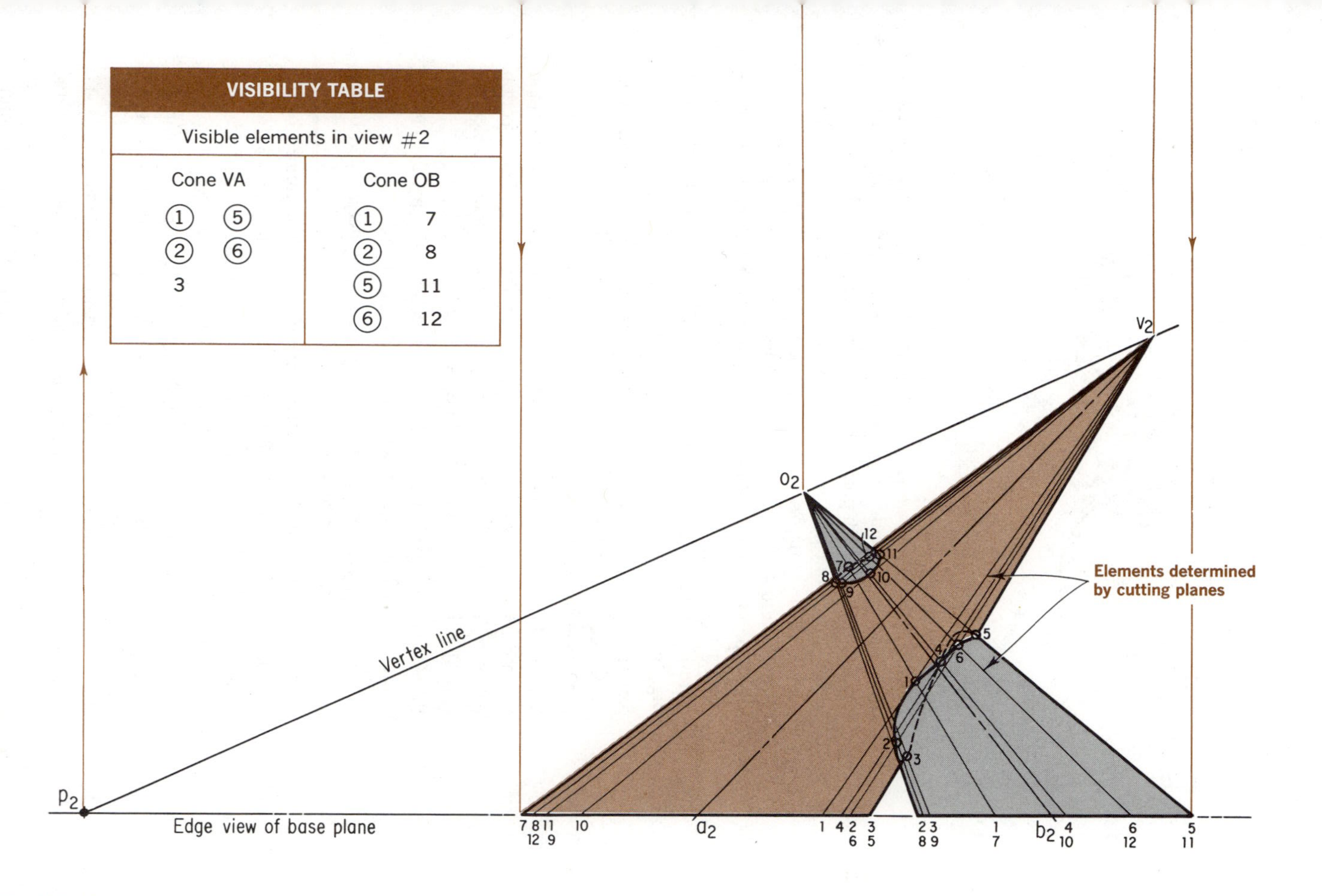
VISIBILITY TABLE
Visible elements in view #2
Cone VA
① ⑤
② ⑥
3
Cone OB
① 7
② 8
⑤ 11
⑥ 12
Elements determined by cutting planes
Vertex line
Edge view of base plane
v_2
o_2
a_2
b_2
p_2

FIGURE 170

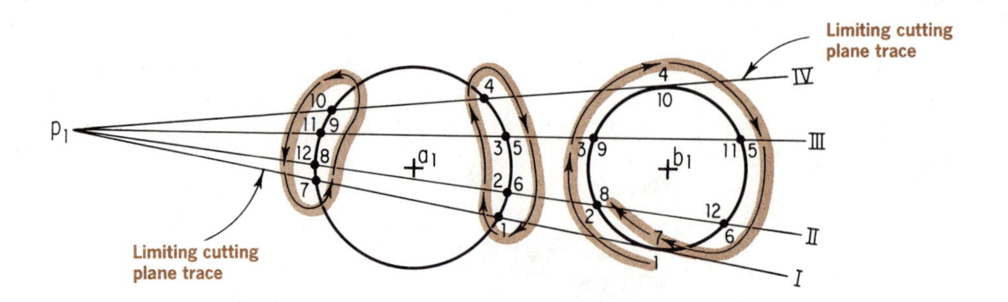

FIGURE 171

tions may be used), you will notice that the points determined by each trace on the outer edge of the base are numbered consecutively through #6, at which point the numbering then proceeds to the inside of the base from points #7–#12. The secant points determined on the cone base center a_1 start with the first point and form a complete loop of numbers, #1–#6. Another loop of numbers at the other secant points is numbered consecutively from #7–#12. On the basis of this numbering system, you can conclude that there will be two *separate* lines of intersection between the two given cones (see Fig. 168).

e. Referring to Fig. 170, construct elements from vertex o_1 to each point on the cone base center b_1 and from vertex v_1 to each point on the cone base center a_1.

f. Where like-numbered elements intersect, determine points common to both cones. These points are located on the lines of intersection between the two cones. (For example, element o_1–5 intersects element v_1–5 at point #5 on the line of intersection.)

g. Project the tangent and secant points determined by the cutting planes on the base of each cone from view #1 to view #2.

h. From vertex v_2, construct elements to each point determined by the cutting planes on the cone base center a_2.

i. From vertex o_2, construct elements to each point determined by the cutting traces on the cone base center b_2.

j. Where like-numbered elements intersect, determine points on the line of intersection between the two cones in view #2.

k. Construct smooth curves to connect all points in views #1 and #2 in numerical order, thus determining the "shape" of the lines of intersection. Determine the visibility of the lines of intersection by considering the visibility of *all* elements on *both* cones. The visible elements in view #1 include points #1, and #7 through #12 on the cone base center a_1. The visible elements for the cone base center b_1 include points #1, #2, #5–#8, #11, and #12. From the tabular summary, you can see that the visible points on the lines of intersection in view #1 are the circled numbers #1, #7, #8, #11, and #12 (see the visibility table in view #1 of Fig. 170). Therefore, the portion of the line of intersection passing through these points is also visible.

To determine the visible elements in view #2, you must refer to the bases of both cones in view #1, considering the front half of each base (nearest the frontal plane). In cone base center a_1 you see that in view #2, points #1, #2, #3, #5, and #6 are visible and that in cone base center b_1, points #1, #2, #5,–#8, #11, and #12 are visible. From the tabular summary, you can see that the circled numbers #1, #2, #5, and #6 are visible in view #2 (see visibility table in view #2 of Fig. 170). Therefore, the portion of the line of intersection passing through these points is also visible.

70 Determination of the line of intersection between two oblique cones having bases in different planes

Cutting planes that pass through the vertices of two intersecting cones having bases in different planes must be utilized in order to determine the line of intersection between these cones. The planes in which the bases of each cone lie, should appear as edges *and* other than edges in the given or constructed orthographic views. In the views in which the base planes do appear as edges, you can determine the points at which the vertex line intersects the base planes. In the views in which the base planes do not appear as edges, you can locate the *traces* of the cutting planes on each base plane. Using the cutting plane traces, you can determine straight line elements on both cones. The intersection of these straight line elements establishes points on the line of intersection between the two cones (see Fig. 172).

The cutting planes must hinge on the vertex line and must also intersect the base planes of both cones. Figure 173 depicts three cutting planes and their traces on base plane A of the cone having vertex V, and on base plane B of the cone having vertex O. Note that the *traces* of the cutting planes intersect on the *line of intersection* between the *base planes.* Note also that the traces *radiate* from points S and R, which are, respectively, the points of intersection between the vertex line and the base planes A and B.

Orthographic Example: Determination of the line of intersection of two oblique cones, the bases of which lie in different planes (Figs. 174–177, inclusive).

Given Data: Horizontal, frontal, and profile projections of two intersecting cones having vertices V and O and bases in two planes that make 90° with each other.

Required:

a. The line of intersection between the two cones in all given views.

b. The visibility of the line of intersection and the two cones in all given views.

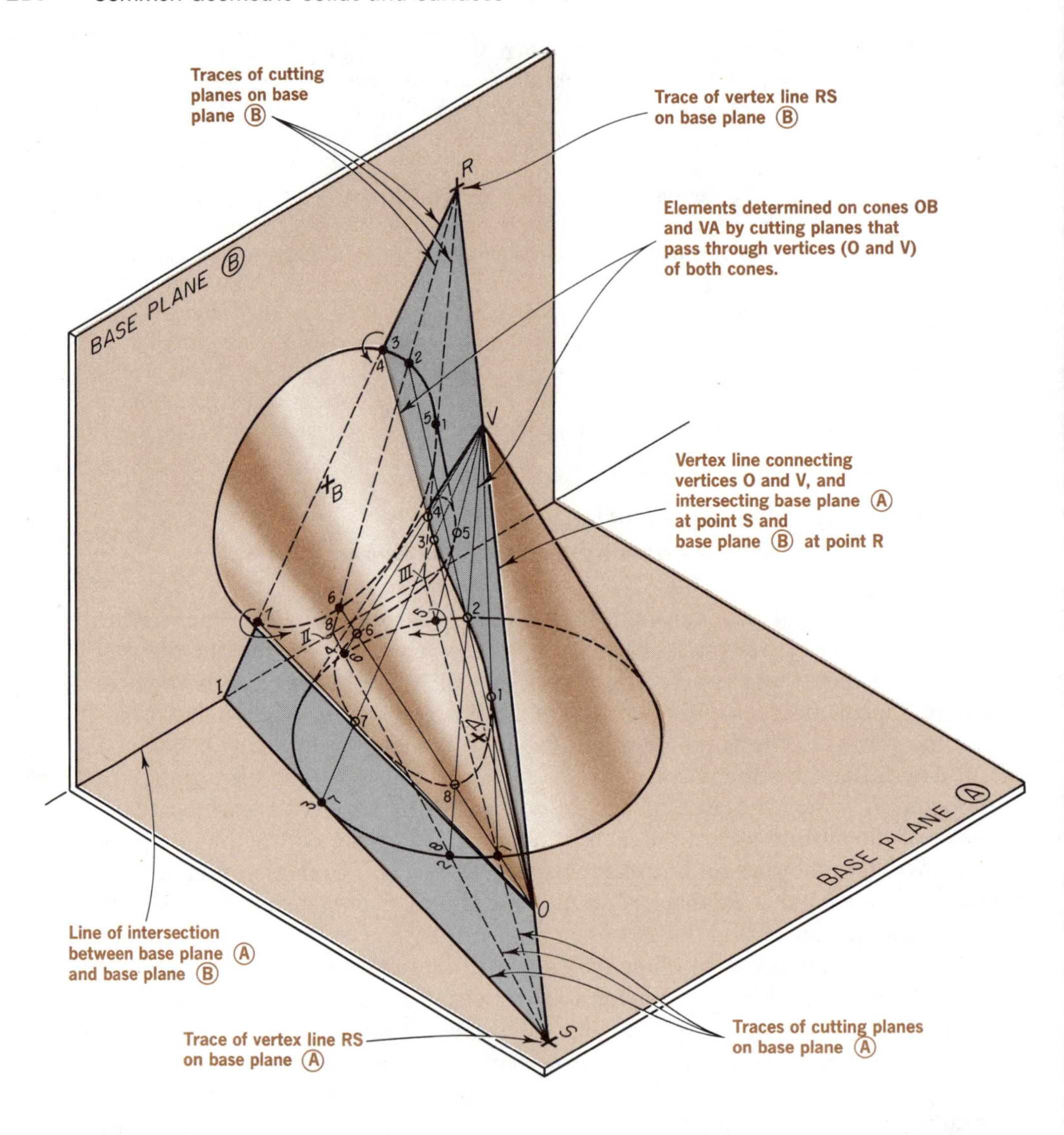

FIGURE 172

Construction Program:

(NOTE: The method of solving this problem has been divided into a series of steps and sequential figures in order to facilitate its presentation. Each step presented should be carefully related to the appropriate figure.)

a. Starting with view #2 in Fig. 174, connect vertices o_2 and v_2 with a straight (vertex) line, extending it until it intersects the base planes of each cone at point s_2 on base plane A, and at point r_2 on base plane B.

FIGURE 173

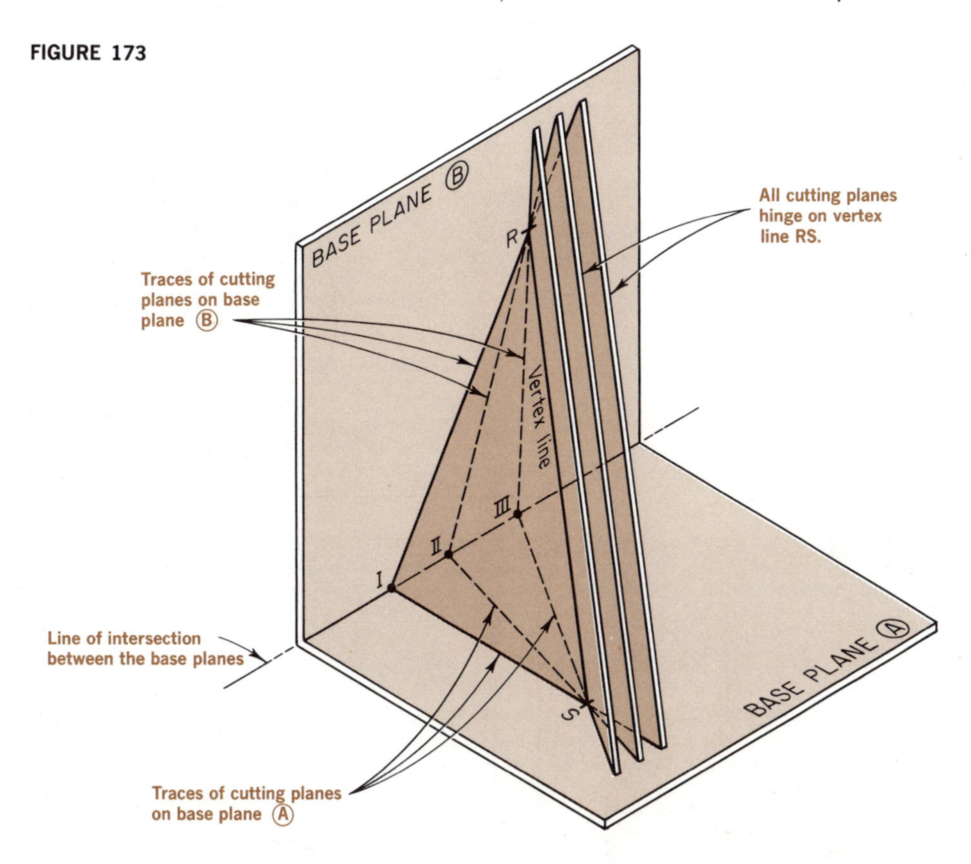

b. Determine, by direct projection from view #2, the vertex line in views #1 and #2, as well as points s_1 and r_1 in view #1, and s_3 and r_3 in view #3 (see Fig. 174).

c. In view #3, the base of the cone with vertex o_3 appears as a circle. The base center is b_3. Draw a trial cutting plane trace tangent to the base at x_3; the trace should start at r_3 and intersect base plane A at w_3. Locate the trace of this trial cutting plane in view #1 by measuring distance W in view #3, and then transferring this distance to view #1 (Fig. 175). The trial trace on plane A in view #1 is s_1w_1, and you can see from this that it does *not* touch the base of the cone having vertex v_1 and cone base with center a_1. The trial trace indicates that this cutting plane does *not* establish straight line elements on *both* cones; therefore it *cannot* be used.

d. In view #3, starting from r_3, draw a trace of a limiting cutting plane tangent at point #1 to the base with center b_3 of the cone having vertex o_3. This trace intersects base plane A at e_3.

e. Locate the trace of the limiting cutting plane on base plane A in view #1 by transferring distance X from view #3 to view #1, thus locating e_1. The trace of this limiting cutting plane on base plane A is s_1e_1.

FIGURE 174

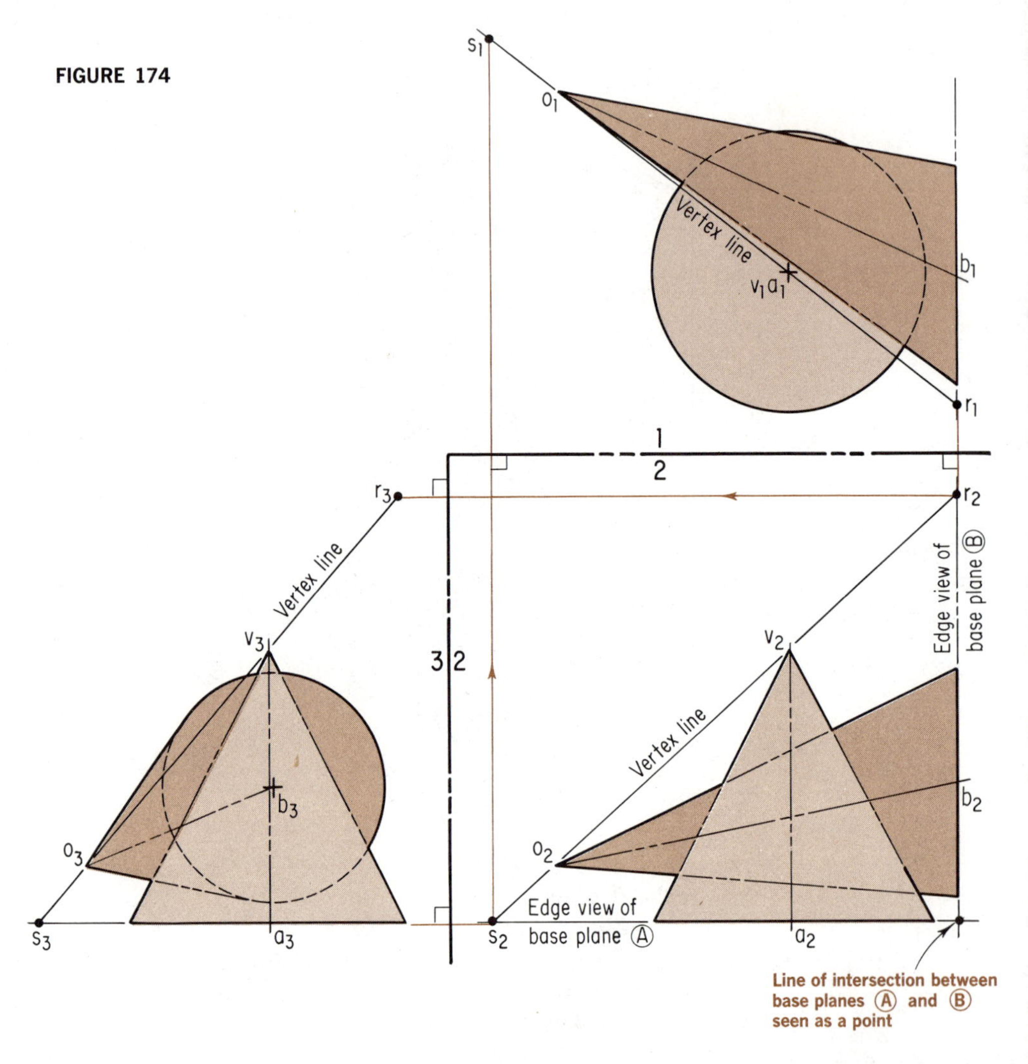

You can now see that it is *secant* to the cone base with center a_1. The cutting plane therefore establishes straight line elements on *both* cones.

f. In view #1, draw a trace from s_1 tangent at point #3 to the base with center a_1 of the cone with vertex v_1. This trace intersects the edge view of base plane B at g_1.

g. Locate the trace of this cutting plane in view #3 by transferring distance Y from view #1 to view #3, thus locating g_3. The trace of this limiting cutting plane on base plane B is r_3g_3. The trace is *secant* to the base with cone base with center b_3. Therefore, this is a valid *limiting cutting plane,* since it establishes straight line elements on *both* cones.

FIGURE 175

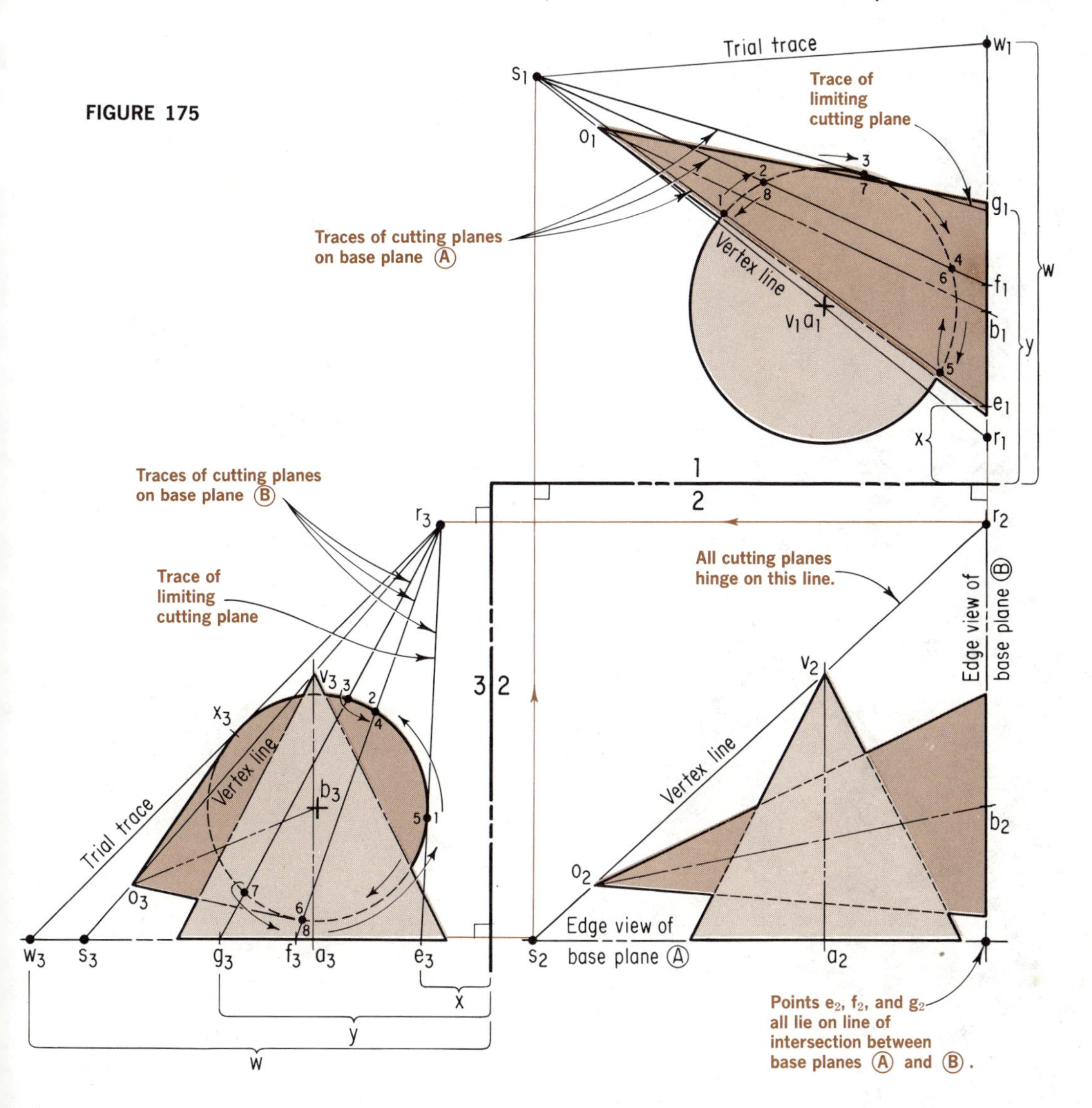

h. In view #3, draw a convenient number of cutting plane traces between traces r_3e_3 and r_3g_3. (We used only one additional cutting plane trace, r_3f_3, in our figure in order to maintain clarity of presentation. Normally, six to ten cutting planes would be required for a more accurate result.)

i. Locate the trace of this last cutting plane in view #1 on the base plane *A* of the cone with vertex v_1. This trace is s_1f_1. In view #3, starting at the first tangent point, number all the traces consecutively (from #1–#8, and then back to #1).

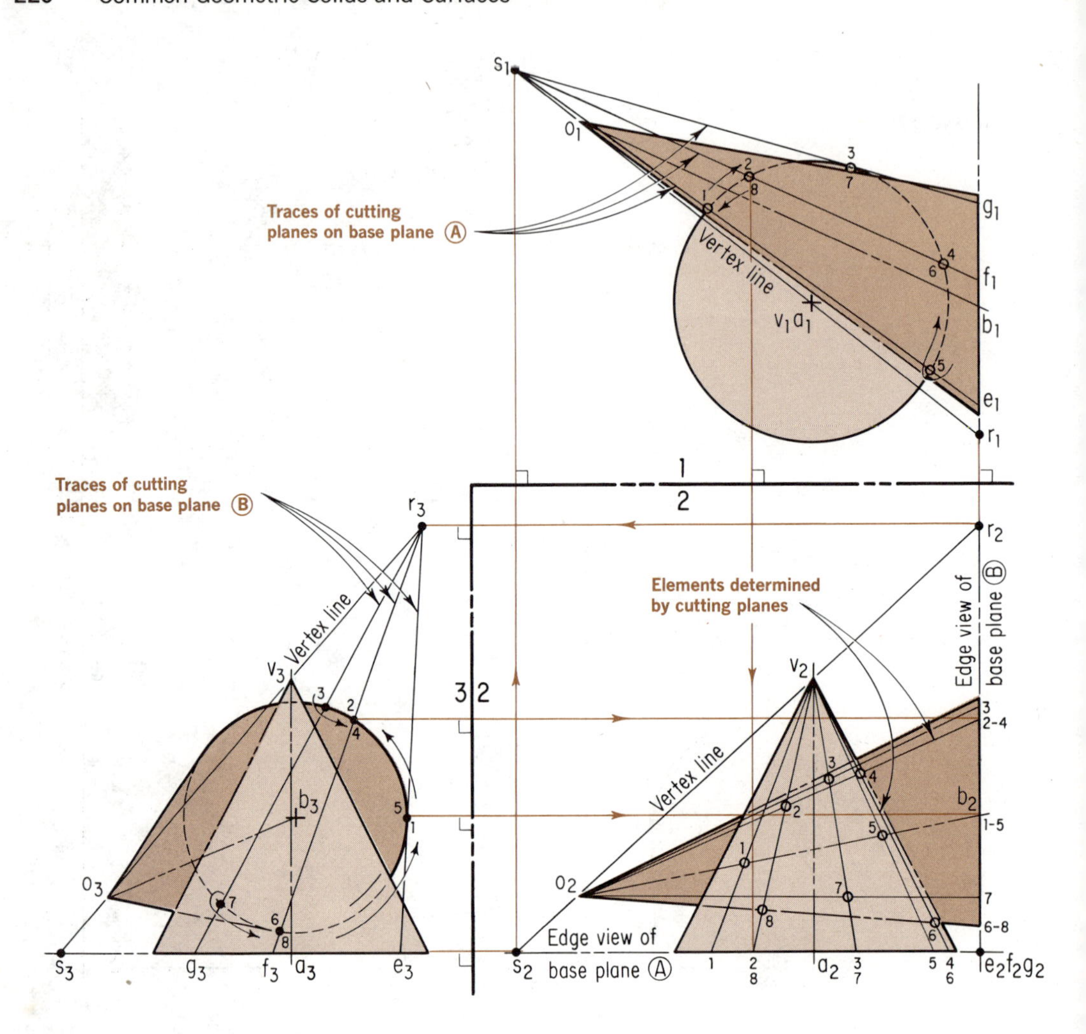

FIGURE 176

Since point #1 is on the trace containing e_3, it must also be on the trace containing e_1 in view #1 (point e_1 is common to both traces on both base planes). Therefore, point #1 in view #1 is a *secant* point, as indicated in Fig. 175. Continue numbering as indicated (from #1–#8, and then back to #1).

(NOTE: In solving this problem, you should constantly distinguish between traces on base planes, the vertex line, and straight line elements.)

j. From view #3, project points #1–#8 on cone base with center b_3 to view #2. From view #1, project points #1–#8 on the cone base with center a_1 to view #2 (see Fig. 176).

FIGURE 177(a)

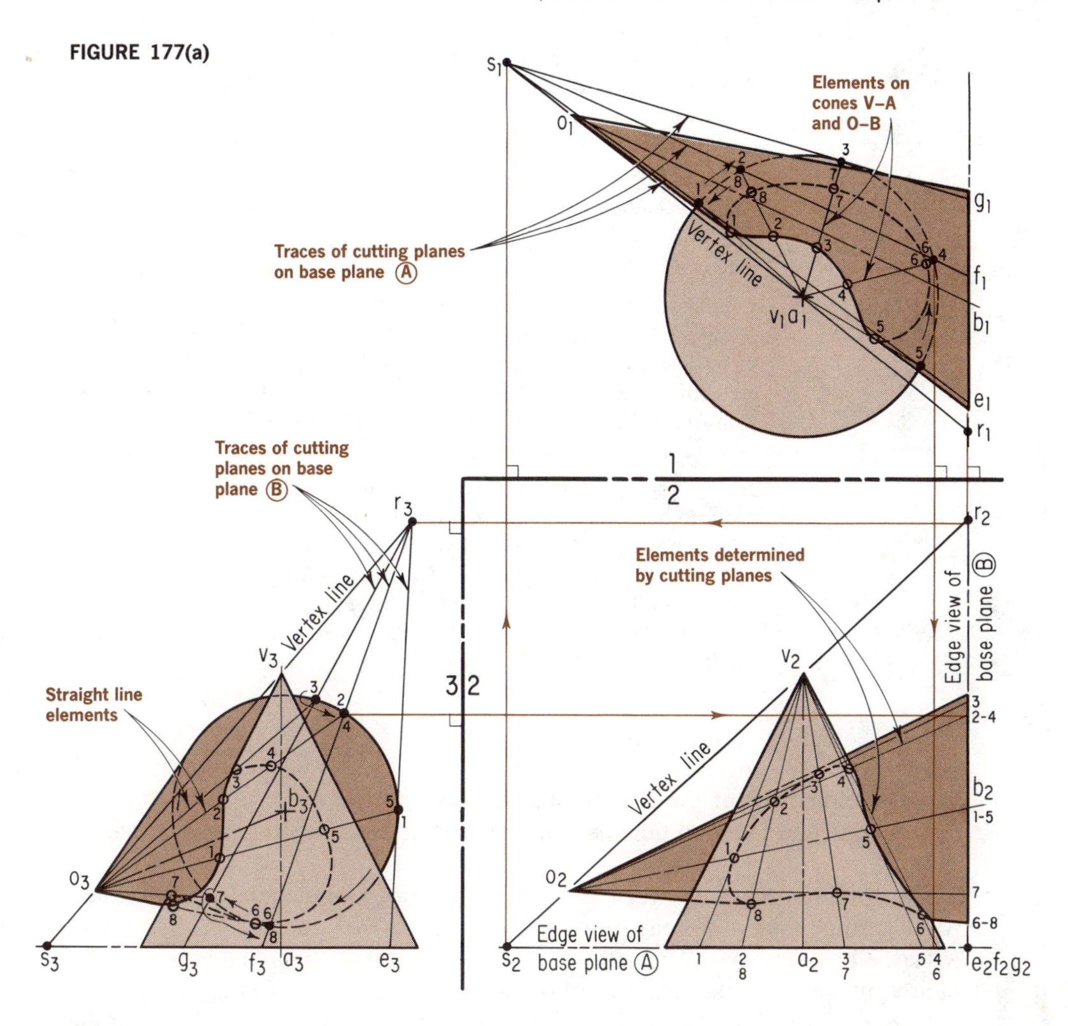

FIGURE 177(b)

VISIBILITY TABLE	
Visible elements in view #1	
Cone OB (See view #3)	Cone VA (See view #1)
① ④	① ⑤
② ⑤	② 6
③	③ 7
	④ 8

VISIBILITY TABLE	
Visible elements in view #2	
Cone OB (See view #3)	Cone VA (See view #1)
1	⑤
2	
3	
4	
⑤	

VISIBILITY TABLE	
Visible elements in view #3	
Cone OB (See view #3)	Cone VA (See view #1)
① 5	①
② 6	②
3 7	⑧
4 ⑧	

k. From both cone vertices (o_2 and v_2, respectively), construct straight line elements to the numbered points on each respective cone base, as shown in view #2. Where like-numbered elements intersect, points on the line of intersection between the given cones are established. These points are numbered and circled (Fig. 176).

l. Connect circled points #1 through #8 with a light, smooth curved line, thus defining the shape of the line of intersection between the two cones (visibility to be determined).

m. Establish, by projection from view #2, points on the line of intersection between the two cones in views #1 and #3. In each of these views, elements are established on each cone. Where like-numbered elements intersect, points on the line of intersection are established. Connect these points with a light, smooth curved line.

n. Determine the visibility of the line of intersection in all views by establishing visibility tables as indicated in Fig. 177(b). (REMEMBER: in order to be visible, a point must be determined by visible elements on *both* cones.) From the tables you can note that in view #1, points #1–#5 are visible on the line of intersection. In view #2, only point #5 is visible. In view #3, points #1, #2, and #8 are visible.

Determine the visibility of the cones by inspection. This completes the requirements of the problem.

71 Determination of the line of intersection between an oblique cone and an oblique cylinder having bases in the same plane

The line of intersection between an intersecting cone and cylinder is determined by the cutting planes that pass through the *vertex* of the cone and that are *parallel* to the cylinder *axis.* The intersection of the straight line elements that the planes cut on both the cylinder and the cone establishes points common to both solids. These points, therefore, are on the line of intersection between the cone and the cylinder.

Cutting planes that cut straight line elements on both the cylinder and the cone hinge on the line that passes through the *vertex* of the cone *parallel* to the cylinder *axis.* Their traces on the base plane radiate from the point of intersection of the hinge line on the base plane (see Fig. 178 for a pictorial representation of this condition).

Orthographic Example: Determination of the line of intersection between an oblique cone and an oblique cylinder, the bases of which lie in the same plane [Figs. 179(a) and 179(b)].

Given Data: Horizontal and frontal projections of an oblique cone having a vertex V and a base center O and intersecting an oblique cylinder having an axis AB and a base center B.

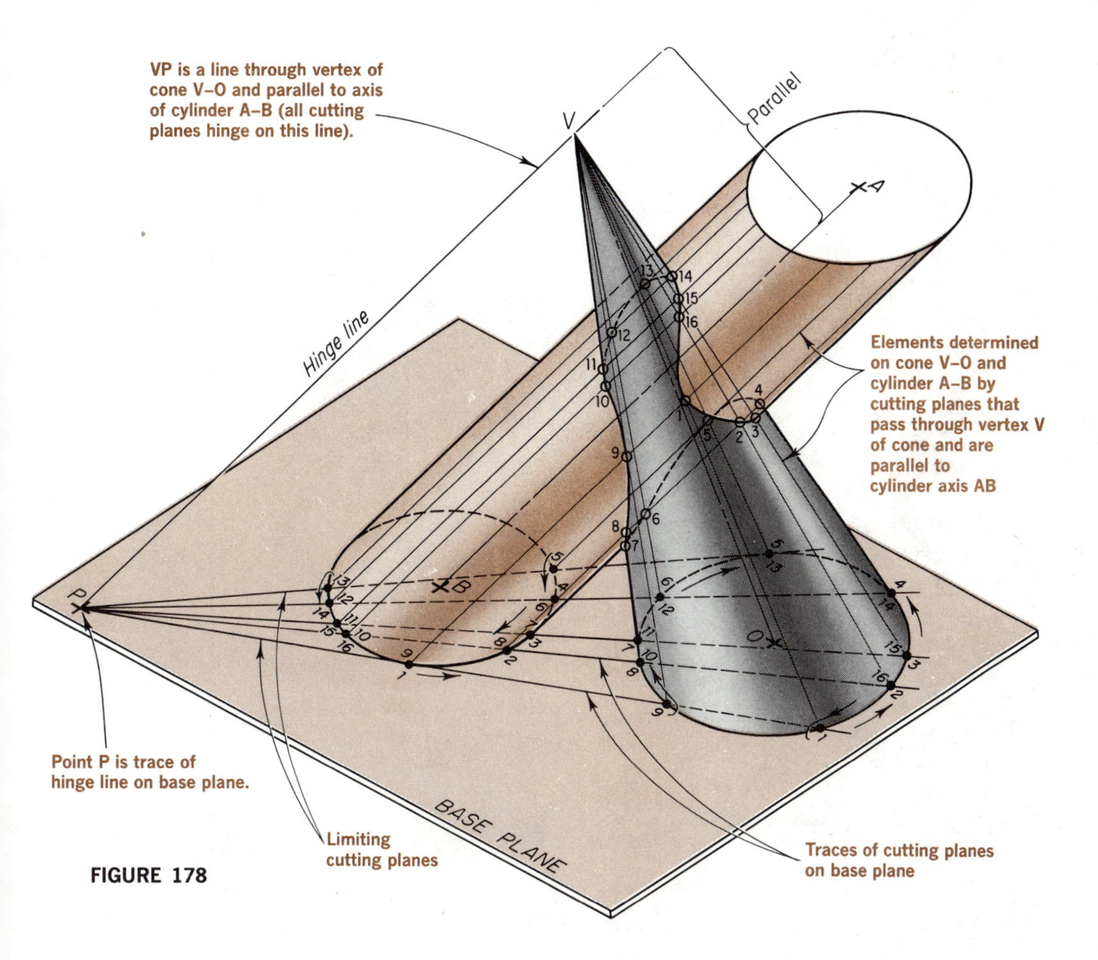

FIGURE 178

Required:

a. The line of intersection between the intersecting cone and cylinder.

b. The visibility of the line of intersection, the cone, and the cylinder.

Construction Program:

a. In view #2, construct a hinge line through vertex v_2 of the given cone parallel to the cylinder axis a_2b_2. Extend the line until it intersects the *edge view* of the base plane at point p_2.

b. In view #1, draw the horizontal projection of the hinge line through vertex v_1 parallel to cylinder axis a_1b_1.

c. Project p_2 upward to view #1 in order to determine point p_1 on the hinge line.

d. From p_1, construct a convenient number of cutting plane traces on the base plane. Note that one limiting trace is tangent to the cylinder with base center b_1, while the other limiting trace is tangent to the cone with base center o_1. [Five cutting plane traces have been shown in Fig. 179(a).]

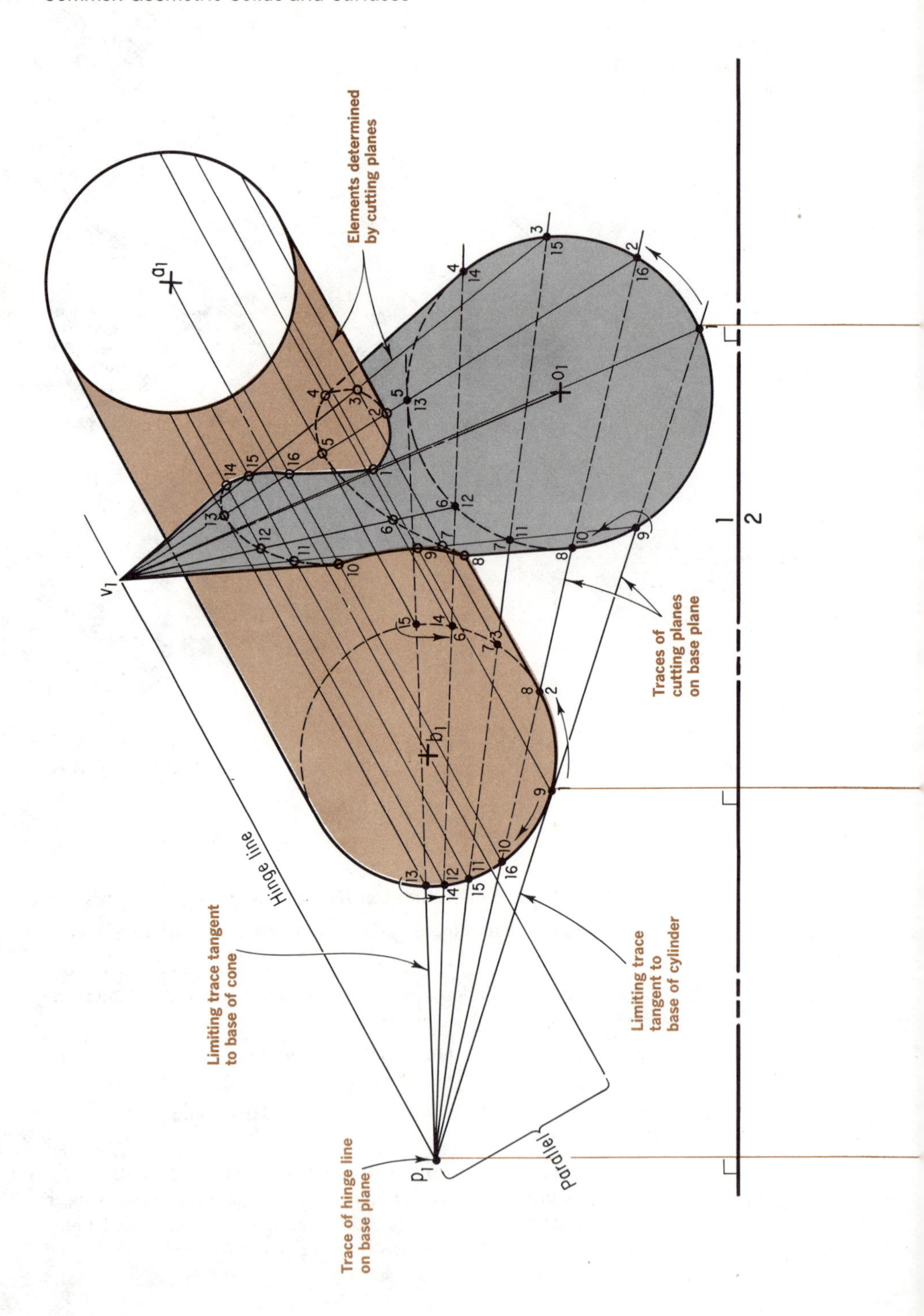
Elements determined by cutting planes
Traces of cutting planes on base plane
Limiting trace tangent to base of cylinder
Limiting trace tangent to base of cone
Trace of hinge line on base plane
Hinge line
Parallel

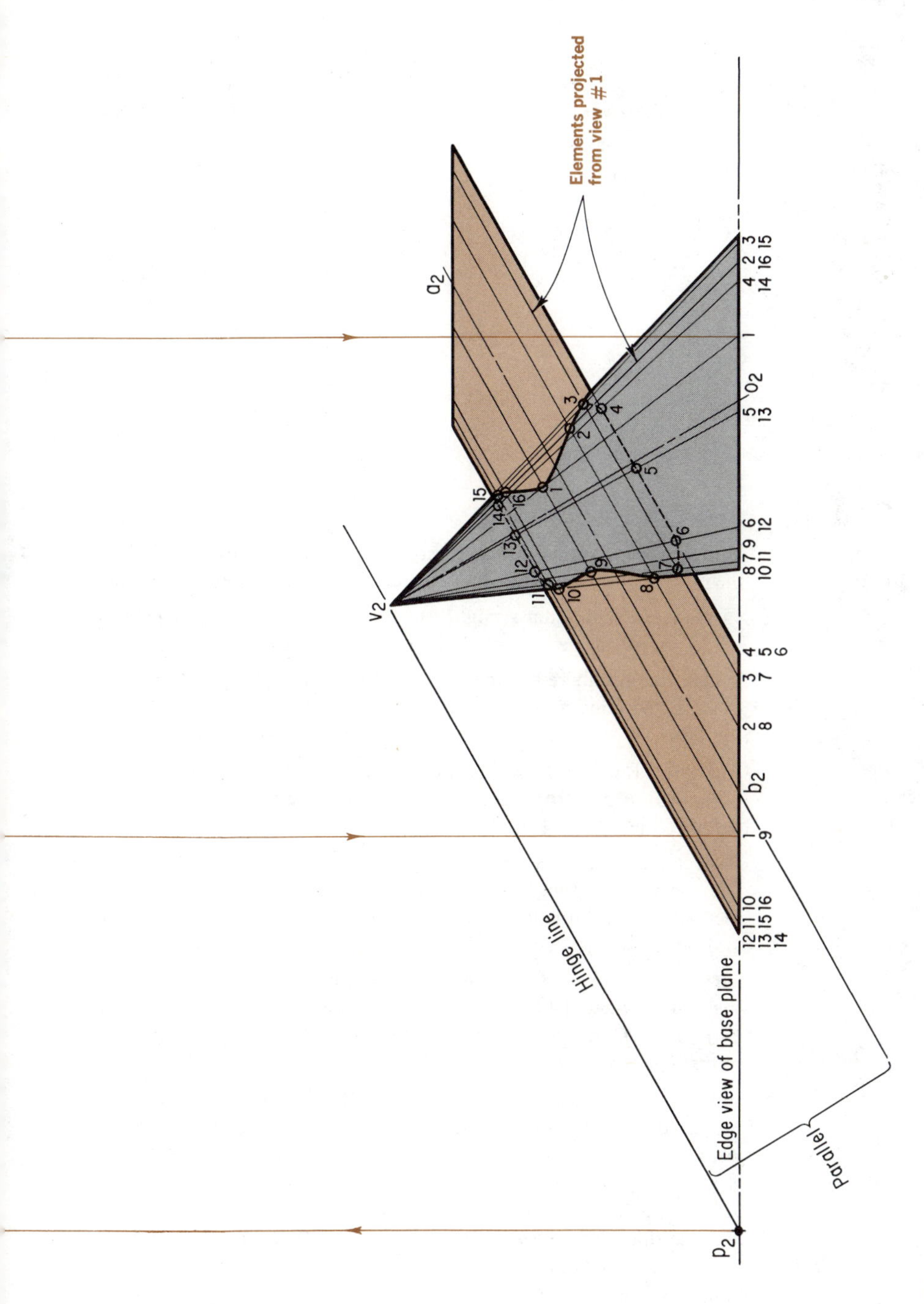
Elements projected from view #1
a2
v2
b2
o2
p2
Hinge line
Edge view of base plane
parallel

FIGURE 179(a)

VISIBILITY TABLE			
Visible elements in view #1			
Cylinder AB		Cone VO	
①	12	①	⑨
②	13	②	⑩
⑧	⑭	3	⑭
⑨	⑮	4	⑮
⑩	⑯	⑧	⑯
11			

FIGURE 179(b)

VISIBILITY TABLE			
Visible elements in view #2			
Cylinder AB		Cone VO	
①	⑩	①	⑨
②	11	②	⑩
3	12	⑧	⑯
4	13		
6	14		
7	15		
⑧	⑯		
⑨			

e. Number the tangent and secant points cut by the traces on both bases in numerical order, according to the established numbering system.

f. In view #1, construct straight line elements from each of the numbered points on the base of the cylinder and of the cone.

g. Where like-numbered elements intersect in view #1, points common to both the cone and the cylinder are determined. These points are therefore on the line of intersection between the two solids.

   In view #1, connect these points with a smooth curve, proceeding in numerical order from #1–#16, and then returning to #1, thus defining the shape of the line of intersection.

h. Project the numbered points on the base of the cylinder and the cone from view #1 to view #2. In view #2, construct straight line elements from each of the numbered points on the base of the cylinder and of the cone.

i. The intersection of like-numbered elements in view #2 determines the frontal view of the points on the line of intersection between the two solids. Connect the points with a smooth curve, proceeding in numerical order.

j. In view #1, establish the visibility of the line of intersection by determining the visible elements on the cylinder and on the cone. From the visibility table in Fig. 179(b), for view #1 you can note that elements #1, #2, #8–#16 are visible on the cylinder and that elements #1–#4, #8–#10, and #14–#16 are visible on the cone. You can also see that elements #1, #2, #8, #9, #10, #14, #15, and #16 are visible on *both* the cylinder and the cone as indicated by the circled numbers. Therefore, the points determined by the intersecting straight line elements occur on the *visible* portion of the line of intersection in view #1.

k. In view #2, establish the visibility of the line of intersection by determining the visible elements on the cylinder and on the cone. From the visibility table in Fig. 179(b), for view #2 you can note that elements #1–#4 and #6–#16 are visible on the cylinder and that elements #1, #2, #8, #9, #10, and #16 are visible on the cone. You can also see that elements #1, #2, #8, #9, #10, and #16 are visible on both the cylinder and the cone as indicated by the circled numbers. Therefore, these elements occur on the visible portion of the line of intersection in view #2.

l. Determine by inspection the visibility of the cone and the cylinder in both views. This completes the requirements of the problem.

72 General summary of possible numbering systems for intersections of cylinders, cones, and cylinders with cones

Figures 180 through 184 pictorially present the various types of possible intersection between two cones. The same numbering system shown in these figures can be applied to intersections of cylinders, cones, and cylinders with cones. In each type of intersection, you must properly determine the cutting plane traces. When a cone intersects another cone, all the cutting planes hinge on the vertex line. When a cylinder intersects another cylinder, the cutting planes must be parallel to both cylinder axes. When a cone intersects a cylinder, the cutting planes hinge on a line that passes through the vertex of the cone and that is parallel to the cylinder axis.

Note particularly that from the numbering system you can predict what type of lines of intersection will result. For example, in Fig. 180, there are two separate lines of intersection where the limiting cutting planes are both secant to one base and tangent to the other.

The numbering system indicated in Fig. 181 shows two separate lines of intersection where the limiting planes are *both* tangent to one base and secant to the other.

In Fig. 182, one of the limiting traces is tangent to one base and secant to the other, while the other limiting trace is secant to one base and tangent to the other. The result is a continuous line of intersection, indicated in the figure by a continuous loop of numbers.

In Fig. 183, the limiting traces are *both* tangent to one base and tangent and secant to the other. This results in a continuous line of intersection that crosses itself at the point of tangency of the two cones. (The point of tangency is established by the limiting traces tangent to both cones.) This is indicated in the figure by a continuous loop of numbers.

Figure 184 illustrates limiting traces that are tangent to both bases. Note that the numbering system is consecutive in the *same* direction on one base, and also consecutive in the *reverse* direction on the other base. The result is two separate lines of intersection that cross each other at the two points of tangency of the two cones. (The points of tangency are determined by the limiting traces tangent to both cones.)

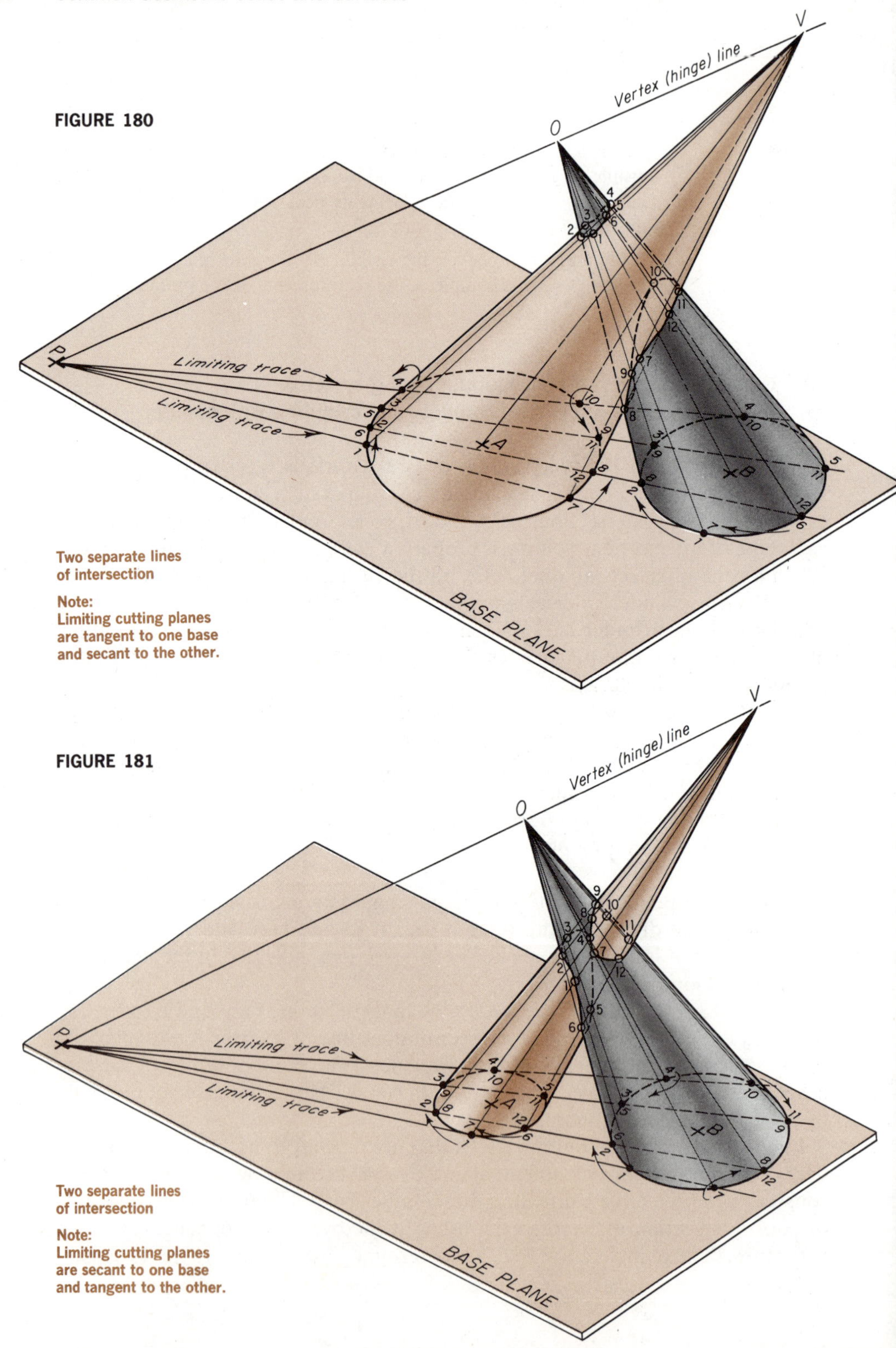

FIGURE 180

Two separate lines of intersection

Note:
Limiting cutting planes are tangent to one base and secant to the other.

FIGURE 181

Two separate lines of intersection

Note:
Limiting cutting planes are secant to one base and tangent to the other.

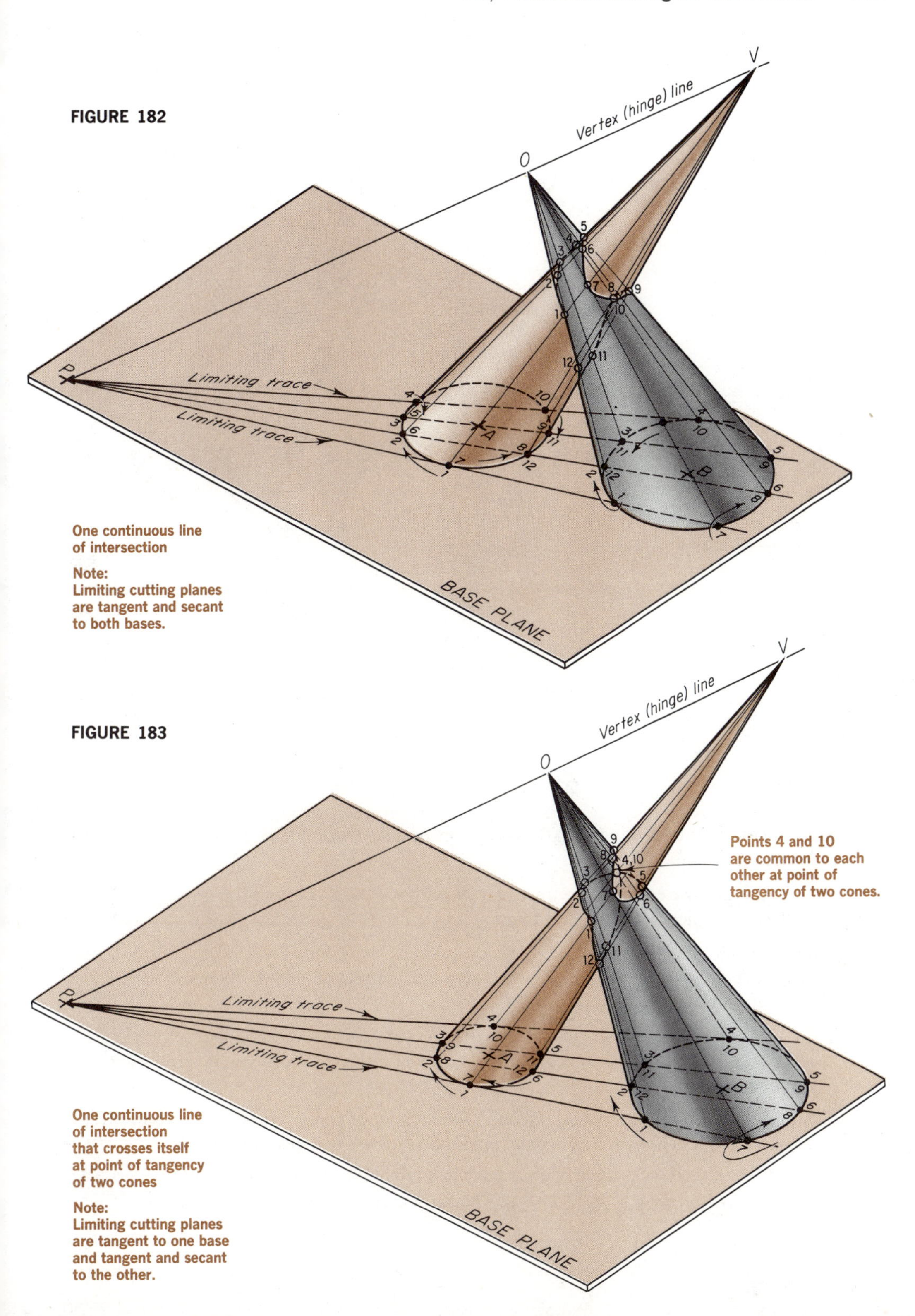

FIGURE 182

One continuous line of intersection

Note:
Limiting cutting planes are tangent and secant to both bases.

FIGURE 183

One continuous line of intersection that crosses itself at point of tangency of two cones

Note:
Limiting cutting planes are tangent to one base and tangent and secant to the other.

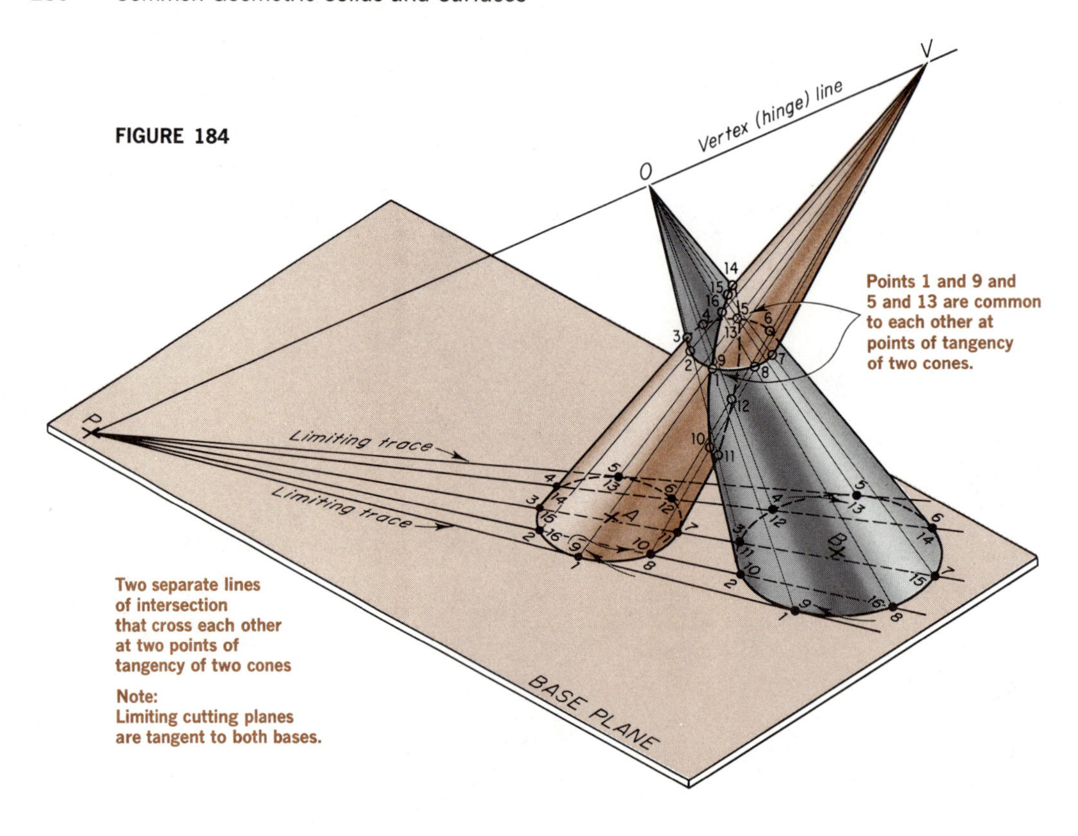

FIGURE 184

SAMPLE QUIZZES

1. Describe the view in which the true shape of the curve cut from an oblique cone by an intersecting plane is seen.

2. How may we establish the trace of a plane that is parallel to two intersecting cylinders, the bases of which lie in the same plane?

3. Explain briefly (without a sketch) how to set up a plane that will indicate the direction of cutting plane traces in order to determine the intersection of two oblique cylinders, the bases of which lie in a common frontal plane.

4. Draw freehand top and front views of two cylinders of revolution that form a figure 8 when they intersect. Show by actual construction how this intersection is determined. Label all points.

5. GIVEN: Two intersecting right cones, the vertices of which are *A* and *B*, respectively. The base of cone *A* lies in a horizontal plane and the base of cone *B* in a profile plane.

a. How can you establish planes that cut straight line elements from both right cones?

b. Illustrate your answer with an orthographic sketch.

6. Sketch two circles that will represent bases of two cylinders. Show the limiting cutting plane traces that occur when two limiting cutting planes are tangent to two cylinders. Apply the established numbering system and state what type of intersection will result under these conditions.

7. Sketch the horizontal and frontal views of a triangular prism that intersects a right cone; then determine the line of intersection between the two solids, using an end view of the prism.

8. Sketch the horizontal and frontal views of two intersecting cylinders, the base planes of which are parallel to each other and to the horizontal projection plane. (Assume that the position of one base plane is higher than that of the other.) Establish limiting cutting plane traces in both views.

PRACTICE PROBLEMS

1. Given Data: Horizontal and frontal projections of a triangular oblique prism and an intersecting plane *WXY* (visibility incomplete).

Required: A. Using only the given views, draw the horizontal and frontal projections of the line of intersection between the prism and the plane.

PROBLEM 1

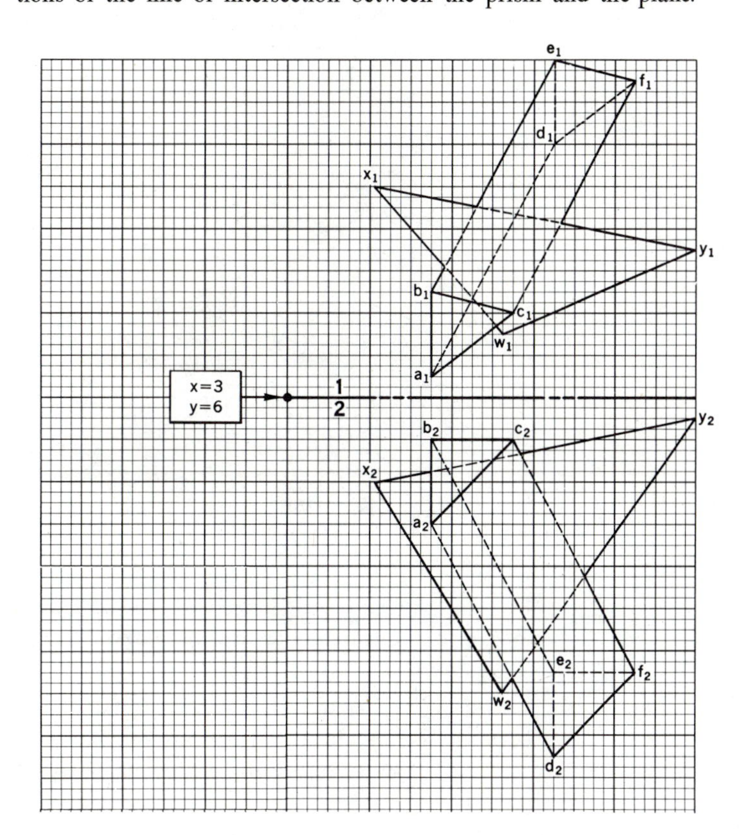

B. Indicate the visibility of the line of intersection in the given views.

C. Indicate the visibility of the prism and the plane in the given views.

D. Check your results with the edge view method.

E. Label and show all construction.

2. Given Data: Horizontal and frontal projections of an oblique triangular pyramid and an intersecting plane *WXY* (visibility incomplete).

Required: A. Using only the given views, draw the horizontal and frontal projections of the line of intersection between the given pyramid and plane.

B. Indicate the visibility of the line of intersection in the given views.

C. Indicate the visibility of the pyramid and the plane in the given views.

D. Check your results with the edge view method.

E. Label and show all construction.

PROBLEM 2

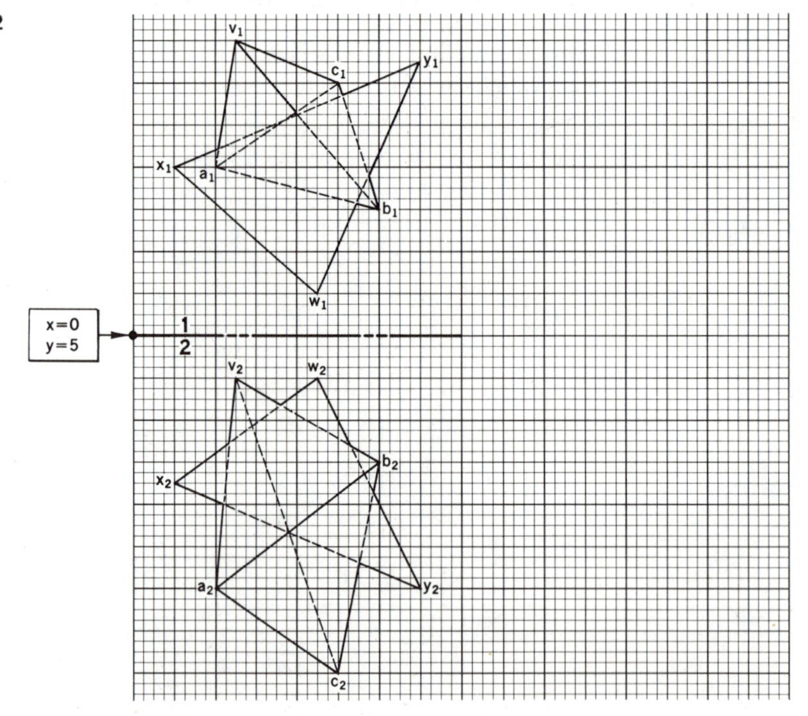

3. Given Data: Horizontal and frontal projections of an oblique cone and an intersecting triangular prism. Assume that the cone and the prism are opaque and of one piece.

Required: A. Using the end view method, determine the horizontal and frontal projections of the line (or lines) of intersection between the given cone and prism. (HINT: Consider the cone to be a multi-sided pyramid.)

B. Indicate the visibility of the line of intersection in all views.

C. Indicate the visibility of the cone and the prism in all views.

D. Label and show all construction.

PROBLEM 3

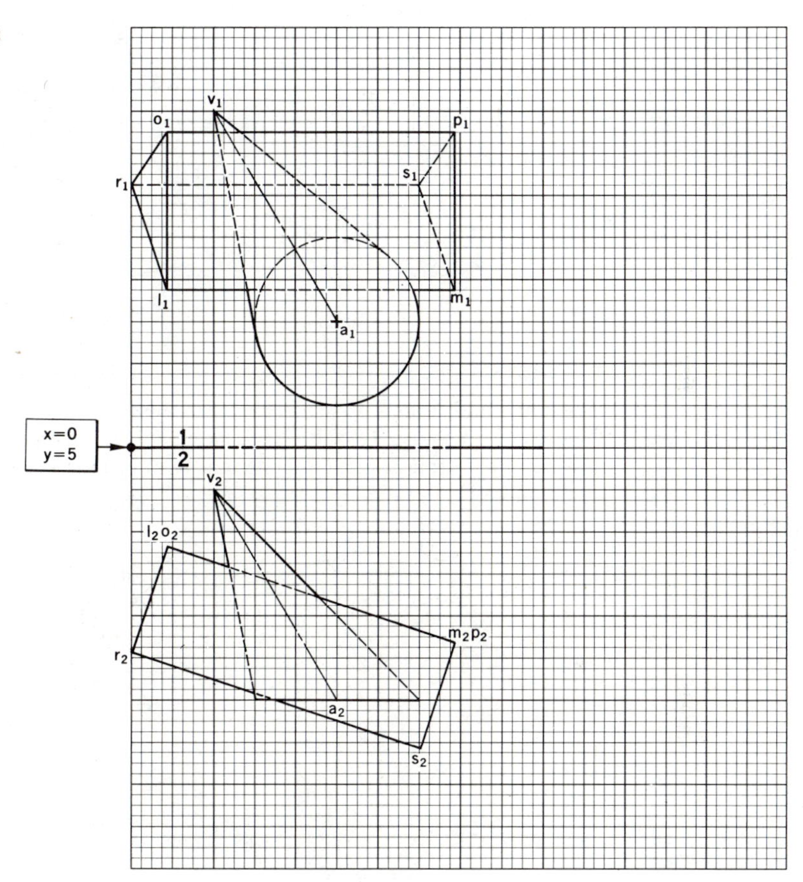

4. Given Data: Horizontal and frontal projections of an oblique triangular pyramid and an intersecting oblique triangular prism. (The visibility of both solids is incomplete.) Assume that the pyramid and prism are opaque and of one piece.

Required:

A. Using the end view method, draw the horizontal and frontal projections of the line of intersection between the given pyramid and prism. (SUGGESTION: Complete the general visibility of the given pyramid and prism first.)

B. Indicate the visibility of the line of intersection in all views.

C. Indicate the visibility of the pyramid and the prism in all views.

D. Label and show all construction.

PROBLEM 4

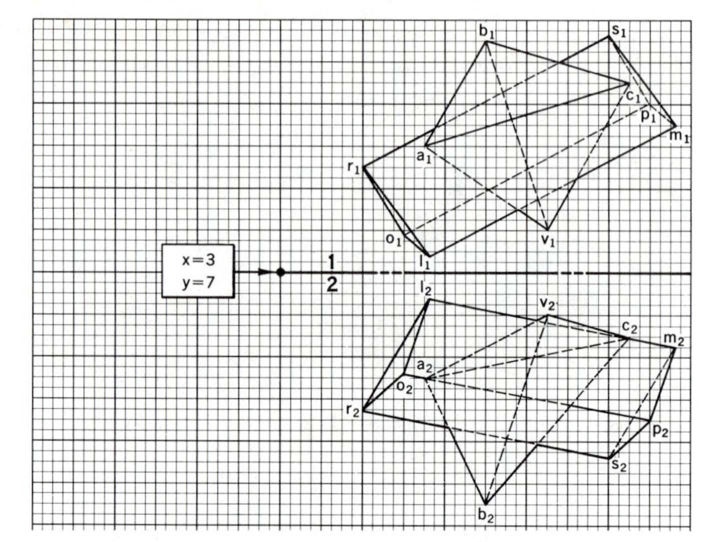

5. Given Data:

Horizontal and frontal projections of two intersecting oblique cylinders, the axes of which are AB and CD, respectively. The bases of both cylinders lie in the same plane. (Assume that both cylinders are opaque and of one piece.)

PROBLEM 5

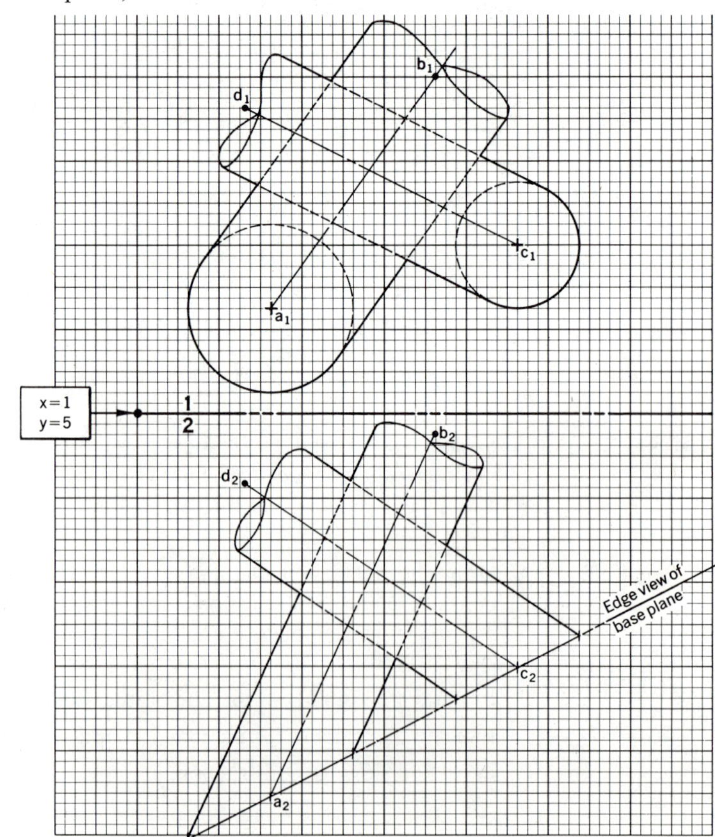

Required:

A. The horizontal and frontal projections of the line (or lines) of intersection between the given cylinders.

B. Indicate the visibility of the line of intersection in the given views.

C. Indicate the visibility of each cylinder in the given views.

D. Label and show all construction, including the numbering system and visibility tables.

6. Given Data: Horizontal and frontal projections of two oblique intersecting cylinders, the axes of which are *AB* and *CD*, respectively. The bases of both cylinders lie in the horizontal projection plane. Assume that both cylinders are opaque and of one piece.

Required:

A. Draw the cutting planes traces that will indicate the straight line elements common to both cylinders. Predict what type of line (or lines) of intersection exists between the given cylinders.

B. Draw the horizontal and frontal projections of the line of intersection between the cylinders.

C. Indicate the visibility of the line of intersection in the given views.

D. Indicate the visibility of each cylinder in the given views.

E. Label and show all construction, including the numbering system and visibility tables.

PROBLEM 6

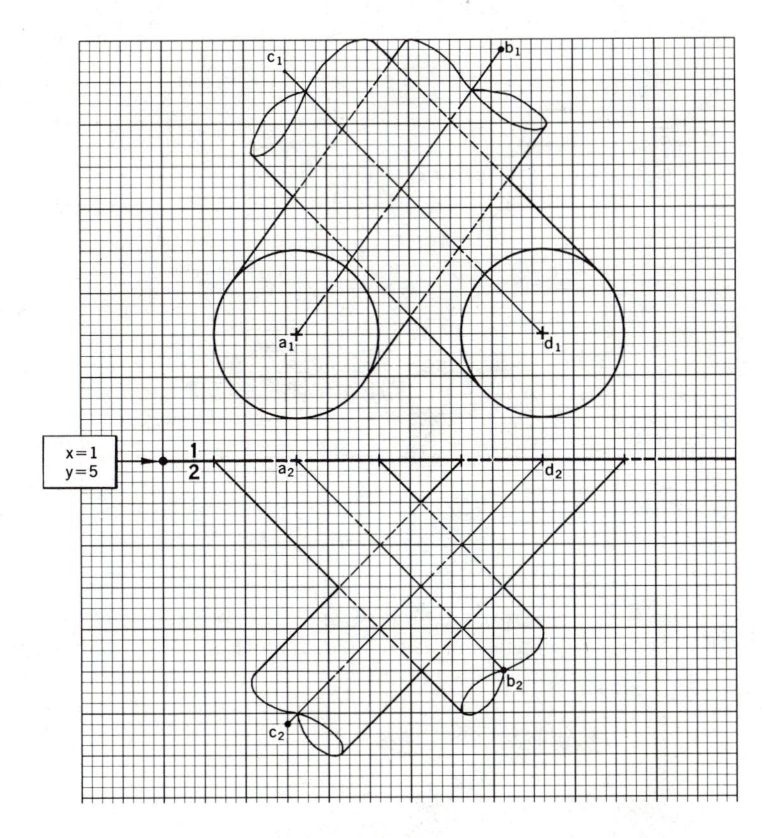

7. Given Data: Horizontal and frontal projections of two oblique intersecting cones, the vertices of which are *V* and *O*, respectively. The bases of both cones lie in the frontal projection plane. (Assume that both cones are opaque and of one piece.)

Required:

A. Draw the cutting plane traces that will indicate the straight line elements common to both cones. Predict what type of line (or lines) of intersection exists between the given cones.

B. Draw the horizontal and frontal projections of the line of intersection between the cones.

C. Indicate the visibility of the line of intersection in the given views.

D. Indicate the visibility of each cone in the given views.

E. Label and show all construction, including the numbering system and visibility tables.

PROBLEM 7

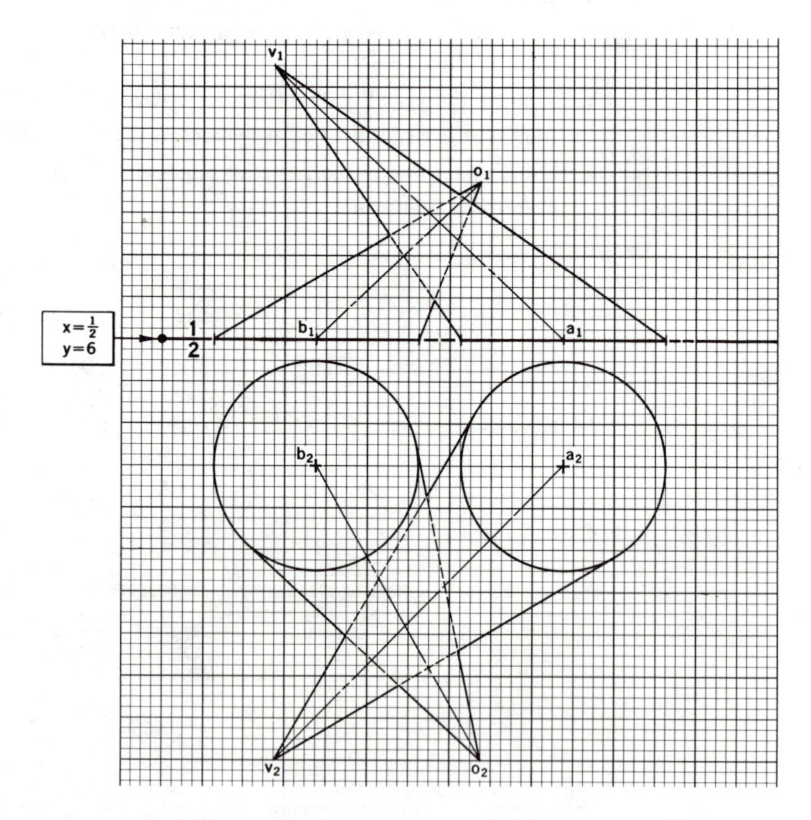

8. Given Data: Horizontal, frontal, and right profile projections of two intersecting oblique cones, the vertices of which are *V* and *O*, respectively. Base *B* of cone *O* is parallel to the horizontal projection plane, and base *A* of cone *V* lies in the right profile projection plane. (Assume that both cones are opaque and of one piece.)

Required:

A. Draw the cutting plane traces that will indicate the straight line elements common to both cones. Predict what type of line (or lines) of intersection exists between the given cones.

B. Draw the horizontal, frontal, and right profile projections of the line of intersection between the cones.

C. Indicate the visibility of the line of intersection in all views.

D. Indicate the visibility of each cone in all views.

E. Label and show all construction, including the numbering system and visibility tables.

PROBLEM 8

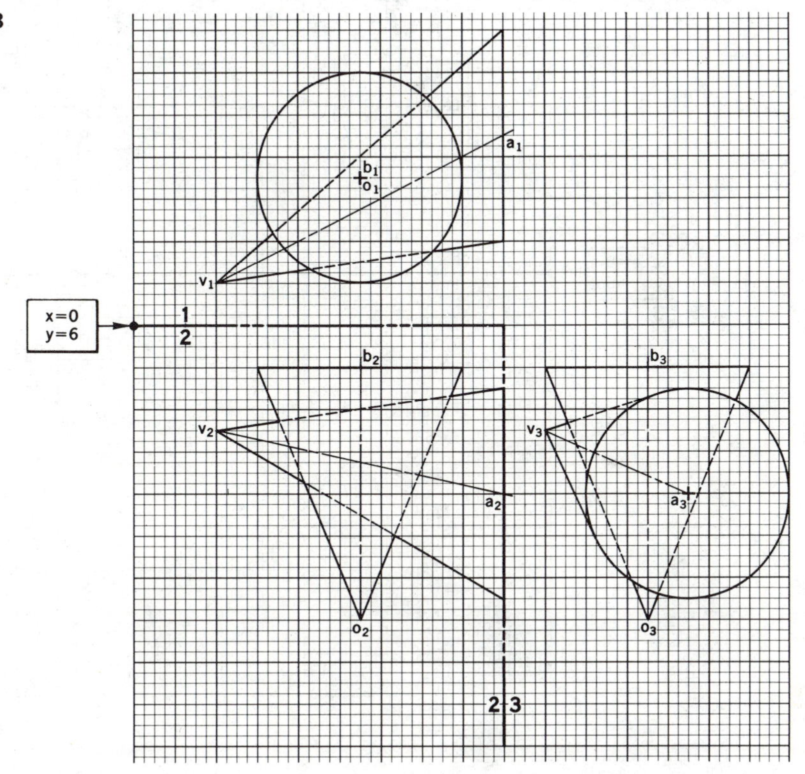

9. Given Data: Horizontal and frontal projections of an oblique cone with vertex *V* and an intersecting cylinder with axis *BC*. Bases of both solids are in the same plane. Assume both the cone and the cylinder to be opaque and of one piece.

Required:

A. Draw the cutting plane traces that will indicate the straight line elements common to the cone and the cylinder. Predict what type of line (or lines) of intersection exists between the given cone and cylinder.

B. Draw the horizontal and frontal projections of the line of intersection between the cone and the cylinder.

C. Indicate the visibility of the line of intersection in the given views.

D. Indicate the visibility of the cone and cylinder in the given views.

E. Label and show all construction, including the numbering system and visibility tables.

PROBLEM 9

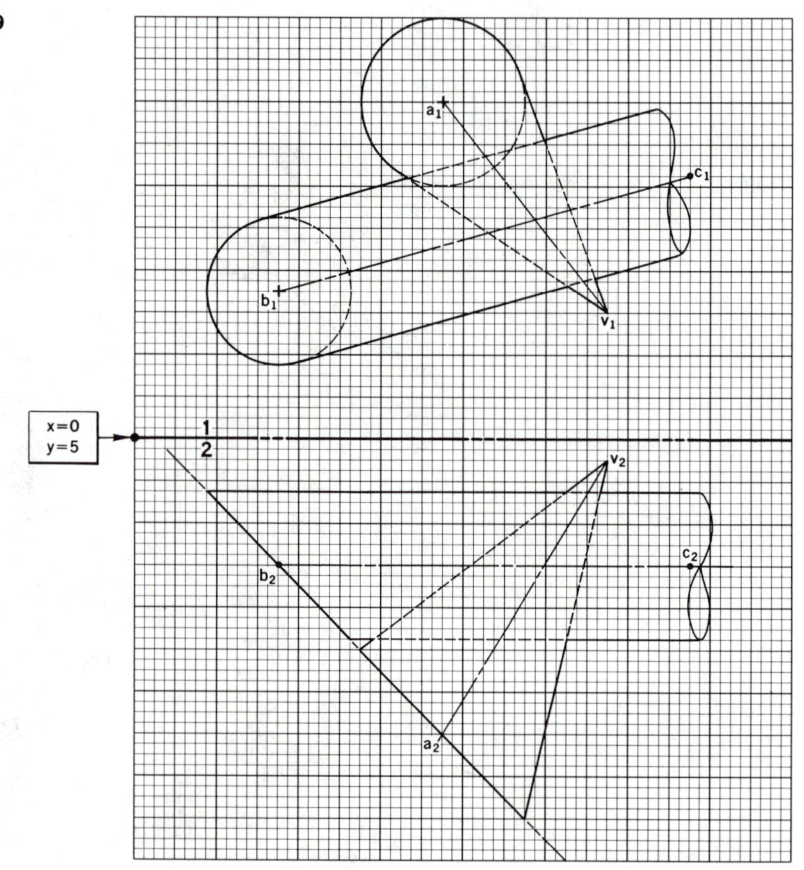

10. Given Data: Horizontal and frontal projections of an oblique cone having a vertex V and an intersecting cylinder having an axis BC. The bases of both solids lie in the same plane. (Assume that the cone and the cylinder are opaque and of one piece.)

Required:

A. Draw the cutting plane traces that will indicate the straight line elements common to the cone and the cylinder. Predict what type of line (or lines) of intersection exists between the given cone and cylinder.

B. Draw the horizontal and frontal projections of the line of intersection between the cone and the cylinder.

C. Indicate the visibility of the line of intersection in the given views.

D. Indicate the visibility of the cone and the cylinder in the given views.

E. Label and show all construction, including the numbering system and visibility tables.

PROBLEM 10

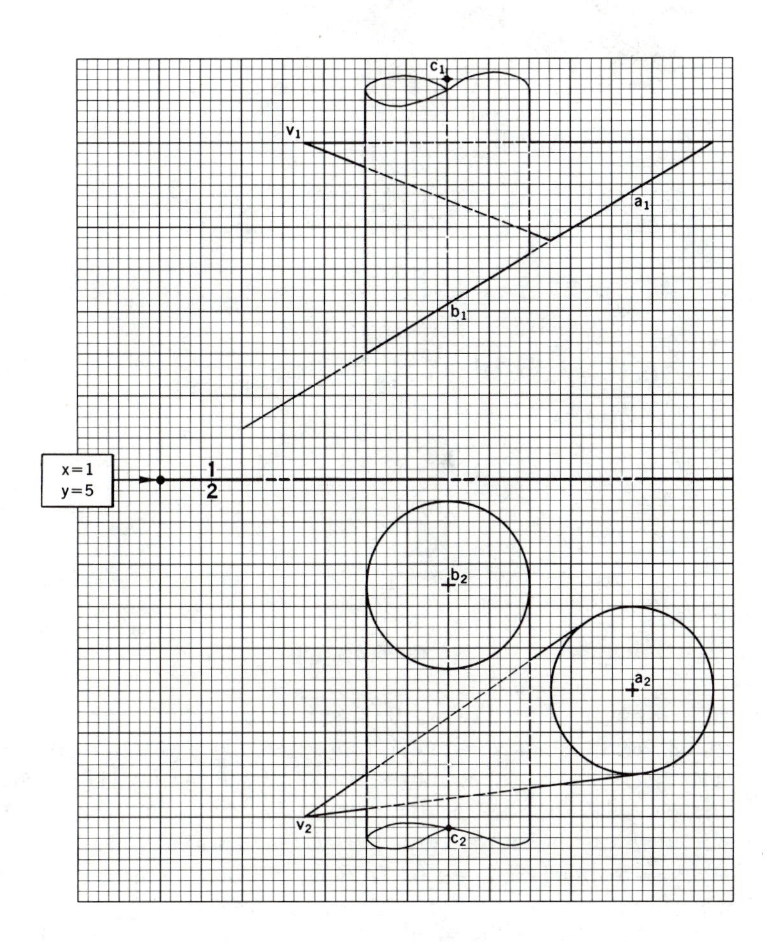

VIII

Development of Surfaces of Basic Geometric Solids

73 Definition of surface development

In addition to determining the line of intersection between various geometric solids, you must also be able to reproduce many geometric solids from *flat plane* surfaces, such as sheet metal, wood, sheet plastic. These surfaces must be "developed," that is, the faces (lateral surfaces) of the geometric solids are unfolded, or laid out, onto a *flat plane.* In this flat plane "state," the surface assumes a shape that can be "rolled" or "bent" into the original shape of the given solid.

An important characteristic of a surface development is that *all lines* appear *true length* in the development. Ruled surfaces, such as cylinders and cones, can be "accurately" developed; that is, during the process *no* distortion of the surface occurs. Certain surfaces, for example, double-curved and warped surfaces, *cannot* be accurately developed since there would have to be actual physical *distortion* of the object's lateral surface. Because all materials are not always easily distorted under controlled conditions, this is normally not desirable. Therefore, developments of such surfaces are approximate.

In this chapter, we will present the development of the surfaces of the several geometric solids. You should note that the development of *cylinders* is analogous to the development of *prisms,* and that the development of *cones* is analogous to the development of *pyramids.*

74 Surface development of a right prism

In order to develop the surface of a right prism the individual faces of the prism must be unfolded (in a consecutive fashion) onto a *flat plane of development.* Figures 185(a) and 185(b) depict a sequential series of illustrations; the complete lateral surface development is indicated in Fig. 185(b). From Fig. 185(b), it can easily be seen that all lines on each face of the prism will appear as *true length* in the surface development.

FIGURE 185(a)

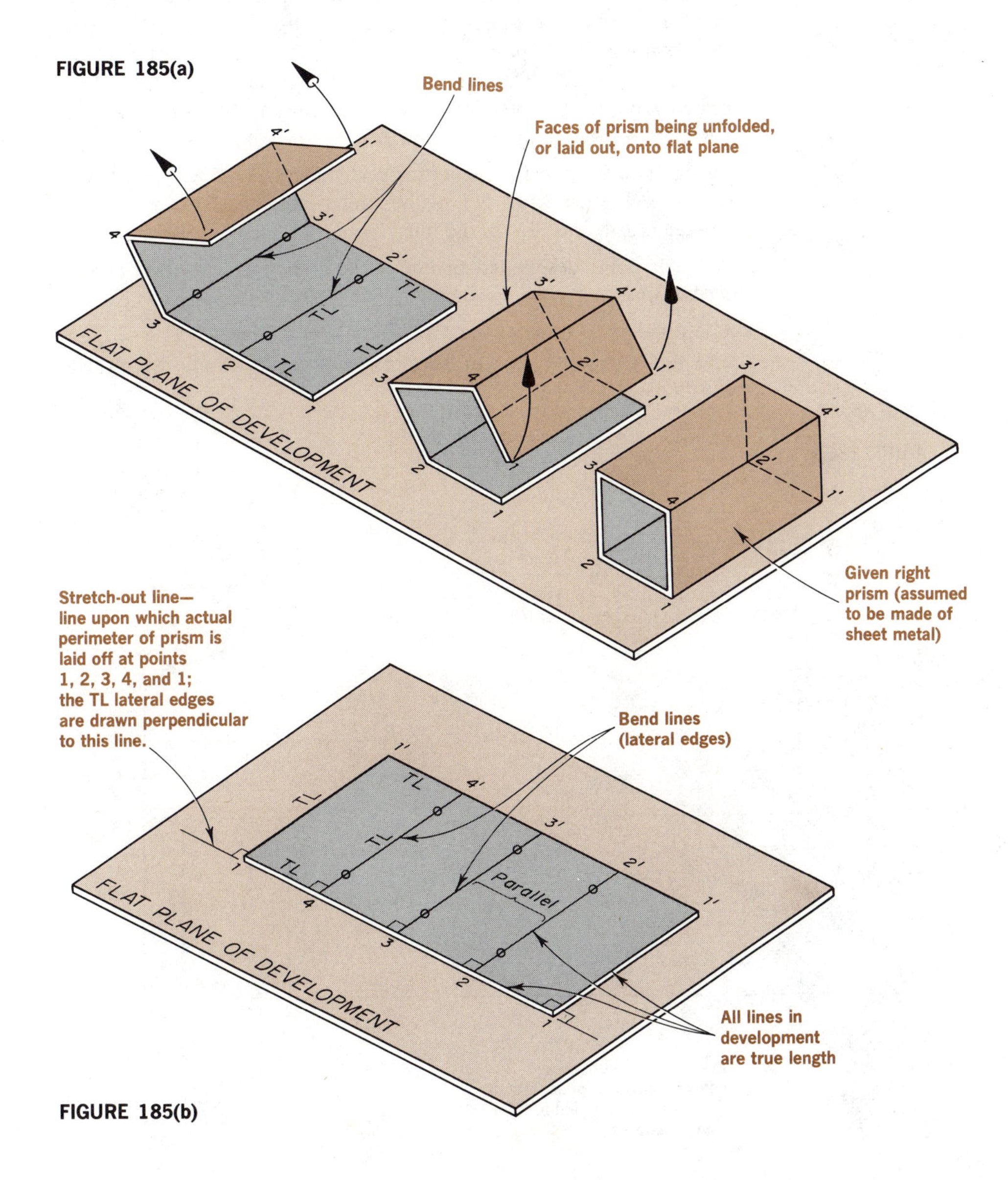

FIGURE 185(b)

Orthographic Example: Surface development of a right prism (Fig. 186).

Given Data: Horizontal and frontal projections of a square right prism.

Required: The lateral surface development of the given right prism, with top and bottom bases attached.

Construction Program:

a. Since the prism faces must be unfolded onto a flat plane of development, establish the actual *length* of the development. It can be seen in view #1, in which the *perimeter,* the actual distance around the prism (for example, from point #1 to #2, #2 to #3, #3 to #4, and #4 to #1) is seen. The flat plane of development is the plane of the drawing paper. On it, construct a "stretch-out" line on which the perimeter of the prism will be unfolded.

b. Using dividers, transfer the perimeter distances #1 to #2, #2 to #3, #3 to #4, and #4 to #1 to the stretch-out line. Note points #1, #2, #3, #4, and #1 on the line.

c. At each point designated on the stretch-out line, construct *perpendicular* lines.

d. Use dividers to transfer the *true length* lateral edges (1–1′, 2–2′, 3–3′, and 4–4′) in view #2 to their respective points on the stretch-out line.

FIGURE 186

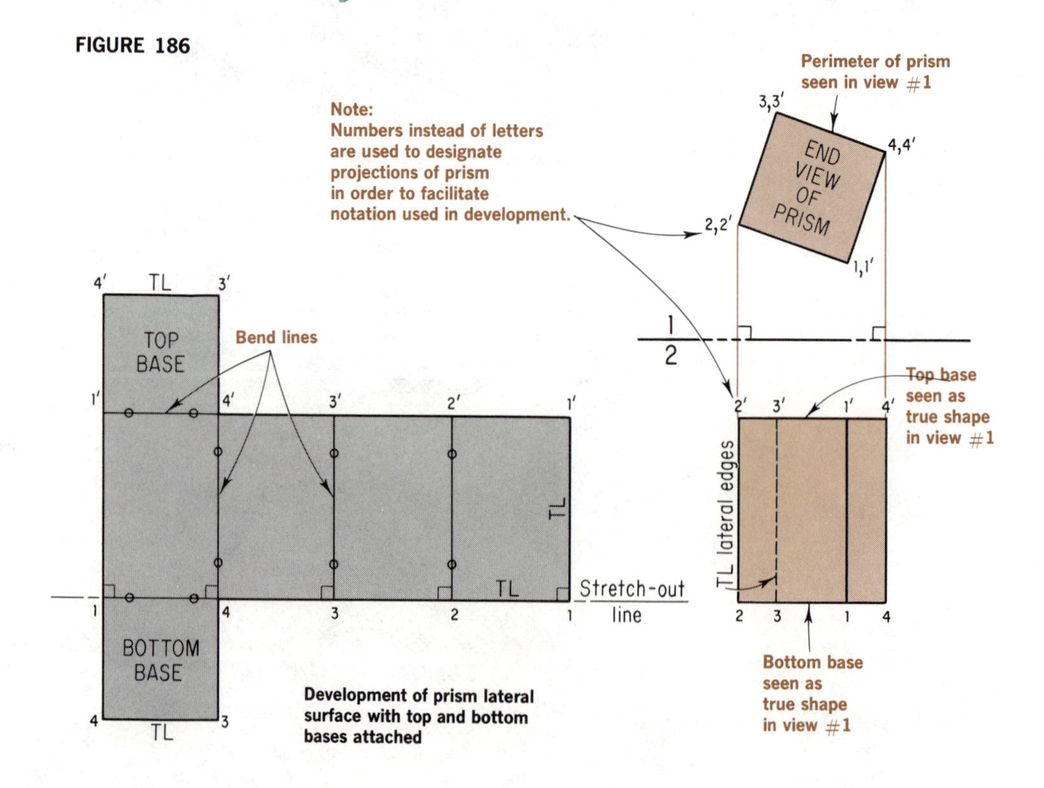

e. Connect points #1′, #2′, #3′, #4′, and #1′ with a straight line.

f. In view #1 you can see that the true shapes of the prism's top and bottom bases appear. Transfer with dividers the bottom base (#1, #2, #3, #4) to the surface development of the prism. Do the same with the top base (#1′, #2′, #3′, #4′). This completes the development of the given prism's lateral surface, with top and bottom bases attached.

The development described above is called a *parallel* line development, since the lines perpendicular to the stretch-out line (and which represent the true lengths of the lateral edges) are *parallel* to each other. (The inner parallel lines are *bend lines,* upon which the surface development would have to be bent in order to re-form the original shape of the prism. Bend lines also occur between the lateral surface development of the prism and its top and bottom bases.) The parallel line development approach is also applied to the development of cylindrical surfaces, as we shall illustrate later.

75 Surface development of a truncated right prism

A truncated prism is one that has been cut by an oblique plane so that its top base is *not* parallel to its bottom base. To develop the surface of such a prism, construct views in which appear the *true lengths of all the lateral edges* (these may vary depending upon how the prism was truncated) and the *true shapes* of the *top base and bottom base.* Likewise you need an *end view* of the prism to see and measure its *perimeter distance.*

Orthographic Example: Surface development of a truncated right prism (Fig. 187).

Given Data: Horizontal and frontal projections of a square truncated right prism.

Required: The lateral surface development of the given prism, with the top and bottom bases attached.

Construction Program:

a. The horizontal and frontal projections of the given truncated right prism show, respectively, the *end view* and the *true length* lateral edges of the prism. Draw a stretch-out line, along which you should measure the perimeter distance around the given prism (start at point #1, where the *shortest lateral edge* is located). Points #1, #2, #3, #4, and #1 are noted on the stretch-out line.

b. Construct perpendicular lines to the stretch-out line at points #1, #2, #3, #4, and #1.

c. On these perpendiculars, start with the *shortest* true length lateral edge 1–1′ (measure with dividers), in view #2 or #3, in which the true length of the lateral edges appear, and lay out the true length lateral edges 1–1′, 2–2′, 3–3′, 4–4′, and back to 1–1′.

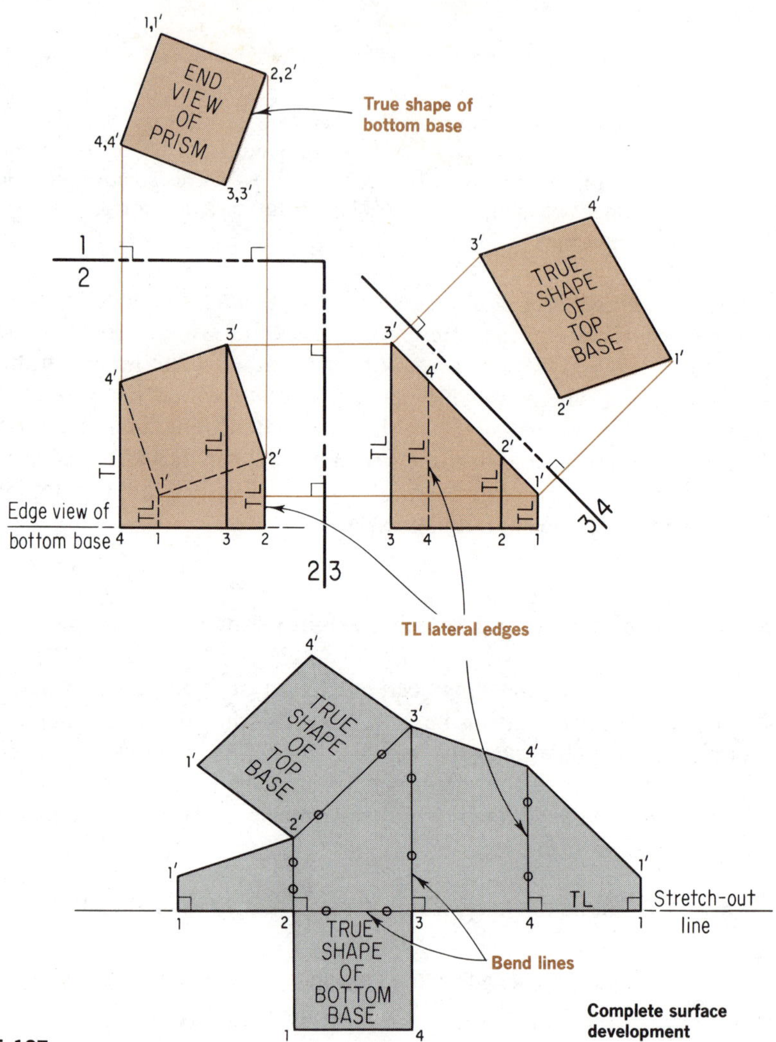

FIGURE 187

d. Connect #1′ to #2′, #2′ to #3′, #3′ to #4′, and #4′ to #1′ with straight lines. This completes the lateral surface development of the prism.

e. Attach the true shape of the bottom base (seen in view #1) to the surface development by transferring the measurements of the true length distances #1 to #2, #2 to #3, #3 to #4, and #4 to #1 with dividers.

f. Determine the true shape of the top base by drawing its edge view (seen in view #3) and then constructing a plane parallel to this edge view (plane #4 is represented by RL 3–4). The true shape of the top base appears in view #4.

g. Attach the true shape of the top base to the surface development along the top base's longest line (2′–3′). Transfer its size and shape with dividers from view #4 to the surface development. This completes the required surface development with the top and bottom bases attached.

The surface development of the truncated right prism was started at the *shortest lateral edge.* In industry this is the most economical procedure, since when the surface of the flat plane is bent to form the required geometric shape, the lateral edges thus meeting could then be attached by either riveting, soldering, or welding. This would occur along the shortest lateral edges. Likewise, the top and bottom bases would be attached along their *longest* lines (see step g above). Only the shortest distance, therefore, would remain and could economically be riveted, soldered, or welded along the remaining lines or edges.

76 Surface development of an oblique prism

The development of an oblique prism requires a view of the prism (either given or constructed) in which all its lateral edges appear as true lengths and in which an end view of the prism appears where the perimeter is seen. In addition, if the top and bottom bases are to be attached to the surface development, their true shape views must also be constructed.

Figures 188(a) and 188(b) pictorially present an oblique prism being unfolded onto a flat plane of development. (The top and bottom bases have been omitted in these figures.)

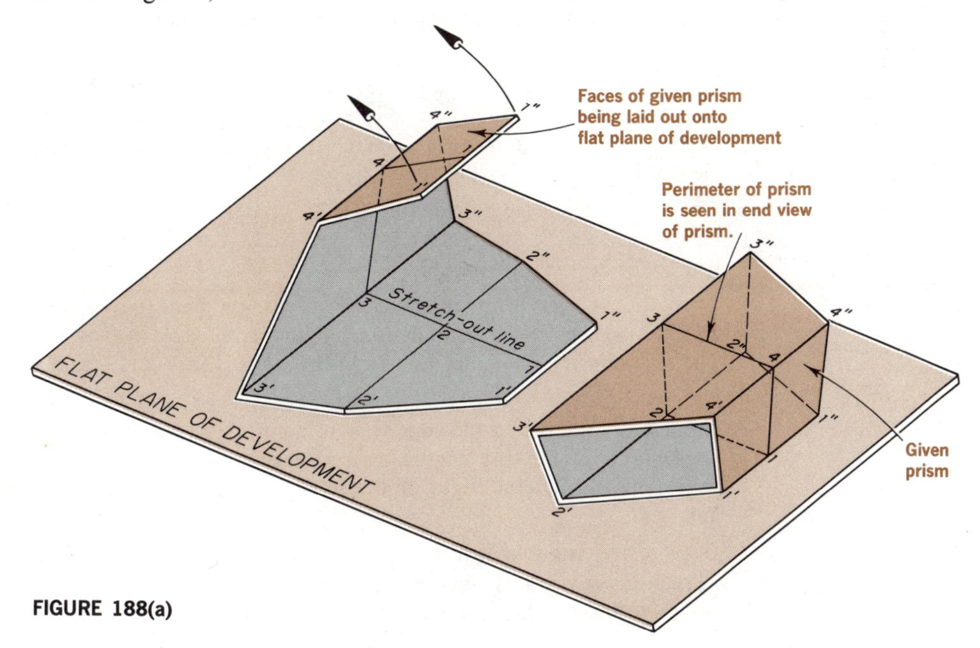

FIGURE 188(a)

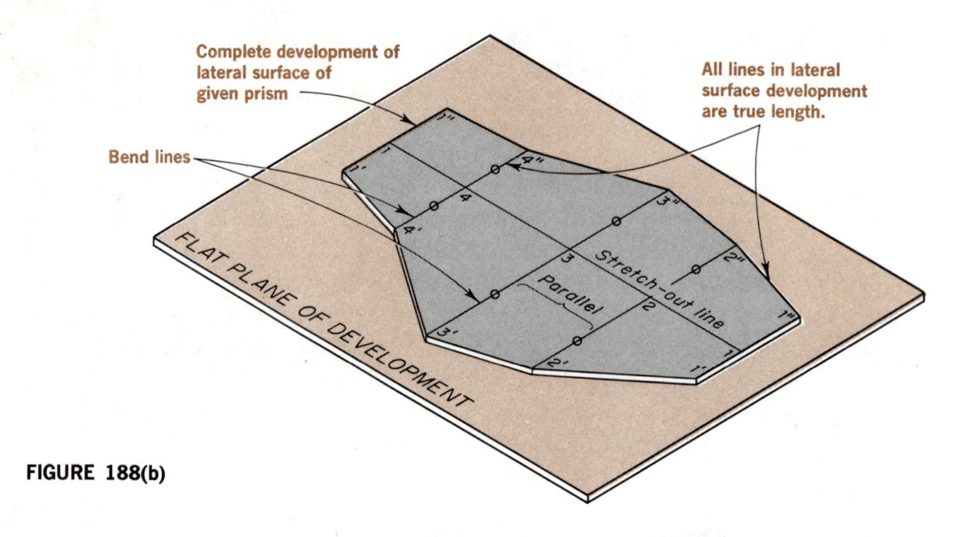

FIGURE 188(b)

Orthographic Example: Surface development of an oblique prism (Fig. 189).

Given Data: Horizontal and frontal projections of an oblique prism.

Required: The lateral surface development of the given prism, with top and bottom bases attached.

Construction Program:

a. Construct a view of the given prism in which all the lateral edges appear as true length. (In the given horizontal projection view #1, you can see that the lateral edges are all parallel to the frontal projection plane; therefore, they appear as true length in frontal projection view #2.)

b. The top and bottom bases appear as edges in view #2. Note that the edge view of the bottom base (#1′, #2′, #3′, #4′) is parallel to the horizontal projection plane and, therefore, it appears as true shape in view #1.

c. Construct a plane (represented by RL 2–3) parallel to the edge view (#1″, #2″, #3″, and #4″) of the top base, thus determining its true shape as seen in view #3.

d. Draw the *end view* of the prism by constructing a plane (represented by RL 2–4) perpendicular to the true length lateral edges in view #2. The end view is seen in view #4 and is noted as points #1, #2, #3, and #4.

e. In view #2, construct the edge view of a cutting plane *S–O* that is *perpendicular* to the true length lateral edges of the prism. This cutting plane cuts the lateral edges of the prism at points #1, #2, #3, and #4.

f. Draw a stretch-out line; measure the perimeter distance of the prism as seen in view #4. (Points #1, #2, #3, #4, and #1 will be indicated on the stretch-out line as shown in Fig. 189.)

g. Construct perpendicular lines from points #1, #2, #3, #4, and #1 on the stretch-out line. (See development in Fig. 189.)

h. Starting with the *shortest lateral edge* (seen as true length in view #2), measure the distance *X* above the stretch-out line and the distance *Y* below it. This locates points #1″–1′ on the perpendiculars.

i. Repeat this procedure and locate points #2″–2′, #3″–3′, #4″–4′, and #1″–1′, measure from points #1, #2, #3, #4, and #1, respectively, on the stretch-out line.

j. Connect points #1′, #2′, #3′, #4′, and #1′ with straight lines. Do the same for points #1″, #2″, #3″, #4″, and #1″. This completes the surface development of the given oblique prism.

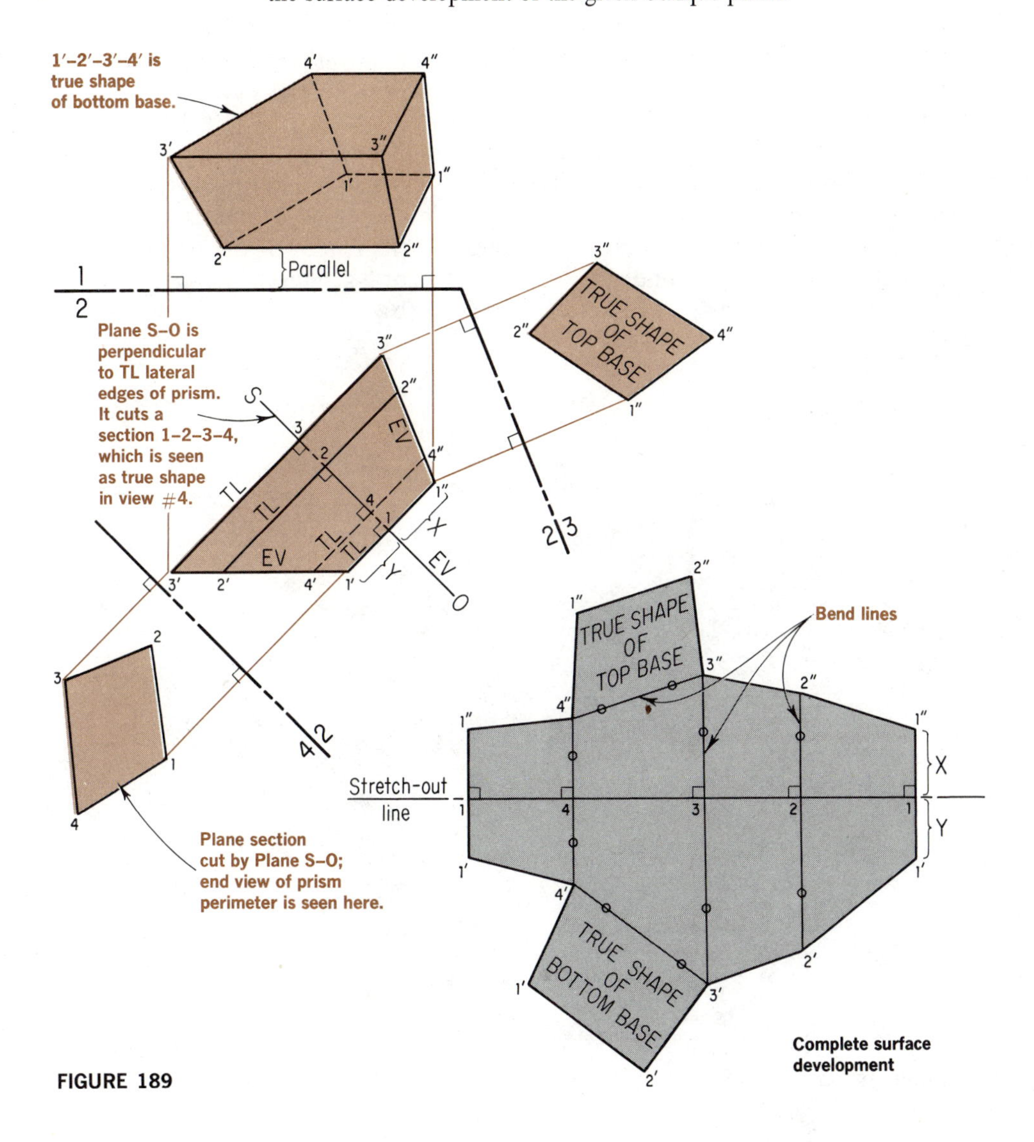

FIGURE 189

k. From view #1, transfer with dividers the *true* shape of the *bottom base* (#1′, #2′, #3′, #4′) and attach it to the longest line (3′–4′) in the surface development.

l. From view #3, transfer the *true shape* of the *top base* (#1″, #2″, #3″, #4″) and attach it to the longest line (3″–4″) of the surface development. This completes the development of the prism surface with bases attached.

77 Surface development of a right cylinder

The development of a cylindrical surface is analogous to the development of a prism. Consider a cylinder to be a prism with an infinite number of faces. In practice, the number of faces may be reduced to a *finite* number, which you can then use graphically. The *elements* of the cylindrical surface are analogous to the lateral edges of a prism. They must be viewed as true lengths that can be used to develop the cylinder surface. The perimeter of a cylinder appears in its end view, as does that of a prism. The perimeter of a cylinder of revolution (in which the cross section is a circle with its perimeter equal to the circumference πD) may be determined. You can approximate the length of the perimeter (or circumference) of the cylinder by measuring chord distances between the elements in the end view [see Figs. 190(a) and 190(b)].

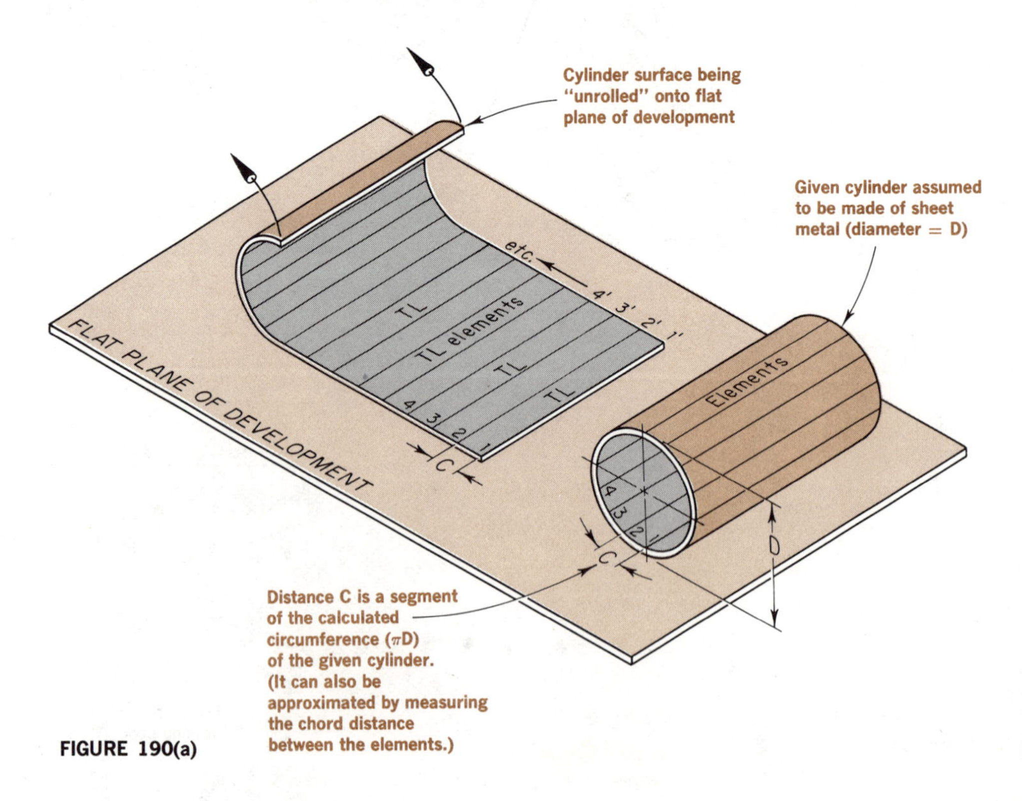

FIGURE 190(a)

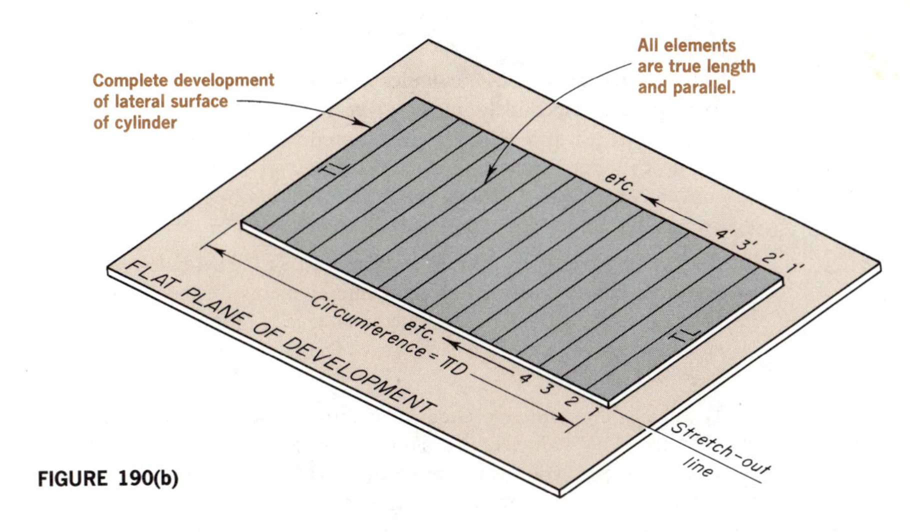

FIGURE 190(b)

Orthographic Example: Surface development of a right cylinder (Fig. 191).

Given Data: Horizontal and frontal projections of a right cylinder of revolution, the end view of which is a circle.

Required: The surface development of the given cylinder, with top and bottom bases attached.

FIGURE 191

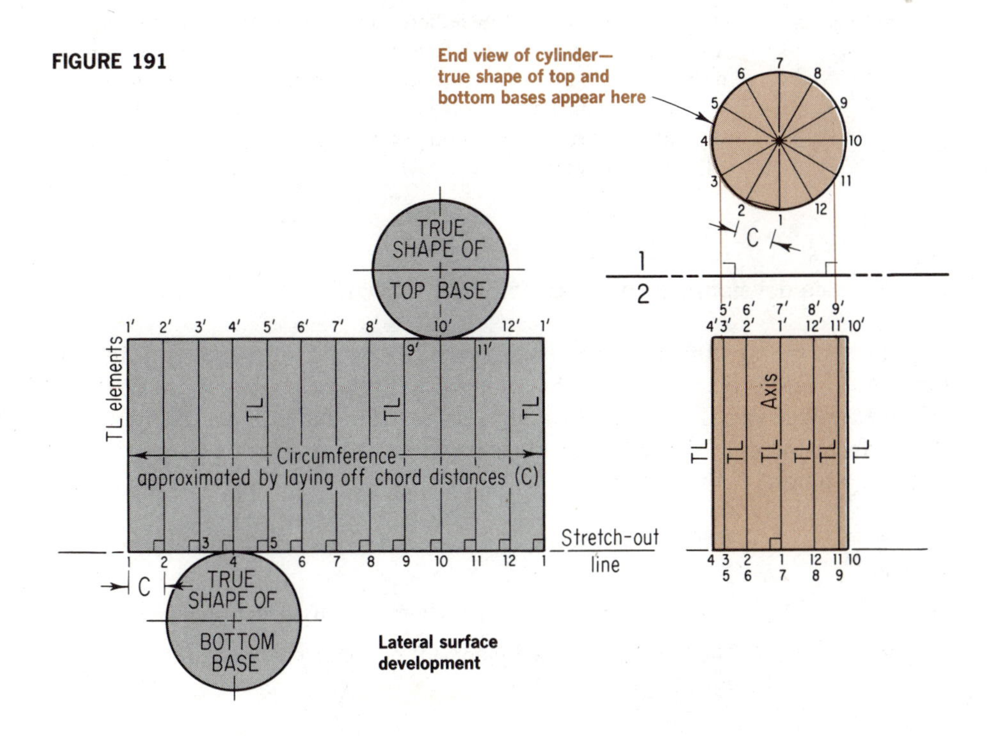

Construction Program:

a. The *end view* of the given cylinder and the *true length* of the cylinder *axis* appear, respectively, in the given horizontal and frontal projection views (#1 and #2). On the circumference, divide the end view of the cylinder into a convenient number of radial divisions. (For convenience, these divisions should be uniform in spacing.) Number each of the units (#1–#12) as indicated in view #1.

b. Project points #1–#12 to the bottom base of the cylinder (view #2). At these points construct straight line elements on the cylinder surface. Since the cylinder axis appears as true length in view #2, these elements also appear as true lengths. The elements are noted as 1–1′, 2–2′, 3–3′, and so forth.

c. Draw a stretch-out line equal in length to the circumference of the cylinder. You can see the actual length of the circumference in view #1 and you can approximate it by measuring chord distances (C) on the stretch-out line. Number the distances consecutively, starting with point #1 on the stretch-out line. Continue through #12, repeating #1 at the extreme right end.

d. At each numbered point on the stretch-out line construct perpendicular lines whose lengths are equal to the *true length elements* of the cylinders, as measured in view #2. This locates points #1′–#12′ and #1′.

e. Connect points #1′–#12′ and #1′ with a straight line. This completes the lateral surface development of the cylinder.

f. The true shapes of the top and bottom bases (which are circles) appear in view #1. Attach the top base to any convenient tangent point on the upper part of the surface development, and attach the bottom base to any convenient tangent point on the stretch-out line. This completes the required lateral surface development of the cylinder, with top and bottom bases attached.

78 Surface development of a truncated right cylinder

A truncated right cylinder is one in which the top base is *not* parallel to its bottom base. Therefore, the true lengths of the straight line elements of the cylindrical surface will vary in length. In addition the true shape of the bottom base differs from the true shape of the top base since the truncating plane cuts an elliptical section on the cylinder.

Orthographic Example: Surface development of a truncated right cylinder (Fig. 192).

Given Data: Horizontal and frontal projections of a truncated right cylinder.

Required: The lateral surface development of the given cylinder, with top and bottom bases attached.

FIGURE 192

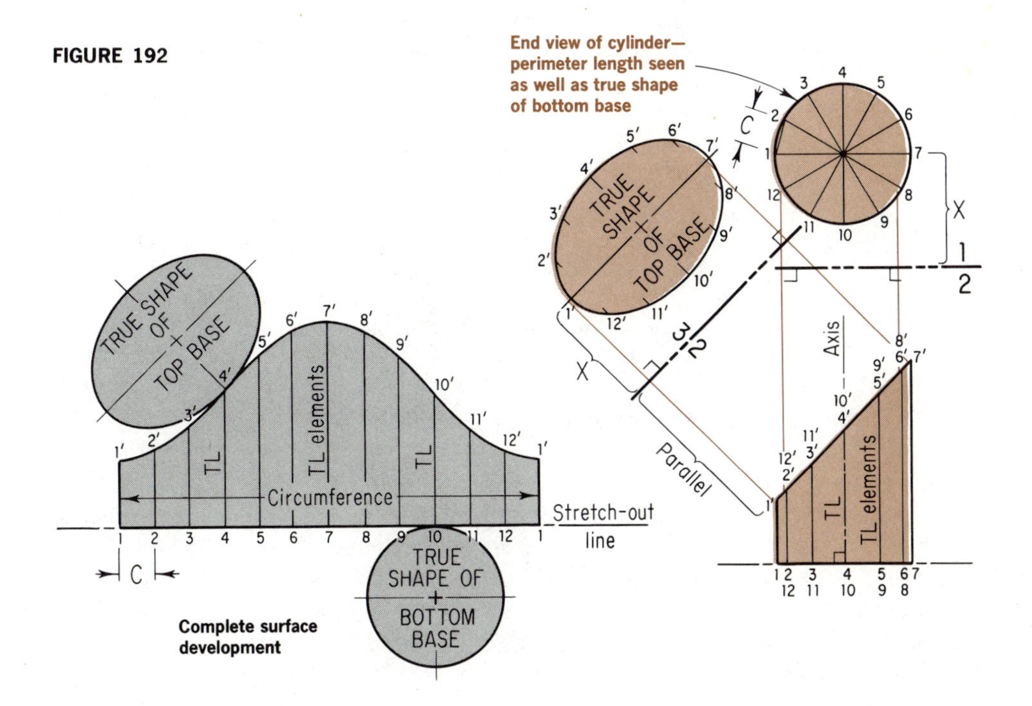

Construction Program:

a. On the circumference, divide the end view of the cylinder (seen in view #1) into a convenient number of radial divisions. (The divisions should be uniform in spacing.) Number each of the units—#1 through #12—as indicated in view #1.

b. Project points #1–#12 to the bottom base of the cylinder (view #2). At these points construct straight line elements on the cylinder surface. Since the cylinder axis appears as true length in view #2, these elements also appear as true lengths. The elements terminate at the truncated end of the cylinder at points #1′–#12′, and are noted as #1-1′, #2-2′, #3-3′, and so forth.

c. Draw a stretch-out line equal in length to the circumference of the cylinder. You can see the actual length of the circumference in view #1, and can approximate it by measuring chord distances (C) on the stretch-out line. Number the distances consecutively, starting from point #1 on the *shortest element.* Continue through #12, repeating number #1 on the stretch-out line. The distances between these numbers represent the chord distances measured in view #1.

d. At each of the numbered points on the stretch-out line, construct perpendicular lines whose lengths are equal to the *true length elements* of the cylinder as measured in view #2, thus establishing points #1′–#12′ and #1′ in the development.

e. Connect points #1′–#12′ and #1′ with a *smooth curved line.* This completes the lateral surface development of the truncated cylinder.

The true shape of the bottom base appears in view #1 and is attached to a convenient tangent point on the surface development.

f. Determine the true shape of the top base by constructing projection plane #3 (represented by RL 2–3) parallel to the edge view of the top base as seen in view #2. (The true shape of the top base appears in view #3.)

g. With dividers, transfer to the surface development the points that describe the true shape of the top base in view #3. (Attach the top base to any convenient tangent point on the upper part of the development.) This completes the required lateral surface development of the cylinder, with top and bottom bases attached.

79 Surface development of an oblique cylinder

An oblique cylinder is one in which the *bottom base* makes an angle *other* than 90° with the *cylinder axis.* To develop an oblique cylinder, you can apply the same basic approach used to develop a right cylinder. Locate the stretch-out distance by constructing a *cutting plane perpendicular* to the *true length elements* of the cylinder. This plane cuts a section which, when seen as true shape, indicates the *actual circumference,* or perimeter, around the cylinder.

Figures 193(a) and 193(b) show the sequential development of an oblique cylinder. (Note the stretch-out line.)

Orthographic Example: Surface development of an oblique cylinder [Figs. 194(a) and 194(b)].

Given Data: Horizontal and frontal projections of an oblique cylinder of revolution.

Required: The surface development of the given oblique cylinder, with top and bottom bases attached.

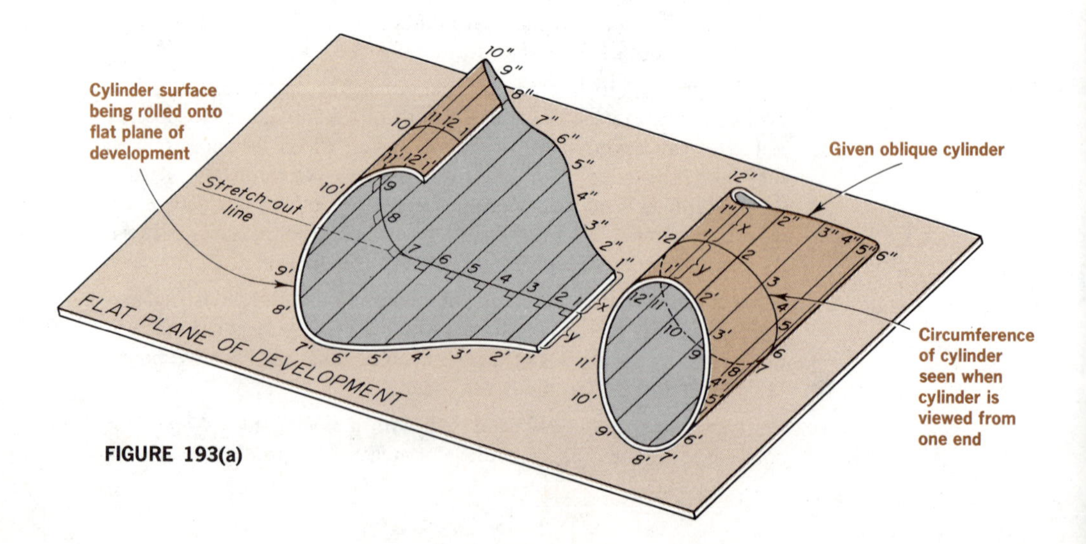

FIGURE 193(a)

FIGURE 193(b)

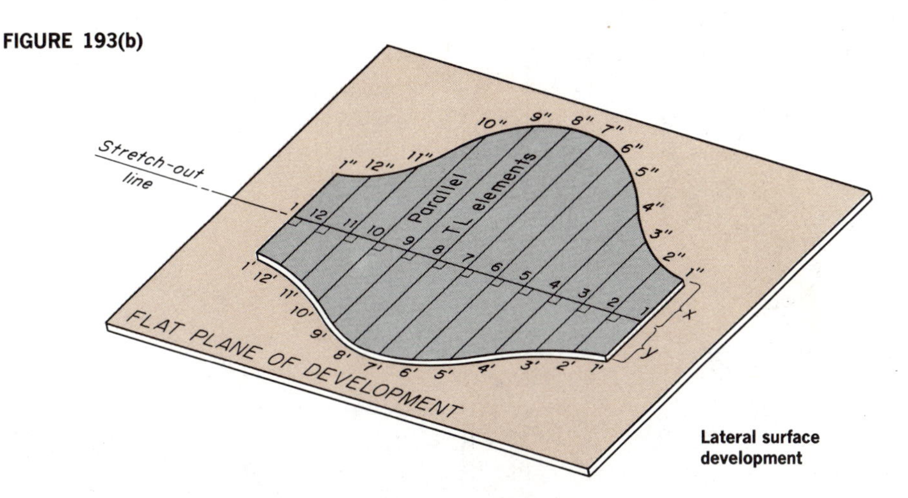

Lateral surface development

Construction Program:

a. In the given views the cylinder axis appears to be vertical and, therefore, it appears as a point in view #1. Consequently, the *end view* of the cylinder must also appear in this view. In view #2, the *true length* of the cylinder axis appears. Divide the *end view* of the cylinder into a convenient number of equally spaced divisions. (Number from #1–#12.)

b. From view #1, project points #1–#12 onto a plane (*S–O*) in view #2. Plane *S–O* is *perpendicular* to the *true length cylinder axis* and, therefore, to the true length elements that appear in this view. Plane *S–O* cuts a plane section—the end view—of the cylinder as seen in its true shape.

c. Through points #1–#12 in view #2, construct *parallel* straight line elements that intersect the *edge view* of the *top base* and the *edge view* of the *bottom base.* (Elements that intersect the edge view of the bottom base are #1′–#12′ and elements that intersect the edge view of the top base are #1″–#12″.)

d. Determine the true shape view of the bottom base by constructing plane #4 (represented by RL 2–4) parallel to the edge view of the bottom base plane.

e. Determine the true shape of the top base by constructing plane #3 (represented by RL 2–3) parallel to the edge view of the top base plane. The true shape of the bottom base appears in view #4, the true shape of the top base in view #3.

f. Draw a stretch-out line equal in length to the circumference of the given cylinder (view #1). This may be approximated by measuring the chord distances between the points on the circumference. Number the chord distance points on the stretch-out line #1–#12, and #1 [Fig. 194(b)].

g. At each point on the stretch-out line construct *perpendicular* lines. In view #2, where the true lengths of all the cylinder elements appear,

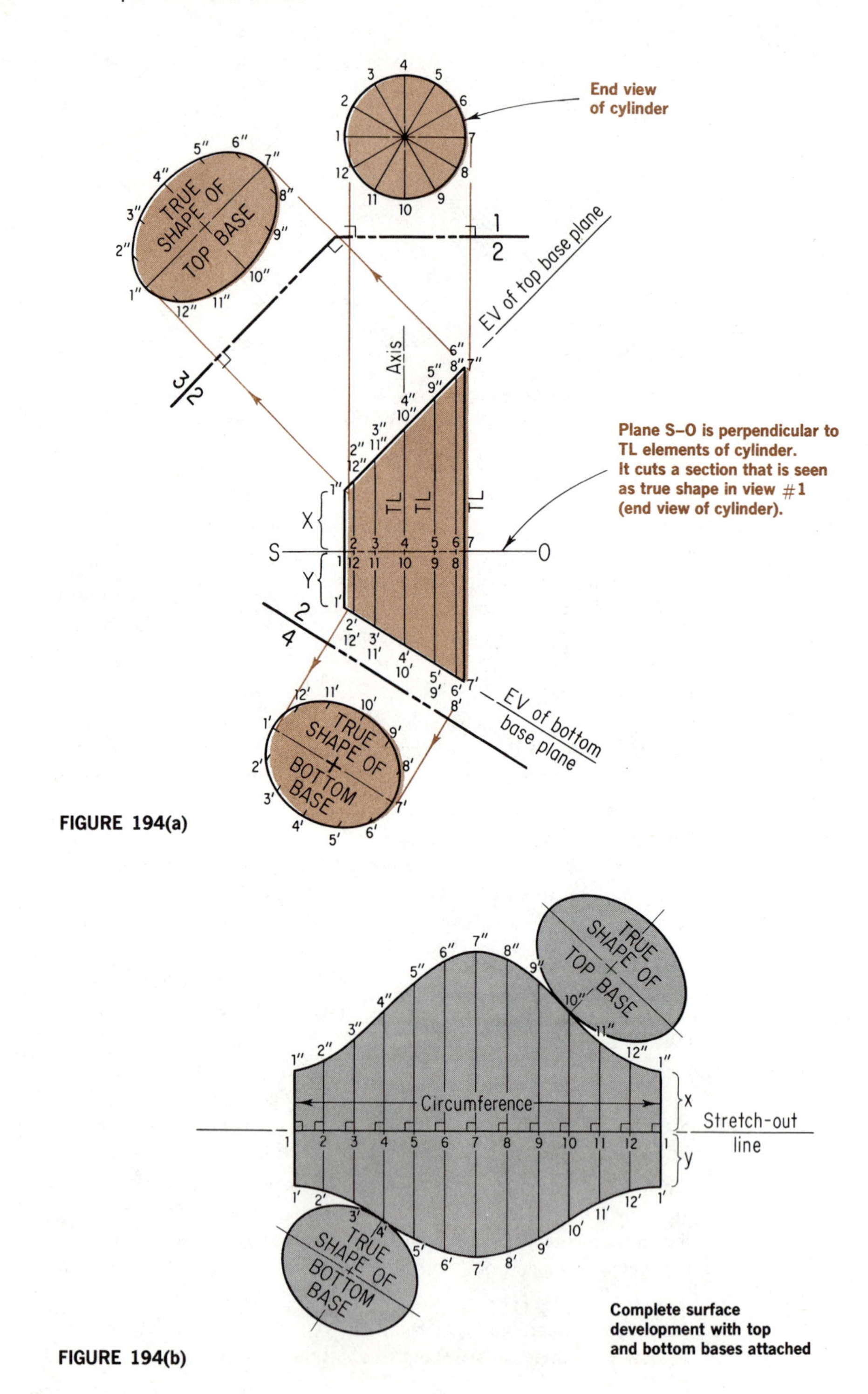

FIGURE 194(a)

FIGURE 194(b)

measure the distance *X above* the plane *S–O* for point #1″ and transfer this distance to the stretch-out line, measuring from point #1.

h. Measure the distance *Y below* the plane *S–O* in view #2, to point #1′. Transfer this distance to the stretch-out line, measuring *below* the line in order to locate point #1′.

i. Repeat the above process for all single-primed numbers *below* the stretch-out line and all double-primed numbers *above* the stretch-out line.

j. Connect points #1′–#12′ and #1′ with a *smooth curved line.* Do the same for points #1″–#12″ and #1″. This is the required lateral surface development of the given oblique cylinder.

k. Transfer with dividers the true shape of the bottom base to the surface development, attaching it to any convenient tangent point.

l. Transfer the true shape of the top base to the surface development, attaching it to any convenient tangent point [Fig. 194(b)]. This completes the required lateral surface development, with top and bottom bases attached.

80 Surface development of a right pyramid

To develop a pyramid, determine the *true shape* of *each face* of the pyramid. Figures 195(a) and 195(b) present a pyramid being unfolded onto a flat plane of development. Note that each face is a triangle, and each has a common vertex *V*, which is the vertex of the pyramid. When you develop a *right* pyramid, note that *all lateral edges* of the pyramid are of *equal length,* a fact that simplifies its surface development.

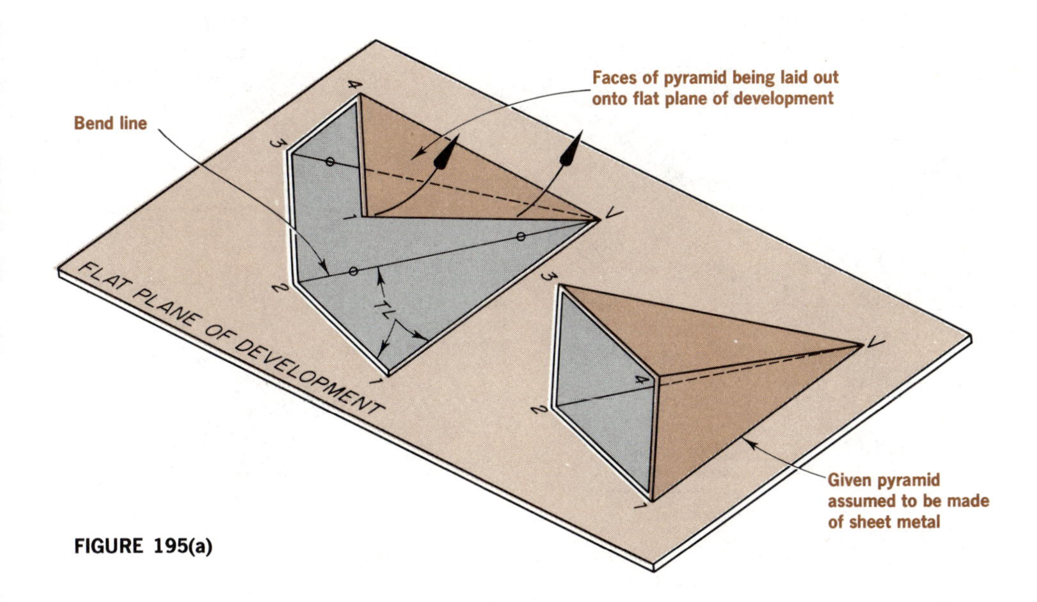

FIGURE 195(a)

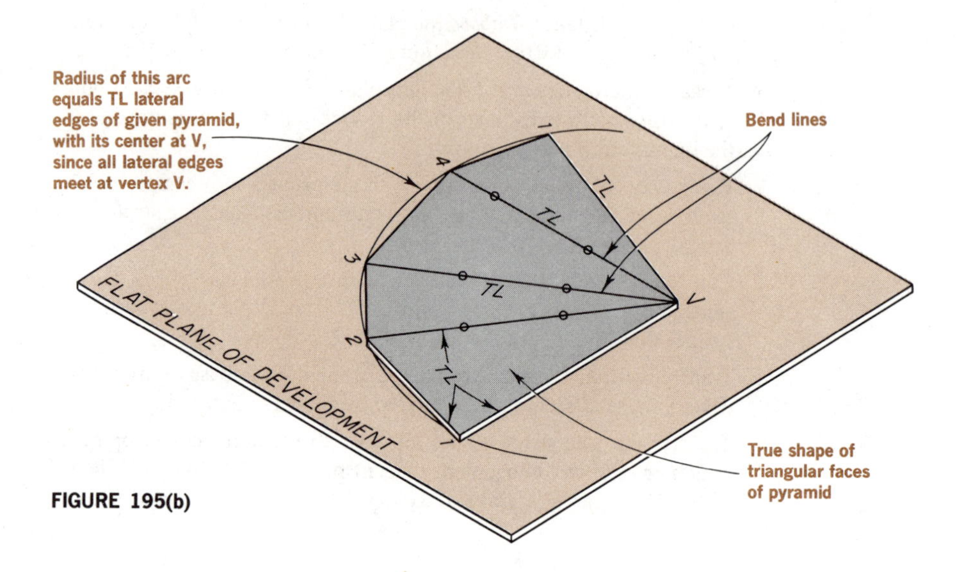

FIGURE 195(b)

Orthographic Example: Surface development of a right pyramid (Fig. 196).

Given Data: Horizontal and frontal projections of a rectangular right pyramid, the vertex of which is V.

Required: The complete surface development of the given pyramid, with base attached.

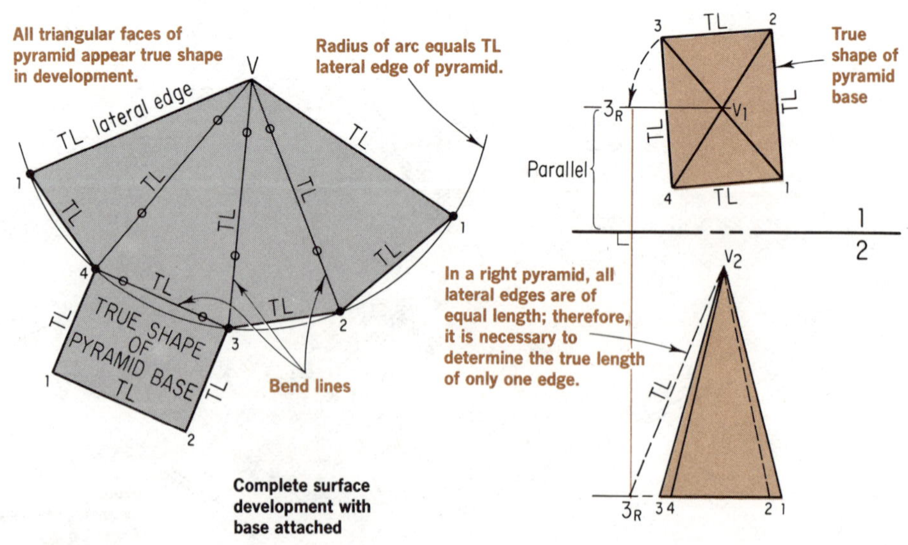

FIGURE 196

Construction Program:

a. Since the development of the surface of a right pyramid involves finding the true shape of all its lateral faces, you must determine the true lengths of all lateral edges and of each side of the pyramid base. The lateral edges of a right pyramid are all the same length. Therefore, if you determine the true length of one lateral edge, you have determined the true lengths of all. Beginning in view #1, rotate lateral edge v_1–3 so that it is parallel to the frontal projection plane. (The revolved position is v_1–3_R.) The true length of this lateral edge appears in view #2 and is indicated as v_2–3_R. From a point noted as *V* in the development, strike an arc having a radius equal to the true length v_2–3_R of lateral edge v_2–3.

b. On this arc, measure with dividers the *true lengths* of the *sides* of the *pyramid base,* which appears as true shape in view #1. (This locates points #1–#4 and #1 on the arc.)

c. Connect point *V* in the development to points #1–#4 and #1.

d. Connect points #1–#4 and #1 on the arc with straight lines. This completes the surface development of the given pyramid.

e. With dividers, transfer the true shape of the base as it appears in view #1 to the surface development, attaching the base along one of its *longest* sides (3–4). This completes the required surface development with base attached. Note that since all the true length lateral edges in the development are concurrent (that is, they meet at one point *V*), the development is called a *radial* development.

81 Surface development of a truncated right pyramid

When developing the lateral surface of a truncated right pyramid, follow a procedure identical to the one used to develop a complete right pyramid. The only difference is that the lateral edges of the truncated pyramid vary in length because of the truncation. Therefore, you will have to determine the true lengths of *each* of the pyramid's lateral edges.

Orthographic Example: Surface development of a truncated right pyramid (Fig. 197).

Given Data: Horizontal and frontal projections of a rectangular truncated right pyramid, the vertex of which is *V*.

Required: The complete surface development of the given pyramid, with bases attached.

Construction Program:

a. When developing any truncated pyramid you should consider the pyramid *before* truncation. Begin by developing a lateral surface as though you were dealing with a *complete* pyramid.

Lightly complete the given truncated right pyramid by extending all the given lateral edges until they intersect at the common vertex *V* in both given views.

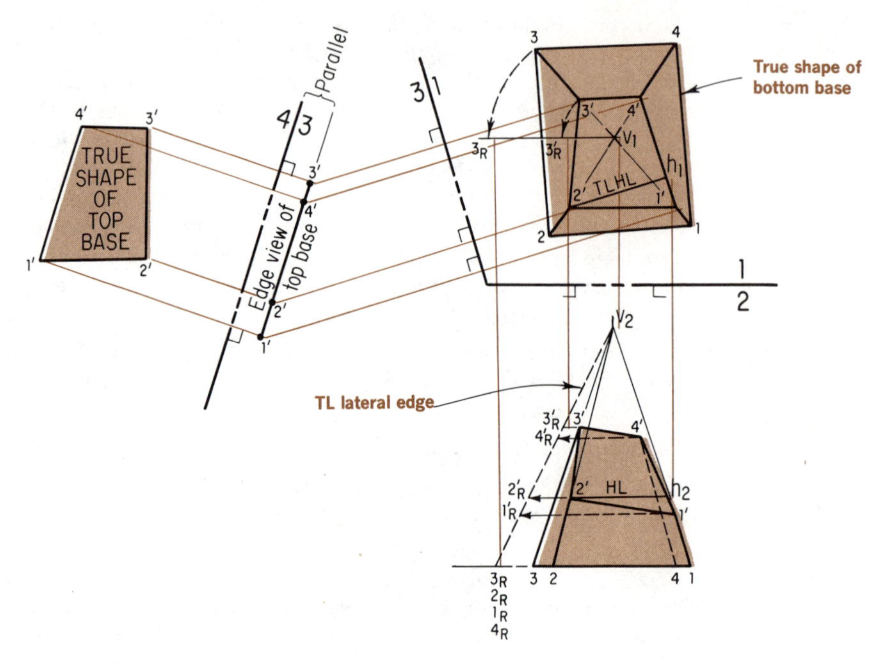

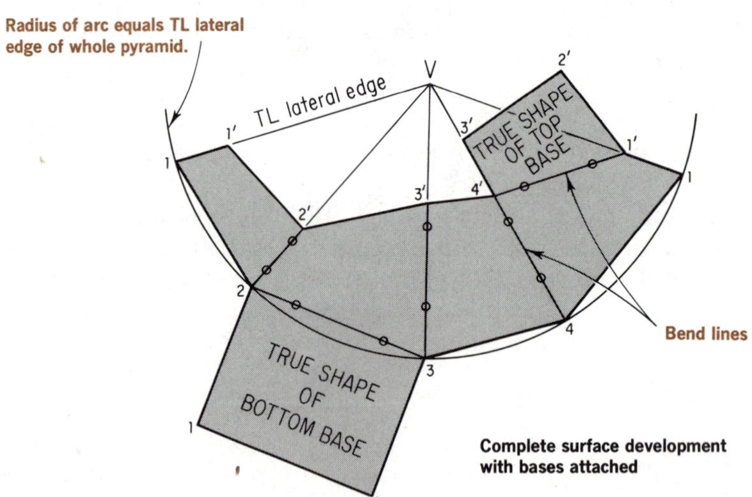

FIGURE 197

b. Determine the true length of one of the extended lateral edges. In our example, lateral edge v_1–3 has been revolved into a position parallel to that of RL 1–2, so that the true length of this lateral edge appears in view #2 as v_2–3_R. Since any point on a rotated line remains on the line, the true distance from vertex V to every point on the top base appears on the true length lateral edge v_2–3_R. (This is true because all the lateral edges of a *complete* right pyramid are of equal length.) Locate points $\#1_R'$, $\#2_R'$, $\#3_R'$, and $\#4_R'$ on the true length lateral edge v_2–3_R. The distances from v_2 to the individual revolved points appear as true length in view #2.

c. Starting at some convenient point V, construct an arc having a radius equal to the true length lateral edge v_2–3_R. On this arc, starting with point #1 (on the *shortest* lateral edge of the given pyramid), measure with dividers the true length distances of the bottom base (from #1 to #2, from #2 to #3, and so on, back to #1). Connect these points with straight lines to V in the development. (This represents the surface of a right pyramid *before* truncation.)

d. In view #2, transfer with dividers the distances from v_2 to points $\#1_R'$, $\#2_R'$, $\#3_R'$, and $\#4_R'$, on their respective lines V–1, V–2, V–3, V–4, and V–1. This locates points #1′, #2′, #3′, #4′, and #1′.

e. Connect points #1′–#4′ and #1′ with straight lines. This is the surface development of the given truncated right pyramid.

f. Attach the *true shape* of the bottom base to the surface development by transferring the true shape measurements from view #1 to the surface development. Attach the base along one of its *longest* sides (2–3).

g. Determine the true shape of the top base (#1′–#4′) by constructing a true length horizontal line $2'$–h_2 in the plane of the top base (see view #2) and by viewing this line as a point so that the plane appears as an edge (see view #3).

h. Construct plane #4 (represented by RL 3–4) parallel to the edge view of the top base in order to determine the true shape of the top base in view #4.

i. Using dividers, transfer the true shape of the top base (seen in view #4) to the surface development. Attach the base along its longest side (4′–1′). This completes the required surface development, with bases attached, of the given truncated right pyramid.

82 Surface development of a truncated oblique pyramid

To develop an oblique pyramid, determine the *true shape* of *each* face of the pyramid as you did in the development of a right pyramid. Note, however, that the *lateral edges* of an oblique pyramid are not necessarily of equal length. Therefore, you must determine the *true length* of *each* of the pyramid's lateral edges. In the case of a truncated oblique pyramid, the development begins by a consideration of the truncated pyramid as a *complete* oblique pyramid.

Figures 198(a) and 198(b) illustrate an oblique pyramid with vertex V being unfolded onto a flat plane of development. (Note that the lateral edges of this truncated oblique pyramid have been extended so that they intersect at vertex V.)

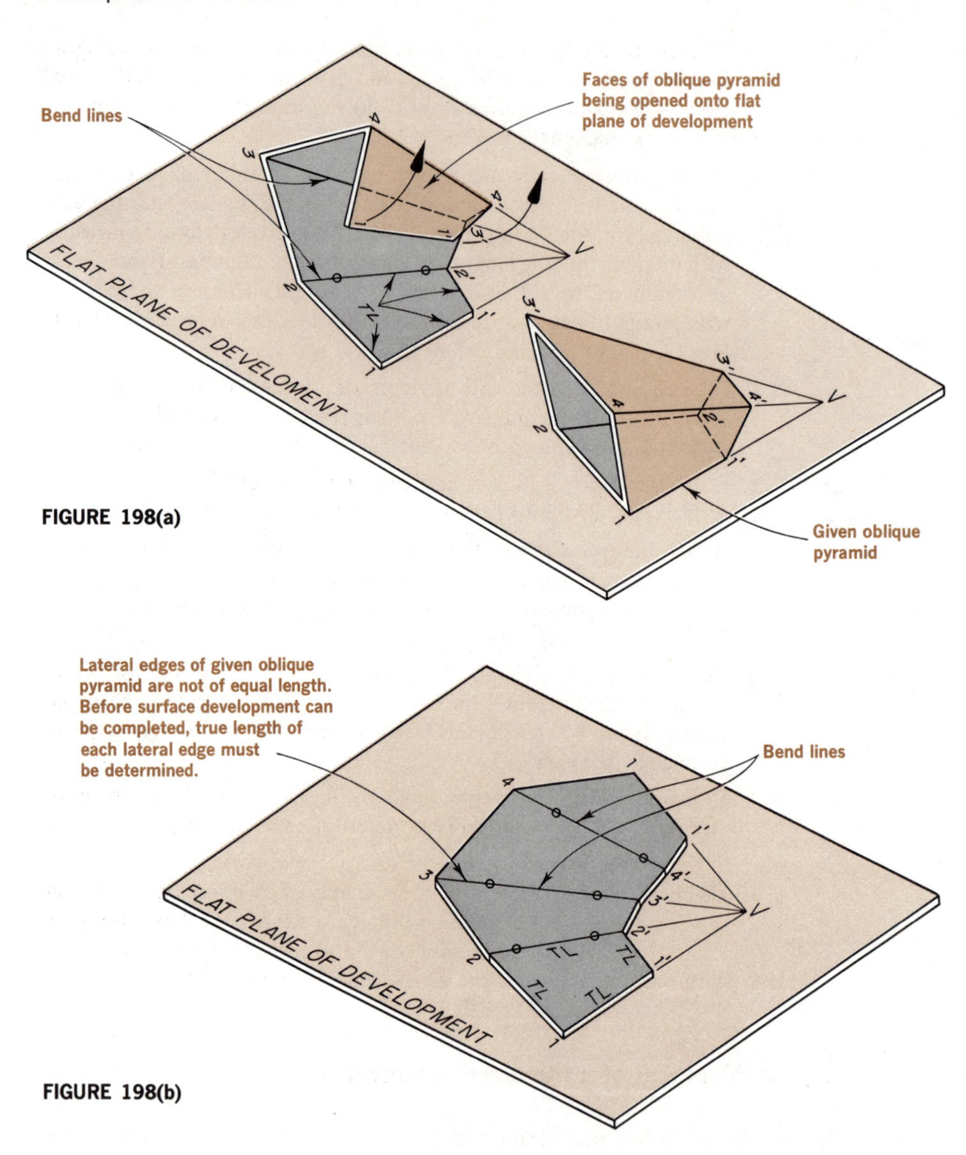

FIGURE 198(a)

FIGURE 198(b)

Orthographic Example: Surface development of a truncated oblique pyramid (Fig. 199).

Given Data: Horizontal and frontal projections of a rectangular oblique pyramid, the vertex of which is V (when the lateral edges are extended).

Required: The complete surface development of the given truncated oblique pyramid, with the bases attached.

Construction Program:

a. Lightly extend the lateral edges of the given pyramid in views #1 and #2 until they intersect at v_1 in view #1, and at v_2 in view #2.

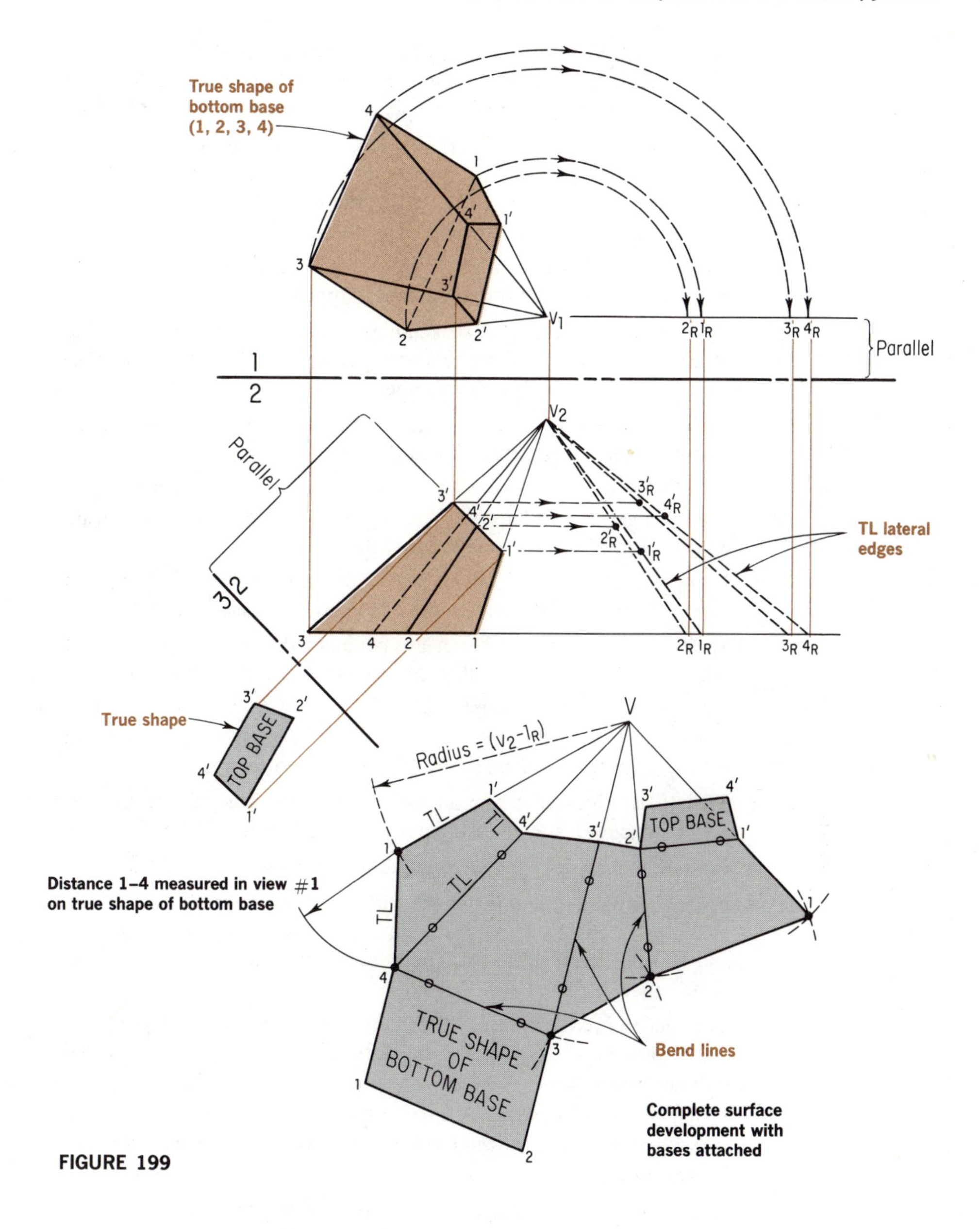

FIGURE 199

These would be the horizontal and frontal projections of the vertex of the oblique pyramid *if* it were complete.

b. Using v_1 in view #1 as an axis of rotation, determine the true length of each lateral edge of the given pyramid. (This means revolving v_1–1, v_1–2, v_1–3, v_1–4 parallel to the frontal projection plane.) The true lengths appear in view #2 as v_2–1_R, v_2–2_R, v_2–3_R, and v_2–4_R.

c. In view #2, by applying the principles of rotation, determine points $\#1_R'$, $\#2_R'$, $\#3_R'$, and $\#4_R'$ on their respective true length lateral edges. Note that the points move perpendicular to the axis of rotation.

d. Using any convenient point V as a center, begin the surface development of the *complete* oblique pyramid (with lateral edges extended to vertex v_2), and strike an arc equal to the true length v_2–1_R of lateral edge v_2–1. Point #1 in the development is on this arc.

e. From point #1, strike an arc having a radius equal to the side 1–4 of the true shape of the bottom base as it appears in view #1.

f. Again using V as a center, strike an arc equal to the true length v_2–4_R of lateral edge v_2–4. Where this arc intersects arc 1–4, you can locate point #4 on the surface development.

g. From point #4, strike an arc equal to the true length side 4–3 of the bottom base, which appears as true shape in view #1.

h. With V as a center, strike an arc equal to the true length v_2–3_R of lateral edge v_2–3. Where this arc intersects arc 4–3, you can locate point #3 on the surface development.

i. From point #3 strike an arc 3–2, which appears as true length in view #1, in which the true shape of the bottom base is seen.

j. With V as a center, strike an arc equal to the true length v_2–2_R of lateral edge v_2–2. Where this arc intersects arc 3–2 you can locate point #2 on the surface development.

k. From point #2, strike an arc 2–1, which appears as true length in view #1, in which the true shape of the bottom base is seen.

l. With V as a center, strike an arc equal to the true length v_2–1_R of lateral edge v_2–1, thus locating point #1 on the surface development.

m. Connect points #1, #4, #3, #2, and #1 with straight lines.

n. From v_2 in view #2, transfer the true length distances of v_2–1′, v_2–4′, v_2–3′, v_2–2′, and v_2–1′ on their respective lines V–1, V–4, V–3, V–2, and V–1 in the development. These true lengths appear in view #2 as v_2–$1_R'$, v_2–$2_R'$, v_2–$3_R'$, and v_2–$4_R'$.

o. Connect points #1′, #4′, #3′, #2′, and #1′ with straight lines. This completes the surface development of the given truncated oblique pyramid.

p. Using dividers, transfer the true shape of the bottom base (seen in view #1) to the surface development. Attach the base along its longest side (4–3).

q. Determine the true shape of the top base by constructing a plane parallel to its edge view (seen in view #2). The true shape of the top base appears in view #3.

r. Using dividers, transfer the true shape of the top base to the surface development, attaching the base along its longest side (1′–2′). This completes the required surface development, with bases attached, of the given truncated oblique pyramid.

83 Surface development of a right cone

You can consider a cone to be a multisided pyramid composed of an infinite number of triangles (lateral faces). The development of its lateral surface is analogous to that of the lateral surface of a pyramid. Actually, the cone's surface can only be divided into a given, or finite, number of lateral faces, or triangles. The *true shapes* of these triangles are laid out in sequential order. This results in an approximate development of the cone, and for most purposes, such a development is sufficiently accurate. (A more accurate development of a cone's surface may be achieved when the circumference of the cone base is calculated.) This system of development is referred to as development by *triangulation*. Figures 200(a) and 200(b) illustrate such a development of a right circular cone. Note that the cone's surface has been divided into a convenient number of triangular sections, and that it appears in its completely flat condition in the plane of development.

Orthographic Example: Surface development of a right cone (Fig. 201).

Given Data: Horizontal and frontal projections of a right circular cone, the vertex of which is *V*.

Required: The complete surface development of the given cone, with base attached.

Construction Program:

a. Starting in view #1, where you see the *true shape* of the *base* of the cone, divide the base into a convenient number of equal divisions. (Number consecutively from #1–#12.)

b. Draw elements from v_1 (in view #1) to each numbered point on the cone's surface.

c. From view #1, project points #1–#12 to the base of the cone in view #2.

d. In view #2, draw elements from vertex v_2 to each point on the base.

e. Determine the *true lengths* of all the *elements* of the given cone. (Since the elements in a right cone are all equal in length, you need only determine the true length of one of them.) In view #1, element v_1–10 is parallel to RL 1–2 and, therefore, it appears as true length v_2–10 in view #2.

f. With *V* as a center, strike an arc having a radius equal to the true length element v_2–10. (See development.)

g. On this radius, measure the *true length chord distances* (they appear in view #1, in which the true shape view of the cone base is seen). Measure from points #1 to #2, #2 to #3, . . . , and so forth, returning to #1.

h. Connect point *V* in the development to each chord distance point on the arc with straight lines. This is the surface development of the cone.

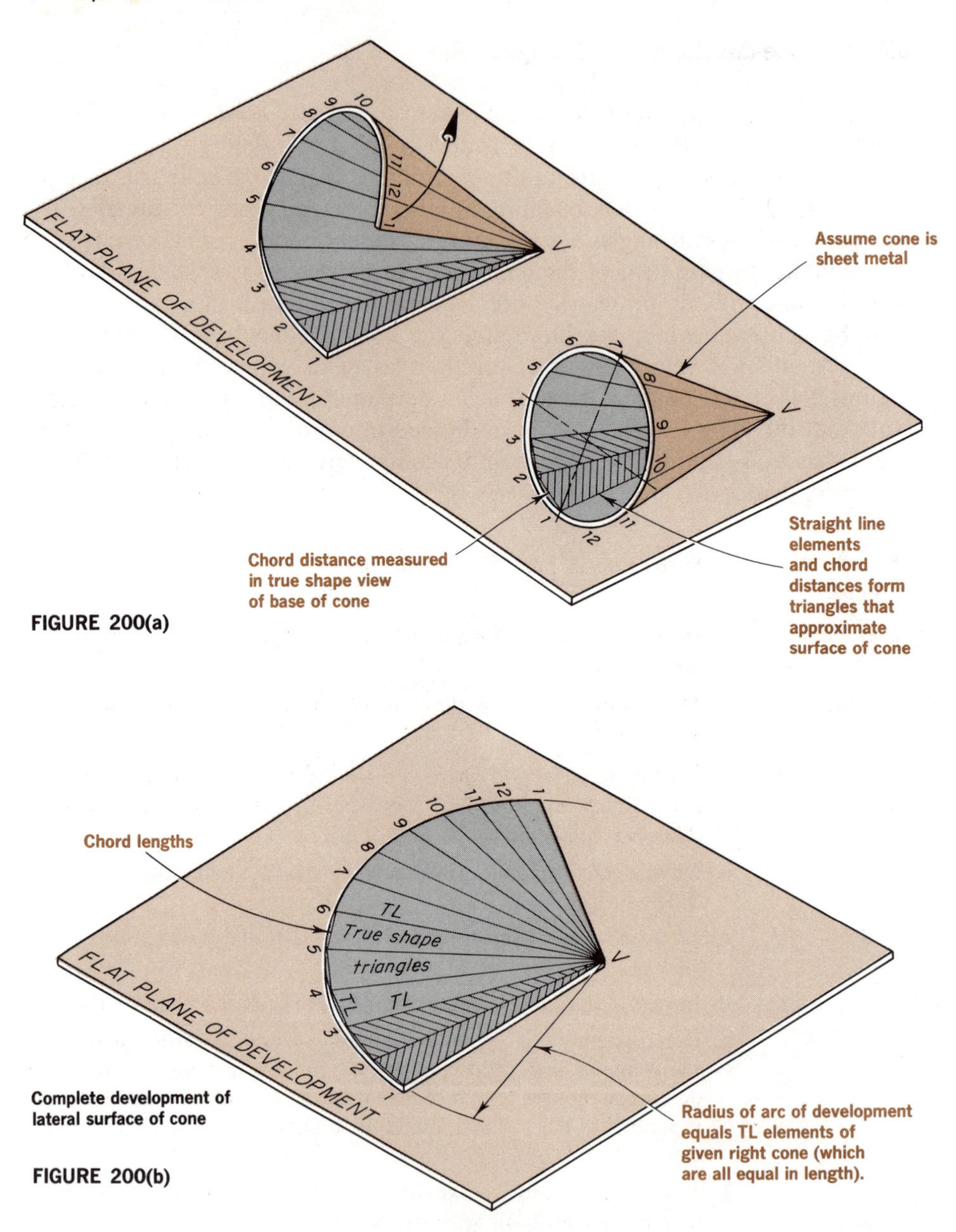

FIGURE 200(a)

FIGURE 200(b)

(Note that all lines are true lengths and that the individual triangles, into which the cone surface of the cone has been divided, appear as true shapes in the development.)

i. The true shape of the cone base appears as a circle in view #1. With a compass, transfer this base to the surface development attaching it at any convenient tangent point. This completes the required surface development, with base attached, of the right cone.

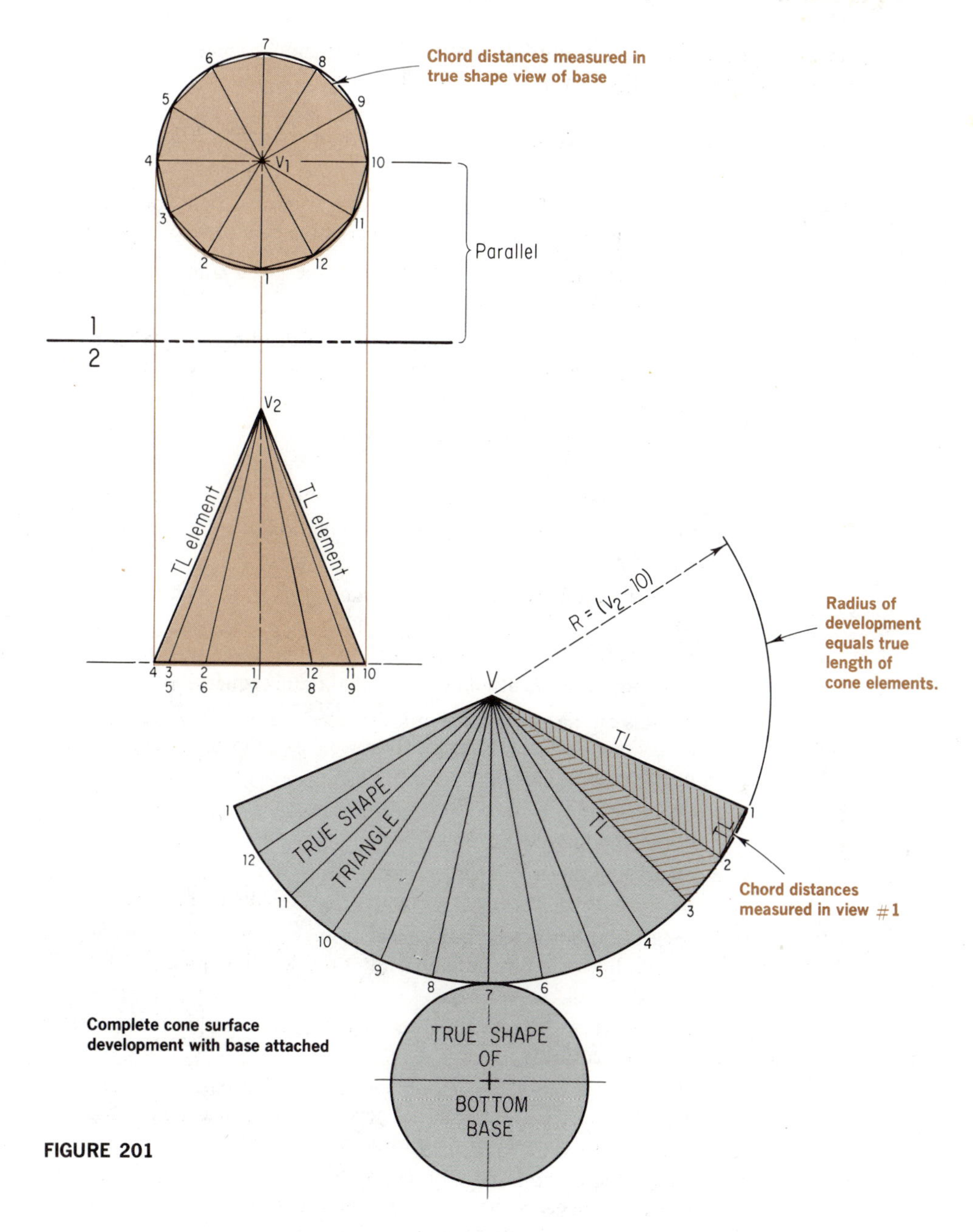

FIGURE 201

84 Surface development of a truncated right cone

The procedure you use to develop the surface of a truncated right cone is similar to that of a complete right cone. Begin the development by considering the truncated right cone to be a complete right cone and then proceeding to determine the individual lengths of the truncated elements.

Orthographic Example: Surface development of a truncated right cone (Fig. 202).

Given Data: Horizontal and frontal projections of a truncated right cone, the vertex of which is V (when the elements of the cone are extended).

Required: The complete surface development of the given cone, with top and bottom bases attached.

Construction Program:

a. Lightly complete the truncated cone by extending its outer elements until they intersect at vertex v_2 in view #2. (Determine v_1 by projection.)

b. In view #1, equally divide the true shape view of the bottom base (#1–#12).

c. From v_1, construct elements to each point on the base in view #1.

d. In view #1, project points #1–#12 downward to the base in view #2.

e. In view #2, construct elements from vertex v_2 to each point on the base.

f. Using a true length element such as v_2–1 in view #2, strike an arc from any convenient point V having a radius equal to v_2–1 (or v_2–7).

g. On this arc measure the chord distances #1 to #2, #2 to #3, #3 to #4, and so forth. Measure from view #1, in which the true shape of the bottom base appears.

h. From point V in the development, construct straight lines to each numbered point on the arc (V–1, V–2, V–3, and so forth).

i. Determine the true length distances from V to points #1′–#12′ on the *top* base of the cone (for example, v_2–$8_R'$ is the true length of v_2–8′).

j. Measure the true length distances from V to #1′–#12′ on the true length lines of the development, thus locating points #1′–#12′ and #1′.

k. Connect these points with a *smooth curve.* This completes the surface development of the cone.

l. The true shape of the bottom base, a circle, appears in view #1. Transfer this base to the surface development with a compass, attaching it at any convenient tangent point.

m. Determine the true shape of the top base by constructing a plane #3 parallel to the edge view of the top base as seen in view #2. The true shape of the top base appears in view #3.

n. Transfer with dividers the true shape of the top base to the surface development, attaching it at any convenient tangent point. This completes the required surface development, with top and bottom bases attached, of the given truncated right cone.

(NOTE: The surface development of a right cone and of a truncated right cone are *radial* developments and are therefore similar in principle to the development of a right pyramid and of a truncated right pyramid.)

FIGURE 202

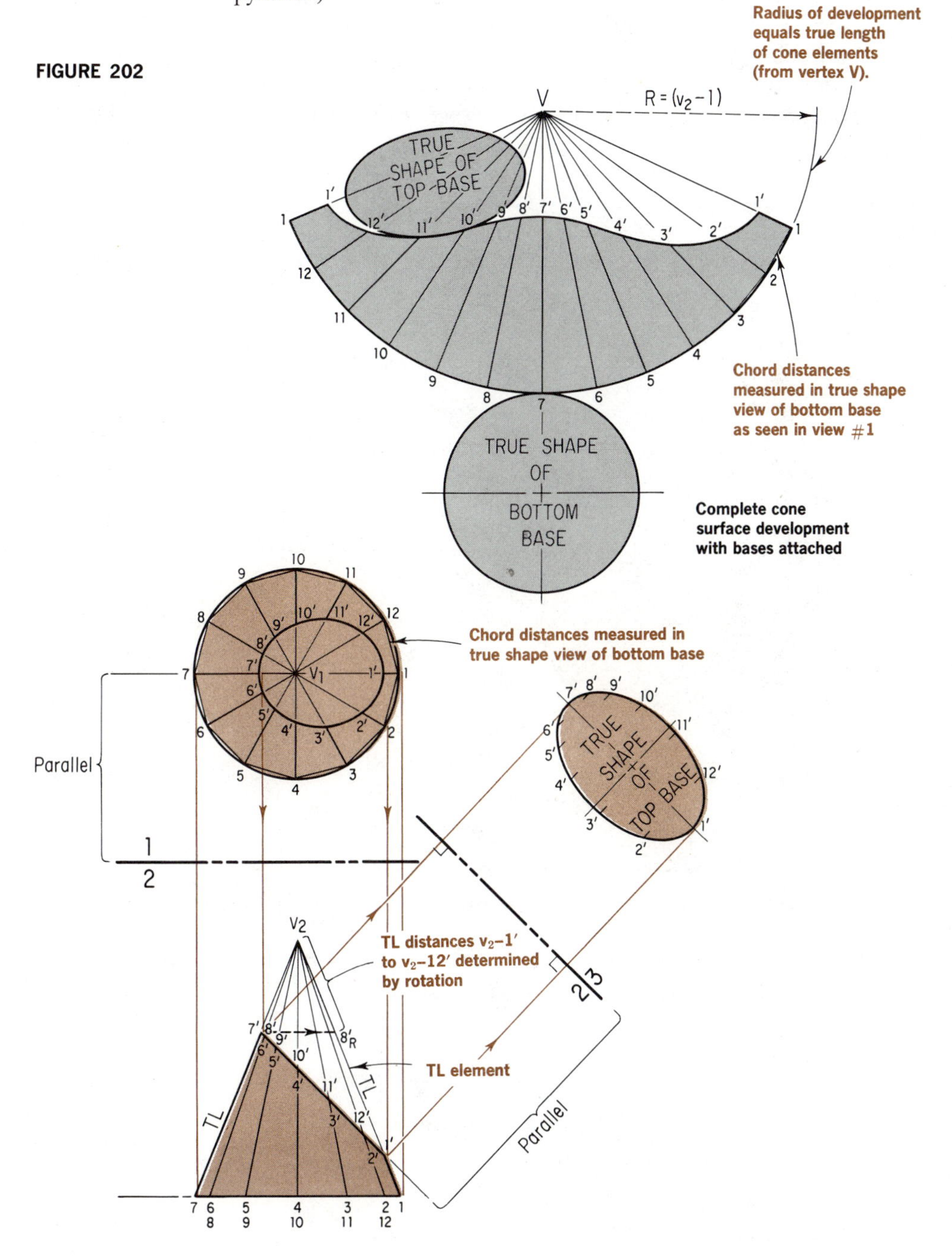

85 Surface development of a truncated oblique cone

The elements of an oblique cone are *not* all of equal length. Therefore, you must determine the true lengths of each element before you can develop the surface of an oblique cone. First, consider a truncated oblique cone to be a *complete* oblique cone. Divide the cone's surface into a finite number of triangular parts, which are used to approximately develop it. (The rotation procedure is the most

FIGURE 203(a)

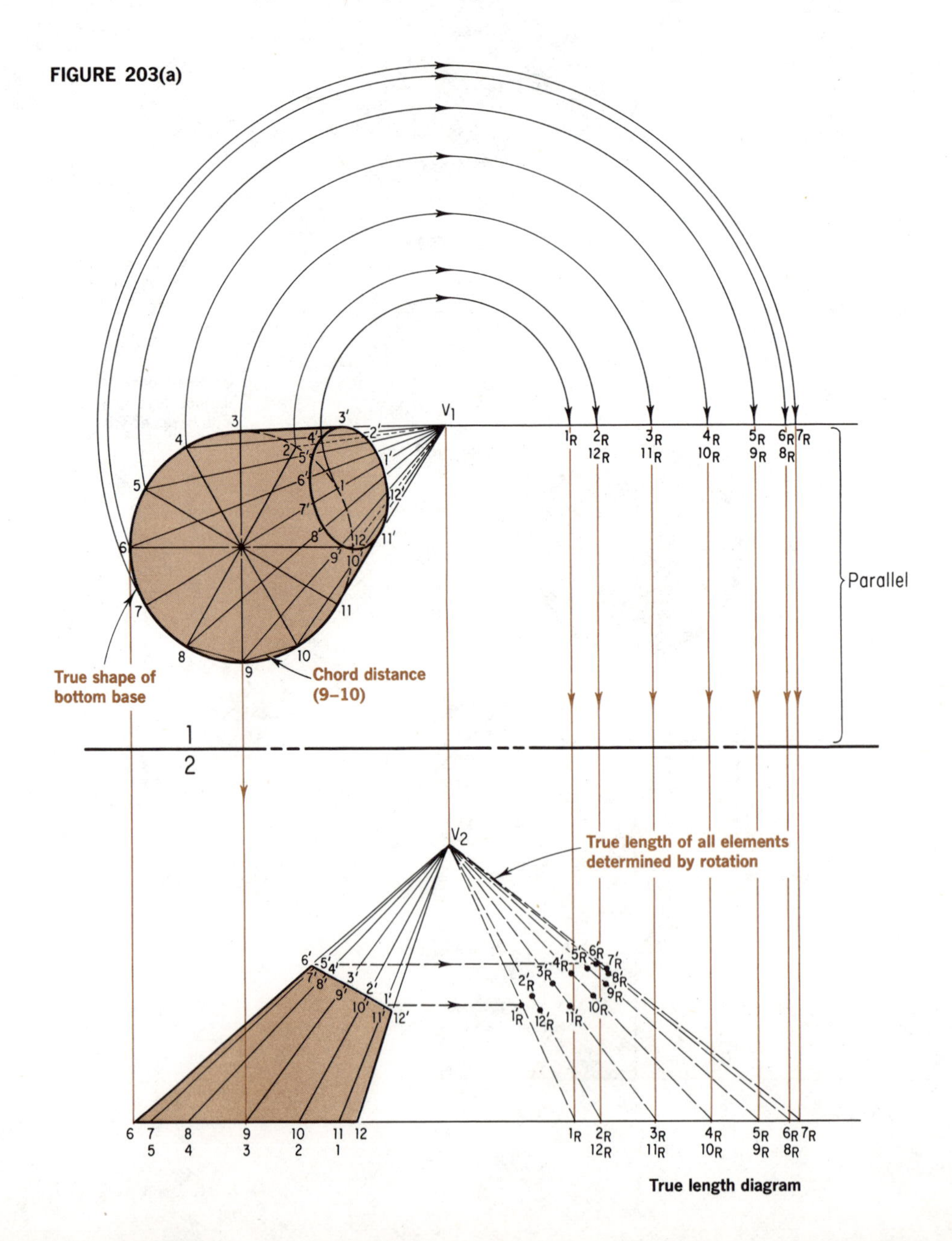

convenient method to use in determining the true lengths of the elements of the oblique cone.)

Orthographic Example: Surface development of a truncated oblique cone (Fig. 203).

Given Data: Horizontal and frontal projections of an oblique cone, the vertex of which is V (when its elements are extended).

Required: The lateral surface development of the given cone.

Construction Program:

a. In views #1 and #2, extend the outer elements of the given cone in order to locate vertex v_1 in view #1 and vertex v_2 in view #2.

b. Divide the true shape of the bottom base of the cone as seen in view #1 into a convenient number of equal divisions (#1–#12).

c. From v_1 in view #1, construct elements to each numbered point on the bottom base.

d. Project the points on the bottom base in view #1 downward to view #2.

e. From v_2 in view #2, construct elements to each numbered point on the bottom base.

f. Using the rotation procedure, determine the true lengths of the elements v_1–1 through v_1–12 (with v_1 serving as an axis of rotation). Note in view #1 that the direction of rotation is clockwise and that all the elements have been rotated into a position parallel to RL 1–2. For this reason, the true lengths of these elements are summarized in a "true length diagram" (view #2).

g. Determine by rotation the location of points #1′–#12′ on their respective lines in the "true length diagram." (Distances from v_2 to each prime-numbered point are true length distances such as v_2–$1_R'$, v_2–$2_R'$, etc.)

FIGURE 203(b)

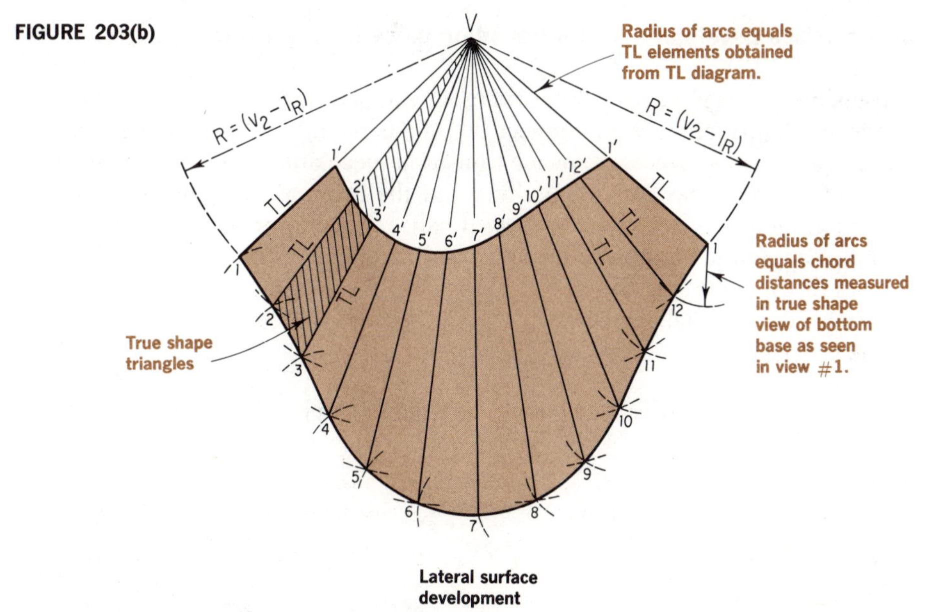

Lateral surface development

h. Starting at any convenient point V, begin the development by striking an arc having a radius equal to the *shortest element* v_2–1_R. Note point #1 on the arc.

i. From point #1, strike an arc having a radius equal to the chord distance 1–12 as it appears in the true shape view of the bottom base (view #1).

j. From V, strike the true length v_2–12_R of element v_2–12. Point #12 on the development can be found at the intersection of this arc with arc 1–12.

k. Using point #12 as a center, strike an arc having a radius equal to the chord distance 12–11, as seen in the true shape view of the bottom base (view #1).

l. From V, strike an arc equal to the true length v_2–11_R of element v_2–11. The intersection of these two arcs determines point #11 on the development.

m. Repeat the procedure until all the points have been located on the development. Connect these points with a *smooth curve.*

n. Referring to the true length diagram (in which the true lengths appear), measure the true length distances from v_2 to the prime-numbered points on the top base (v_2–$1_R'$, v_2–$2_R'$, etc.), on their respective elements in the development, thereby locating points #1′–#12′ and #1′. Connect these prime-numbered points with a *smooth curve.* This completes the required lateral surface development of the given truncated oblique cone.

(NOTE: In this procedure we are again determining true shape triangles into which the surface of the truncated oblique cone has been divided.)

86 Surface development of a transition piece having a warped surface

A transition piece is a geometric form used widely in sheet metal work. It is used in sheet metal duct work to connect two ducts having different cross sections; that is, the openings, or ends, of the transition piece differ from each other. One opening, for example, may be a circle and the other a rectangle. This type of transition piece has a *warped surface:* it can only be approximately developed. You also apply the method of triangulation here, that is, divide the surface into a series of triangles and find the true shapes of these triangles.

Figure 204 pictorially presents a transition piece that has two openings, both of them circles.

Orthographic Example: Surface development of a transition piece having a warped surface (Figs. 204–209, inclusive).

Given Data: Horizontal and frontal projections of a transition piece having a circular horizontal base and a circular top base that is not parallel to the bottom base. (The center of the bottom base is point *B*, and the center of the top base is point *A*.)

Required: The lateral surface development of the given transition piece.

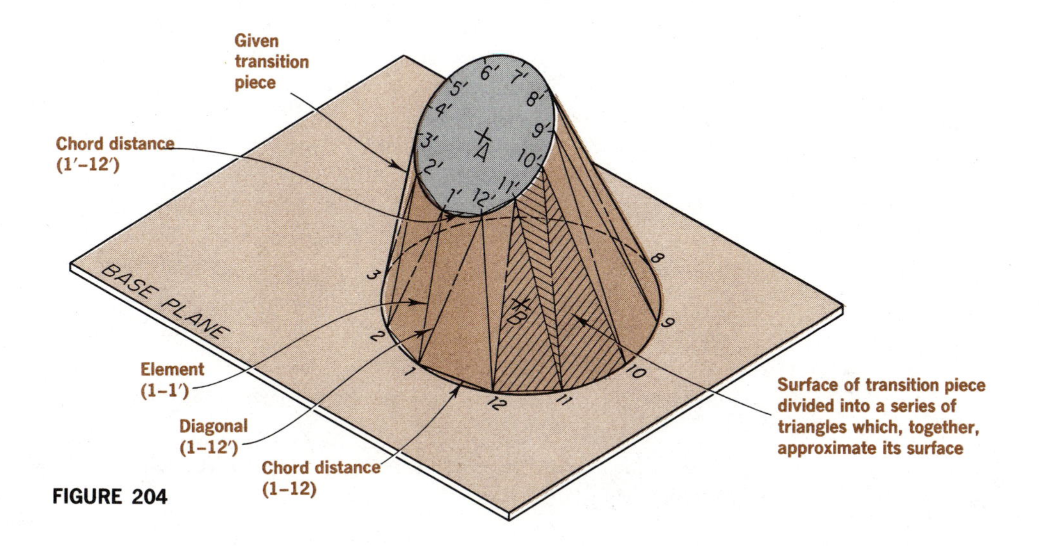

FIGURE 204

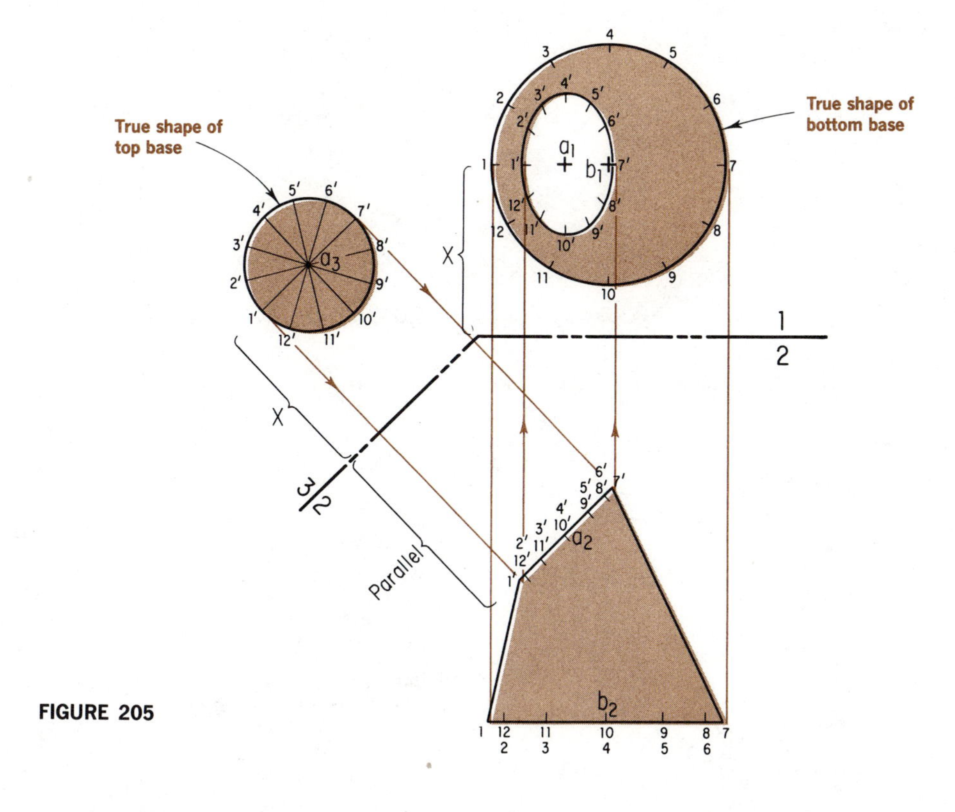

FIGURE 205

Construction Program:

NOTE: Carefully relate each step with the proper figure.

The given transition piece appears similar to a truncated cone. What, however, makes it different? The development involves two types of lines: *elements* and *diagonals,* located on the surface of the transition piece (see Fig. 204). Each of these lines will be clearly distinguished and must be separately identified. Figure 204 pictorially presents *elements* connecting *like-numbered* divisions on each base of the transition piece. The diagonals connect the numbers immediately *opposite* the counterparts (such as #1–#12′). Chord distances on both the bottom and top base planes (such as 1–12 and 1′–12′, respectively) are utilized.

a. Referring to Fig. 205, the true shape of the bottom base, a circle, appears in view #1. Divide it into a convenient number of equal parts (#1–#12).

b. Draw the true shape of the top base by constructing a plane #3 (represented by RL 2–3) parallel to the edge view of the top base. The true shape of the top base appears in view #3.

c. Divide the top base into the same number of parts as the bottom base.

(NOTE: Point #1 on the bottom base and point #1′ on the top base are both located at the same distance *X* from the frontal projection plane.)

FIGURE 206

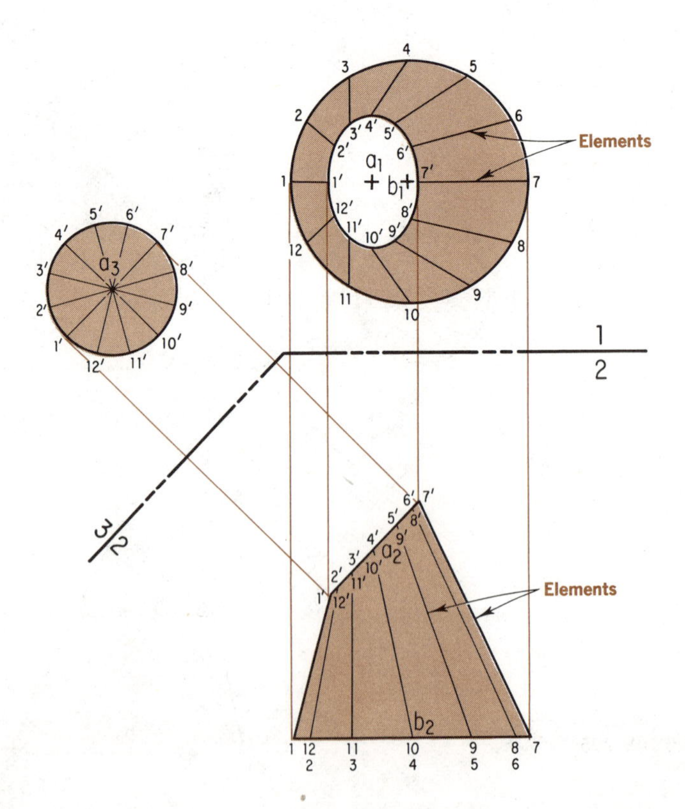

d. Referring to Fig. 206, establish straight line *elements* on the surface of the transition piece by connecting *like-numbered* points on the bottom and top bases, such as #1 with #1′, #2 with #2′, #3 with #3′, and so forth (see views #1 and #2 in Fig. 206).

e. Referring to Fig. 207, draw dotted lines to connect *opposite* points from the bottom base to the top base, such as #1 with #12′, #2 with #1′, #3 with #2′, and so forth. These are the straight line *diagonals* on the surface of the transition piece.

f. In Fig. 207, the chord distances #1 to #2, #2 to #3, #3 to #4, and so forth, are drawn in the true shape view of the bottom base (view #1). The chord distances #1′ to #2′, #2′ to #3′, #3′ to #4′, and so forth, are drawn in the true shape view of the top base (view #3).

g. Determine the true lengths of *elements* 1–1′, 2–2′, 3–3′, and so forth, and the true lengths of *diagonals* 1–12′, 2–1′, 3–2′, and so forth, by rotation. To facilitate this, a short-cut method has been illustrated in Fig. 208. In this figure, element 8–8′ is shown in its horizontal and frontal projections. To find its true length by rotation, define the *axis of rotation* and move it to a convenient location, so as to prevent the overlapping of the given views and the true length views. The horizontal projection of element 8–8′ is indicated as distance *H*. By transferring, with dividers, distance *H* to the new location of the axis of

FIGURE 207

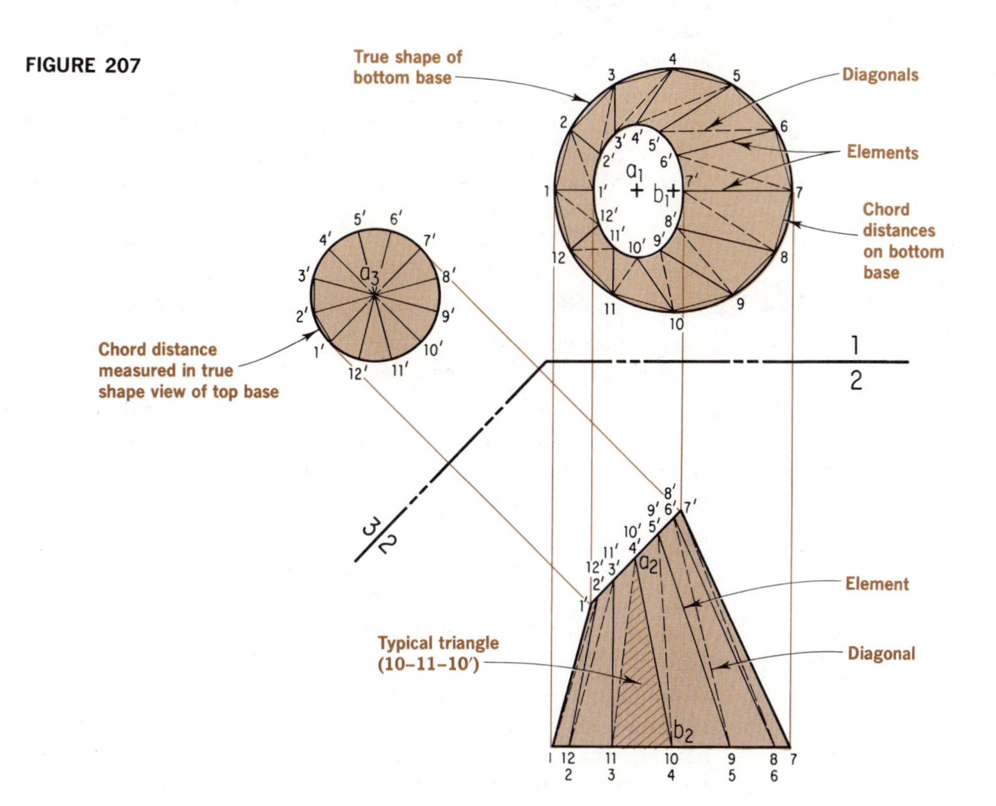

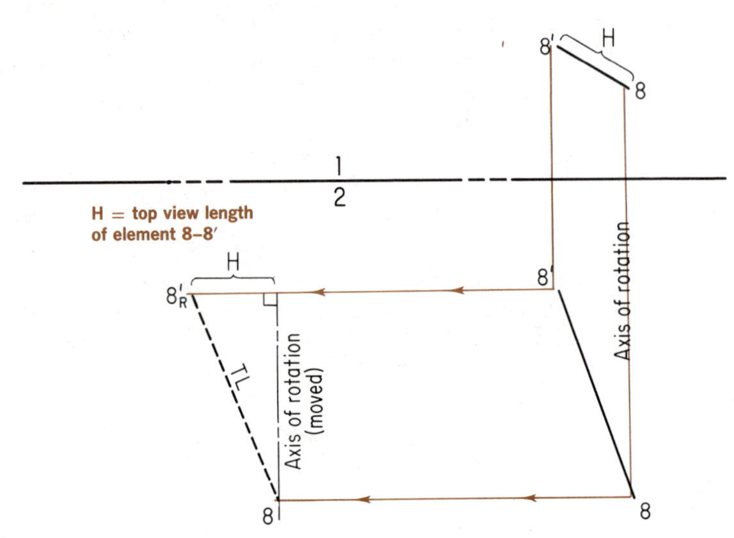

FIGURE 208

rotation, and by measuring it *perpendicular* to the axis of rotation, you will locate $8_R'$ (the revolved position of 8').

The line 8–$8_R'$ is the true length of the element 8–8'. This procedure is used to determine the true lengths of all the elements and diagonals indicated on the surface of the transition piece (summarized in the true length diagram in Fig. 209).

h. Referring to Fig. 209, start at any convenient place and note it as #1'. Strike an arc having a radius equal to the shortest true length element 1'–1. From #1', strike an arc having a radius equal to the true length chord distance 1'–2', as seen in the true shape view of the top base (view #3). From point #1 strike an arc equal to the true length of diagonal 1–2'. The intersection of this arc and chord distance 1'–2' locates #2' on the development.

i. From #2' strike an arc equal to the true length of element 2'–2.

j. From point #1, strike the chord distance 1–2, as seen in the true shape view of the bottom base (view #1). The intersection of element 2'–2 and chord distance 1–2 locates point #2 on the development.

k. From #2', construct an arc equal to the chord distance 2'–3', as seen in the true shape view of the top base (view #3).

l. From #2', strike an arc equal to the true length of the diagonal distance 2'–3.

m. From point #2, strike an arc equal to the chord distance 2–3, as seen in the true shape view of the bottom base (view #1). The intersection of chord distance 2–3 and diagonal 2'–3 locates point #3 on the development.

n. Continue this procedure for all elements, diagonals, and chord distances on the bottom base, and for the chord distances on the top base, thereby completing the development. Each line in this development appears as *true length* and, therefore, the chord distances,

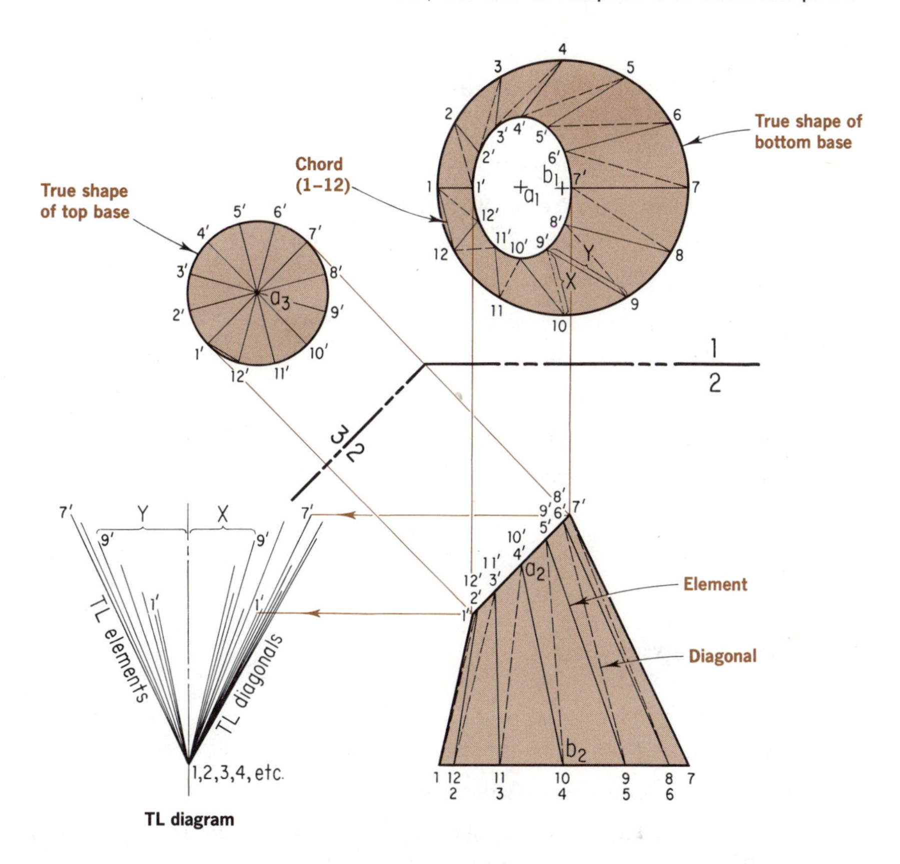

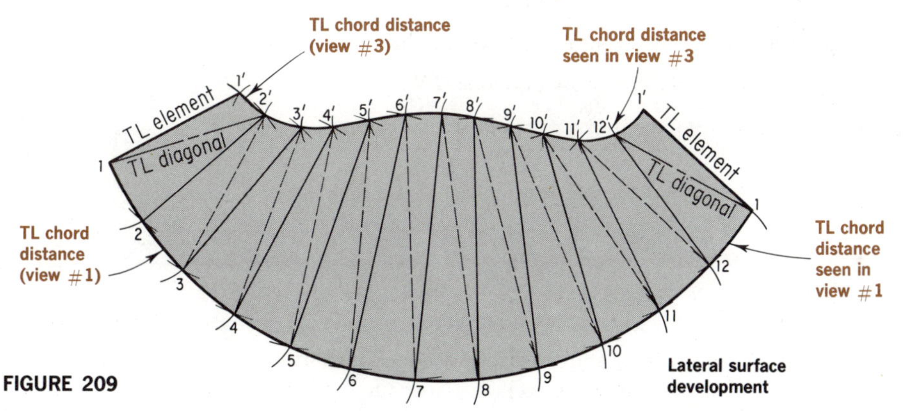

FIGURE 209

together with the diagonals and elements, form *true shapes* of the triangles that approximate the surface of the transition piece. Referring to the lateral surface development in Fig. 209, connect points #1′–#12′ and #1′ with a smooth curve and points #1–#12 and #1 with a smooth curve. This is the required lateral surface development of the given transition piece.

SAMPLE QUIZZES

1. What is meant by the "development" of a surface?

2. In developing an oblique cone of revolution, is triangulation

   (a) necessary? (b) sometimes necessary? (c) never necessary?

Verify your answer with a sketch.

3. In developing a right cone of revolution, is triangulation

   (a) necessary? (b) sometimes necessary? (c) never necessary?

Verify your answer with a sketch.

4. In developing the right cone of an elliptical cross section, is triangulation

   (a) necessary? (b) sometimes necessary? (c) never necessary?

Verify your answer with a sketch.

5. Without using a sketch briefly define the following:

   a. Element.
   b. Ruled surface.
   c. Double-curved surface.
   d. Cylindrical surface.
   e. Warped surface.

6. How is a cone of revolution usually developed?

7. How is a cylinder of revolution usually developed?

8. Explain briefly how to use the method of triangulation to develop the surface of a truncated cone that is *not* a cone of revolution.

9. How is a cylinder of revolution formed and how is it developed?

10. Can a warped surface be truly developed? Why?

11. Why is the development of a pyramid similar to that of a cone?

PRACTICE PROBLEMS

1. Given Data: Horizontal and frontal projections of a truncated right prism having a pentagonal cross section. (Note the double truncation.)

Required:

A. Starting from the shortest lateral edge, develop the lateral surface of the given prism.

B. Attach the true shape of the prism's top and bottom bases to the development. (Note that the top base consists of two intersecting planes.)

C. Label and show all construction.

PROBLEM 1

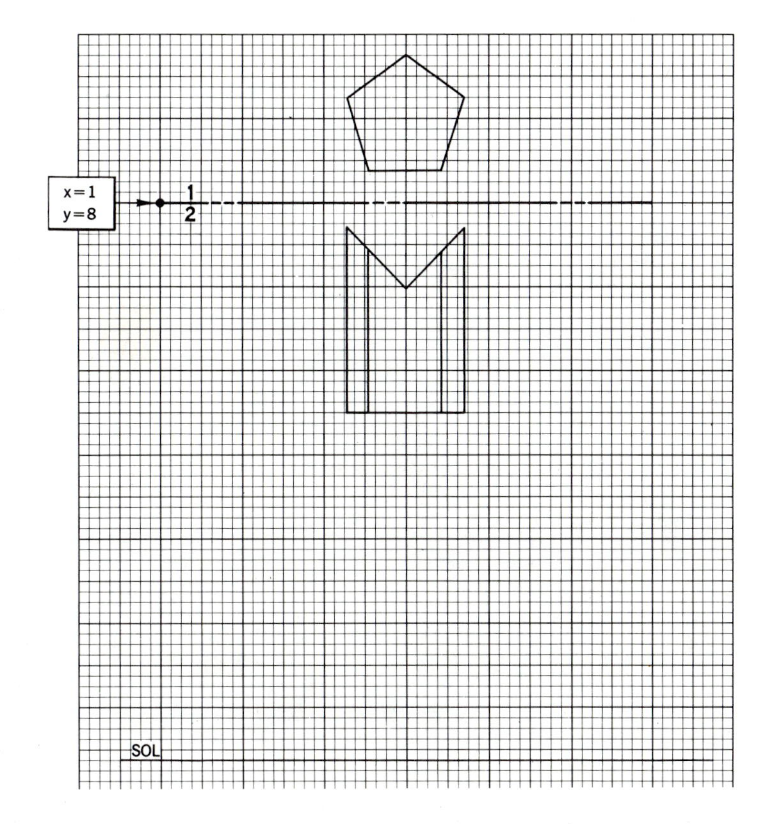

2. Given Data: Horizontal and frontal projections of an oblique prism having a pentagonal cross section.

PROBLEM 2

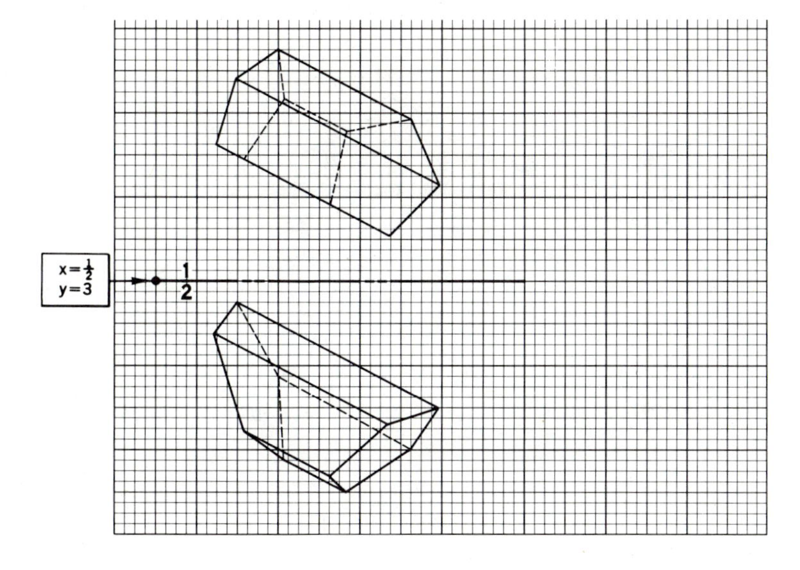

Required:

A. Starting from the shortest lateral edge, develop the lateral surface of the given prism.

B. Attach the true shape of the top base to the development. (SUGGESTION: Use separate sheet of cross-section paper for the development.)

C. Label and show all construction.

3. Given Data: Horizontal and frontal projections of a truncated circular right cylinder. (Note the double truncation.)

Required:

A. Starting from the shortest element, develop the lateral surface of the given cylinder.

B. Attach the true shape of the bottom base and the left and right halves of the top base to the development.

C. Label and show all construction.

PROBLEM 3

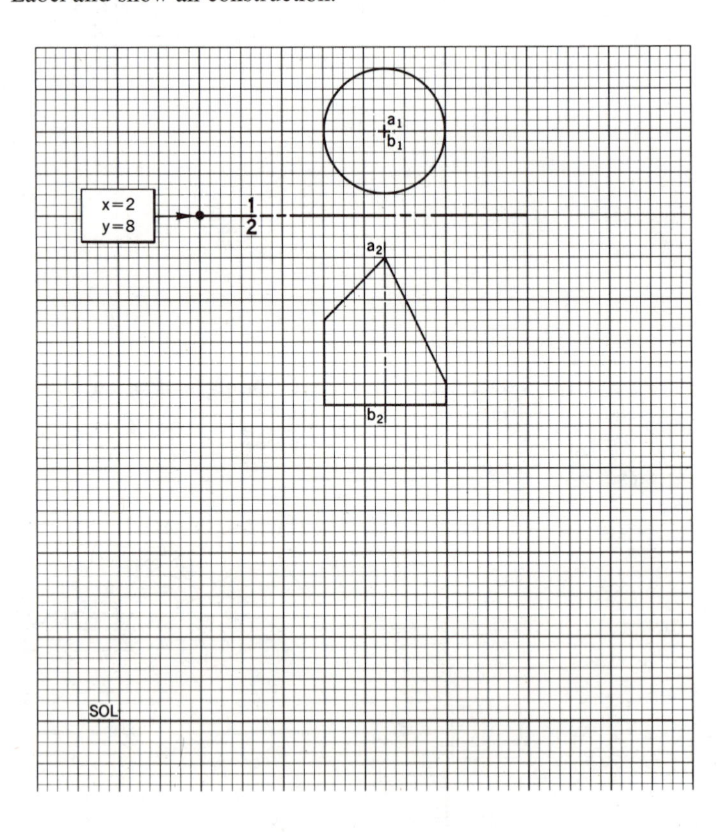

4. Given Data: Horizontal and frontal projections of a truncated oblique cylinder.

A. Starting with the shortest element, develop the lateral surface of the given cylinder.

B. Label and show all construction.

PROBLEM 4

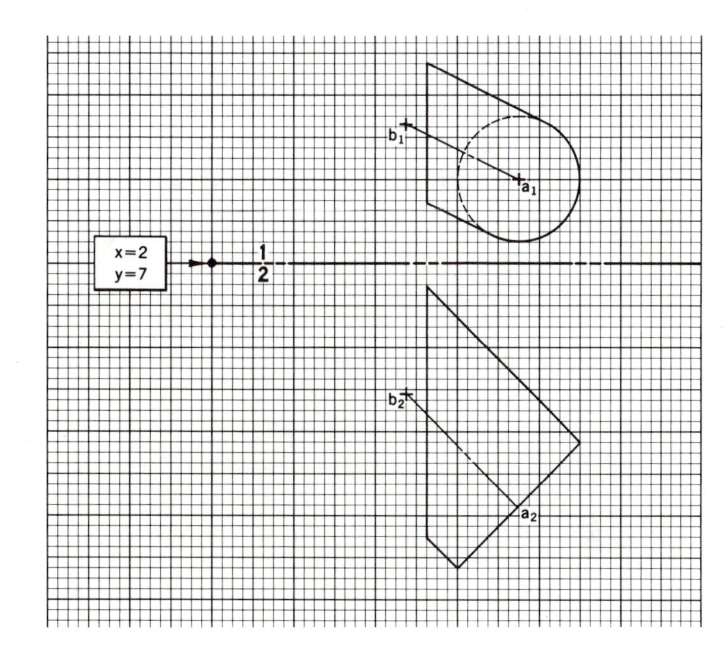

5. Given Data: Horizontal, frontal, and right profile projections of a truncated right pyramid having a vertex *V* and a base that is parallel to the right profile projection plane.

Required:

A. Starting from the shortest lateral edge, develop the lateral surface of the given pyramid.

B. Attach the true shapes of the top and bottom bases to the development.

C. Label and show all construction.

PROBLEM 5

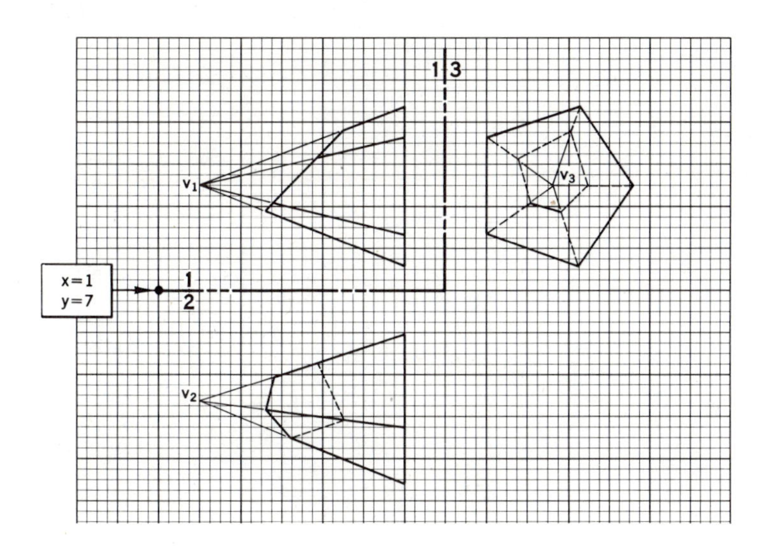

6. Given Data: Horizontal and frontal projections of a truncated oblique pyramid having a vertex *V* and an inclined base.

Required:

A. Starting from the shortest edge, develop the lateral surface of the given pyramid.

B. Label and show all construction.

PROBLEM 6

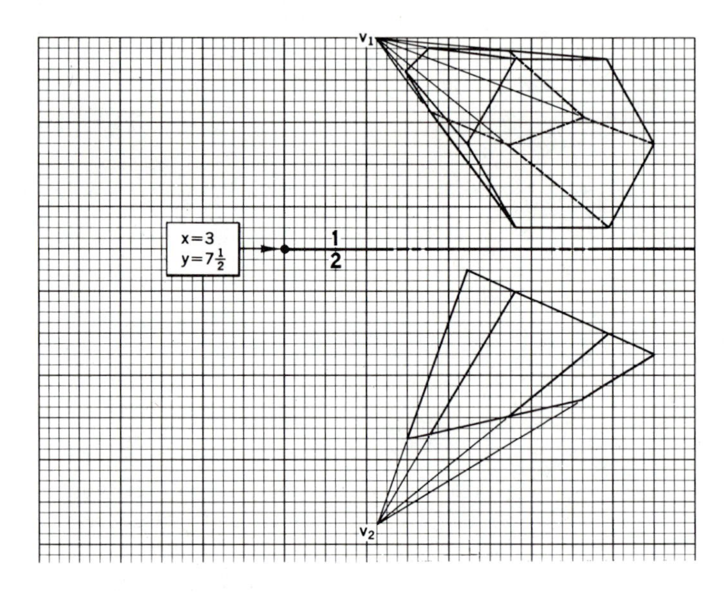

7. Given Data: Horizontal, frontal, and partial right profile projections of a truncated right circular cone having a vertex *V* and a base that is parallel to the right profile projection plane.

PROBLEM 7

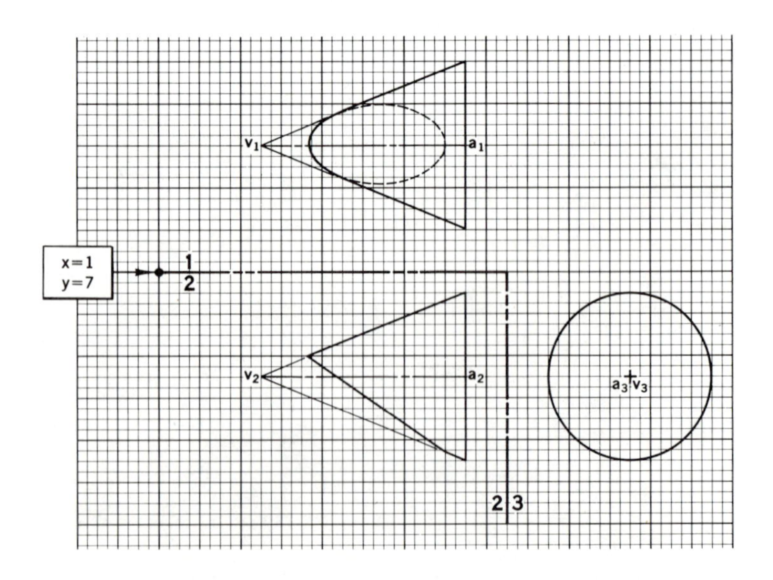

Required:

A. Starting from the shortest element, develop the lateral surface of the given cone.

B. Attach the true shapes of the top and bottom bases to the development.

C. Label and show all construction.

8. Given Data: Horizontal and frontal projections of a truncated oblique cone having a vertex *V* and a base that is parallel to the horizontal projection plane.

Required:

A. Starting from the shortest lateral element, develop the lateral surface of the given cone.

B. Label and show all construction.

PROBLEM 8

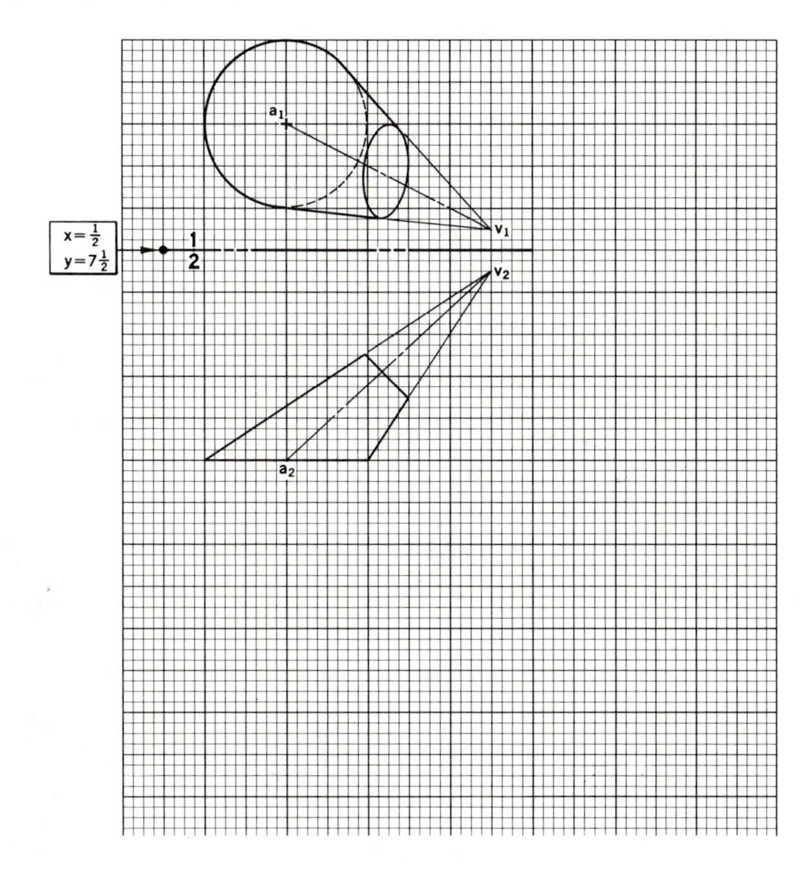

9. Given Data: Horizontal, frontal, and a partial left profile projection of a transition piece having circular openings in different planes.

Required:

A. Starting from the shortest element, develop the lateral surface of the given transition piece. (SUGGESTION: Use a separate sheet of cross-section paper for the surface development.)

B. Label and show all construction.

PROBLEM 9

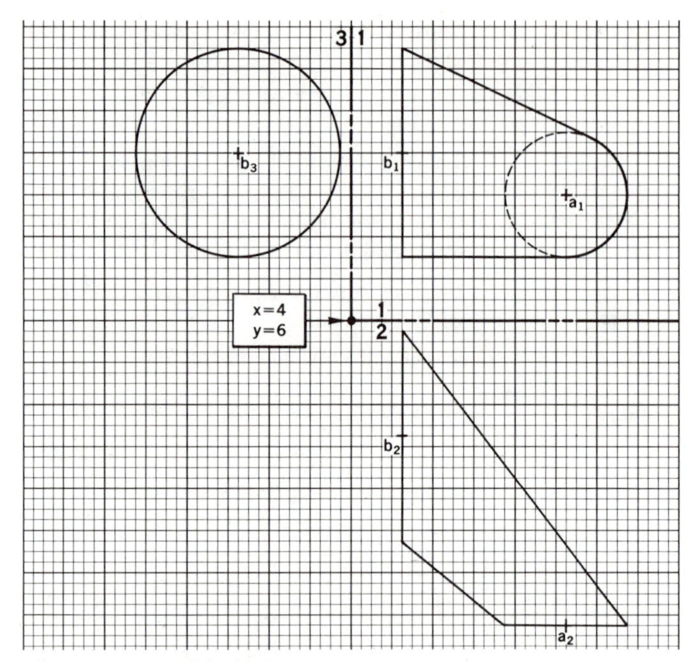

10. Given Data: Horizontal and frontal projections of a transition piece having a top circular opening and a bottom square opening.

Required: A. Starting from the shortest element, develop the lateral surface of the given transition piece. (SUGGESTION: Use a separate sheet of cross-section paper for the surface development.)

B. Label and show all construction.

PROBLEM 10

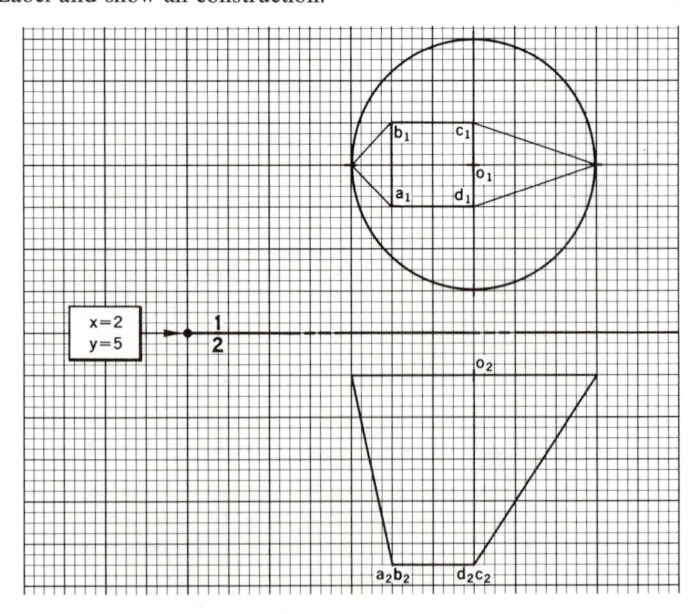

Principles of Descriptive Geometry Applied to Three-Dimensional Space Vectors

87 Introduction

Science, engineering, and architecture are basically concerned with the effects of forces in and on natural bodies and man-made structures. These forces in nature include nuclear forces, gravitational forces, wind forces, snow forces (such as snow loads on the roof of a building), hydraulic forces due to water (such as the force of water acting against a dam). Forces that act on man-made structures include so-called "dead" and "live" loads. Dead loads (such as the weight of furniture) remain *stationary* on the floor of a building. Live loads *travel* across the floor (for example, a person's weight as he walks across the floor or the movement of trucks in factories). There are dead and live loads in machinery, as well as dynamic loads that involve inertia forces, vibrational forces, and so forth.

In order to analyze efficiently the various types of force systems that occur in nature and that act on man-made structures, various methods must be devised. Much force analysis is done through graphical and mathematical methods. In many cases, graphical methods can be used to check the results of mathematical analyses. In certain problems, graphical methods prove to be more direct and, therefore, more efficient.

We present in this chapter a number of force systems that can be analyzed by applying the principles of descriptive geometry. They can also be analyzed mathematically, using the algebraic–trigonometric approach, or the approach of vector notation (see the Bibliography).

88 The concept of a force

Force is the *effect* of one body on another, tending to change the state of motion, or rest, of the two bodies, for example, push and pull.

In Fig. 210 we see the gravitational forces of the earth acting on a weight hanging from a rope that is attached to a structure that rests on the earth's surface. The gravitational forces *pull* the weight *downward,* while the effect of the rope *pulls* the weight *upward.* Figure 211 illustrates the effect of one body on another; a rope is *pulling upward* against the weight and the weight is *pulling downward* against the rope.

In Fig. 212 a beam that rests on two knife-edge supports is carrying a load of weight W. Because of gravitational forces, the weight of the beam *acts* downward, while the supports at points *A* and *B* push upward (see Fig. 213).

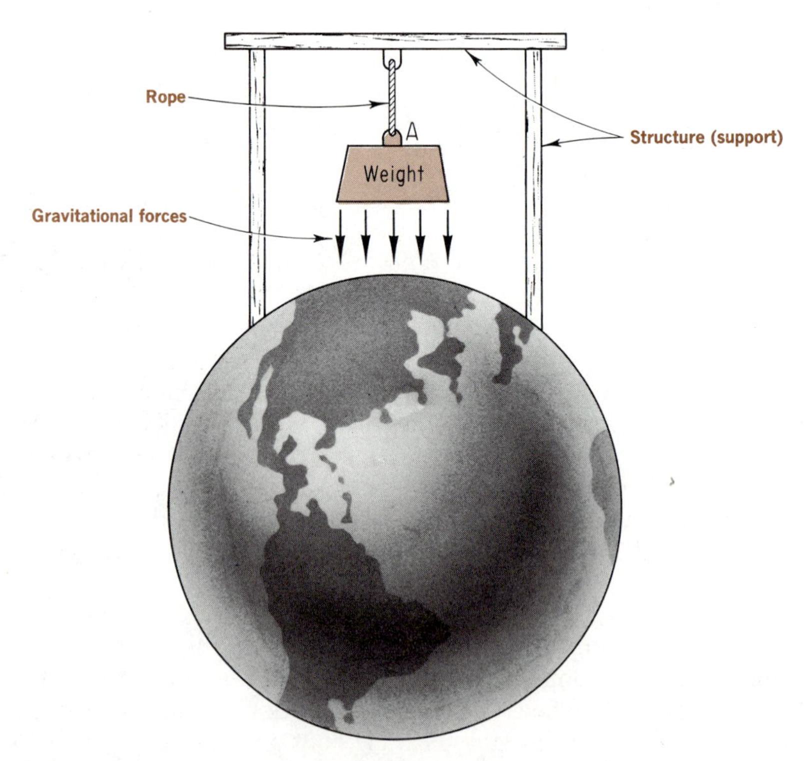

FIGURE 210

FIGURE 211

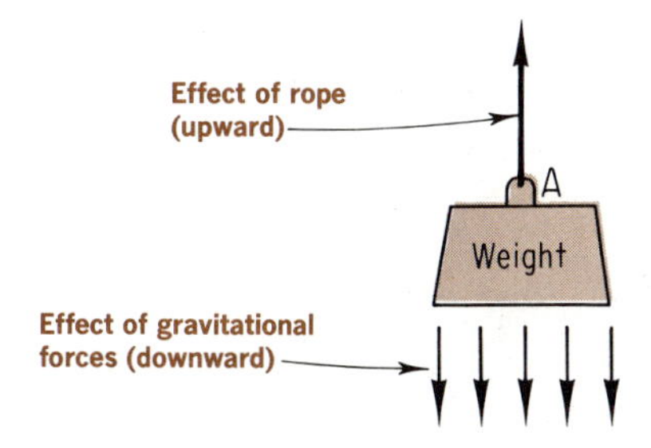

FIGURE 212

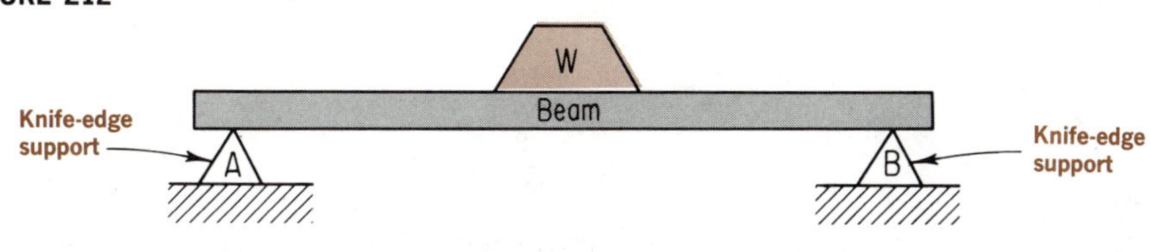

FIGURE 213

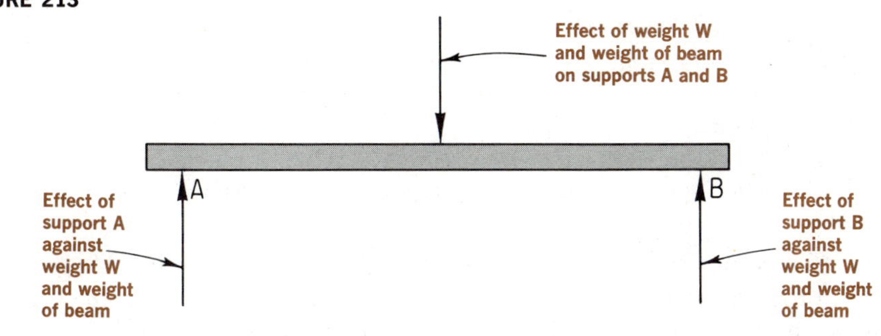

89 The vector as a geometric concept

You must know the magnitude, sense, and direction of a force in order to describe it physically.

In Figs. 211 and 214 we can see that the effect of the rope supporting weight W is such that it acts against the total effect of gravity on the given weight. The force in the rope has a specific direction (that is, a line of action with the *total* effect of gravity of the given weight), a specific sense (that is, its upward action as indicated by the arrow), and a definite point of application (A). From the

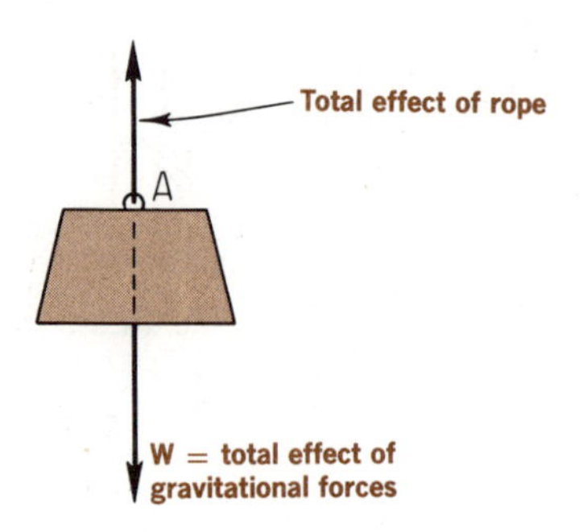

FIGURE 214

physical description we can determine certain basic factors that will allow us to completely describe a force:

A. A FORCE MUST HAVE MAGNITUDE.

The magnitude of a force is the measure of its intensity. The magnitude of a force, for example, can be measured in pounds.

B. A FORCE MUST HAVE A POINT OF APPLICATION WHEN IT ACTS ON A BODY.

The point of application of a force is the point of contact between the force and the body upon which it acts.

C. A FORCE MUST HAVE A LINE OF ACTION.

Such a line of action is the direction of the rope in Fig. 214.

D. A FORCE MUST HAVE A SENSE.

The sense indicates whether the forces are pushing or pulling on a body. This is usually denoted by an arrow as in Figs. 213 and 214.

The foregoing list defines the geometric concept of a force. This geometric concept may be represented graphically by a *vector.* A vector is a line (in a specific position with respect to a frame of reference) drawn to a definite length (a given scale). An arrowhead at one end of the vector indicates sense. Therefore a vector can be considered as a physical *model* of a force, represented in geometric or graphical form. Thus, a force acting on a body is a *vector quantity* having a specific point of application.

In Fig. 215, a vector AB has been drawn to a scale in which 2 in. = 200 lb.* This means that vector AB represents a force of 200 lb and has a sense indicated by the arrowhead at B, a position relative to a frame of reference (the horizontal projection plane), and a direction of N30°W. Thus you can see that all the specifications of a vector have been satisfied (except a point of application).

* Vector scales used in this chapter are not full size due to the reduction of figures.

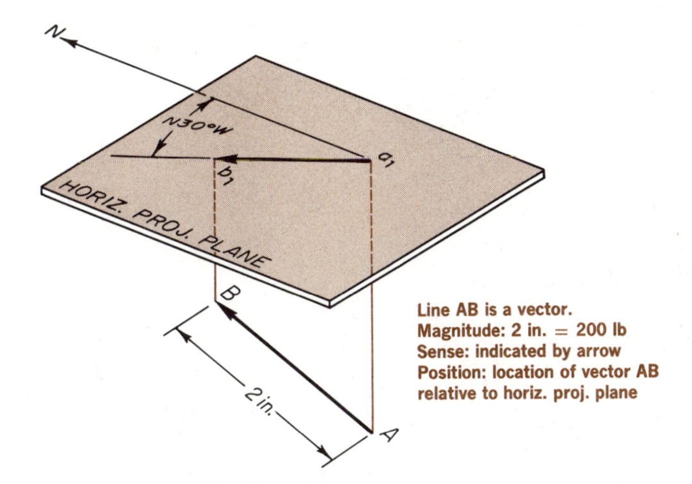

FIGURE 215

90 Force system

When a group of forces, whatever their nature may be, acts on an object, it is defined as a *force system.* For example, Fig. 211 shows a force system that involves (1) the gravitational forces of the earth acting on a weight, (2) the force in a rope supporting this weight and resisting the gravitational forces, and (3) a structure that rests on the surface of the earth and to which the rope is attached. (This structure has forces in it that resist both the force in the rope and the gravitational forces of the earth.)

The general types of force systems are as follows:

A. COPLANAR FORCE SYSTEMS

The lines of action of the forces in these systems lie in one plane [see Figs. 216(a) and 216(b)].

B. NONCOPLANAR FORCE SYSTEMS

The forces of the lines of action in these systems do not lie in one plane (see Figs. 217–219, 223, 227, and 228).

C. CONCURRENT FORCE SYSTEMS

The lines of action of the forces in these systems meet at one point. Such force systems may be either coplanar or noncoplanar [see Figs. 216(a), 216(b), 217, 220, 222, and 227].

D. NONCONCURRENT FORCE SYSTEMS

The lines of action of the forces in these systems do not meet at one point. Such force systems may be coplanar or noncoplanar (see Figs. 218 and 219).

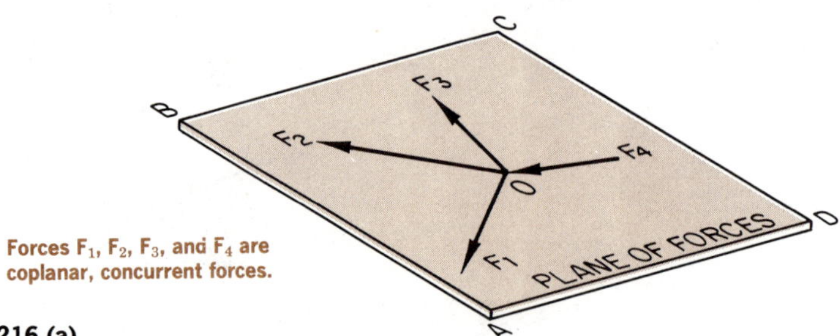

Forces F_1, F_2, F_3, and F_4 are coplanar, concurrent forces.

FIGURE 216 (a)

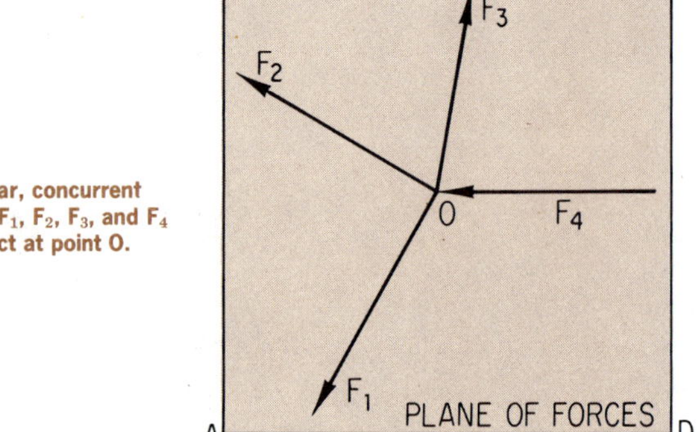

Coplanar, concurrent forces F_1, F_2, F_3, and F_4 intersect at point O.

FIGURE 216 (b)

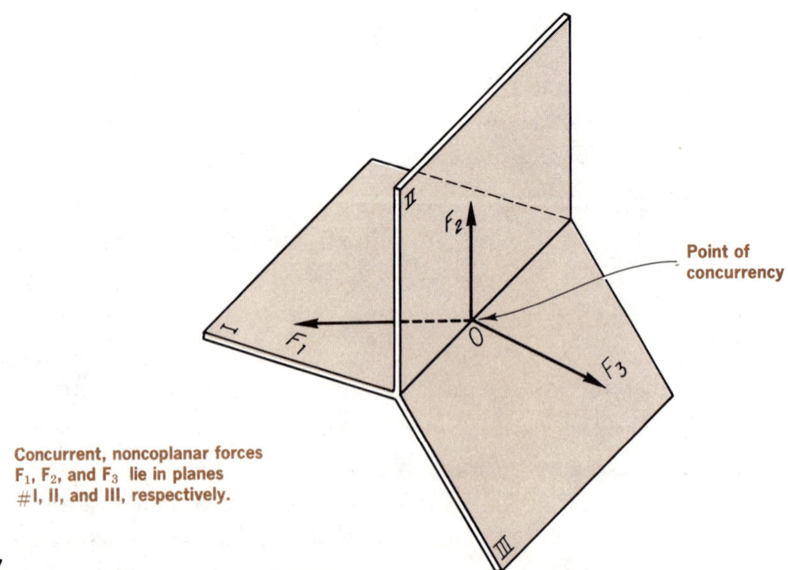

Concurrent, noncoplanar forces F_1, F_2, and F_3 lie in planes #I, II, and III, respectively.

FIGURE 217

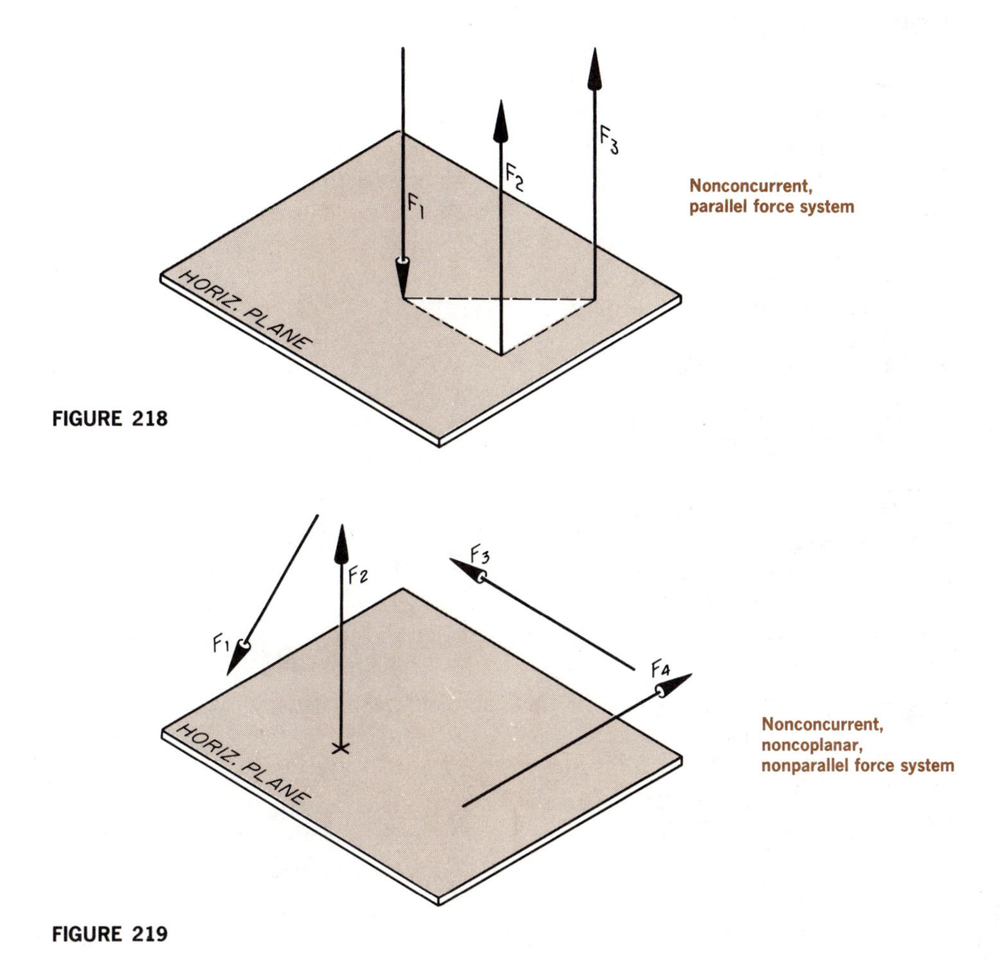

FIGURE 218

FIGURE 219

In this chapter we present methods of solving problems involving concurrent, noncoplanar force systems as well as a nonconcurrent, noncoplanar parallel force system. Sections 91 and 92 include references to simple *coplanar* force systems, making it easier for you to understand the transition from two-dimensional (coplanar) force systems to three-dimensional (noncoplanar) force systems.

91 Equilibrium of a force system

When we deal with force systems acting on structures that are in a state of rest relative to a given frame of reference (such as a building standing on the ground or a bridge suspended between two points across a river), we are concerned with *balanced* force systems. If a given external force system is applied to a given structure in a state of rest, we will discover whether the structure will remain in a state of rest. If it moves from its original position, we must determine what additional force can be applied to the structure in order to keep it at rest. If the structure remains at rest when we apply the initial force system, we must conclude that the structure has forces acting on and resisting it, creating a *balanced* condition. The structure can then be said to be in a state of *static equilibrium.*

92 Resultant of a force system

You must know the *total effect* of a given force system on a structure in order to determine its state of equilibrium. This total effect is called the *resultant* of a force system, and can be represented by a *vector.* The resultant, the simplest form of a given force system, is determined by adding the forces in the system.

Figure 220 shows two concurrent, coplanar forces, represented by vectors $\overrightarrow{F_1}$ and $\overrightarrow{F_2}$, acting at a point O. $\overrightarrow{F_1}$ and $\overrightarrow{F_2}$ form a plane since they intersect at O. By completing the parallelogram, the sides of which are equal in length to the vectors representing F_1 and F_2, and by drawing the *diagonal* from O (the point of application of the two forces), you can determine the resultant of the two forces. The resultant is a vector, the length of which starts at point O and terminates at point R. The length of this vector $\overrightarrow{O\text{–}R}$, measured to a vector scale, defines the *magnitude* of the resultant.

In Fig. 221, three concurrent, noncoplanar forces are represented by vectors $\overrightarrow{F_1}$, $\overrightarrow{F_2}$, and $\overrightarrow{F_3}$. The *resultant* vector $\overrightarrow{O\text{–}R}$ is determined by completing the parallelepiped, the sides of which are equal in length to the vectors representing F_1, F_2, and F_3, and by drawing the diagonal O–R of the parallelepiped.

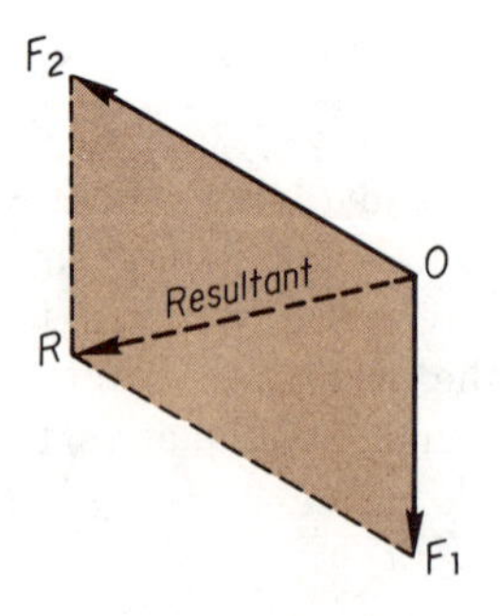

Vectors F_1 and F_2 represent two concurrent, coplanar forces.

FIGURE 220

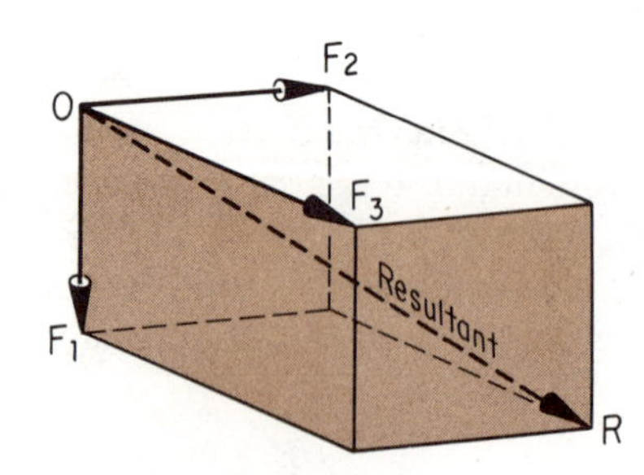

Vectors F_1, F_2, and F_3 represent three concurrent, noncoplanar forces.

FIGURE 221

93 Equilibrant of a force system

If a force system is *balanced,* or in a state of rest, the *resultant* of the total force system has a magnitude equal to *zero;* that is, the total effect of all the forces in the system is zero. If a *balanced* force system acts on a structure that is at rest, it will have no effect on the equilibrium of the structure. The resultant of the entire force system must be equal to zero. On the other hand, if an *unbalanced* force system acts on a structure that is in a state of rest, it may change the position or the equilibrium of the structure. If the structure remains at rest and in its original position, it is *counterbalancing* the effect of the *resultant* of the applied unbalanced force system. The counterbalancing effect of the structure is actually another force equal in magnitude, with the same line of action and a sense opposite to that of the *resultant* of the applied force system. This counterbalancing force is called the *equilibrant* of the entire force system.

In Fig. 222(a) we see two coplanar forces, represented by the vectors $\overrightarrow{F_1}$ and $\overrightarrow{F_2}$, that are concurrent at a point O. The resultant of these two forces (as we have seen before) is represented by the diagonal of the rectangle O–R.

For the two forces acting at point O to have no effect on point O, they must be resisted by an equilibrant equal and opposite to that of their resultant. Figure 222(b) shows the equilibrant. Note that the magnitude and line of action of the equilibrant are the same as that of the resultant, but that its sense is *opposite.*

Let us apply the concept of equilibrium to a concurrent, noncoplanar force system. In Fig. 223(a) we see three forces pictorially represented by their vectors $\overrightarrow{F_1}$, $\overrightarrow{F_2}$, and $\overrightarrow{F_3}$, all acting from point O. The *resultant* of these forces is represented by a vector $\overrightarrow{O\text{–}R}$. $\overrightarrow{O\text{–}R}$ is the diagonal of the parallelepiped having sides equal in length to $\overrightarrow{F_1}$, $\overrightarrow{F_2}$, and $\overrightarrow{F_3}$.

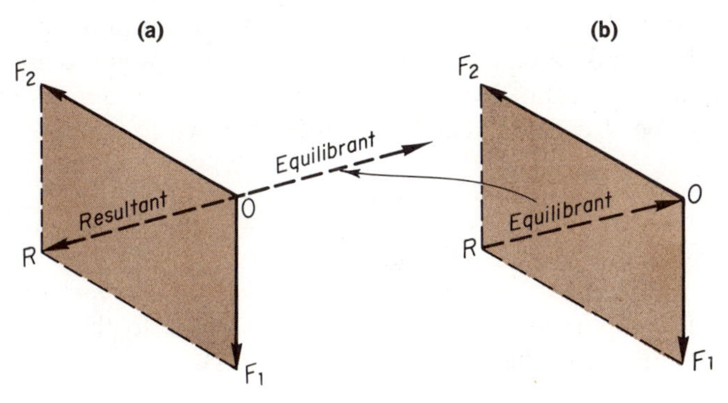

Vectors F_1 and F_2 represent two concurrent, coplanar forces.

FIGURE 222

The equilibrant $\overrightarrow{R\text{–}O}$ [see Fig. 223(b)] counterbalances the resultant effect of the three given forces in the system. From Fig. 223(b) we see that the equilibrant $\overrightarrow{R\text{–}O}$ has the same magnitude and line of action as the resultant $\overrightarrow{O\text{–}R}$ of Fig. 223(a), but that its sense is *opposite.*

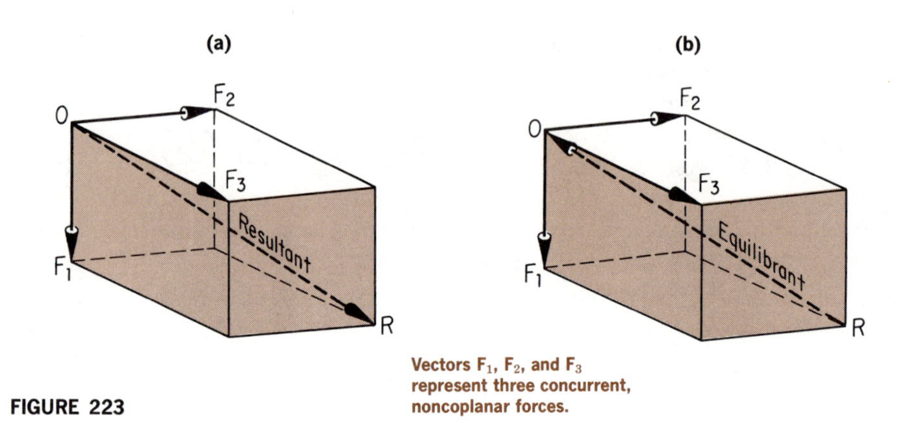

FIGURE 223

Vectors F_1, F_2, and F_3 represent three concurrent, noncoplanar forces.

94 Structure, space, and vector diagrams

To solve problems of force systems acting on structures, you should represent graphically (1) the structure itself, indicating the lines of action of the applied forces, (2) the forces in the structure itself that resist the applied forces, and (3) the vectors that represent all the forces in the given system.

A. STRUCTURE DIAGRAM

Figure 224 is a *structure diagram* that shows two links *A–O* and *B–O* supporting a 100-lb load that is suspended from point *O* (the point at which the links are pinned to each other). Points *A* and *B* are fixed points on a ceiling beam. Because the structure diagram defines the *geometry* of a given structure, it can be accurately drawn to scale.

B. SPACE DIAGRAM

Figure 225 is a *space diagram* that shows lines of action of the forces acting in and on the structure (refer to Fig. 224). The space diagram need not be drawn to any scale, but the *lines of action* of the forces indicated in this type of diagram *must* be parallel to the members of the structure and to the applied load (or loads).

C. VECTOR DIAGRAM

The *vector diagram* illustrates the applied forces acting on the structure and the forces in the structure that resist the applied forces. These are represented by vectors drawn consecutively to a specific vector scale. The vectors must be parallel to the forces they represent in the *space diagram.*

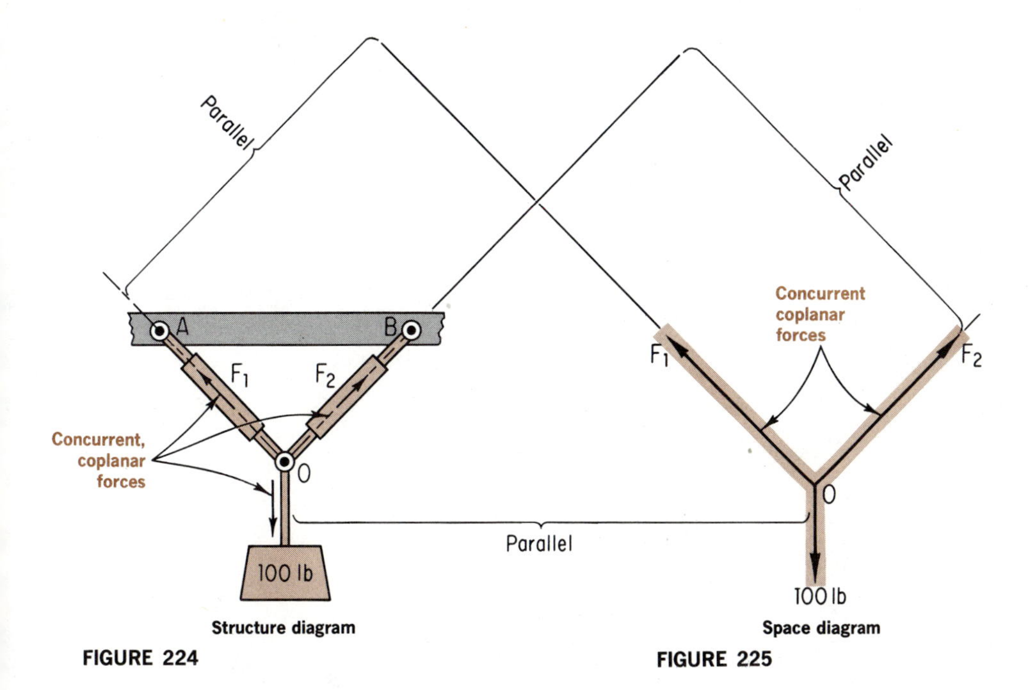

FIGURE 224

FIGURE 225

Figure 226(a) is a space diagram that shows the concurrent, coplanar forces of the structure diagram in Fig. 224. Figure 226(b) is a *vector diagram* in which the vectors are parallel to the respective forces indicated in the space diagram in Fig. 226(a) and in which the vectors have been added to each other in consecutive order.

In Fig. 226(b), the vector diagram indicates that the 100-lb load, as applied at point O in the structure diagram (see Fig. 224), is counterbalanced by the *vector summation* of $\overrightarrow{F_1}$ and $\overrightarrow{F_2}$; that is, the 100-lb force is the equilibrant of $\overrightarrow{F_1} + \overrightarrow{F_2}$ (or vice versa).

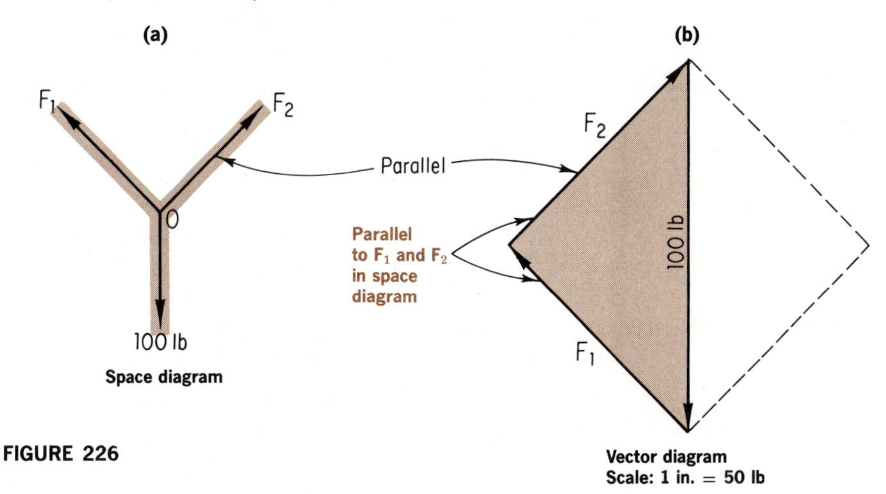

FIGURE 226

Before you can construct the space and vector diagrams of nonconcurrent force systems (Fig. 227), you must draw a minimum of *two* orthographic projection views (horizontal and frontal, for example) of the given force system. This is necessary because noncoplanar force systems are three-dimensional spatial systems.

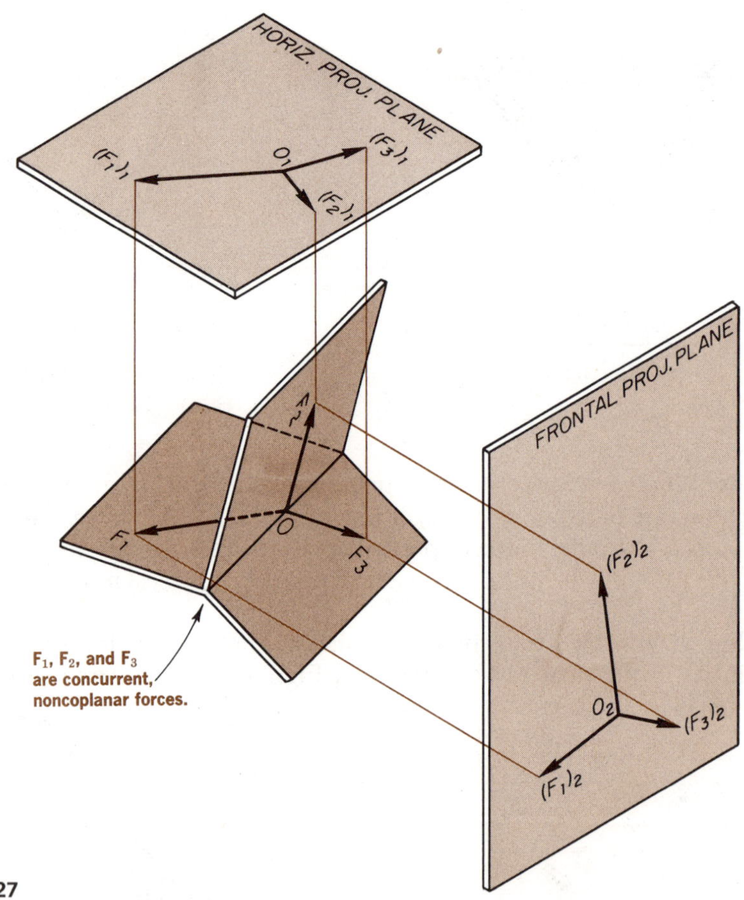

FIGURE 227

Forces F_1, F_2, and F_3, shown in Fig. 227, are noncoplanar forces that are concurrent at point O. Figure 228(a) is the space diagram of this force system, which is defined by the horizontal and frontal projections of the given forces. Figure 228(b) is the vector diagram of the horizontal and frontal projections that represent the given forces. Note especially in the vector diagram that *each vector is parallel to its respective force in its respective projection.* Also note that the vector diagram is drawn to a specified *vector scale* of 1 in. = 50 lb.

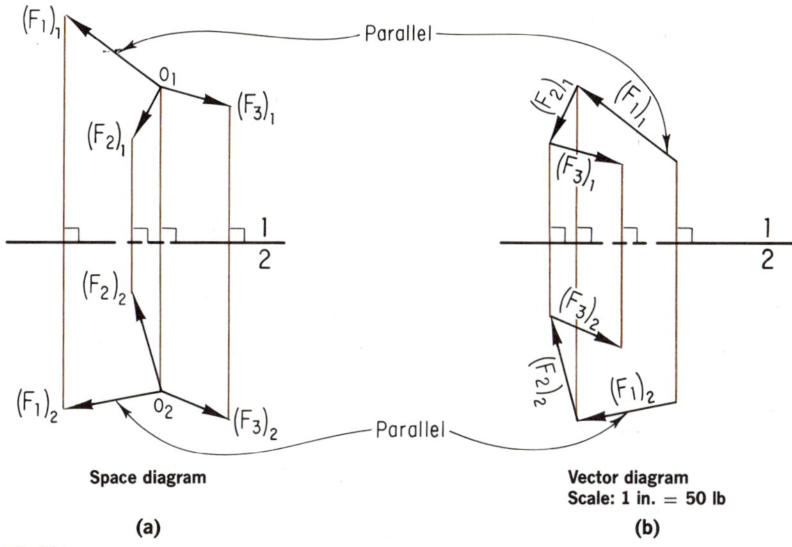

FIGURE 228

95 Determination of the resultant and the equilibrant of a concurrent, noncoplanar force system

We can now apply the principles of descriptive geometry to the solution of problems involving forces and vectors. This is illustrated in the following examples:

Orthographic Example: Determination of the resultant and equilibrant of a concurrent, noncoplanar force system (Figs. 229–232).

Given Data: Horizontal and frontal projections of the space diagram of three concurrent, noncoplanar forces: $F_1 = 120$ lb, $F_2 = 115$ lb, and $F_3 = 80$ lb (Fig. 229).

Required: The resultant and the equilibrant of the given force system.

Construction Program:

a. The space diagram in Fig. 229 of the given force system F_1, F_2, and F_3 shows only the *positions* of these forces in space, their *lines of action,* and their *senses.* Their magnitudes are *not* represented by the length of the lines; they have been indicated by a *number* (for example, $F_1 = 120$ lb) in the given data. Before you can determine the resultant of the given force system, you must represent the given *forces* by vectors, the *lengths* of which should be drawn to a convenient vector scale and which indicate the magnitude of each of the forces.

Figure 230(a) is a space diagram of force F_1. In order to determine the vector that represents F_1, you must find the *true length view* of the *line of action* of the force and on this view measure (according to a vector scale) $F_1 = 120$ lb. Figure 230(b) illustrates how an *assumed*

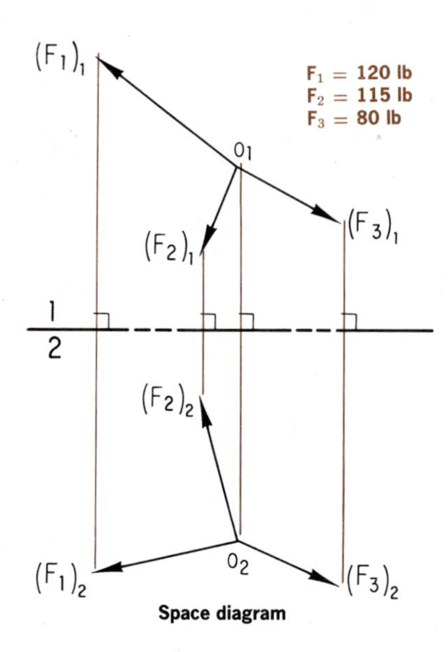

FIGURE 229

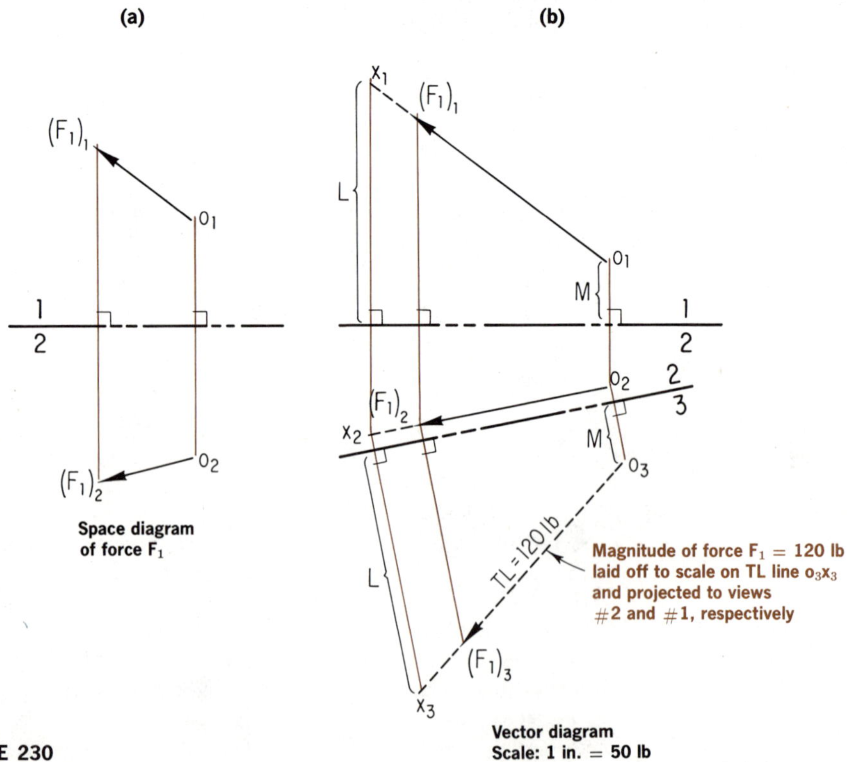

FIGURE 230

length o_1–x_1 in view #1 of force F_1 is first determined as true length. On this true length (o_3–x_3 in view #3), the *magnitude* of $F_1 = 120$ lb is measured. (We have chosen a vector scale of 1 in. = 50 lb.) This true length, o_3–$(F_1)_3$, is then projected to frontal view #2 and horizontal view #1, thus determining the *projection lengths* of the given vector $\overrightarrow{F_1}$.

b. Repeat this procedure to find the *projection lengths* of vectors $\overrightarrow{F_2}$ and $\overrightarrow{F_3}$.

[NOTE: You may use the rotation procedure, or a version of it, to determine the lengths illustrated in Fig. 208. Figure 231(b) shows how this method may be adapted.]

From Fig. 231(b) we can see that RL 1–2 was moved *parallel* to *itself* (see view #1) and through point o_1 of the line of action o_1–x_1 of force F_1. RL 2–3 has been placed on o_2–x_2 in view #2. The true length of the assumed length (o_1–x_1 in view #1) of the line representing force F_1 is determined by measuring the *perpendicular* distance D from the new location of RL 1–2 to point x_1. Distance D is then transferred with dividers to RL 2–3 and measured from x_2 (in view #2) perpendicular to RL 2–3, thus locating x_3. Therefore, in view #3, the distance from o_3 to x_3 is the true length of the assumed length of the line representing force F_1. *On this true length, measure, according to a vector scale,* $F_1 = 120$ lb, *and project this length to views #2 and #1.* (Repeat this procedure for F_2 and F_3.)

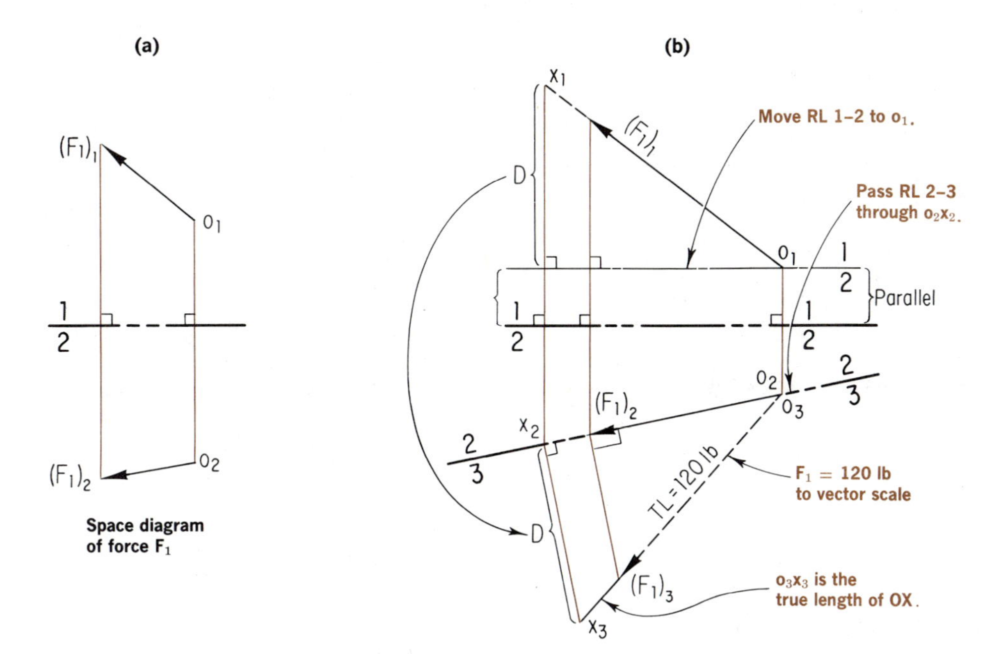

FIGURE 231

c. In Fig. 232(b), the horizontal and frontal projections of the vectors representing forces F_1, F_2, and F_3 in the vector diagram are laid out consecutively from points o_1 and o_2. (The true lengths and projection lengths of the vectors for each force were determined by the preceding method.) The vector representing force F_1 starts at point O and ends at point P; the vector representing force F_2 starts at point P and ends at point Q; and the vector representing force F_3 starts at point Q and ends at point R. The resultant is the *closing vector* that starts at point O and ends at point R (from o_1 to r_1 in view #1, from o_2 to r_2 in view #2).

d. Determine the true length of the resultant $\overrightarrow{O\text{–}R}$. (The true length of O–R, according to a vector diagram with a scale of 1 in. = 50 lb, is 75 lb.)

e. The equilibrant has the same direction and magnitude, but it is *opposite* in sense to the vector representing the resultant. Therefore, the arrow representing the sense of the equilibrant would be at point O rather than at point R; that is, for equilibrium the vector diagram must *close* and the sense of all the vectors in the diagram must follow each other.

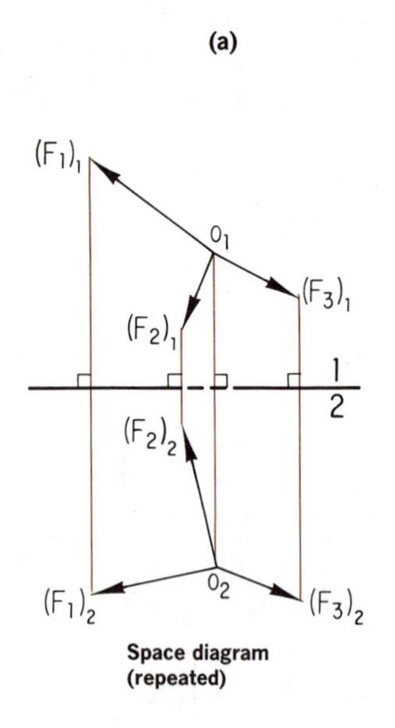

Space diagram (repeated)

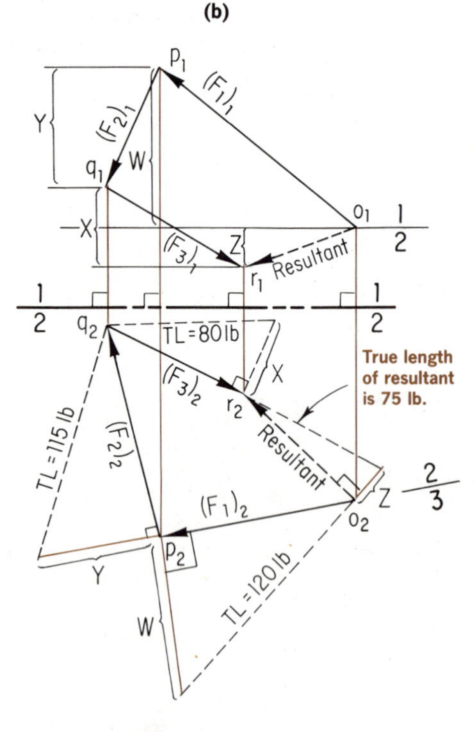

Vector diagram
Scale: 1 in. = 50 lb

FIGURE 232

96 Analysis of loads in a structure involving concurrent, noncoplanar forces

For a structure that is carrying an external load (or loads) to be in a state of equilibrium, the total force system, including the applied forces and the reactive forces of the structure, must add up to a resultant of zero. By utilizing the concept of vectors and equilibrium, as well as the principles of descriptive geometry, you can solve problems involving *three unknown* concurrent forces. Such forces are present, for example, in a tripod that is loaded at its apex (common intersecting point of the legs). Figure 233 is a pictorial representation of a tripod, the legs (*OA*, *OB*, and *OC*) of which meet at point *O*. The tripod is subjected to an external *vertical load* (100 lb) acting on it at point *O*. The load is "distributed," or resolved, into each leg of the tripod. The *angle* each individual leg makes with the line of action of the applied load determines the distribution and magnitude of the loads in each leg. The *length* of the tripod's legs has *no effect* on the amount of load they carry. In this type of analysis we assume that no deflection, or bending, of the legs occurs. In other words, the tripod is considered to be a *rigid body*. (This is an idealistic assumption created for the initial stages of the analysis.)

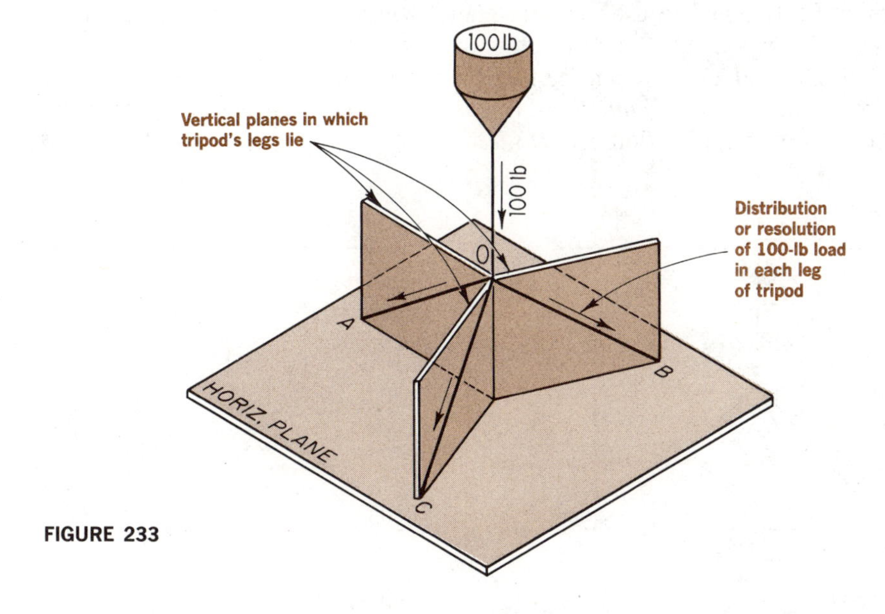

FIGURE 233

In Fig. 234, the line of action of the 100-lb vertical load is indicated with a line and an arrowhead aimed toward point *O* of the tripod. Each leg *reacts* to the load to maintain the original condition of equilibrium that existed before its application. Each leg is thus reacting *against* the sense of the applied load, and is, therefore, under a condition of *compression,* that is, it is *pushing against* the applied load.

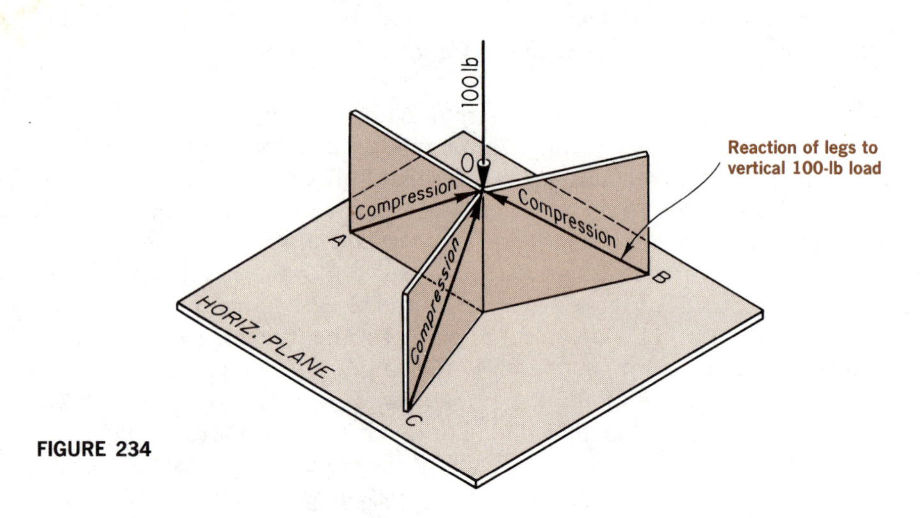

FIGURE 234

Figure 235 shows the frontal projection (view #2) of a tripod subjected to a 100-lb vertical load. Here again, each leg is pushing against the applied load.

If the 100-lb vertical load has an opposite sense (see Fig. 236), all the legs of the tripod (if attached to a horizontal plane) will *pull* against the applied load and cause the tripod legs to be in a state of *tension.*

In Figs. 235 and 236 you can see that the forces that act toward the common point of application are *compressive* forces, and forces that act away from the common point of application are *tensile* forces.

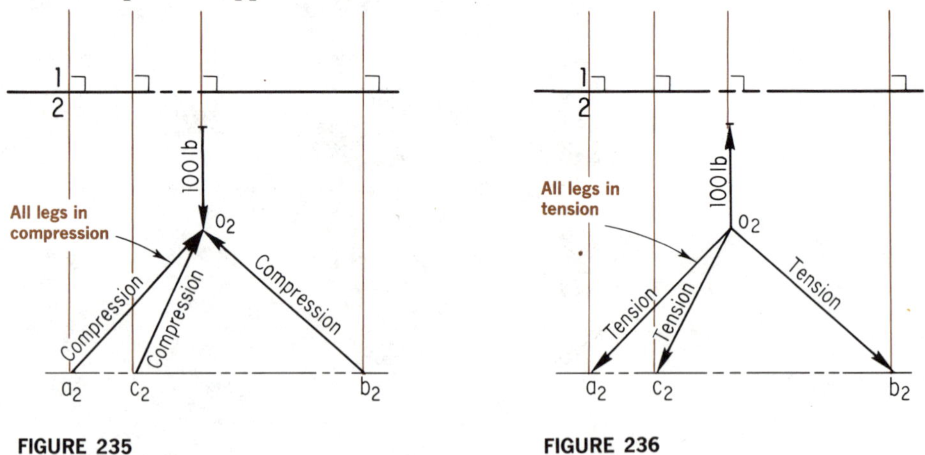

FIGURE 235

FIGURE 236

If the applied load is a horizontal one (see Fig. 237), the types of forces acting in each leg of the tripod may vary. For example, the frontal projection of the tripod indicates that leg *OA* is in a state of tension and that legs *OB* and *OC* are in a state of compression. It is not always possible to determine by inspection whether the members of a structure are in tension or compression. For example, there are times when a tripod may be loaded with a series of external loads, some of which may be both vertical and horizontal, in addition to being inclined. Under these conditions a detailed analysis is necessary to determine the types and magnitudes of the forces acting in the structure. Only then can it be determined whether they are tensile or compressive. (You will perform this analysis in a later example.)

FIGURE 237

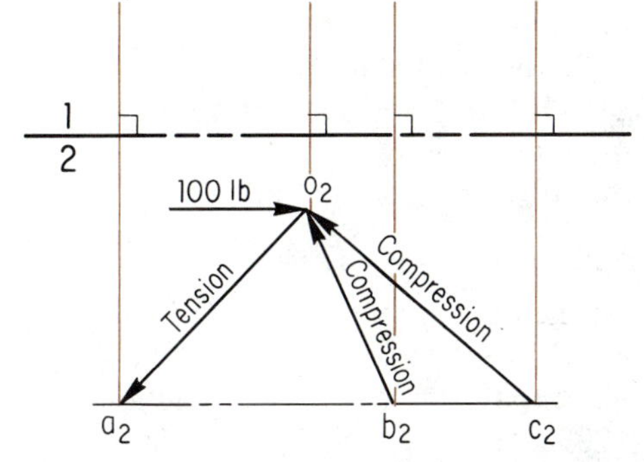

Figure 238 depicts the horizontal and frontal projections of the space diagram of a tripod, the legs (*OA*, *OB*, and *OC*) of which are subjected to an external 100-lb vertical load. The plane formed by legs *OB* and *OC* appears as an edge in view #2. This necessary condition must be fulfilled before a graphic solution involving three unknown forces can be initiated. A single force may be resolved into just two components if only the lines of action of the components are known. If a force were to be resolved into three components, and its lines of action (but not their magnitudes) were known, you could not determine magnitudes for these components.

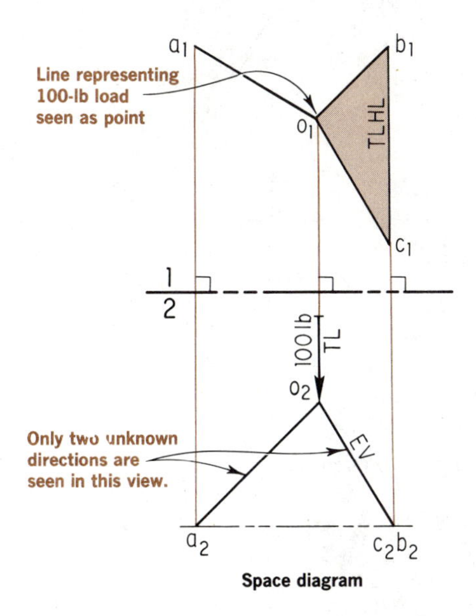

FIGURE 238

Another useful procedure in solving this type of problem is to record methodically the order in which vectors appear in the vector diagram. *Bow's notation,* a widely used method, involves numbering the *spaces* on *either side* of the known or applied forces and on *either side* of *each member* of the structure. (In a three-dimensional structure, such as a tripod, you must use *two views* of the structure.) For the tripod you need: a view in which the plane of two members or two legs of the tripod appear as an edge and an adjacent view.

Figure 239 shows the horizontal and frontal projections of the space diagram of a tripod, the legs (*OA*, *OB*, and *OC*) of which are subjected to an external 100-lb vertical load acting at point *O*.

To apply Bow's notation, follow this procedure:

a. Choose one of the views of the tripod (view #2 was chosen in Fig. 239).

b. Starting with the known applied 100-lb load (*this is important*), place, in consecutive order, #1 in the space to the right of the line of action and #2 in the space to the left of the line of action. (The direction of numbering in Fig. 239 is counterclockwise, as indicated by the curved arrow, but it also could have been clockwise. The *direction* of the numbering is arbitrary; it is necessary only that the numbers be *consecutive* from space to space.)

c. Before placing #3 in the space on the right side of leg o_2a_2, decide whether this space is enclosed by legs o_2a_2 and o_2c_2 or by legs o_2a_2 and

FIGURE 239

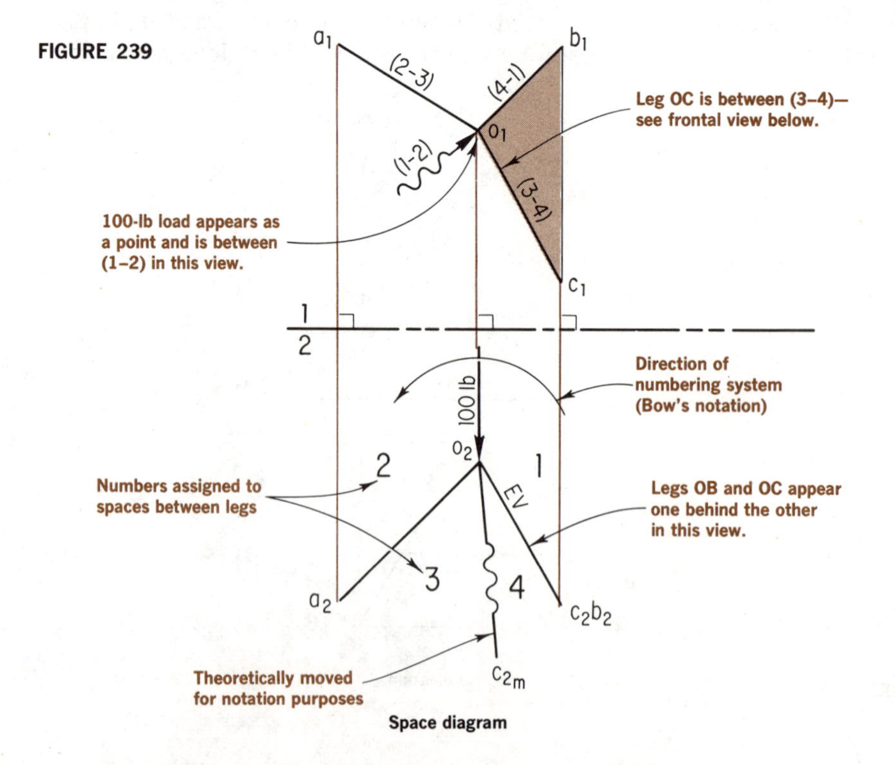

Space diagram

o_2b_2 (legs o_2b_2 and o_2c_2 are behind each other in this view). For purposes of notation, arbitrarily "move" one of these two legs (o_2b_2 *or* o_2c_2). In Fig. 239, leg o_2c_2 has been symbolically moved; it is therefore indicated as a *zigzag* line $o_2c_{2_m}$. Now the space indicated by #3 is enclosed by legs o_2a_2 and $o_2c_{2_m}$.

d. Place #4 in the space enclosed by legs $o_2c_{2_m}$ and o_2b_2. *This completes Bow's notation.*

e. In view #1 of the space diagram, *summarize* Bow's notation by indicating, on the individual tripod legs, the numbers you have placed on either side of each leg in view #2. (Include the applied 100-lb load.) In view #1 you see that the 100-lb load, which appears here as a point, is noted as a zigzag line between #1 and #2; leg o_1a_1 is between numbers #2 and #3, leg o_1b_1 between #4 and #1, and leg o_1c_1 between #3 and #4. This completes Bow's notation as *applied* in view #2 and as *summarized* in view #1.

Bow's notation names the vectors representing the known 100-lb load and the loads carried by each leg of the tripod. For example, the vector representing the 100-lb load in the vector diagram will *begin* at #1 and *end* at #2. The distance between the two points is equal to the *magnitude* of the 100 lb, according to a vector scale. The next vector to be drawn will be the vector representing the load in leg o_2a_2. It will begin at #2 and end at #3. The vector representing the load in leg *OC* will begin at #3 and end at #4. The vector representing the load in *OB* will begin at #4 and end at #1, thus closing the vector diagram and indicating a condition of equilibrium.

The general procedure for determining the loads on the legs of an externally loaded tripod is as follows:

a. Draw the horizontal and frontal projections of the given loaded tripod to a convenient *space scale.* These views fix the three-dimensional geometry of the tripod and its external loading in space. (They also represent two views of the *space diagram.*)

b. Construct a view in which the plane formed by two legs of the tripod appears as an edge.

c. Apply Bow's notation in the view in which this plane appears as an edge *or* in an adjacent view. Start numbering in the space on either side of the *known load* (or loads), then, continue numbering consecutively (either clockwise or counterclockwise) on either side of each leg of the tripod.

d. Construct a vector diagram in which the vectors are *parallel* to the *legs* of the tripod in the space diagram. They will be parallel to the legs of the tripod in the view in which the plane of two legs appears as an edge and in the adjacent view. The result is two views of the vector diagram that are necessary in order to define vectors in a three-dimensional space. To indicate a condition of *equilibrium,* the vector diagram must close since the resultant force is equal to zero.

Orthographic Example: Determination of the loads in the legs of an externally loaded tripod by viewing the plane formed by two legs of the tripod as an edge (Figs. 240–246 inclusive).

Given Data: Horizontal and frontal projections of an externally loaded tripod, the legs of which are *OA*, *OB*, and *OC*. The external load is a downward vertical load with a 100-lb magnitude.

Required: The magnitude and type of the load (tensile or compressive) carried by each leg of the tripod.

Construction Program:

a. In the space diagram in Fig. 240, the plane formed by legs *OB* and *OC* appears as an edge $o_2b_2c_2$ (view #2). Apply Bow's notation in view #2. (Number consecutively the spaces on either side of the known load, and then number the spaces on either side of each tripod leg.)

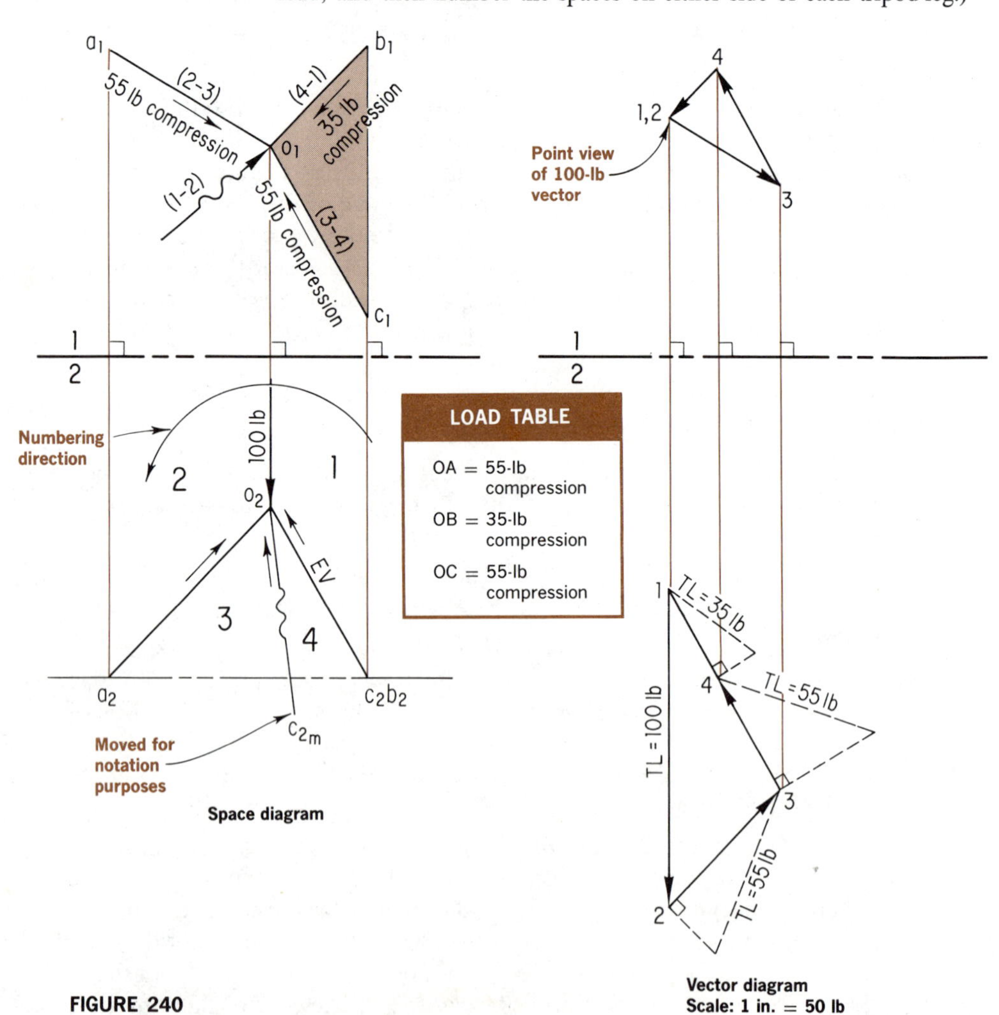

FIGURE 240

For purposes of notation, leg o_2c_2 was moved to a general position indicated by $o_2c_{2_m}$.

b. Summarize Bow's notation in view #1, indicating o_1a_1 between #2 and #3, o_1b_1 between #4 and #1, and o_1c_1 between #3 and #4.

c. Using a convenient scale (1 in. = 50 lb has been indicated in Fig. 240) construct a vector diagram in which the vectors are parallel to the given external load and to the respective legs of the tripod.

[NOTE: The following steps sequentially develop the procedure for drawing the complete vector diagram (Figs. 241–246).]

d. Referring to Fig. 241, construct a vector representing the applied 100-lb load. This vector will start at some convenient point #1 (as indicated in view #2) and end at point #2. Its horizontal projection is a point indicated as #1,2.

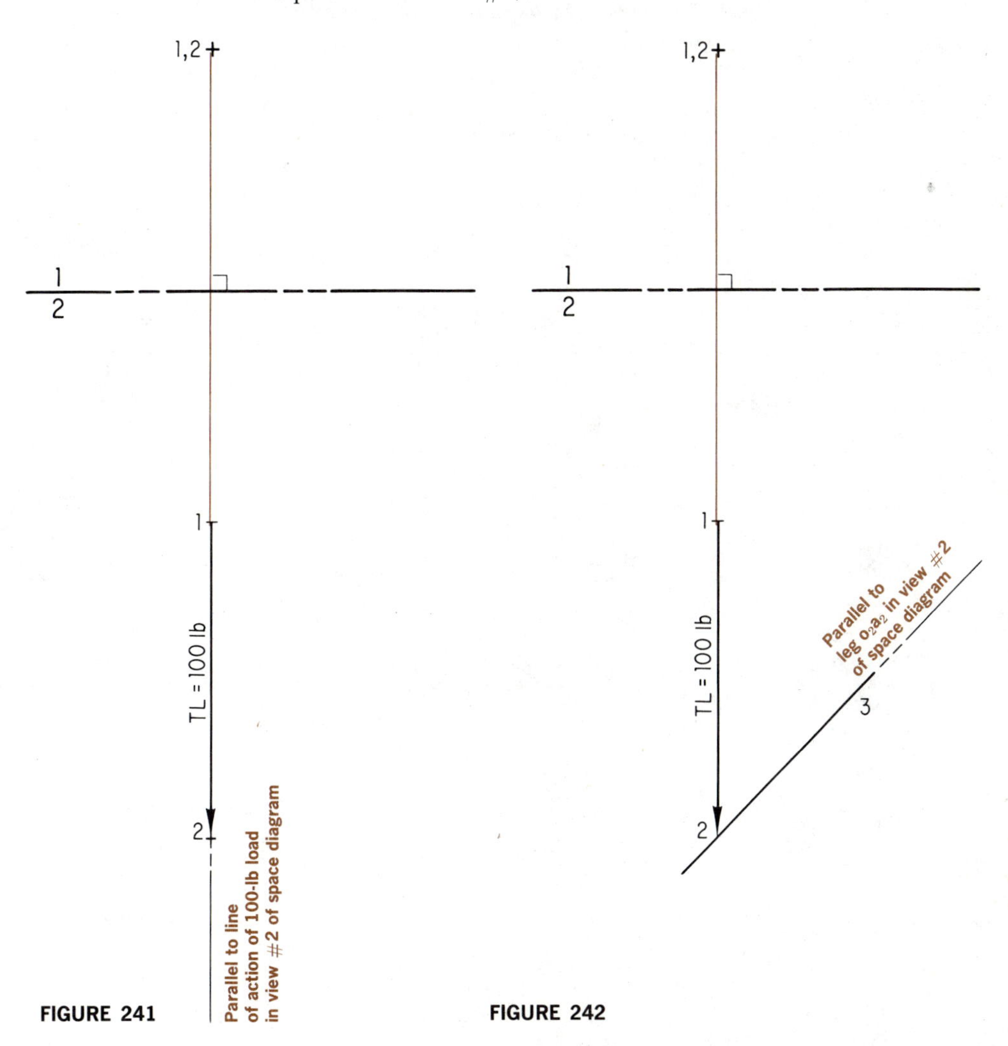

FIGURE 241

FIGURE 242

e. Referring to Fig. 242, in view #2, construct a line from point #2 parallel to leg o_2a_2. (The length of this line is at this stage *indefinite*.)

f. Referring to Fig. 243, in view #2, construct a line from point #1 parallel to legs o_2b_2 and o_2c_2 (in this view they are behind each other and therefore have the same direction). Extend this line until it intersects the line 2–3 (which is parallel to leg o_2a_2 in view #2). The point of intersection of these two lines locates #3 in the vector diagram. In view #2, note that the vector diagram has *closed,* in spite of the fact that #4 has not yet been located as a specific point. You must obtain additional information from view #1 in order to locate #4.

g. Referring to Fig. 244, in view #1, construct a line from the point view of the 100-lb force (indicated as #1,2) parallel to leg o_1a_1. (The length of this line is *indefinite*.)

h. Referring to Fig. 244, from view #2, project #3 upward to view #1 to determine point #3.

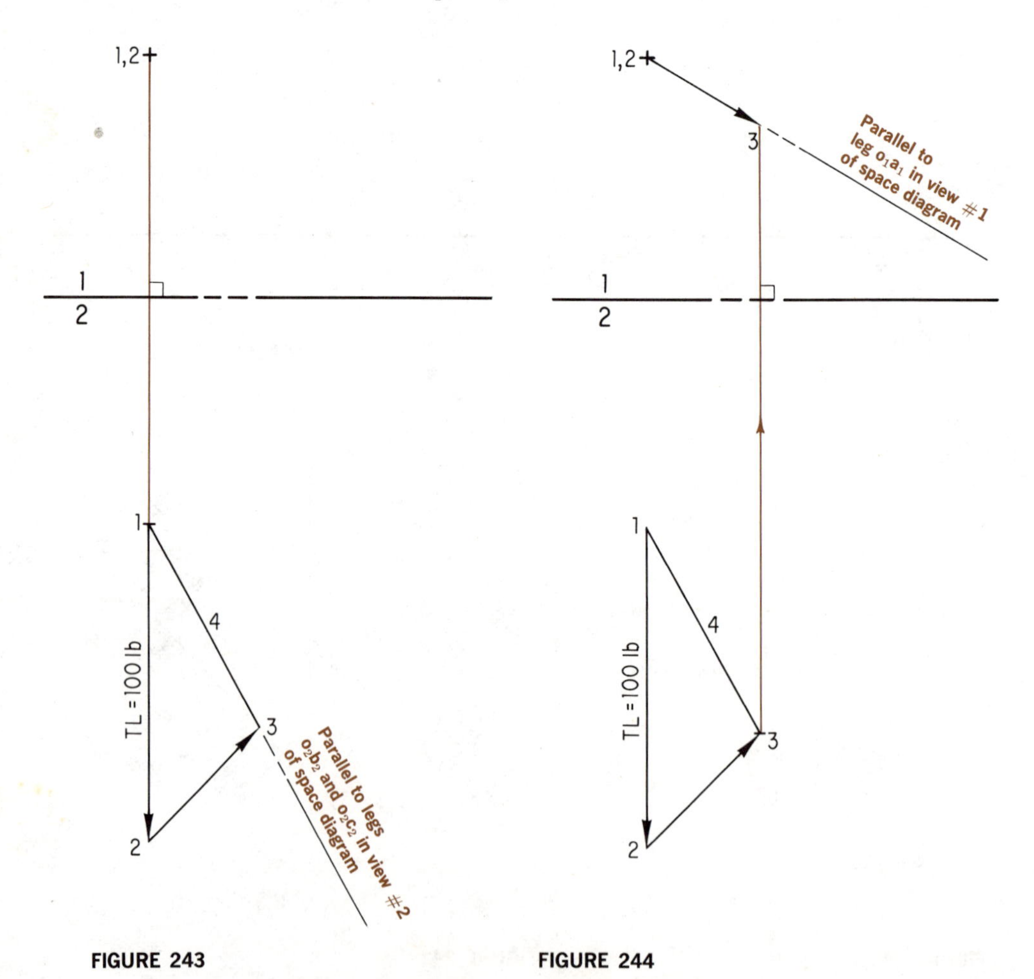

FIGURE 243

FIGURE 244

i. Referring to Fig. 245, in view #1, construct a line from point #3 parallel to leg o_1c_1.

j. Referring to Fig. 246, from the point view of the applied 100-lb load (indicated as #1,2 in view #1 of the vector diagram) construct a line parallel to leg o_1b_1 in the space diagram. The point of intersection of the lines parallel to legs o_1c_1 and o_1b_1 locates point #4 in view #1 of the vector diagram. Project point #4 from view #1 downward to view #2 in the vector diagram.

k. Referring to Fig. 246, the sense of the applied 100-lb load determines the sense of the remaining vectors. To indicate a condition of *equilibrium*, the sense of the vectors *must follow each other*. This means that the arrows will be pointing toward #2, #3, #4, and #1. You can see that the vector diagram closes, indicating a resultant of zero magnitude and, therefore, a condition of equilibrium for the given tripod. (Show the sense of the vectors in both views.)

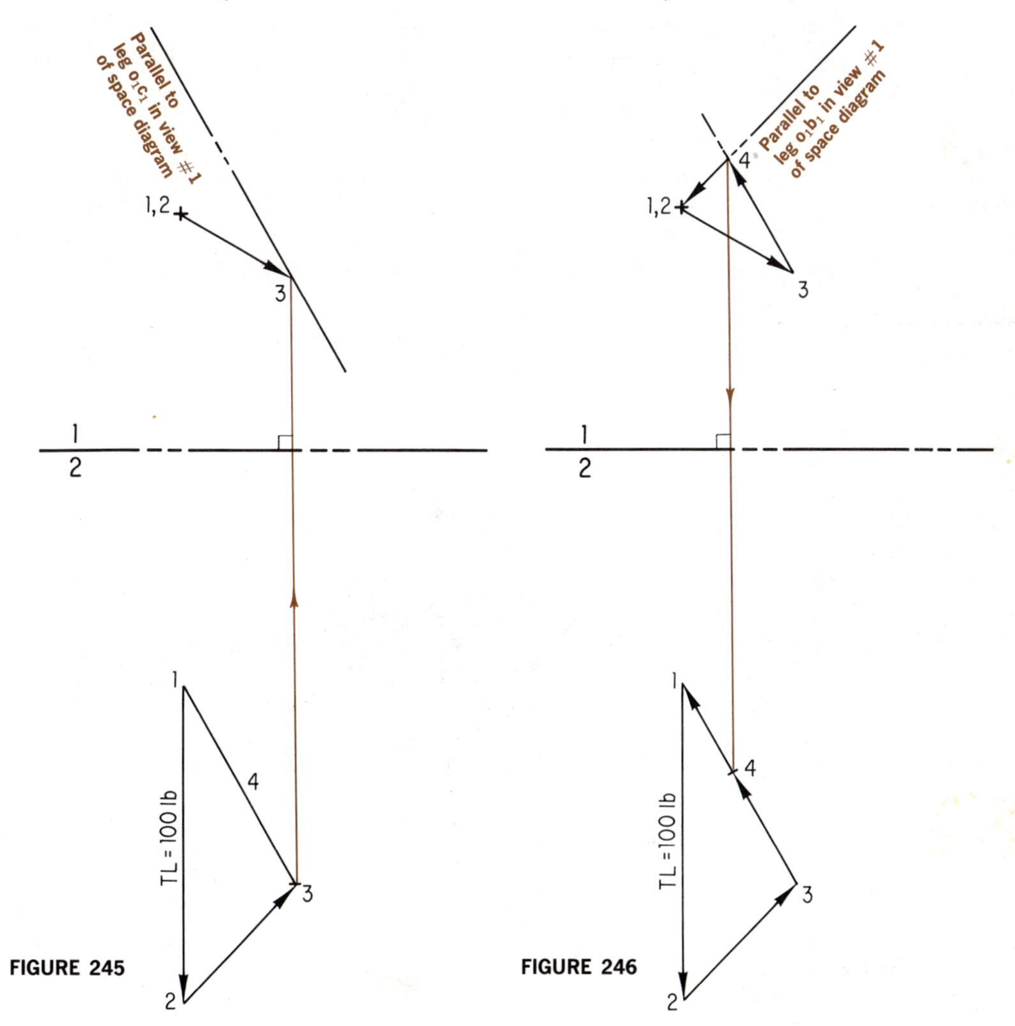

FIGURE 245

FIGURE 246

l. Referring again to Fig. 240, in the vector diagram, determine the true length of each vector using the short-cut method. The loads carried by each leg of the tripod are summarized as: leg $OA = 55$ lb, leg $OB = 35$ lb, and leg $OC = 55$ lb.

m. To determine whether the loads are tensile or compressive, transfer the sense of each vector to its *respective leg* in either view #1 or #2. The sense of a vector acting *toward* the point of application of the external 100-lb vertical load indicates that the leg is in *compression.* From the space diagram in Fig. 240, you can see that the sense of all the vectors act toward the point of application; therefore, all the tripod legs are in compression.

Orthographic Example: Determination of the loads carried by the legs of a tripod that has a vertical and a horizontal force acting at its apex [Figs. 247, 248(a), and 248(b)].

Given Data: Horizontal and frontal projections of a tripod, the legs of which are OA, OB, and OC. A 100-lb vertical load and a 175-lb horizontal load act at apex O of the tripod (Fig. 247).

The space diagram in Fig. 248(a) indicates the line of action of each of the applied forces.

(NOTE: The lengths of the lines of action of the applied loads in the space diagram are *arbitrary.* They show only the line of action and the sense of the loads. The vector lengths of these loads will either be established in the vector diagram or will be superimposed on the space diagram.)

Required: The magnitude and type of load (tensile or compressive) carried by each leg of the tripod.

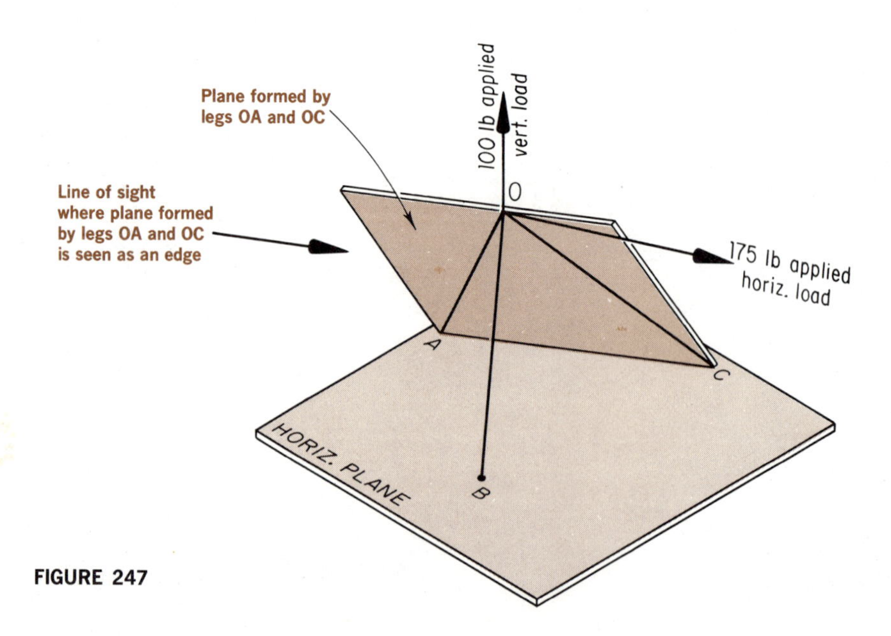

FIGURE 247

a. Construct an edge view of the plane formed by legs OA and OC of the tripod. (In view #2, a_2c_2 is a horizontal line that appears as true length a_1c_1 in view #1. RL 1–3 is perpendicular to a_1c_1, and it establishes view #3, in which the plane formed by legs OA and OC appears as an edge.)

b. Then apply Bow's notation in view #3 by moving point c_3 of leg o_3a_3 to an imaginary position c_{3_m}. Number consecutively the spaces on either side of the known loads. (The 100-lb vertical load has #1 and #2 on either side of it, and the 175-lb horizontal load has #2 and #3 on either side of it. Leg o_3b_3 has #3 and #4 on either side of it; leg o_3a_3 has #4 and #5 on either side of it, and leg $o_3c_{3_m}$ has #5 and #1 on either side of it).

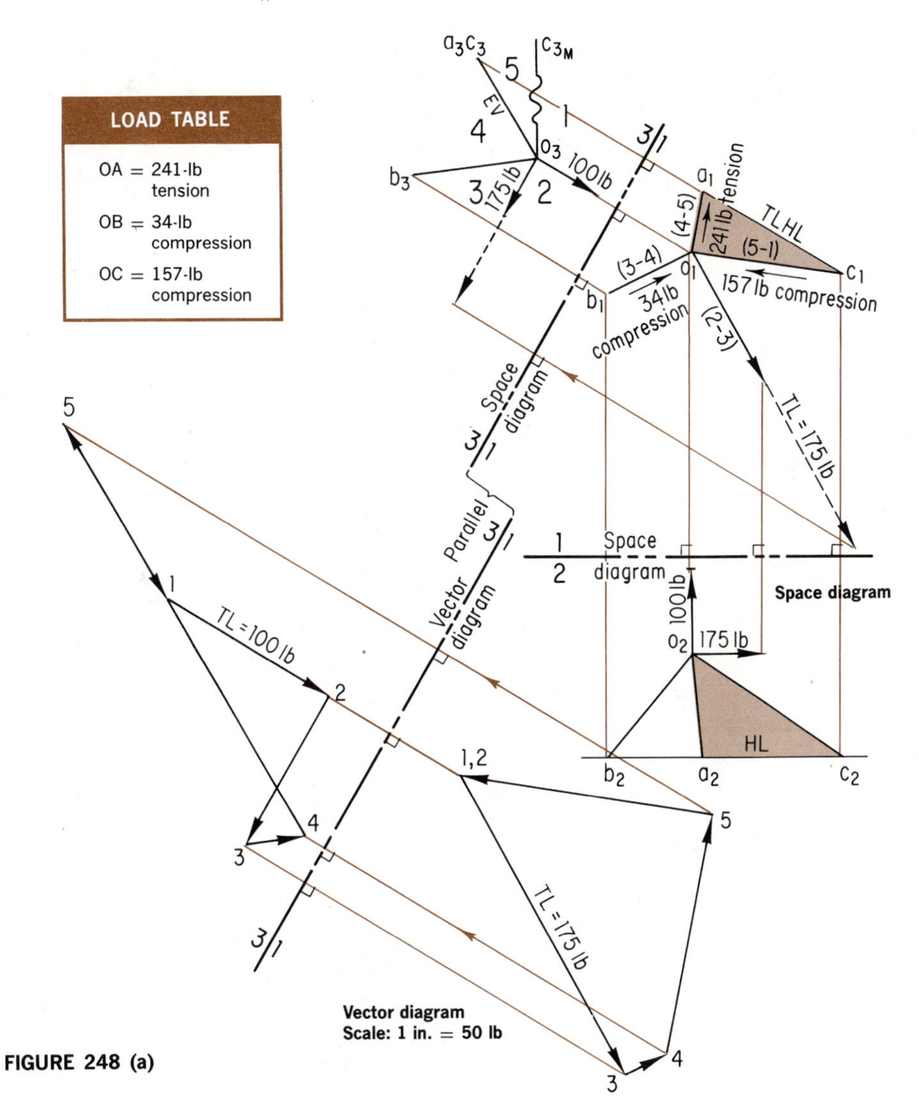

FIGURE 248 (a)

c. In view #1, summarize Bow's notation as it pertains to the legs of the tripod and the lines of action of the applied loads.

d. Using a convenient vector scale (1 in. = 50 lb) start the vector diagram, in which RL 1–3 will be parallel to RL 1–3 of the space diagram. First construct a vector parallel to the 100-lb vertical load. The vector representing this load starts at point #1 and ends at point #2 (in accordance with Bow's notation).

e. Before we can indicate the applied 175-lb horizontal load in the vector diagram, we must determine its *projection length* in view #3. For convenience, this projection length has been constructed in the space diagram (1 in. = 50 lb). Since the load has a horizontal line of action, its horizontal projection in view #1 will appear as true length. Therefore, in view #1 measure the vector length of the load along its line of action from point o_1 as indicated. Project the vector length from view #1 to view #3 in order to determine its projection length.

f. Returning to the vector diagram, measure the projection length of the applied 175-lb load (view #3 in the space diagram) and transfer this length with dividers to the vector diagram. Start from point #2 in the vector diagram and end at point #3 (in accordance with Bow's notation).

g. Complete view #3 of the vector diagram by drawing a line, parallel to the legs of the tripod in view #3 of the *space diagram,* from point #3 parallel to leg o_3b_3; then extend the line. From point #1, construct a line parallel to legs o_3a_3 and o_3c_3 (which have the same direction in view #3), extending it until it intersects the line coming from point #3. This locates point #4 and closes the vector diagram in view #3.

h. Complete horizontal projection view #1 of the vector diagram by first projecting the vectors representing the applied 100-lb and 175-lb loads to this view. (The lines of action of the vectors in view #1 of this diagram are parallel to their lines of action in view #1 of the space diagram). The 100-lb load appears as a point, (indicated as #1,2) and the 175-lb load as a vector (from point #2 to point #3). To complete view #1 of the vector diagram, draw a line from point #3 (in the vector diagram) parallel to leg o_1b_1 (in the space diagram) and extend it.

i. From view #3 of the vector diagram, project point #4 to view #1 and determine point #4 in view #1. From point #4 in view #1, construct a line parallel to leg o_1a_1 (in view #1 of the space diagram) and extend this line.

j. From point #1 in view #1 of the vector diagram, construct a line parallel to leg o_1c_1 (in view #1 of the space diagram). As you extend this line, it intersects the line parallel to leg o_1a_1 and determines point #5 in view #1 of the vector diagram.

k. In the vector diagram, project point #5 in view #1 to view #3, thus locating point #5 in view #3.

l. The senses of the applied 100-lb and 175-lb loads establish the sense direction of all vectors in the vector diagram. Draw an arrow: from #1 to #2, from #2 to #3, from #3 to #4, from #4 to #5, and from #5 to #1. Transfer the arrow senses from view #1 of the vector diagram to view #1 of the space diagram.

m. Referring to Fig. 248(b), using the short-cut method, determine the true lengths of all the vectors in the vector diagram. These are summarized as vector 3–4, which represents the load carried by leg *OB* (34 lb) and which is a compressive load, since the arrow sense is acting *toward* the point of application; vector 4–5, which represents the force carried by leg *OA* (241 lb), and which is a tensile load, since the arrow sense is acting *away* from the point of application; and vector 5–1, which represents the load carried by leg *OC* (157 lb), and which is a compressive load, since the arrow sense in this leg is acting *toward* the point of application. (See Fig. 248a—space diagram.)

This completes the requirements of the problem.

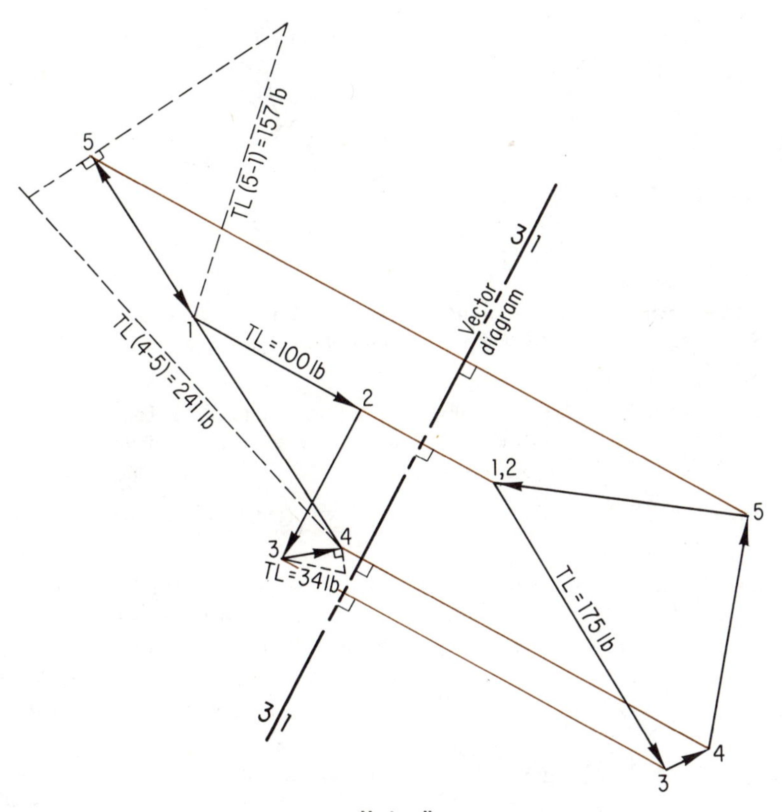

FIGURE 248 (b)

Vector diagram
Scale: 1 in. = 50 lb

You may use another approach to determine a view in which only two directions of the unknown forces in the tripod legs may be seen—view *one* of the legs of the tripod as a *point.* You can see the directions of the other two legs (having unknown forces) and of the known applied loads (Fig. 249).

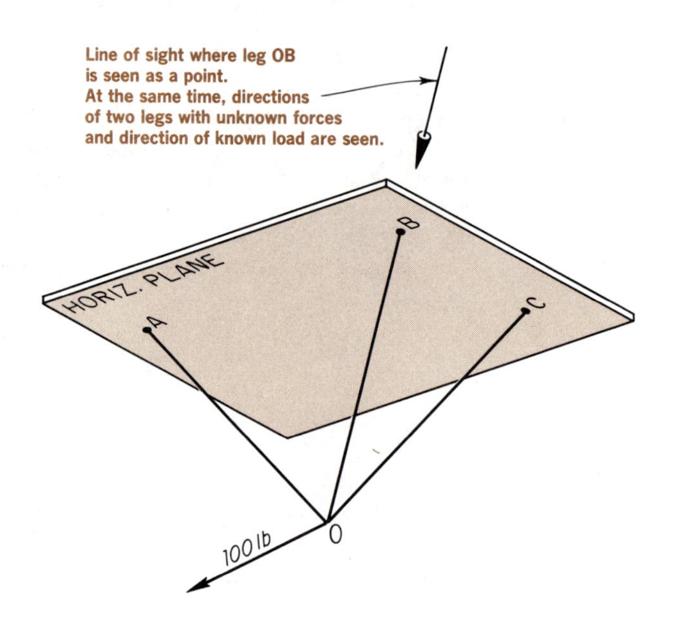

FIGURE 249

Orthographic Example: Determination of the loads in the legs of a tripod from a view in which one leg appears as a point (Fig. 250).

Given Data: Horizontal and frontal projections of a space diagram in which a tripod hangs from a horizontal ceiling. The legs of the tripod are OA, OB, and OC, and the sense and line of action of an applied 100-lb load acts at its apex.

Required: The magnitude and type (tensile or compressive) of the load carried by each leg of the tripod.

Construction Program:

a. Determine the point view of tripod leg OB by constructing a plane (represented by RL 1–3) parallel to its horizontal projection o_1b_1 in view #1. (The true length o_3b_3 of leg OB appears in view #3.)

b. Construct a plane (represented by RL 3–4) perpendicular to true length o_3b_3. The point view o_4b_4 of leg OB appears in view #4.

c. In order to apply Bow's notation, move leg o_4b_4 and represent it by a zigzag line $o_4b_{4_m}$ (see view #4 of the space diagram). Number consecutively the spaces on either side of the applied 100-lb load (the line of action is indicated as o_4f_4). Continue numbering consecutively the spaces on either side of each leg of the tripod.

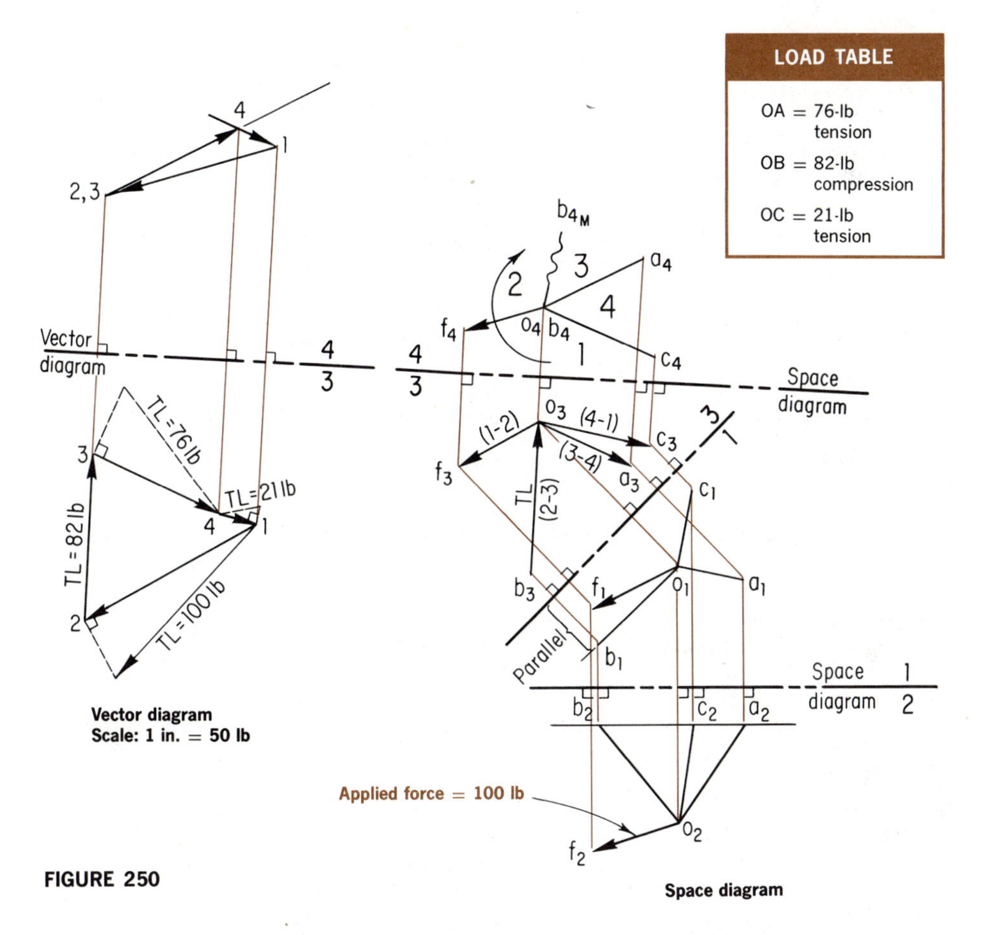

FIGURE 250

d. Construct a vector diagram (1 in. = 50 lb) starting with the applied 100-lb load (the true length of the load was determined and projected to views #3 and #4 of the vector diagram).

  (NOTE: All vectors are parallel to their respective legs in their respective views #3 and #4.)

e. Complete the vector diagram: indicate the sense of the vectors in each view and determine their true lengths by the short-cut method. The true lengths of the vectors are indicated: vector 2–3, which represents the force in leg *OB* (82-lb compression) (see view #3 of the space diagram for tension and compression directions); vector 3–4, which represents the force in leg *OA* (76-lb tension); vector 4–1, which represents the force in leg *OC* (21-lb tension).

You may use another approach to determine the loads in an externally loaded tripod. Combine the forces in two legs of the tripod into one force having a line of action that is *coplanar* with the applied load and with the line of action of the third tripod leg (Fig. 251). The forces on the legs of the tripod and of the applied

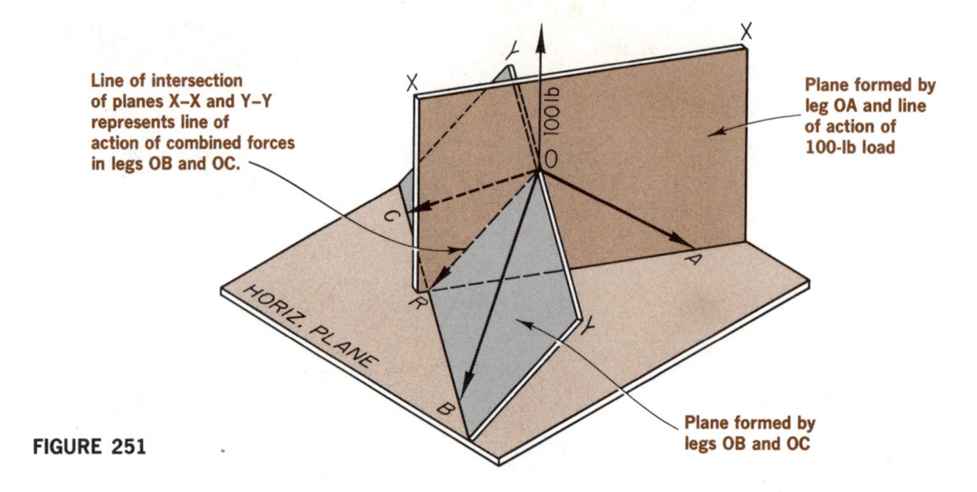

FIGURE 251

load are reduced to *two separate coplanar systems* that can be readily handled to solve unknowns.

Orthographic Example: Determination of the loads in an externally loaded tripod by combining the forces in two tripod legs into one force (Fig. 252).

Given Data: Horizontal and frontal projections of a tripod, the legs *OA*, *OB*, and *OC* of which are subjected to an upward vertical force of 100 lb.

Required: The magnitude and type (tensile or compressive) of the loads acting in each leg of the tripod.

Construction Program:

a. Through leg o_1a_1, construct a vertical plane *X–X* (which appears as an edge in view #1 of the space diagram) that contains the line of action of the applied 100-lb load. (The load appears as a point in view #1.) Plane *X–X* intersects the plane formed by legs o_1c_1 and o_1b_1 (the line of intersection is indicated as o_1r_1 in view #1, and as o_2r_2 in view #2).

b. The line of intersection *OR* is the line of action of the *combined* forces in legs *OB* and *OC*. (They are represented as F_c.) Now determine the magnitudes of the forces F_1 (leg *OA*) and F_c (legs *OB* and *OC*).

   Construct vector diagram #I so that the vectors are parallel to the line of action of forces F_1 (leg *OA*) and F_c (leg *OR*). The result is a *coplanar* vector diagram, which appears as an edge in view #1.

c. Revolve the edge view of vector diagram #I parallel to RL 1–2. The true shape of the diagram—a triangle—appears in view #2. The magnitude of the vectors representing forces $F_1 = 70$ lb and $F_c = 75$ lb can now be measured.

d. In the space diagram, determine the true shape view of the plane formed by legs *OB* and *OC*. (You should view the true length horizontal line b_1c_1 as a point in order to determine the edge view $o_3b_3c_3$ of the plane in view #3.) The true shape of this plane appears in view #4. Included in the true shape is the line of intersection *OR* between the vertical plane X–X and the plane formed by legs OB and OC. (This line of intersection is indicated as o_4r_4 and is the direction of the combined force F_c.) From view #4 of the space diagram, construct vector diagram #II, utilizing the magnitude of the combined force F_c determined in the first vector diagram. The vectors in vector diagram #II are parallel to the direction of F_c (o_4r_4) in view #4 and to the legs o_4b_4 and o_4c_4 in view #4. Since the true shape of the plane formed by legs OB and OC appears, the vector diagram #II is also true shape and therefore all the vectors appear true length. You find by measuring that the vector representing the force in leg OB is equal to 38 lb and the vector representing the force in leg OC is equal to 55 lb.

e. Determine the types of load in the tripod legs by transferring the arrow senses from the vector diagrams to the space diagram (this was done in view #2). From this you can see that all the tripod legs are under tensile loads.

(NOTE: In this example, remember that both vector diagrams #I and #II deal with *coplanar* force systems. This approach is possible because two unknown forces were *combined* into one unknown force.)

FIGURE 252

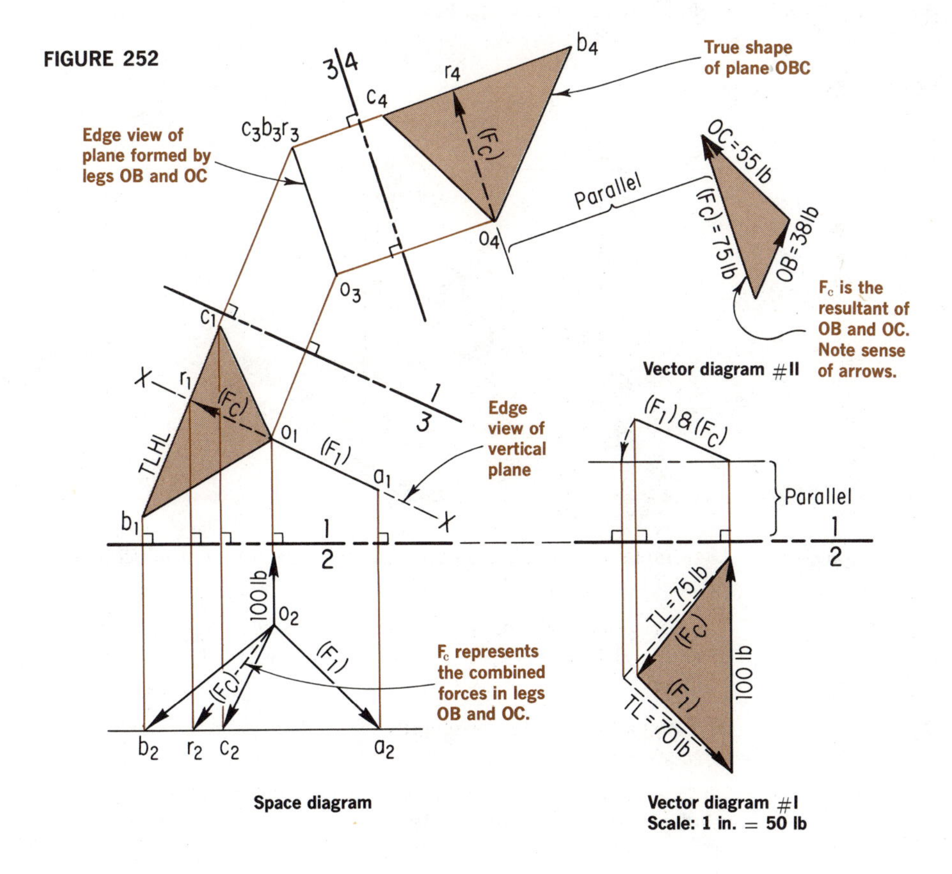

97 Resultant and equilibrant of noncoplanar, parallel forces

You must utilize three concepts to determine the resultant or equilibrant of a noncoplanar, parallel force system:

A. VECTOR DIAGRAM

A *vector diagram* shows the algebraic–vectorial addition of vectors representing the parallel forces.

B. RAY POLYGON

A *ray polygon* indicates the resolution of the forces in a given system into components.

C. STRING POLYGON

A *string polygon* (also called a funicular polygon) replaces the lines of actions of the given parallel forces with the lines of action of their components (as determined in the ray polygon).

Figure 253 depicts a parallel force system that involves three forces. Their positions relative to each other are related to one corner of a plane of reference.

FIGURE 253

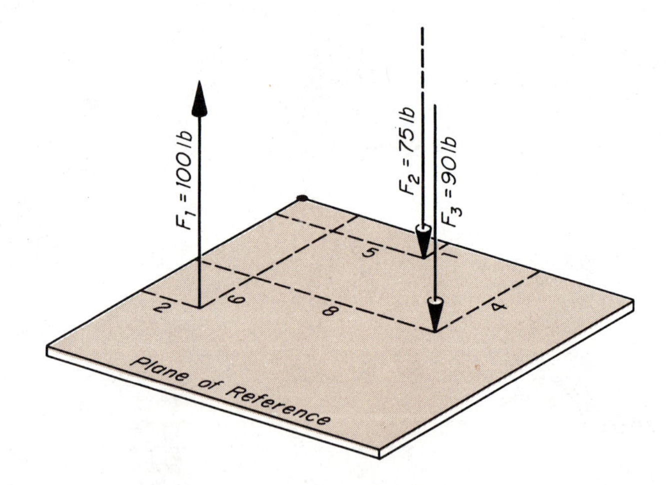

Orthographic Example: Determination of the resultant and equilibrant of a noncoplanar, parallel force system (Figs. 254 and 255).

Given Data: The horizontal and frontal projections of a space diagram illustrating three parallel forces: $F_1 = 100$ lb, $F_2 = 75$ lb, and $F_3 = 90$ lb (Figs. 253 and 254).

Required:

a. The magnitude of the resultant of the given noncoplanar, parallel force system.

b. The location (in the horizontal and frontal views) of the equilibrant relative to the given forces in the system.

Construction Program:

a. Draw the horizontal and frontal projections (views #1 and #2, respectively) of the given noncoplanar, parallel forces based on a convenient scale (1 in. = 3.0 ft). These two views comprise the space diagram of the given force system (Fig. 254).

b. Apply Bow's notation to the lines of action of the given forces by numbering consecutively the spaces on either side of the forces (Figs. 254 and 255).

FIGURE 254

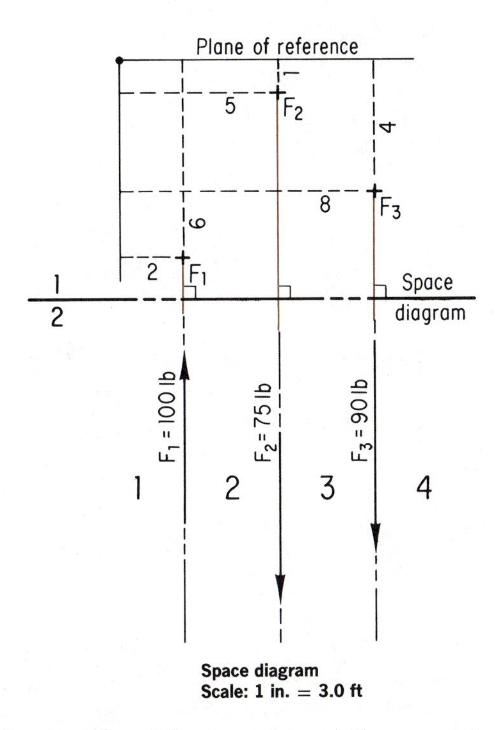

Space diagram
Scale: 1 in. = 3.0 ft

c. Referring to Fig. 255, draw vector diagram #I, showing the algebraic–vectorial addition of the forces (as represented by vectors $\overrightarrow{F_1}$, $\overrightarrow{F_2}$, and $\overrightarrow{F_3}$), and determining the resultant $\vec{R}$ of the given system. (Note that vector 1–2 represents force $F_1 = 100$ lb, vector 2–3 represents force $F_2 = 75$ lb, vector 3–4 represents force $F_3 = 90$ lb, and vector 1–4 represents the *resultant* force of the given system, *65 lb.*)

d. From any convenient point P draw lines to points #1–#4 in vector diagram #I. These lines represent the *ray polygon* and are *components* of each force in the vector diagram. For example, rays P–1 and P–2 are components of vector $\overrightarrow{1\text{–}2}$; rays $\overrightarrow{P\text{–}2}$ and $\overrightarrow{P\text{–}3}$ are components of vector $\overrightarrow{2\text{–}3}$. (Note that in this case, $\overrightarrow{P\text{–}2}$ has a sense *opposite* to its sense as a component of vector $\overrightarrow{1\text{–}2}$. This means that component $\overrightarrow{P\text{–}2}$ has *cancelled* itself.) The components of vector $\overrightarrow{3\text{–}4}$ are rays $\overrightarrow{P\text{–}3}$ and $\overrightarrow{P\text{–}4}$.

FIGURE 255

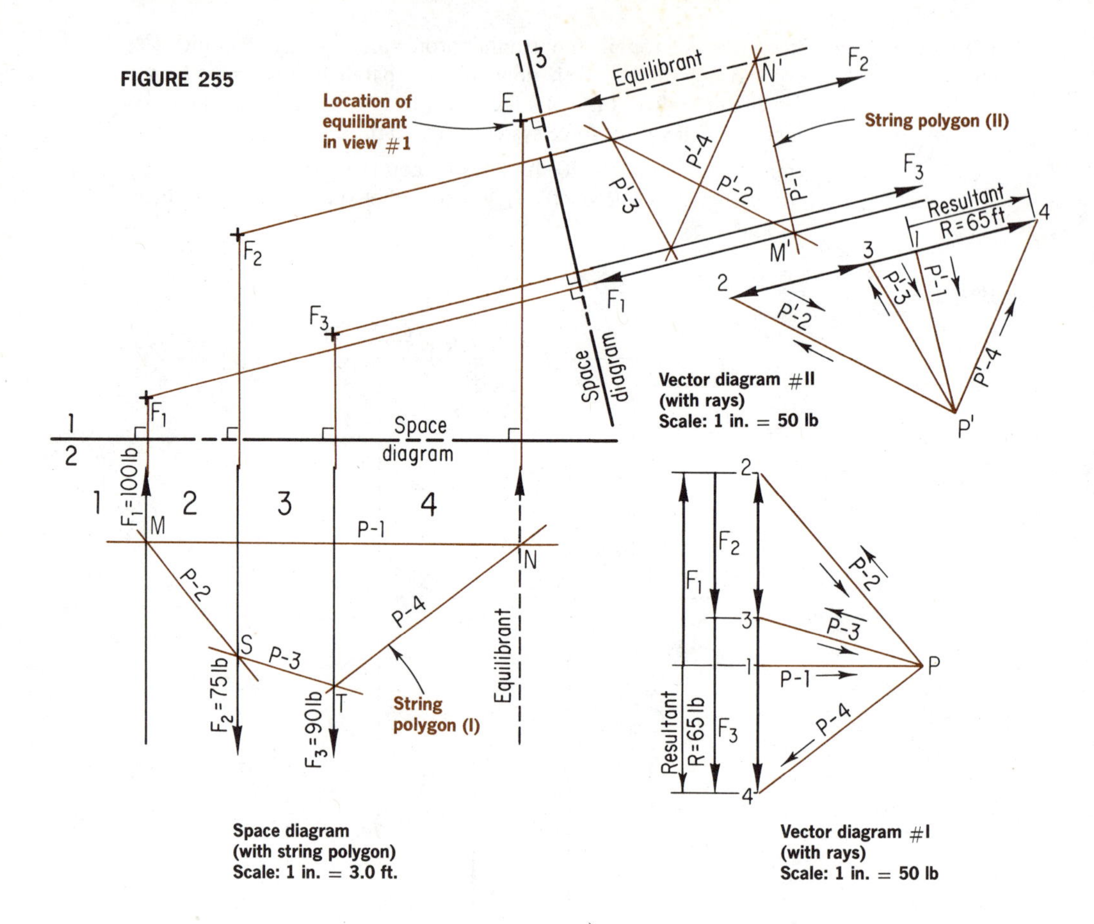

(Note that in this case $\overrightarrow{P-3}$ has a sense *opposite* to its sense as a component of vector $\overrightarrow{2-3}$. $\overrightarrow{P-3}$ has also *cancelled* itself.) The remaining components $\overrightarrow{P-1}$ and $\overrightarrow{P-4}$ are not cancelled. They represent the *components* of the *resultant* vector $\overrightarrow{1-4}$.

e. In the space diagram in Fig. 255, choose any convenient point M on the line of action of force F_1 in view #2. Through this point, construct lines parallel to the lines of action of components $\overrightarrow{P-1}$ and $\overrightarrow{P-2}$, as seen in vector diagram #I. These two component lines of action replace, in effect, the line of action of force F_1 in the space diagram.

f. Extend the line of action of component $\overrightarrow{P-2}$ in view #2 of the space diagram until it intersects the line of action of force F_2 at point S ($\overrightarrow{P-2}$ is a component of forces F_1 and F_2).

g. At point S, construct a line parallel to the line of action of component $\overrightarrow{P-3}$, as seen in vector diagram #I. From the vector diagram you can see that components $\overrightarrow{P-2}$ and $\overrightarrow{P-3}$ are components of force F_2.

h. Extend the line of action of component $\overrightarrow{P-3}$ until it intersects the line of action of force F_3 at point T.

i. From point T, construct a line parallel to the lines of action of component $\overrightarrow{P\text{–}4}$, as seen in vector diagram #I. Extend this line until it intersects the line of action of component $\overrightarrow{P\text{–}1}$ (extended) at point N.

j. Through point N, construct a line *parallel* to the line of action of the *resultant* in vector diagram #I. Place an arrow in the sense *opposite* to that of the resultant. This is the frontal projection of the required equilibrant, the magnitude of which is 65 lb.

k. Project the line of action of the equilibrant to view #1. We require another view of the given parallel force system in order to specifically locate the equilibrant in view #1.

l. Choosing *any* view #3 (coming from view #1 of the space diagram), project the lines of action of forces F_1, F_2, and F_3 into this view.

m. In view #3, construct a vector diagram #II parallel to the lines of action of forces F_1, F_2, and F_3.

n. Choose any convenient point P' in vector diagram #II. Construct rays to points #1–#4 in order to determine the components, $\overrightarrow{P'\text{–}1}$, $\overrightarrow{P'\text{–}2}$, $\overrightarrow{P'\text{–}3}$, and $\overrightarrow{P'\text{–}4}$ of each of the vectors.

o. In view #3, construct the *string polygon* by replacing the lines of action of the given forces by their components. (Start at any convenient point M' on the line of action of F_1.) The equilibrant of the given force system must pass through point N', which is at the intersection of the uncancelled components $\overrightarrow{P'\text{–}1}$ and $\overrightarrow{P'\text{–}4}$. (The equilibrant in view #3 is parallel to the resultant vector $\overrightarrow{1\text{–}4}$ in vector diagram #II.)

p. Project the equilibrant line of action from view #3 to view #1. Point E is determined by the intersection of this projection and the projector from view #2.

Note that the given force system has been projected onto two projection planes (views #2 and #3) in which the lines of action of the forces are seen as parallel lines. Each of these views represents a coplanar "picture" of the given parallel force system. By combining these two coplanar pictures (using orthographic projection) you determined the *third dimension* in space that located the equilibrant of the parallel forces relative to their spatial locations.

SAMPLE QUIZZES

1. Define the following:

a. Concurrent forces
b. Equilibrant
c. Vector diagram
d. Space diagram

2. What is a vector?

3. The vector diagram for three or more concurrent, noncoplanar forces that are not in equilibrium, will *not* be a closed polygon. However, if one vector closes the vector diagram, and if the sense of this vector is *opposite* to the sense of all other vectors around the diagram, what is the significance of this closing vector?

4. What does the ray polygon represent?

5. What does the string, or funicular, polygon represent?

6. How do you determine the type (tensile or compressive) of load in a tripod (assuming that an inclined external force is acting on it)?

7. Describe the use and significance of Bow's notation.

8. What basic condition must you satisfy before you can construct a vector diagram for a problem involving a loaded tripod?

9. How must you represent the known external forces acting on a tripod, in order to draw a vector diagram?

10. Define the following:

a. Structure diagram   b. Space diagram

PRACTICE PROBLEMS *

PROBLEM 1

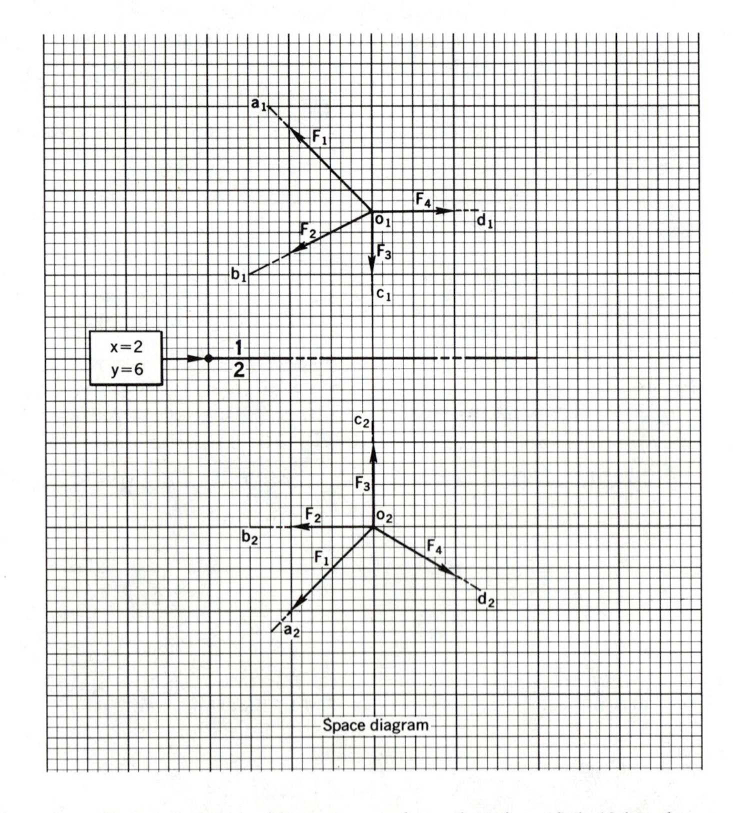

* For all vector problems, use two sheets of cross-section paper taped together along their 10-in. edges.

1. Given Data: Horizontal and frontal projections of a space diagram that illustrates the sense and lines of action of four concurrent, noncoplanar forces (F_1 = 200 lb, F_2 = 120 lb, F_3 = 160 lb, and F_4 = 120 lb).

Required:

A. Graphically determine the magnitude, bearing, and slope of the resultant of the given force system. (*Ans.:* Resultant = 128 lb, bearing = S70°W, slope = 21°.) (SUGGESTION: Use a vector scale in which 1 in. = 80 lb.)

B. The horizontal and frontal projections of the resultant in the space diagram.

C. Label and show all construction.

2. Given Data: Horizontal and frontal projections of a space diagram that illustrates the senses and lines of action of three concurrent, noncoplanar forces (F_1 = 120 lb, F_2 = 200 lb, and F_3 = 160 lb).

Required:

A. Graphically determine the magnitude, bearing, and slope of the equilibrant force that will counterbalance the given force system. (*Ans.:* Equilibrant = 335 lb, bearing = N56°W, slope = 38°.) (SUGGESTION: Use a vector scale in which 1 in. = 80 lb.)

PROBLEM 2

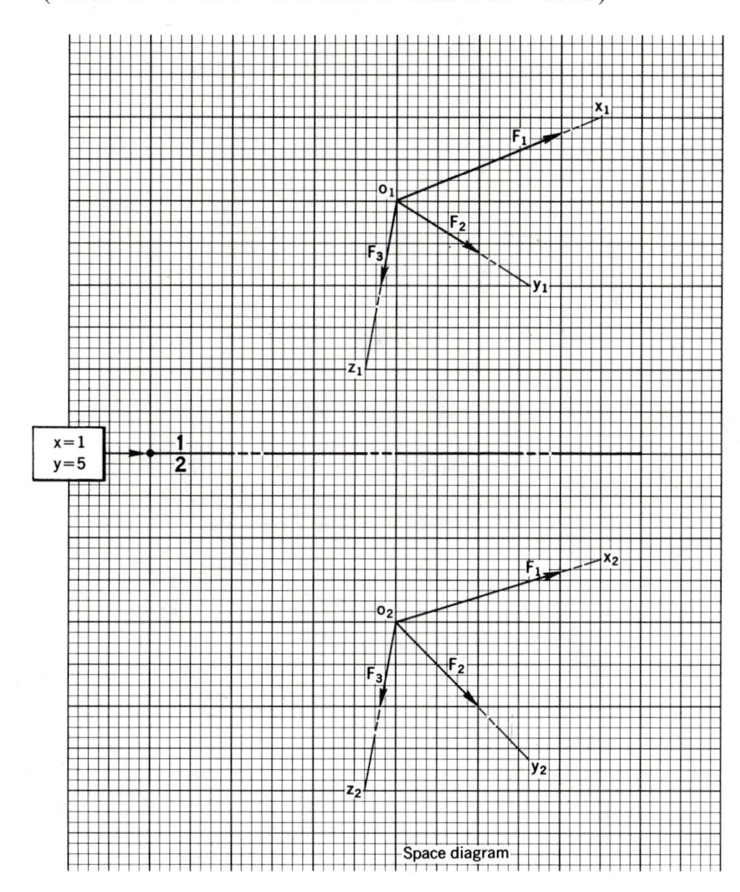

B. The horizontal and frontal projections of the equilibrant force in the space diagram.

C. Label and show all construction.

3. Given Data: Horizontal and frontal projections of a space diagram that illustrates a tripod (the legs of which are OA, OB, and OC) hanging from a horizontal ceiling. The sense and line of action OX of a downward vertical load $F_1 = 200$ lb acting at apex O of the tripod are also shown.

Required: A. Using the edge view method, graphically determine the magnitude and type of load (tensile or compressive) carried by each leg of the tripod in order for it to maintain a condition of equilibrium. (*Ans.:* $OA = 67$-lb tension, $OB = 110$-lb tension, and $OC = 96$-lb tension.) (SUGGESTION: Use a vector scale in which 1 in. = 80 lb.)

B. Summarize the results in view #1 of the space diagram, indicating by arrows the sense of the load carried by each tripod leg.

C. Label and show all construction.

PROBLEM 3

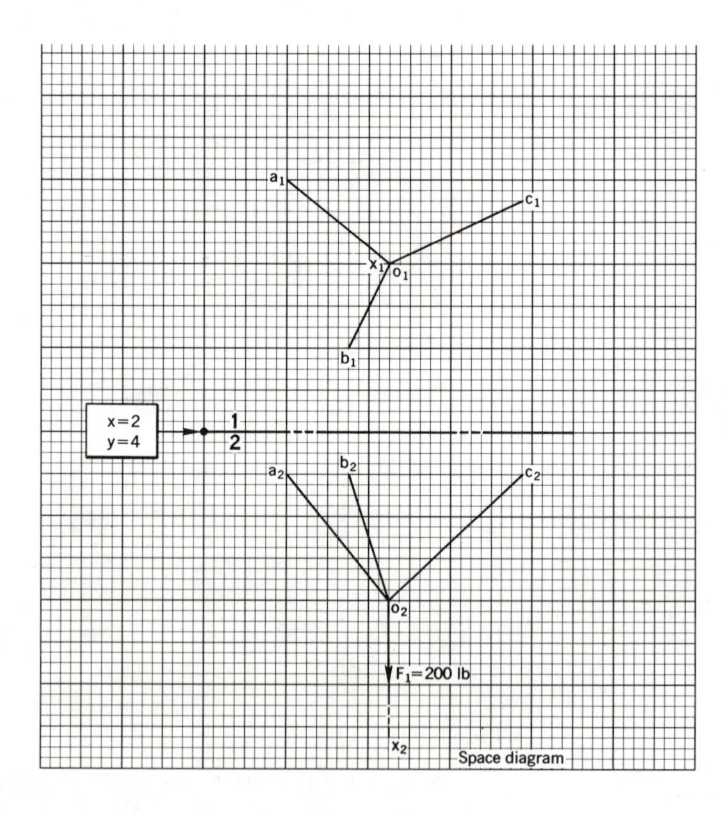

4. Given Data: Horizontal and frontal projections of a space diagram that illustrates a tripod (the legs of which are OA, OB, and OC) and the senses and directions OX and OY, respectively, of two forces ($F_1 = 200$ lb and $F_2 = 160$ lb) acting at apex O of the tripod.

Required:

A. Using the edge view method, graphically determine the magnitude and type of load (tensile or compressive) carried by each leg of the tripod in order for it to maintain a condition of equilibrium. (*Ans.:* OA = 265-lb tension, OB = 14-lb tension, OC = 85-lb compression.) (SUGGESTION: Use a vector scale in which 1 in. = 80 lb.)

B. Summarize the results in view #1 of the space diagram, indicating by arrows the sense of the load carried by each tripod leg.

C. Label and show all construction.

PROBLEM 4

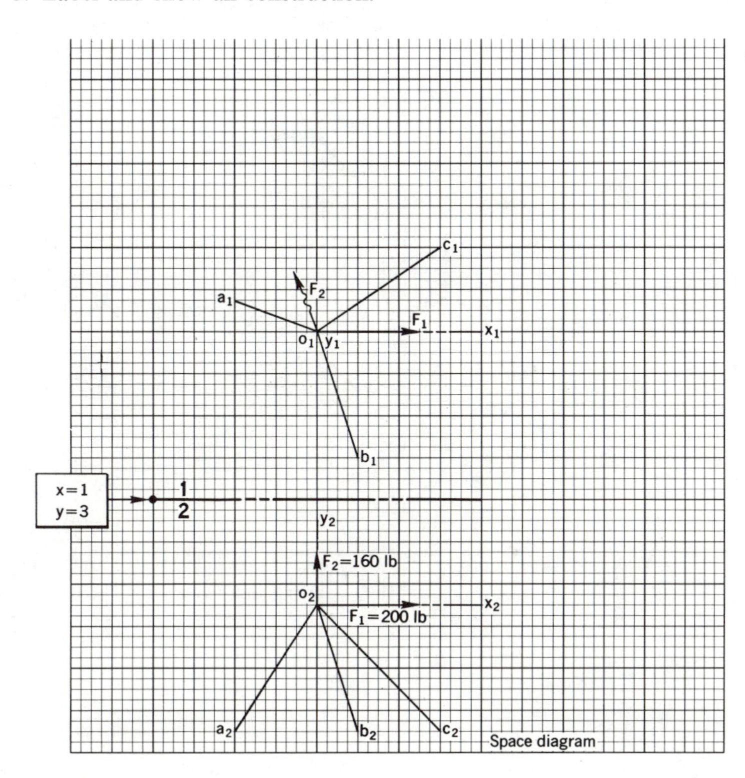

5. Given Data: Horizontal and frontal projections of a space diagram that illustrates a tripod (the legs of which are OA, OB, and OC) and the sense and line of action OX of an inclined force (F = 250 lb) acting at apex O of the tripod.

Required:

A. Using the edge view method, graphically determine the magnitude and type of load (tensile or compressive) carried by each leg of the tripod in order for it to maintain a condition of equilibrium. (*Ans.:* OA = 185-lb compression, OB = 65-lb tension, OC = 230-lb compression.) (SUGGESTION: Use a vector scale in which 1 in. = 80 lb.)

B. Summarize the results in view #1 of the space diagram, indicating by arrows the sense of the load carried by each tripod leg.

C. Label and show all construction.

PROBLEM 5

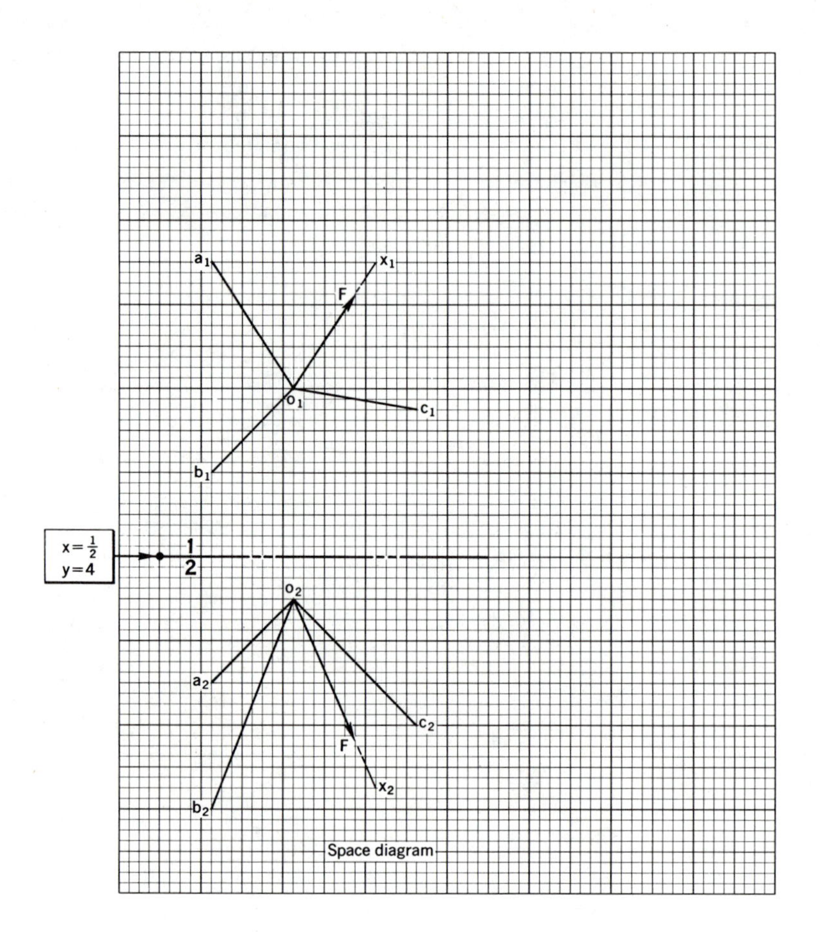

6. Given Data: Horizontal and frontal projections of a space diagram that illustrates a tripod (the legs of which are OA, OB, and OC), hanging from a vertical wall. The sense and line of action of a downward vertical load ($F =$ 200 lb) acting at the apex of the tripod are also shown.

Required:

A. Using the point view method, graphically determine the magnitude and type of load (tensile or compressive) carried by each leg of the tripod in order for it to maintain a condition of equilibrium. (*Ans.:* $OA =$ 210-lb compression, $OB =$ 80-lb compression, $OC =$ 234-lb tension.) (SUGGESTION: Use a vector scale in which 1 in. = 80 lb.)

B. Summarize the results in view #1 of the space diagram, indicating by arrows the sense of the load carried by each tripod leg.

C. Label and show all construction.

PROBLEM 6

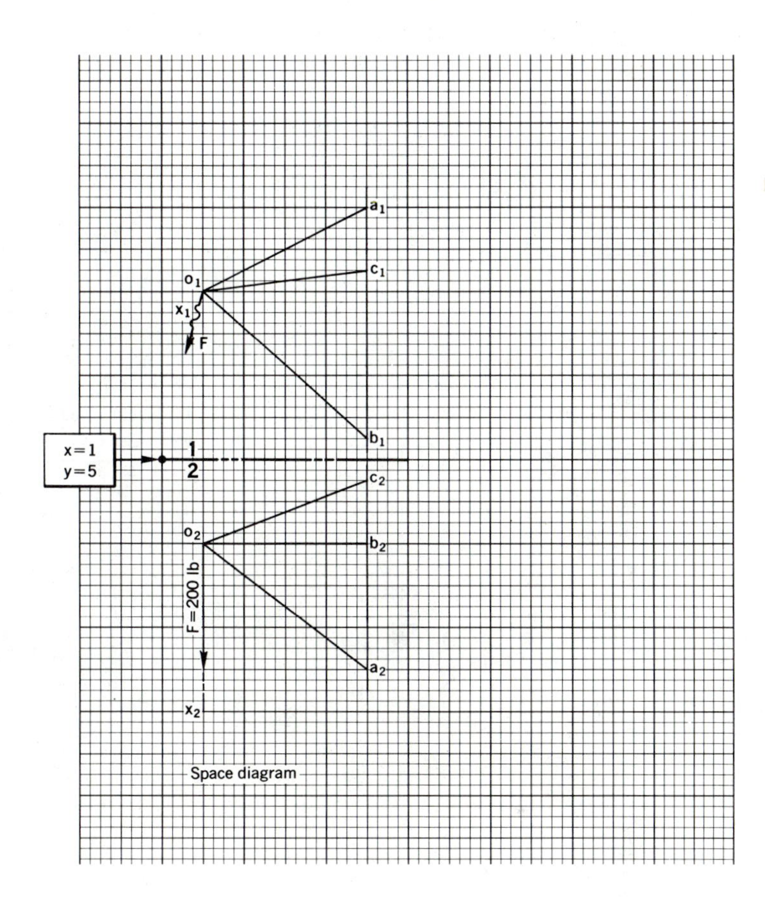

7. Given Data: Horizontal and frontal projections of a space diagram that illustrates a tripod (the legs of which are *OA*, *OB*, and *OC*) and the sense and line of action of a force ($F = 160$ lb) acting at apex *O* of the tripod.

Required:

A. Using the method of combining two unknown forces into one, determine the magnitude and type of load (tensile or compressive) carried by each leg of the tripod in order for it to maintain a condition of equilibrium. (*Ans.:* $OA = 18$-lb tension, $OB = 80$-lb compression, $OC = 155$-lb tension.) (SUGGESTION: Use a vector scale in which 1 in. = 80 lb.)

B. Summarize the results in view #1 of the space diagram, indicating by an arrow the sense of the load carried by each tripod leg.

C. Label and show all construction. (NOTE: Use two sheets of cross-section paper taped together along their 10-in. edges for this problem. Construct a vertical cutting plane through the given force *F* in order to determine the directions of *two* forces that represent the combined loads carried by legs *OA* and *OB and,* respectively, by legs *OA* and *OC*.)

PROBLEM 7

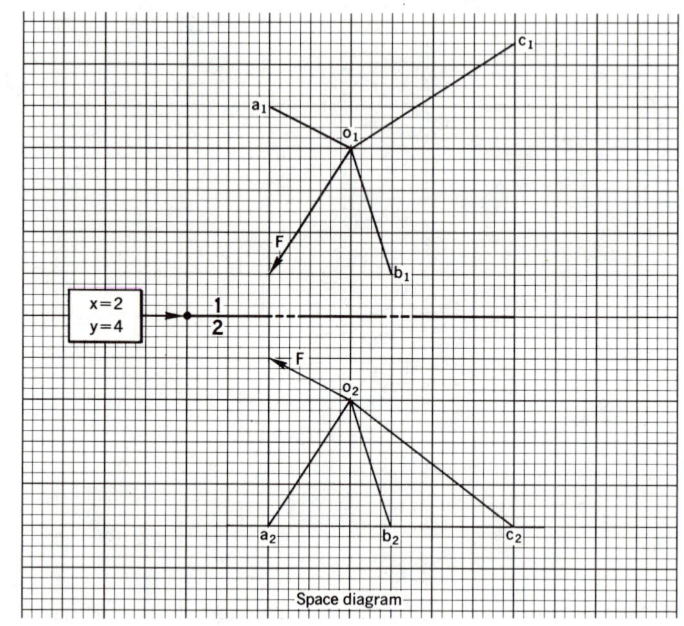

8. Given Data: Horizontal and frontal projections of a space diagram that illustrates the senses and lines of action of four parallel forces ($F_1 = 160$ lb, $F_2 = 80$ lb, $F_3 = 120$ lb, $F_4 = 160$ lb).

PROBLEM 8

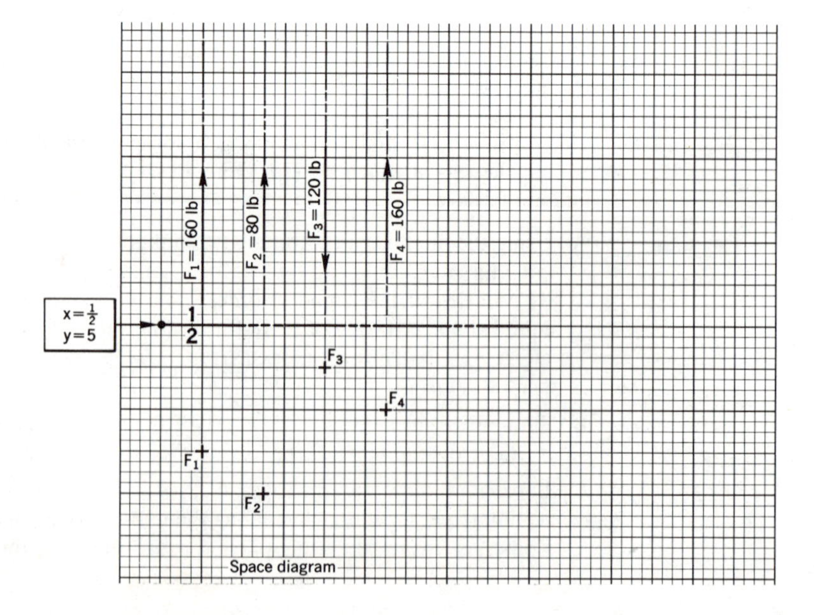

Required: A. Graphically determine the magnitude of the resultant of the given force system. (SUGGESTION: Use a vector scale in which 1 in. = 80 lb.)

B. The horizontal and frontal projections of the sense and line of action of the resultant relative to the forces in the space diagram.

C. Label and show all construction.

9. Given Data: Horizontal and frontal projections of a space diagram that illustrates the senses and lines of action of three parallel forces ($F_1 = 120$ lb, $F_2 = 200$ lb, $F_3 = 80$ lb).

Required: A. Graphically determine the magnitude of the resultant of the given force system. (SUGGESTION: Use a vector scale in which 1 in. = 80 lb.)

B. The horizontal and frontal projections of the sense and line of action of the resultant relative to the given forces in the space diagram.

C. Label and show all construction.

PROBLEM 9

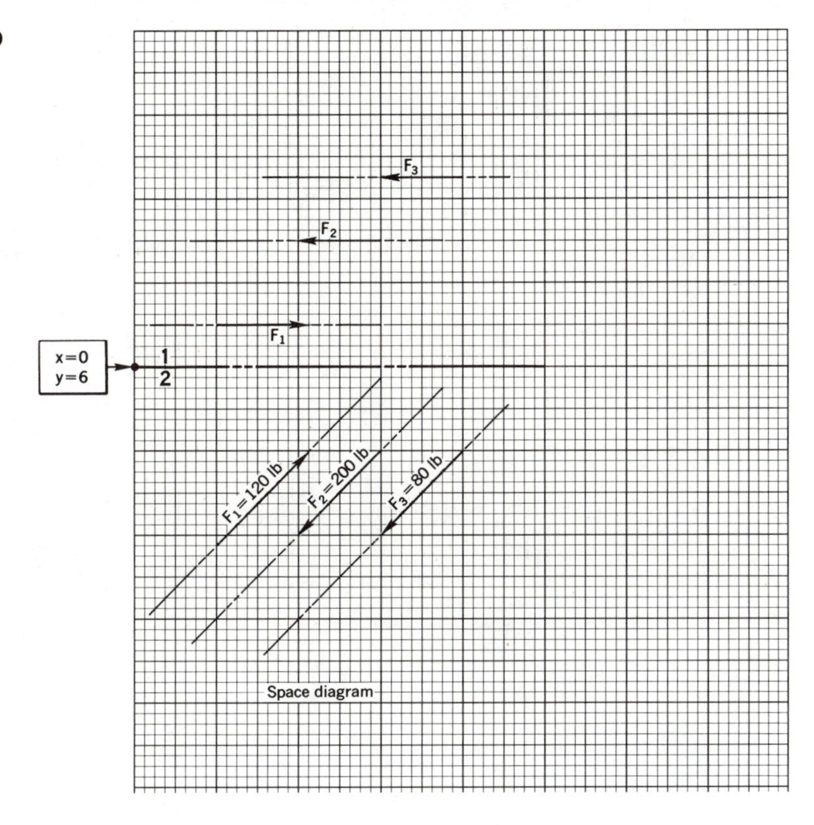

10. Given Data: The horizontal and frontal projections of a space diagram that illustrates an irregular solid block, the flat plane faces of which make right angles with each other. The weight of each section of the block is indicated as $F_1 = 160$ lb, $F_2 = 80$ lb, $F_3 = 90$ lb.

Required: A. The total weight of the block. (SUGGESTION: Use a vector scale in which 1 in. = 80 lb.)

B. The location (horizontal and frontal projections) of a force that will counterbalance the weight of the block.

C. Label and show all construction.

PROBLEM 10

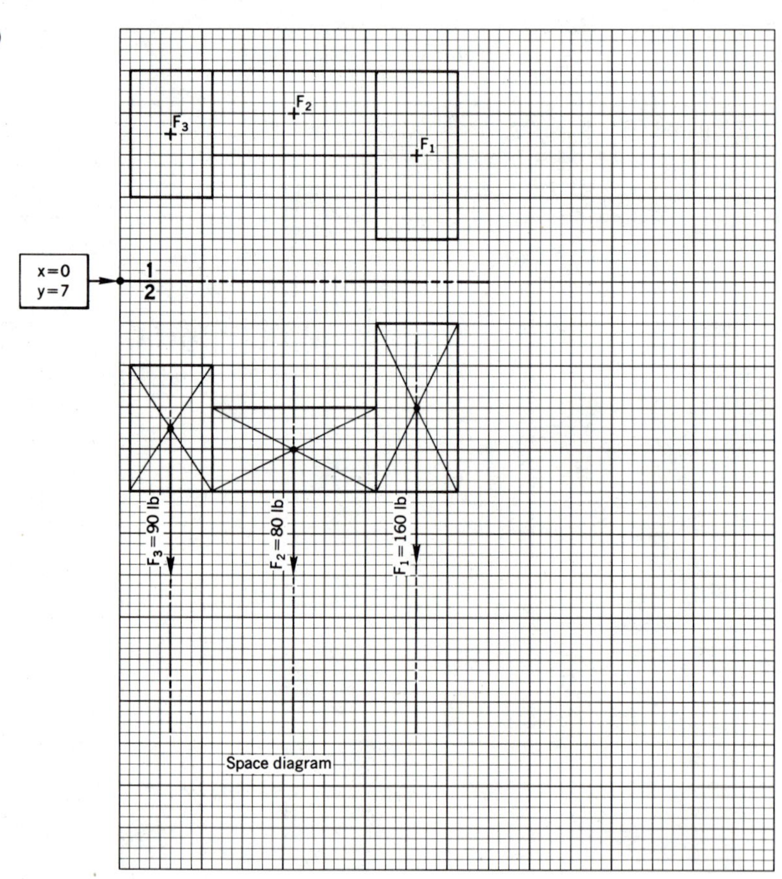

Principles of Descriptive Geometry Applied to Selected Topics

98 Introduction

In this chapter we introduce the application of three-dimensional descriptive geometry to three areas: basic mining problems, projections of simple shades and shadows, and the development of basic perspective projection (central projection). Relate each application to the principles of three-dimensional descriptive geometry and to the interrelationships of the geometric elements of points, lines, and planes. (Sample Quizzes and Practice Problems have been excluded from this chapter.)

99 Basic mining problems

A. PRINCIPLES OF DESCRIPTIVE GEOMETRY

To solve problems of basic mining operations (for example, the location of a vein of ore under the earth's surface), the fundamental concepts of three-dimensional descriptive geometry used are: true length and bearing of a line, edge view and slope of a plane, parallel lines, skew lines, and parallel planes.

B. DEFINITIONS AND TERMINOLOGY USED IN MINING OPERATIONS

Several basic terms and definitions common to mining operations are:

1. *Ore stratum* (vein of ore): generally considered to be a plane having a

uniform thickness (two equidistant parallel surfaces) and usually inclined under the surface of the earth (see Fig. 256).

2. *Headwall:* the upper surface of an ore stratum (represented in Fig. 256 by points *A*, *B*, and *C*).

   Footwall: the lower surface of an ore stratum (represented in Fig. 256 by points *A*′, *B*′, and *C*′).

3. *Boreholes:* vertical or inclined holes that are drilled from the earth's surface to an ore stratum below the surface (Fig. 257). Boreholes are used to determine points on the headwall and footwall of an ore stratum. For example, the location of three points on the headwall provides enough information so that you can describe the position of the ore stratum below the earth's surface. (This is possible because three noncollinear points define a plane, and the distance, or depth, of a borehole from the earth's surface to an ore stratum defines the positions of these points.) In Fig. 257 you see an ore stratum located by an *outcrop* at point *A*, a vertical borehole that intersects the headwall at point *X*, and an inclined borehole that intersects the headwall at point *Y*. Points *A*, *X*, and *Y* define the plane of the headwall. (You can now specify the location of the ore stratum by defining the bearing of a horizontal line in this plane as well as the slope of the plane.)

4. *Outcrop:* a point (sometimes a line) at which a portion of an ore stratum pushes through and lies on or above the earth's surface (see point *A* in Figs. 256 and 257).

5. *Strike:* the bearing of a horizontal line in either the headwall or footwall of the ore stratum (see Fig. 258).

6. *Dip:* the angle that either the headwall or footwall of an ore stratum makes with the horizontal projection plane (see Fig. 258). You will recognize this as the definition of the *slope of a plane.*

FIGURE 256

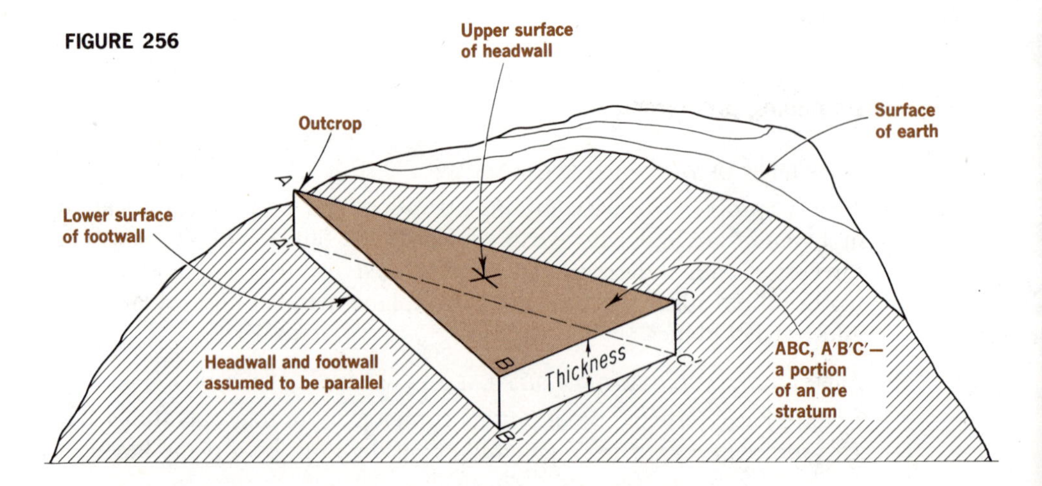

FIGURE 257

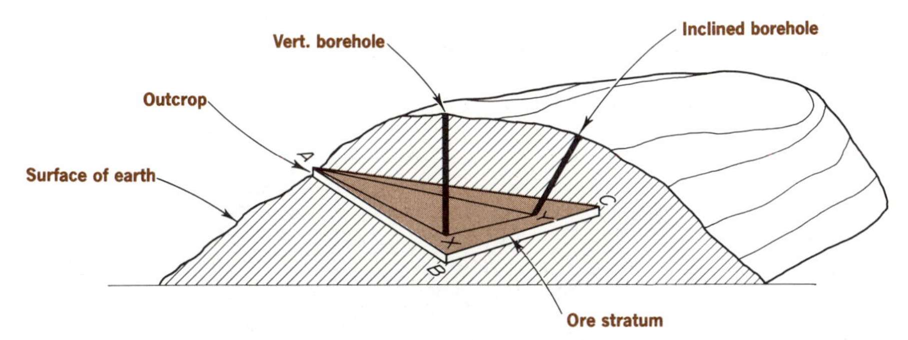

FIGURE 258

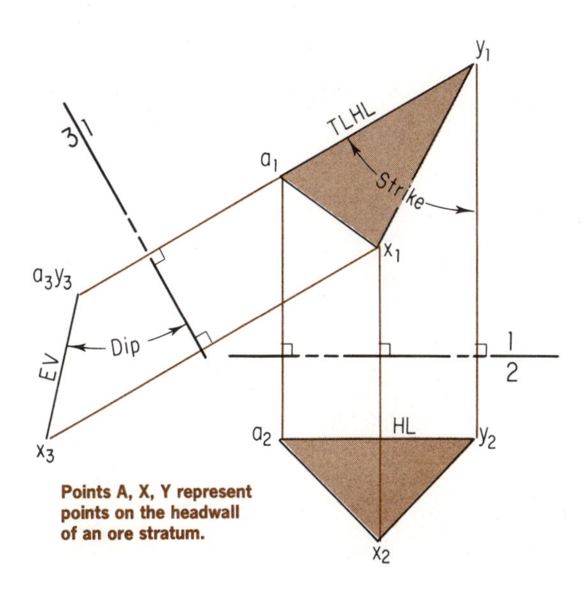

7. *Thickness:* the perpendicular distance between the headwall (upper surface) and footwall (lower surface) of an ore stratum (see Fig. 256).

C. ORE STRATUM LOCATION—PARALLEL BOREHOLES

Figure 259 depicts an ore stratum, located by an outcrop at point *A*, and two vertical parallel boreholes #I and #II. These boreholes locate points *C* and *B*, respectively, on the headwall. The three points *A*, *B*, and *C* define the plane of the headwall. From this information you can establish the strike and dip of the vein of ore.

Orthographic Example: Determination of the strike, dip, and thickness of an ore stratum by two parallel boreholes and an outcrop (Figs. 259 and 260).

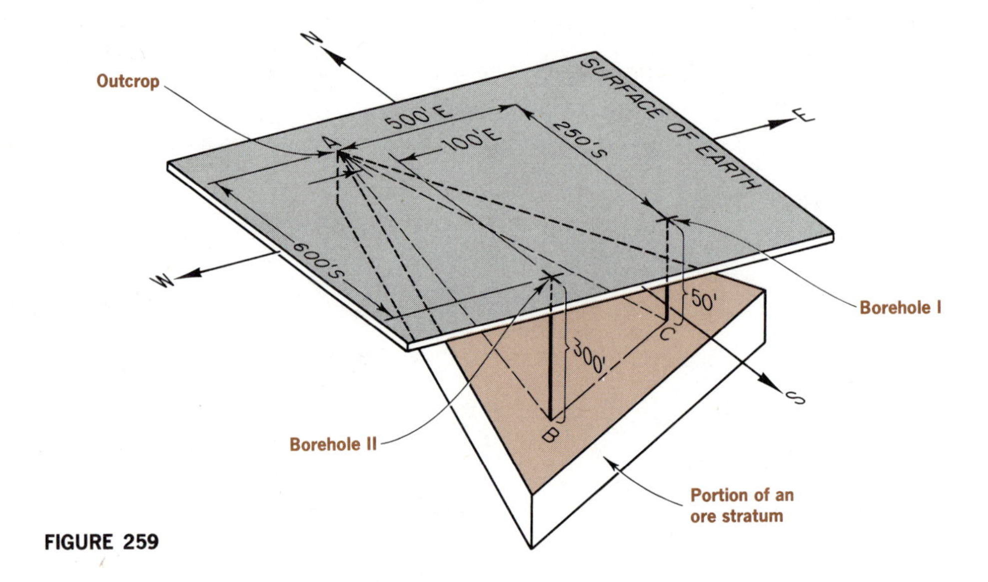

FIGURE 259

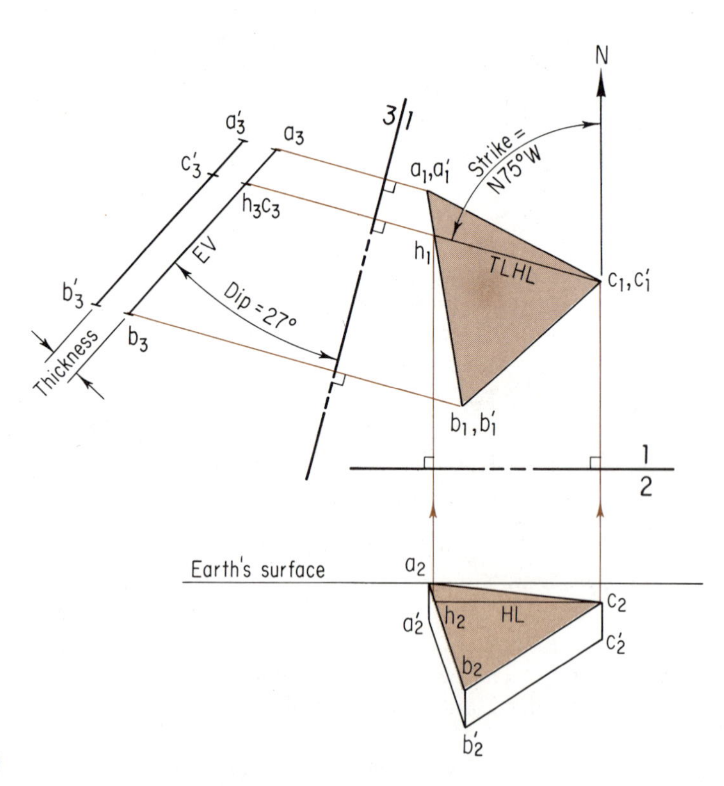

FIGURE 260

Given Data: The horizontal and frontal projections of three points (A, B, and C) on the headwall of a vein of ore (see Fig. 259). Determine points A, B, and C as follows:

a. Point A is an outcrop point of the vein of ore.

b. Point C is established by vertical borehole #I, which intersects the headwall 50 ft below the earth's surface and the footwall 150 ft below the surface. Borehole #I is located 500 ft east and 250 ft south of outcrop point A.

c. Point B is established by vertical borehole #II, which intersects the headwall 300 ft below the earth's surface and the footwall 400 ft below the surface. Borehole #II is located 100 ft east and 600 ft south of outcrop point A.

Required: The strike, dip, and thickness of the vein of ore.

Construction Program:

a. In Fig. 260, the horizontal and frontal projections of outcrop A and points B and C on the headwall are located according to the specifications given above. Points A', B', and C' on the footwall are also located according to these specifications.

b. To determine the strike, construct, in view #2, a horizontal line c_2h_2 in plane $a_2b_2c_2$.

c. Determine the true length c_1h_1 of the horizontal line in view #1 by direct projection.

d. Measure the bearing of horizontal line a_1h_1 (consider the top of your paper to be north) and determine the strike of the given ore stratum. In Fig. 260, the strike has been measured as N75°W.

e. Determine the edge view of the headwall $a_1b_1c_1$ by constructing a projection plane #3 (represented by RL 1–3) perpendicular to the true length horizontal line c_1h_1. In view #3 the headwall $a_3b_3c_3$ appears as an edge. The dip can be measured in view #3; it is equal to 27°.

f. Since the headwall and footwall are assumed to be parallel to each other, the footwall $a_3'b_3'c_3'$ also appears as an edge in view #3. The edge view of both the headwall and footwall appear parallel in this view, and the thickness of the ore stratum can therefore be measured here.

D. ORE STRATUM LOCATION—NONPARALLEL BOREHOLES

Since you assume that the headwall and footwall of an ore stratum are *parallel* to each other, you can determine the strike, dip, and thickness of a vein of ore by using only two *nonparallel* boreholes. In Fig. 261, borehole #I starts from an assumed point A and intersects the headwall at point B and the footwall at point C. Borehole #II starts at an assumed point D and intersects the headwall at point E and the footwall at point F.

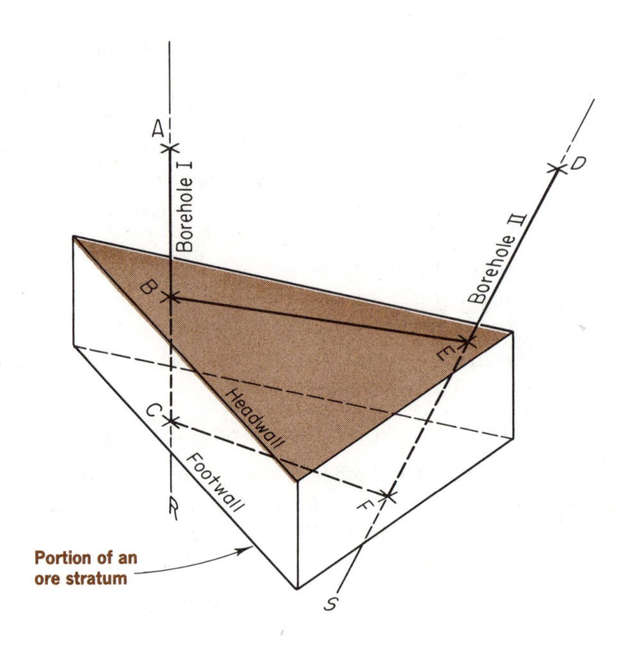

FIGURE 261

Lines *BE* and *CF* are *skew lines* that lie in parallel planes. To determine the thickness of the ore stratum, find the shortest perpendicular distance between skew lines *BE* and *CF*.

Orthographic Example: Determination of the strike, dip, and thickness of an ore stratum by two nonparallel boreholes (Fig. 262).

Given Data: Horizontal and frontal projections of two nonparallel boreholes *AR* (#I) and *DS* (#II). Borehole #I intersects the headwall at point *B* and the footwall at point *C*. Borehole #II intersects the headwall at point *E* and the footwall at point *F*.

Required: The strike, dip, and thickness of the ore stratum.

Construction Program #I: Using the Plane Method—Fig. 262.

a. Construct a plane containing one of two skew lines that is parallel to the other line outside the plane. In this case, one of the skew lines can be the line *BE* on the headwall that contains the plane parallel to the other line *CF* on the footwall. (We can use the plane method, since we will assume that the headwall and footwall are parallel planes.) Create a plane that contains line b_2e_2 on the headwall in view #2, and that is also parallel to line c_2f_2 on the footwall. (In view #2, construct a line through e_2 parallel to line c_2f_2. Assume x_2 on this line to define a portion of the parallel plane $b_2x_2e_2$.)

b. Construct a horizontal line b_2h_2 in plane $b_2x_2e_2$. Determine the true length of this line by direct projection (the true length appears in view #1 as b_1h_1).

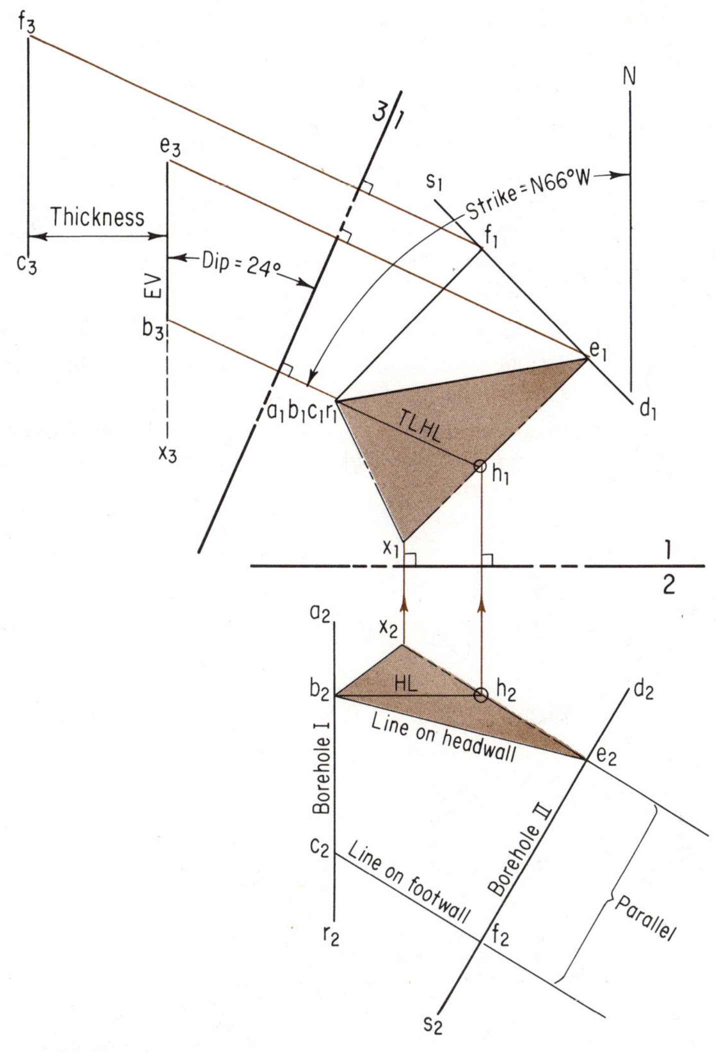

FIGURE 262

c. Determine a view in which plane $b_1x_1e_1$ appears as an edge by constructing plane #3 perpendicular to the true length horizontal line b_1h_1. (In view #3, the constructed plane $b_3x_3e_3$ appears as an edge.)

d. Project line c_1f_1 to view #3 and determine line c_3f_3. In this view, line c_3f_3 appears *parallel* to line e_3b_3. The shortest distance between these two skew lines is now seen in view #3. Since these lines are located in the parallel headwall and footwall planes, the distance represents the thickness of the ore stratum.

e. The dip of the ore stratum, measured in view #3, is equal to 24°.

f. The strike of the ore stratum, measured in view #1, is the bearing of the true length horizontal line b_1h_1. The strike is N66°W.

Construction Program #II:

Using the Line Method—Fig. 263.

a. To find the strike, dip, and thickness of a vein of ore, the line method of determining the shortest distance between two skew lines can be used since we assume the planes of the headwall and footwall to be parallel.

By direct projection, determine the true length of either line *BE* on the headwall or line *CF* on the footwall. (Line *CF* on the footwall was used in our example. RL 1–3 is parallel to c_1f_1 and, therefore, c_3f_3 appears as true length in view #3.)

b. From point c_3 in view #3, construct a *horizontal line* c_3p_3.

c. Assume point x_3 on this horizontal line, and define a plane ($f_3c_3x_3$) that contains point x_3 and line c_3f_3.

d. Determine the point view of line c_3f_3 by constructing projection plane #4 (represented by RL 3–4) perpendicular to the true length c_3f_3 in view #3.

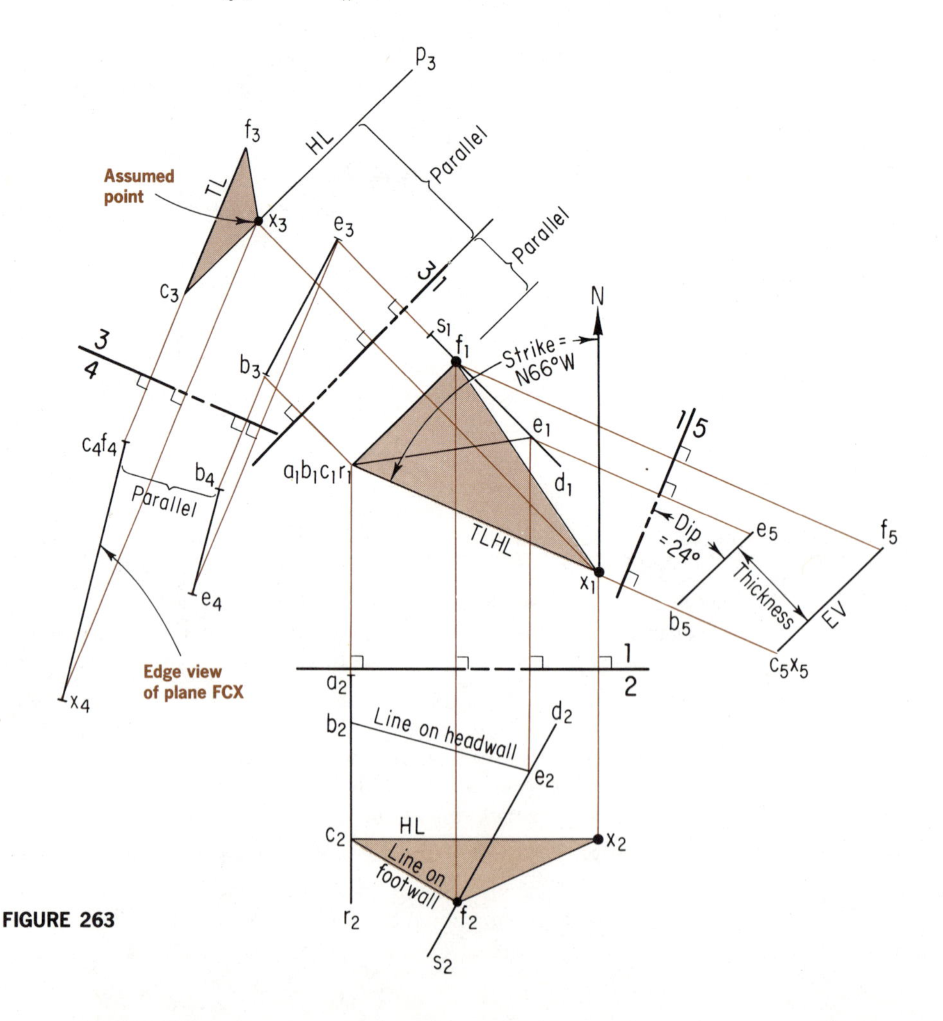

FIGURE 263

e. Project line b_3e_3 to view #4 and determine its projection b_4e_4.

f. From the point view c_4f_4 in view #4, construct the *edge view* $f_4c_4x_4$ of plane FCX parallel to line b_4e_4. (This construction is valid because both lines c_4f_4 and b_4e_4 are located in assumed parallel planes—the footwall and the headwall.)

g. By direct projection, locate point x_1 in view #1 and x_2 in view #2. Since c_1x_1 is a true length horizontal line in the footwall plane, the strike of the ore stratum appears in view #1; it is equal to N66°W.

h. Determine the dip of the ore stratum by constructing an *edge view* of the plane $f_1c_1x_1$ in an *elevation view.* (View #5 is the edge view of this plane $f_5c_5x_5$, and line b_5e_5 is parallel to this edge.) The dip of the ore stratum appears in view #5; it is equal to 24°. The thickness of the ore stratum appears in views #4 and #5.

100 Projection of simple shades and shadows

The projection of simple shades and shadows is a versatile technique, useful in architectural drawing. In this section we present some of the basic principles used to find shades and shadows, applying the descriptive geometry concepts of the piercing point of a line and a plane (flat or curved).

Light rays are assumed to be parallel, and to travel in straight lines. The source of light is at infinity.

When an object intercepts rays of light, the side (or sides) of the object facing the light source are illuminated, while its other side (or sides) remain obscured. The part of the object from which light is obscured is called *shade.* The *shadow* of the object on a plane surface is determined by the light rays, as *intercepted* by the *object* and the plane surface.

DEFINITIONS OF SIMPLE SHADES AND SHADOWS (Fig. 264)

1. *Parallel Light Rays:* parallel lines representing rays of light and having a source at infinity.
2. *Shade:* the part of an object's surface from which light is obscured by the object itself.
3. *Shade Line:* the line that separates the shaded from the illuminated portions of the object.
4. *Umbra:* the space (either a line, plane, or volume) *between* the object and the surface on which its shadow is cast. The umbra is indicated by lines that represent *intercepted light rays.*
5. *Cast Shadow:* the image made on a given surface by an object that intercepts light rays: the intercepted light rays are the umbra, and the shadow is the *intersection* of the umbra with the given surface.

The following paragraphs present methods for determining cast shadows.

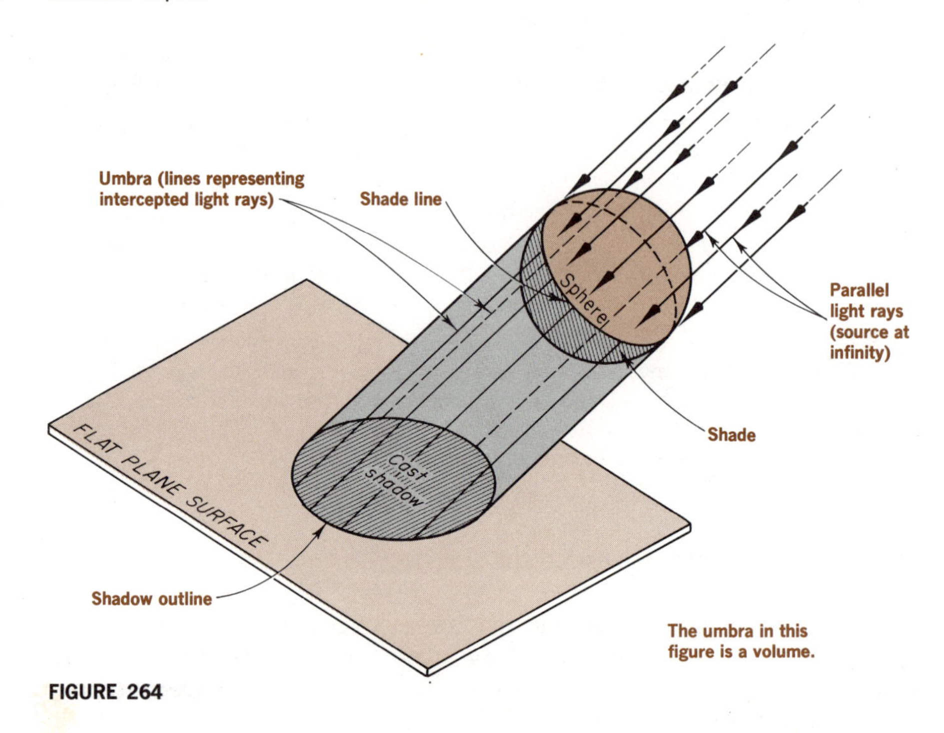

FIGURE 264

A. CAST SHADOW OF A POINT ON A FLAT PLANE SURFACE (Fig. 265)

The cast shadow of a point is determined by locating the piercing point of the *intercepted* light ray that illuminates the point with the plane surface. In Fig. 265, a point P is exposed to a single light ray RA. The intercepted light ray pierces the flat plane surface at point S_p. The umbra is one straight line.

Orthographic Example: Determination of the cast shadow of a point on a flat plane surface (Fig. 266).

Given Data: Horizontal and frontal projections of a point P, directions of a light ray RA, and a flat plane surface $LMNO$.

Required: The cast shadow of point P on the flat plane surface $LMNO$.

Construction Program:

a. Construct a vertical cutting plane X–X through the given direction r_1a_1 of the light ray in view #1. This determines points t_1 and u_1 on plane $l_1m_1n_1o_1$.

b. Project t_1 and u_1 to view #2 and determine points t_2 and u_2. s_{p_2} is located where the line connecting t_2 and u_2 intersects the direction line of the light ray r_2a_2 in view #2. This is the cast shadow of point p_2 on the flat plane surface in view #2.

c. Project s_{p_2} to view #1 and determine s_{p_1}, which is the cast shadow of point p_1 on the given flat plane surface in view #1.

FIGURE 265

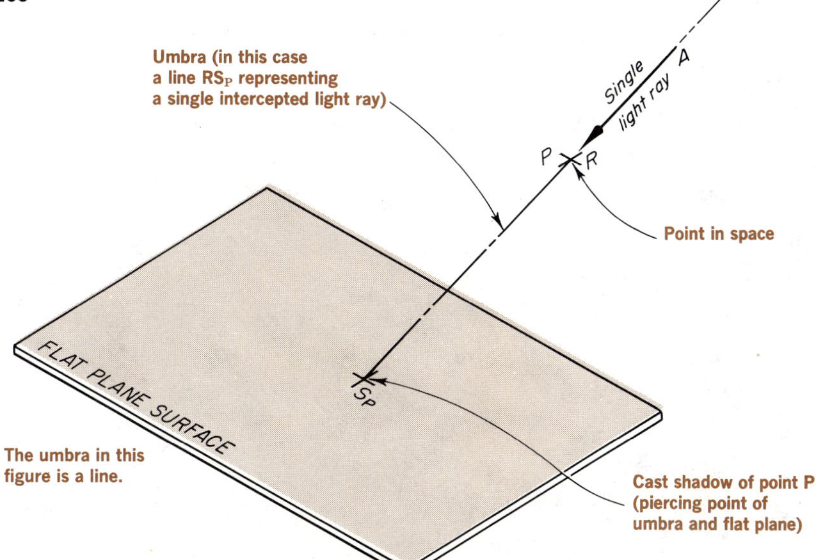

FIGURE 266

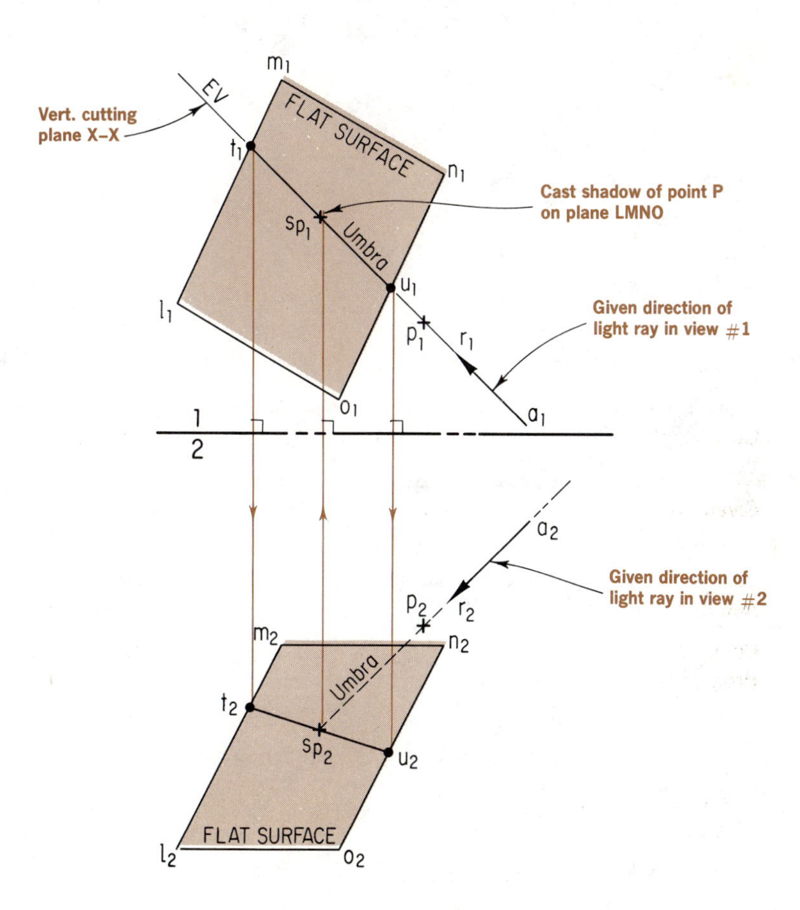

B. CAST SHADOW OF A POINT ON THE SURFACE OF A CYLINDER (Fig. 267) The cast shadow of a point outside the surface of a cylinder, illuminated by a light ray, is the piercing point of the intercepted light ray (umbra) and the cylinder surface. In Fig. 267, a given point P in space is illuminated by light ray RA, which is intercepted by point P. The *intercepted* light ray pierces (intersects) the cylinder surface at S_P, which is the cast shadow of point P on the cylinder surface.

FIGURE 267

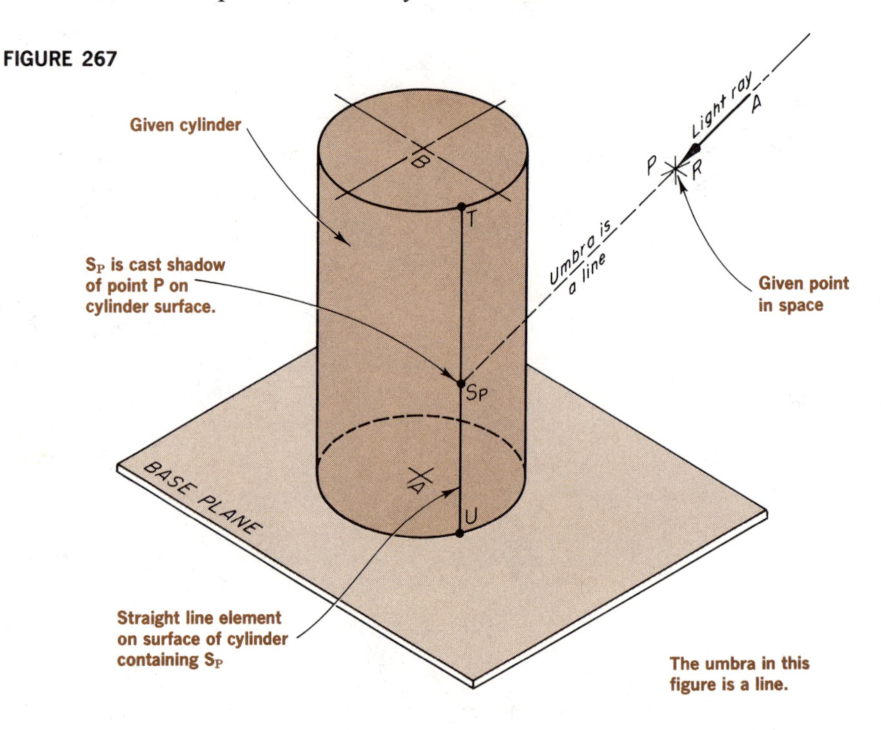

Orthographic Example: Determination of the cast shadow of a point on the surface of a cylinder (Fig. 268).

Given Data: Horizontal and frontal projections of a right circular cylinder, a point P outside the cylinder, and the directions of a light ray RA that illuminates point P.

Required: The cast shadow of point P on the surface of the given cylinder.

Construction Program:

a. Construct a vertical cutting plane X–X through the direction r_1a_1 of light ray RA in view #1. This cutting plane cuts the surface of the cylinder at element t_1u_1.

b. Project element t_1u_1 to view #2 and determine its frontal view t_2u_2.

c. In view #2, extend the direction r_2a_2 of light ray RA until it intersects element t_2u_2 at point s_{p_2}. Point s_{p_2} is the cast shadow of point p_2 on the cylinder surface. (s_{p_1} in view #1, which coincides with element

FIGURE 268

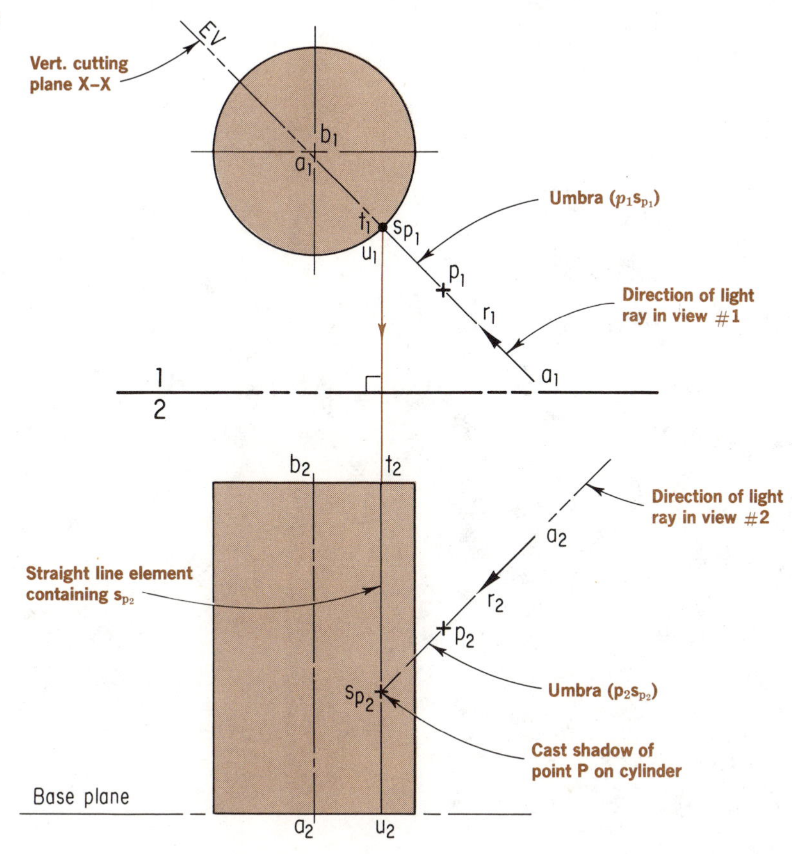

t_1u_1, is the cast shadow of point p_1 in view #1.) Note that the umbra in this example is one line.

C. CAST SHADOW OF A LINE ON A FLAT PLANE SURFACE (Fig. 269)

The parallel light rays that illuminate a line in space are intercepted by the line. Since parallel lines form a plane, the umbra in this case will also be a plane. The line of intersection between the umbra plane and a flat plane is the cast shadow of the given line in space on the flat plane surface.

Figure 269 shows a line *AB* in space that intercepts parallel light rays, forming a plane that intersects a flat plane surface on the line S_AS_B. This line is the cast shadow of the given line *AB* on the flat plane. (Note the plane formed by the umbra.)

Orthographic Example: Cast shadow of a line on a flat plane surface (Fig. 270).

Given Data: Horizontal and frontal projections of a line *AB*, the direction of parallel light rays, and a plane surface *LMNO*.

FIGURE 269

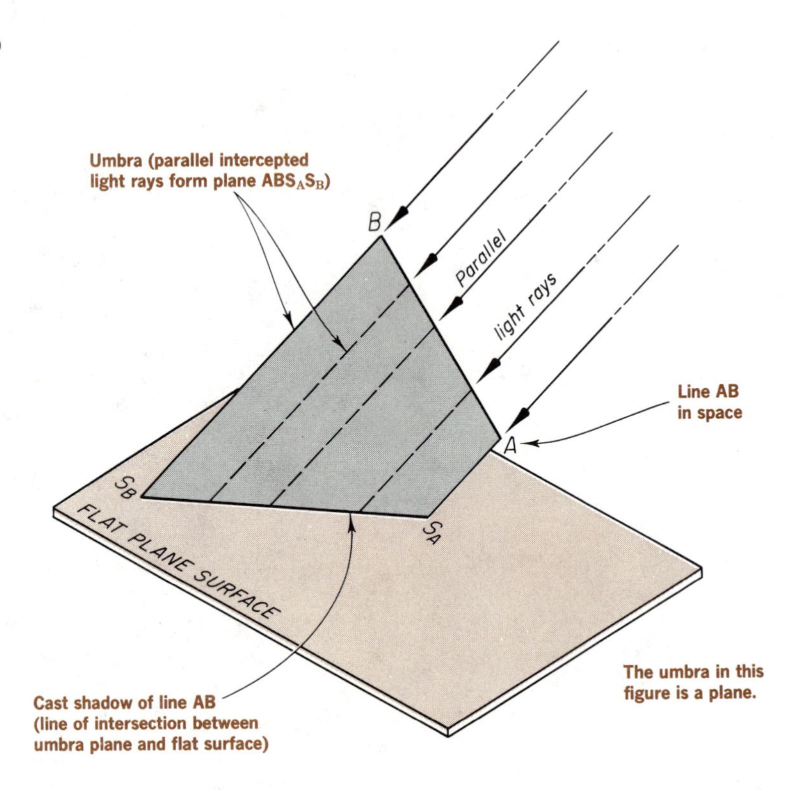

Required: The cast shadow of line AB on the given flat plane $LMNO$.

Construction Program:

a. Determine the line of intersection between the umbra plane and the given flat plane $LMNO$. In view #1, construct vertical cutting planes through the light rays illuminating points a_1 and b_1. This determines points y_1, z_1, w_1, and x_1 on $LMNO$.

b. Project y_1, z_1, w_1, and x_1 to view #2 and determine points y_2, z_2, w_2, and x_2.

c. Connect y_2 and z_2 with a straight line. Point s_{a_2} appears at the intersection of this line and the direction of the light ray illuminating a_2. This point is the cast shadow of point a_2 on the flat plane.

d. Connect points w_2 and x_2 with a straight line. Point s_{b_2} is found at the intersection of this line and direction of the light ray illuminating point b_2. This point is the cast shadow of point b_2 on the flat plane.

e. Connect s_{a_2} and s_{b_2} with a straight line. This line is the cast shadow of line AB on the flat plane in view #2.

f. Project line s_{a_2}–s_{b_2} to view #1 and determine line s_{a_1}–s_{b_1}, which is the cast shadow of line AB on the flat plane surface in view #1.

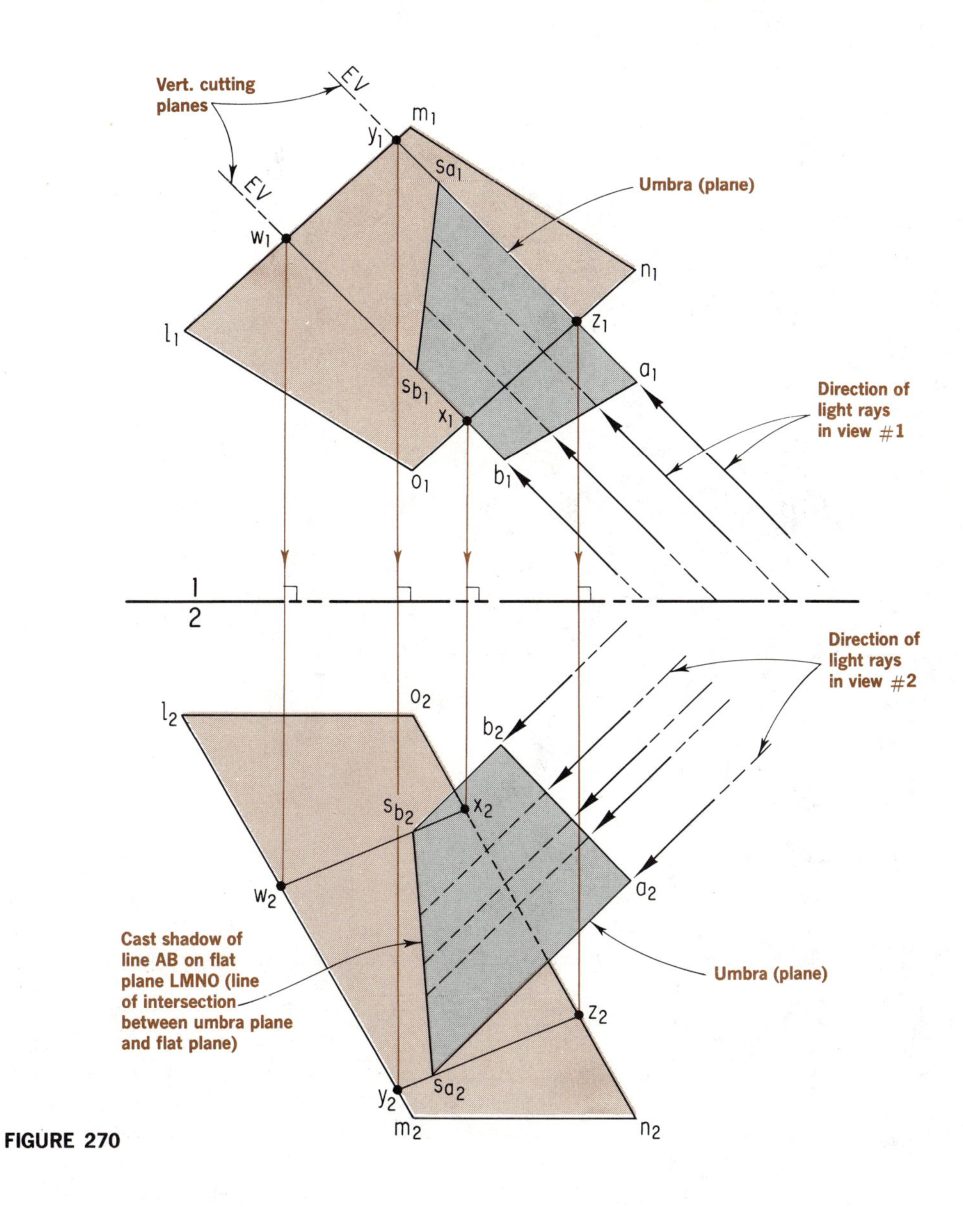

FIGURE 270

D. CAST SHADOW OF A LINE ON THE SURFACE OF A CYLINDER (Fig. 271) Since parallel light rays intercepted by a line in space form an *umbra plane,* the cast shadow of a line in space on a cylinder surface is the line of intersection between the umbra plane and the cylinder surface (see Fig. 271).

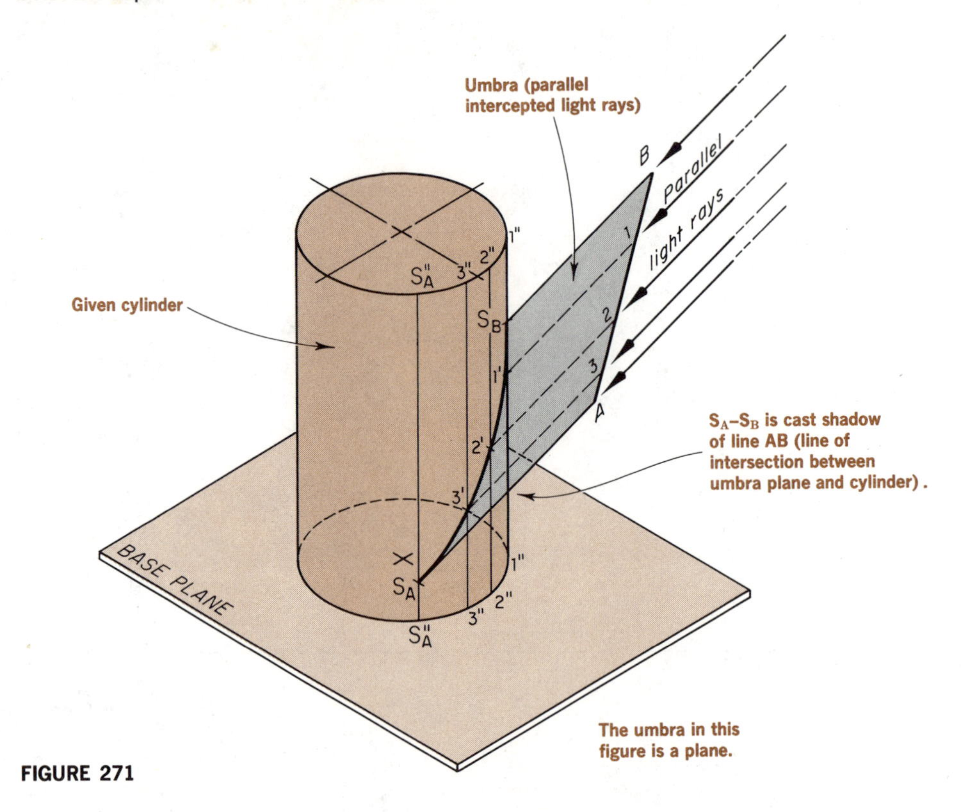

FIGURE 271

Orthographic Example: Determination of the cast shadow of a straight line on the surface of a cylinder (Fig. 272).

Given Data: Horizontal and frontal projections of a right cylinder, a straight line AB outside the cylinder surface, and the direction of parallel light rays.

Required: The cast shadow of line AB on the surface of the given cylinder.

Construction Program:

a. In views #1 and #2, draw a convenient number of lines through A and B and intermediate assumed points #1, #2, #3 on the given line AB. These lines represent extended light rays. Extend them until they intersect the cylinder at points s_{a_1}, s_{b_1}, #1′, #2′, and #3′.

b. Through these lines, construct vertical cutting planes in view #1 that cut *straight line elements* at each intersecting point s_{a_1}, s_{b_1}, #1′, #2′, and #3′. These elements are noted in view #2 as s_{a_2}''–s_{a_2}'', #3″–3″, #2″–2″, #1″–1″, and s_{b_2}''–s_{b_2}''.

c. In view #2, extend the intercepted light ray lines until they intersect their respective elements at points s_{a_2}, #3′, #2′, #1′, and s_{b_2}. Connect these points with a *smooth curve.* The curved line represents the cast shadow (in view #2) of line AB on the given cylinder.

FIGURE 272

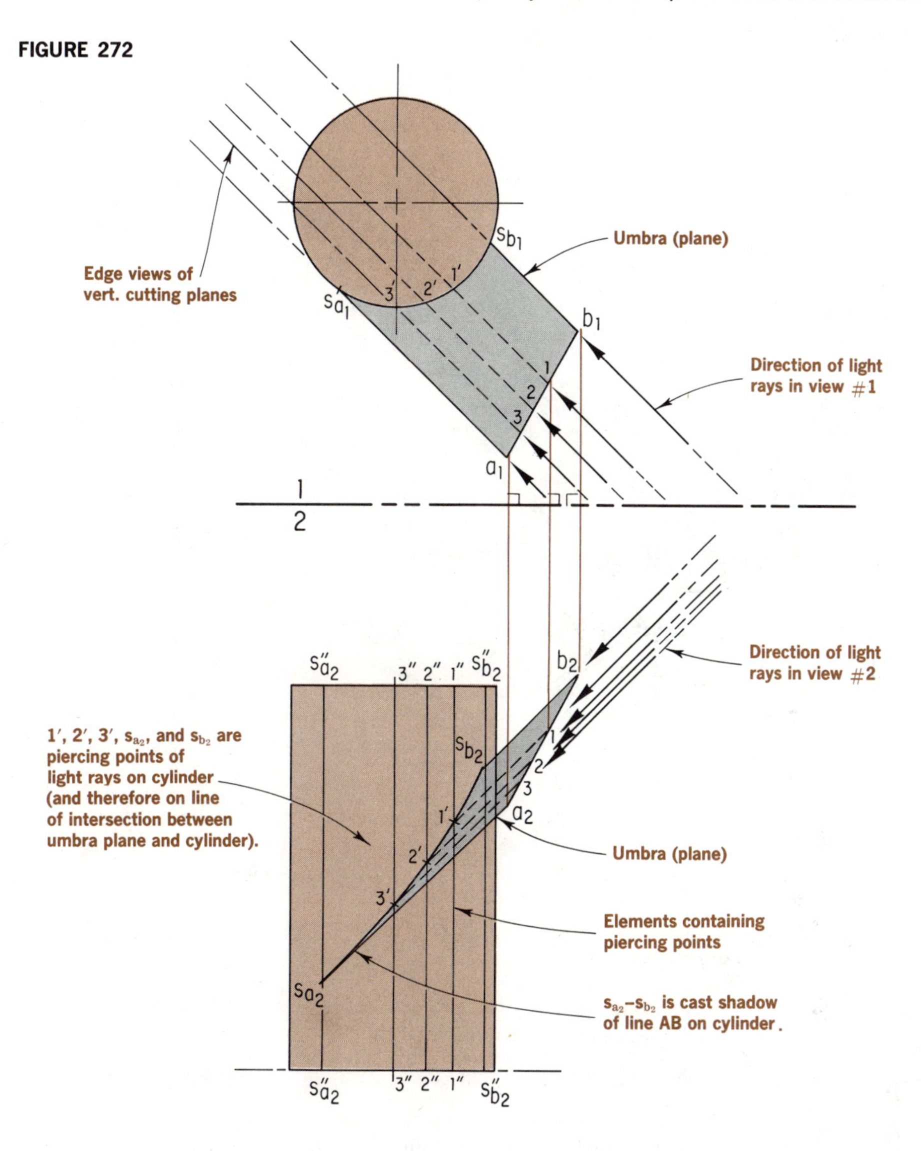

E. CAST SHADOW OF A PLANE ON A FLAT PLANE SURFACE (Fig. 273)

The parallel light rays intercepted by a plane in space form an umbra that has a *volume* (encloses a three-dimensional space). The intersection of this volume with a flat plane surface is a plane section, which is the cast shadow of the given plane on the flat plane surface.

In Fig. 273, a given plane ABC has a cast shadow of S_A, S_B, S_C on a flat plane surface. This shadow was determined by the piercing points of the intercepted light rays through points A, B, and C of the given plane.

FIGURE 273

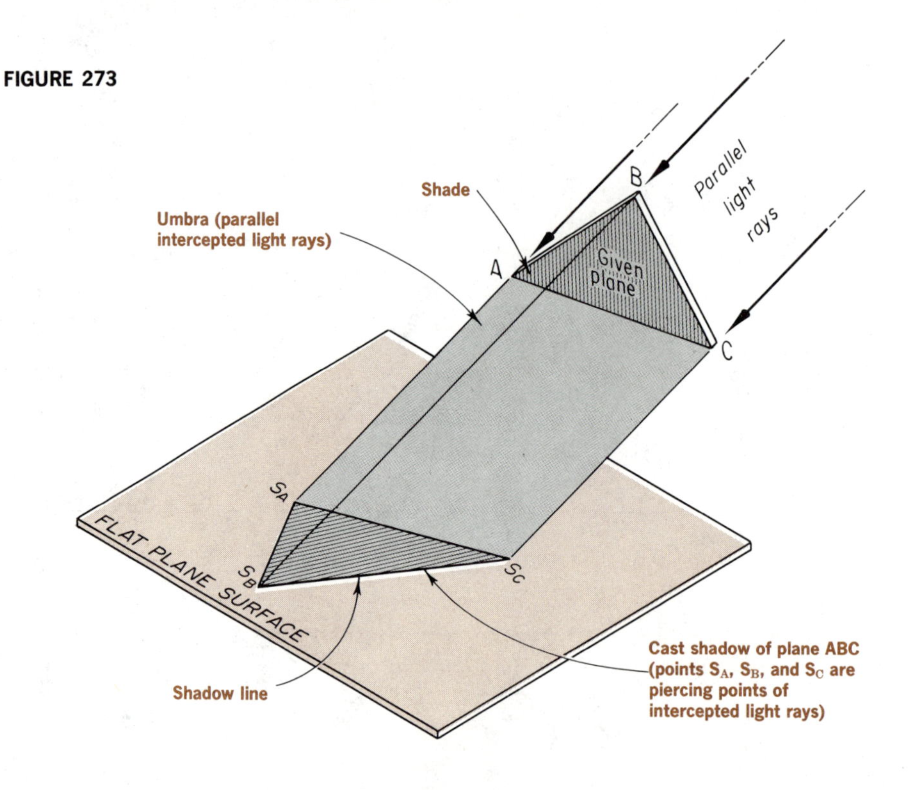

The umbra in this figure is a volume.

Orthographic Example: Determination of the cast shadow of a plane on a flat plane surface (Fig. 274).

Given Data: Horizontal and frontal projections of plane *ABC*, flat plane surface *RSTU*, and the direction of parallel light rays.

Required: The cast shadow of plane *ABC* on the flat plane surface *RSTU*.

Construction Program:

a. In view #1, construct vertical cutting planes #I, #II, and #III through each intercepted light ray. The planes should pass through points a_1, b_1, and c_1 of plane *ABC*, determining points n_1, o_1, p_1, q_1, l_1, and m_1 on the flat plane surface $r_1s_1u_1t_1$. Project these points to view #2 and determine points n_2, o_2, p_2, q_2, l_2, and m_2. Connect these points with lines (as shown in view #2). The lines intersect the directions of their respective light rays in view #2, thus determining points s_{a_2}, s_{b_2}, and s_{c_2}. Connect these points and determine the cast shadow of plane *ABC* on plane *RSTU* in view #2.

b. Project s_{a_2}, s_{b_2}, and s_{c_2} to view #1 and determine points s_{a_1}, s_{b_1}, and s_{c_1}. Connect these points to establish the cast shadow of plane *ABC* on plane *RSTU* in view #1.

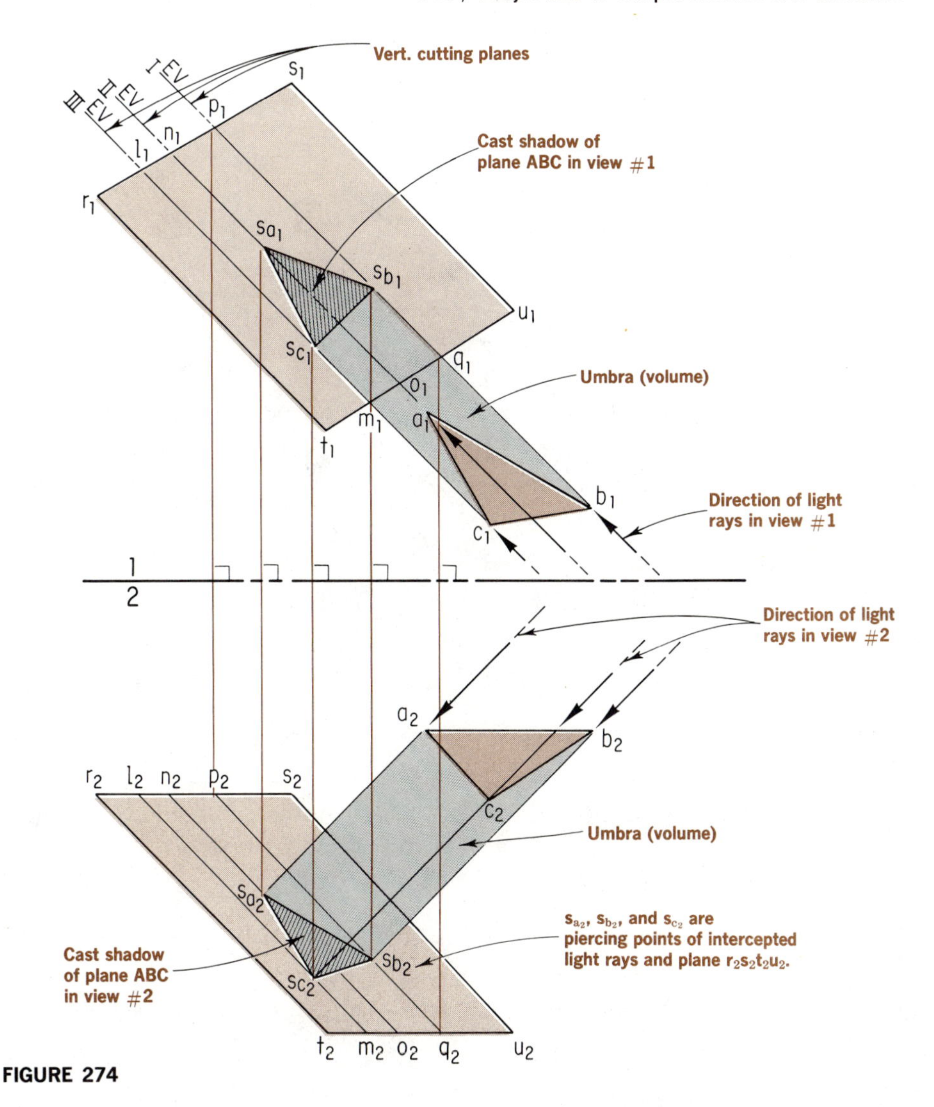

FIGURE 274

F. CAST SHADOW AND SHADE OF A GEOMETRIC SOLID WITH PLANE FACES

The umbra formed when light rays intercept a geometric solid that has plane faces is a volume. The intersection of the umbra volume with a flat surface is a plane section that represents the cast shadow of the geometric solid. The face or faces of the solid that are in the shade are shielded from the light rays by other faces, or sides, of the object. In Fig. 275, a rectangular prism intercepts parallel light rays that form an *umbra volume.* This volume intersects a flat plane surface and casts the prism's shadow on the flat surface. The edges of the prism define the shade line *ABCD.*

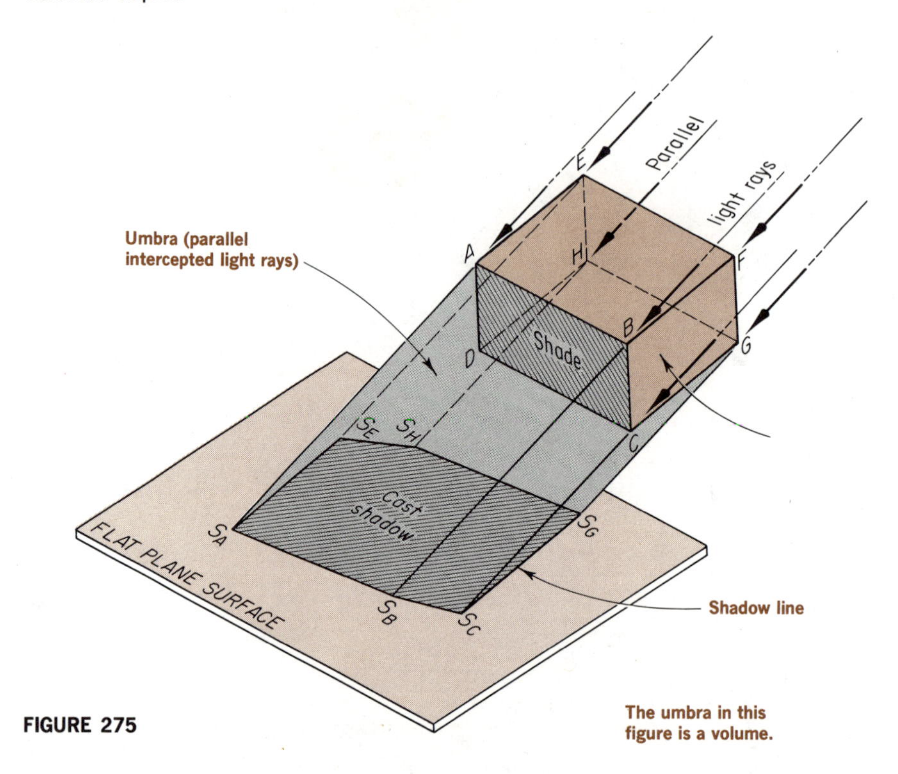

FIGURE 275

The shadow line (indicating the outline of the cast shadow) is defined by the piercing points of the intercepted light rays on the flat plane surface (the plane section of the umbra volume).

Orthographic Example: Determination of the cast shadow and shade of a rectangular prism on a flat plane surface (Fig. 276).

Given Data: Horizontal and frontal projections of a rectangular right prism, the direction of parallel light rays, and the edge view of a flat plane surface.

Required: The cast shadow of the given prism on the given flat plane surface.

Construction Program:

a. In view #2, extend the light rays through each point on the prism (a_2, b_2, e_2, f_2 and c_2, d_2, h_2, g_2) until the intercepted rays intersect the *edge view* of the flat plane surface at points s_{a_2}, s_{b_2}, s_{e_2}, s_{c_2}, s_{h_2}, and s_{g_2}.

b. Extend the light rays through each point on the prism in view #1.

c. Project points s_{a_2}, s_{b_2}, s_{e_2}, s_{c_2}, s_{h_2}, and s_{g_2} to their respective intercepted light rays in view #1, thus determining their horizontal projections s_{a_1}, s_{b_1}, s_{e_1}, s_{c_1}, s_{h_1}, and s_{g_1}. Connect these points with straight lines and define the *shadow line*. The area within the shadow line is the required cast shadow of the given prism on the given flat plane surface.

(NOTE: Points s_{b_1} and s_{f_1} fall within the shadow line and are therefore not indicated.)

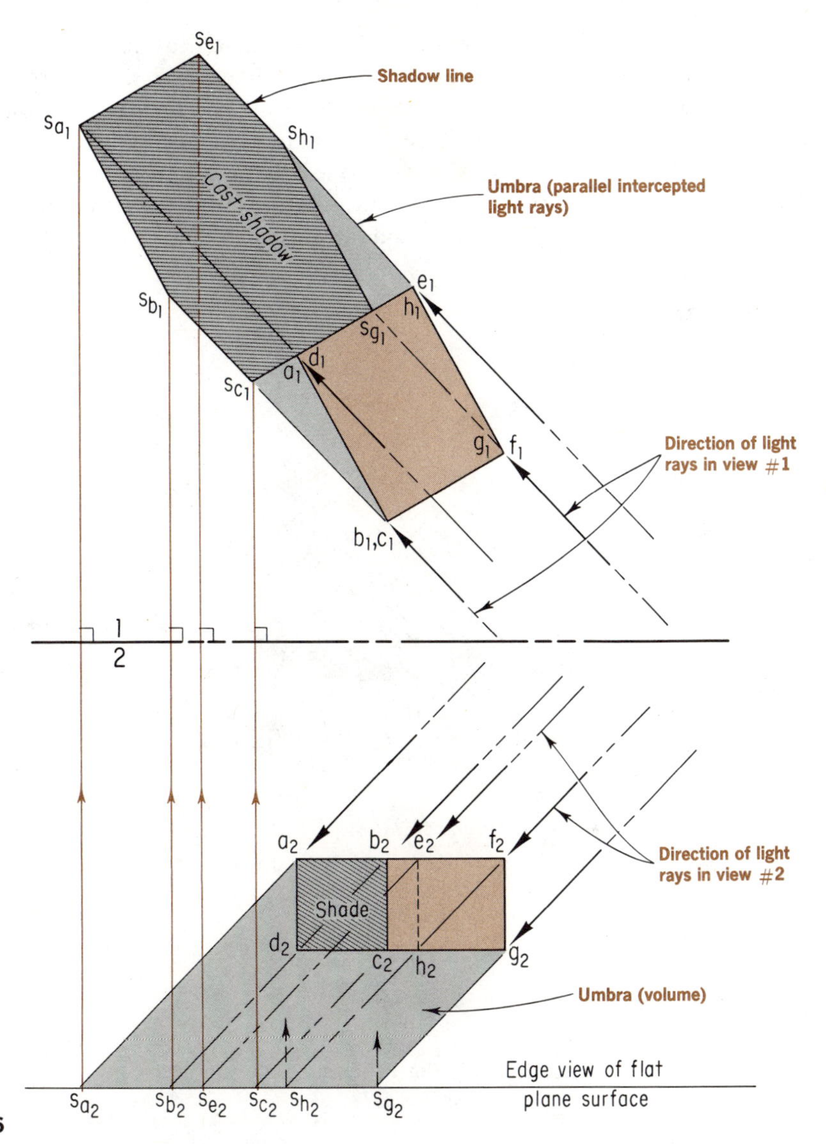

FIGURE 276

d. The shade area is determined by inspection. In view #1, we see that edge b_1c_1 of the prism excludes light from face $a_1b_1c_1d_1$. Therefore, this face is shaded, as shown in view #2.

G. CAST SHADOW AND SHADE OF A RIGHT CONE

The shadow line of the cast shadow of a cone is determined by parallel light rays that are *tangent* to the cone. The umbra, under this condition, is a volume approximating the shape of a tetrahedron with one curved face. In Fig. 277, a right cone is illuminated by parallel light rays that are intercepted so that shade lines *VA* and *VB* are determined by the planes these intercepted light rays form. The shadow lines S_VA and S_VB are defined by the intersection of these planes with the flat plane surface. The cast shadow of the cone exists within these shadow lines. The shade is defined by the surface of the cone between the tangent elements *VB* and *VA* on the side from which light is obscured.

FIGURE 277

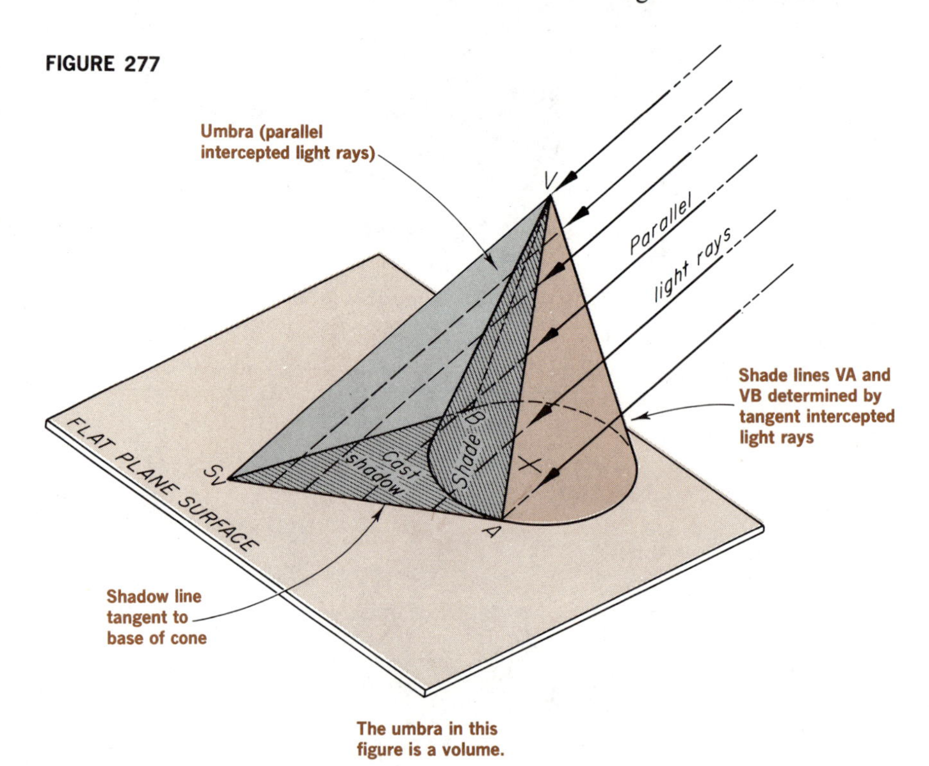

The umbra in this figure is a volume.

Orthographic Example: Determination of the cast shadow and shade of a right cone on a flat plane surface (Fig. 278).

Given Data: Horizontal and frontal projections of a right cone with vertex *V*, direction of parallel light rays, and the edge view of a flat plane surface.

Required: The cast shadow and shade of the given right cone on the flat plane surface.

Construction Program:

a. In view #2, extend the light ray illuminating v_2 until v_2 is intercepted and the ray intersects the flat plane surface at point s_{v_2}.

b. Extend the light ray illuminating v_1 in view #1.

c. Project s_{v_2} to view #1 and determine point s_{v_1}.

FIGURE 278

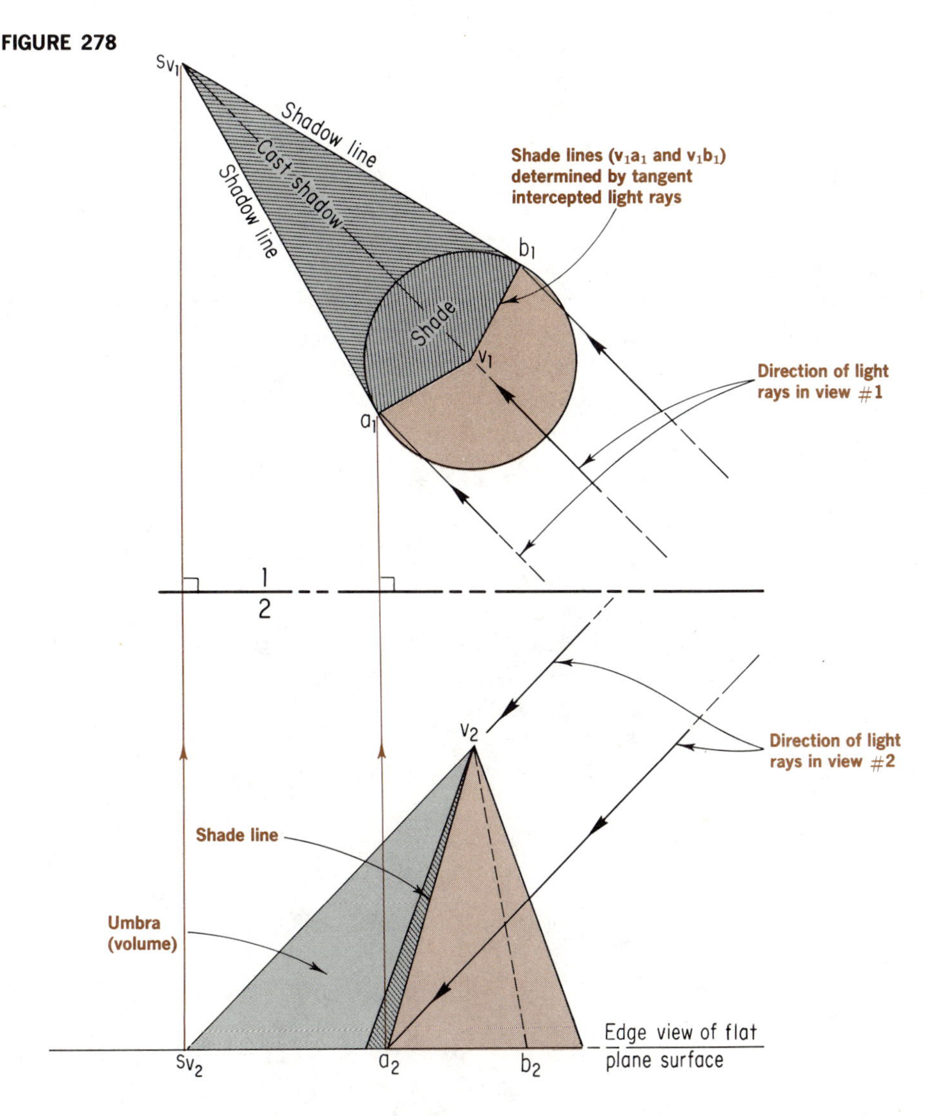

d. From point s_{v_1} in view #1, construct *tangent lines* to the cone base (at points a_1 and b_1, respectively). These two lines are shadow lines, and define the outline of the cast shadow.

e. Connect v_1 and a_1, and v_1 and b_1 with straight lines. These two lines are the shade lines, and represent straight line elements on the cone surface. The shade exists between these two lines and is indicated in view #1.

f. Project a_1 and b_1 to view #2 and determine points a_2 and b_2.

g. In view #2, only part of the shade can be indicated because it is partially hidden behind the cone.

101 Perspective projection

Perspective projection (also known as central projection) involves representations known as *perspective drawing.* Perspective drawing is widely used to portray realistic views of various buildings and structures.

The concept of descriptive geometry used in perspective projection is that of the piercing point of a line and a plane. The basic difference between central projection and orthographic projection is illustrated in Figs. 279 and 281. In orthographic projection, as we have seen, the observer's lines of sight are assumed to be parallel. Theoretically, this means that the observer's position relative to the object he is viewing is at a point at *infinity* (see Fig. 279).

FIGURE 279

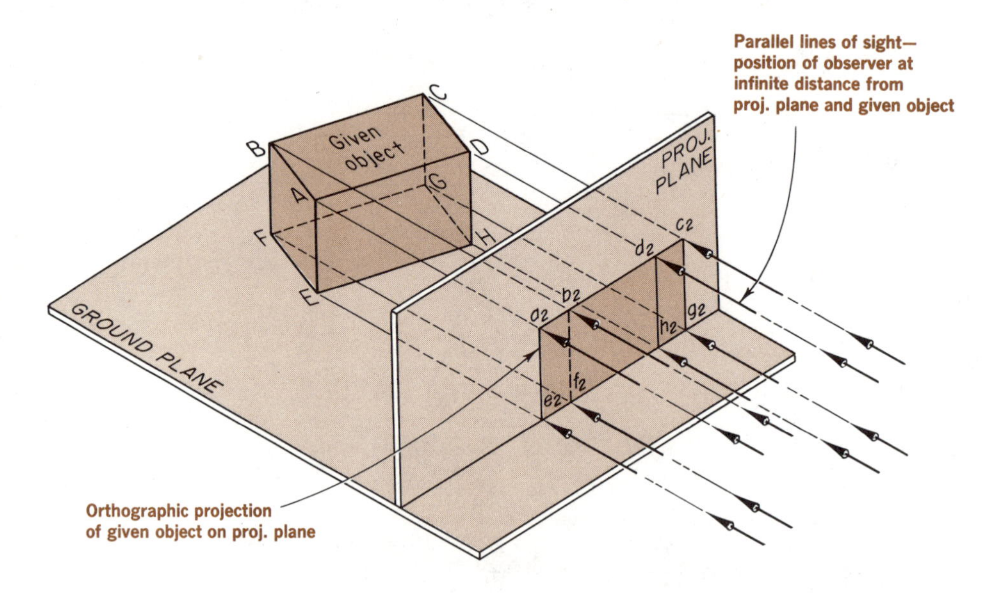

By placing a projection plane between the observer and the object he is viewing, the points at which his parallel lines of sight intersect the projection plane determine an *orthographic projection* of the given object on the plane.

However, we know from experience that when we look at an object we are located at a *finite* distance from it. This means that our lines of sight "radiate" from our eyes to each point on the object we are viewing. In central projection the lines of sight of the observer are *not* parallel (see Figs. 280 and 281).

A. DEFINITIONS (Fig. 282)

1. *Picture plane*—the frontal projection plane located between the observer and the object he is viewing.

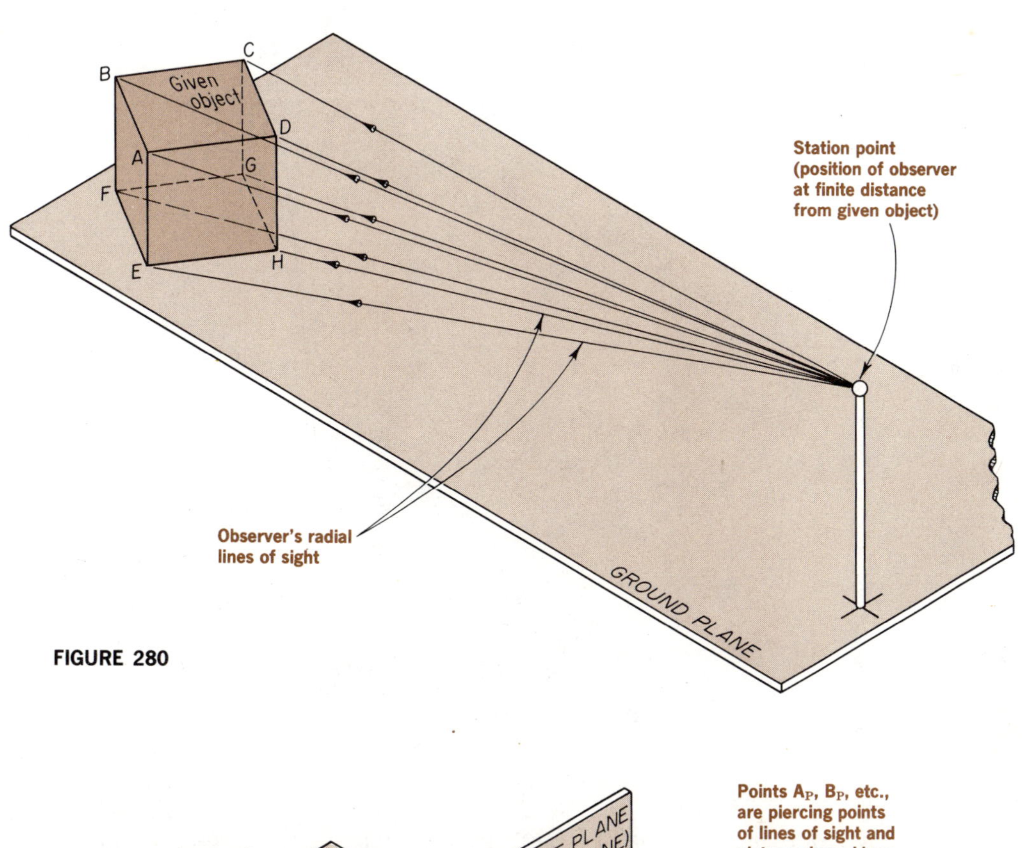

FIGURE 280

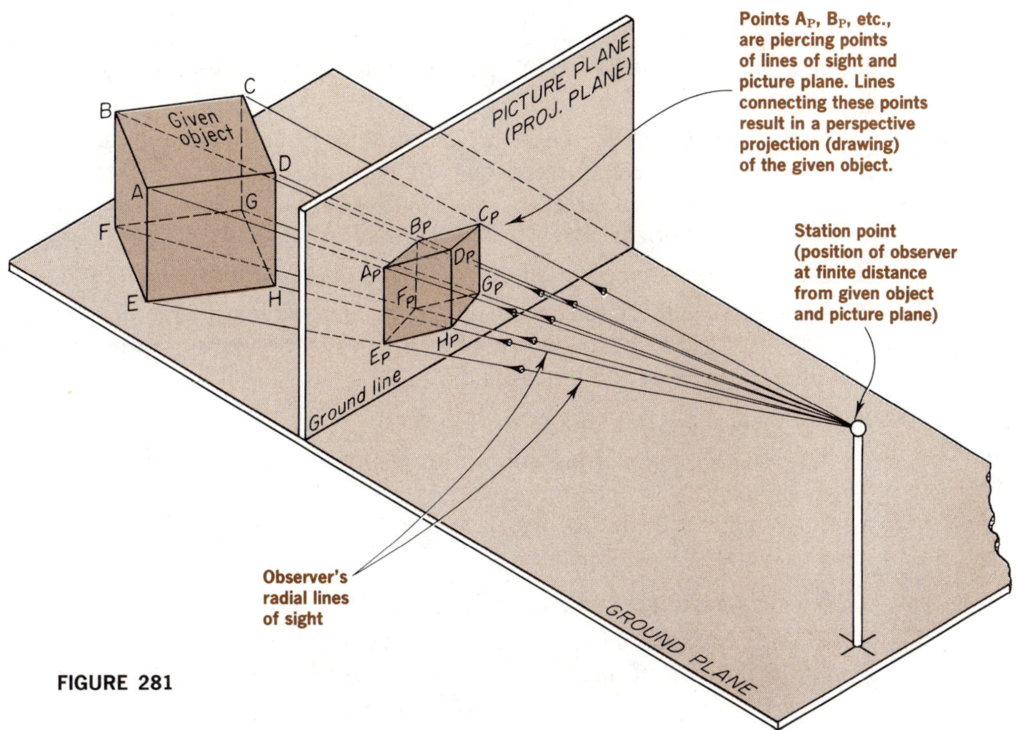

FIGURE 281

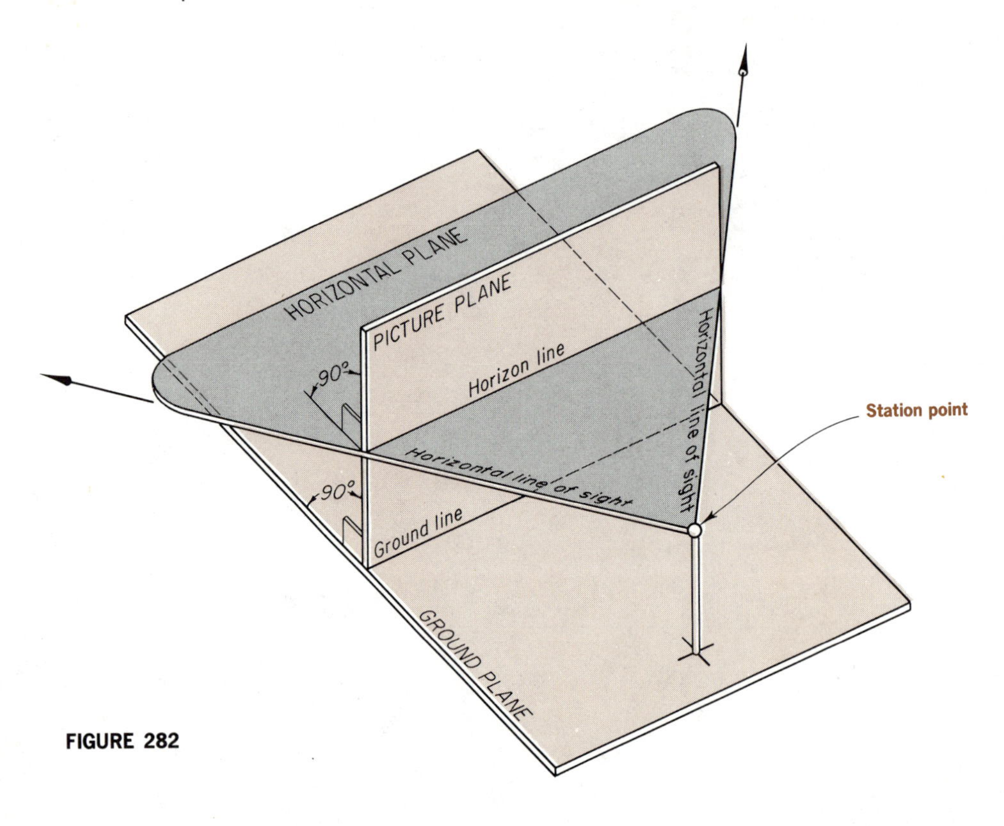

FIGURE 282

2. *Station point*—the observer's position relative to the object and the picture plane.
3. *Horizon line,* or *horizon*—the intersection of a horizontal plane (formed from the horizontal lines of sight at the observer's station point as the observer turns his head) and the picture plane. In Fig. 282 the horizon line is at the eye level of the observer.
4. *Ground plane*—the plane upon which the observed object rests.
5. *Ground line*—the intersection of the ground plane and the picture plane. When viewing a picture plane from a station point, the object will appear to rest on the ground line.

B. PARALLEL PERSPECTIVE DRAWING (Fig. 283)

In developing a parallel perspective projection the object being viewed is positioned so that one of its faces is parallel to the picture plane. This is illustrated in Fig. 283, in which face *ADHE* is parallel to the picture plane.

C. ANGULAR PERSPECTIVE DRAWING (Figs. 284 and 285)

There are two fundamental types of angular perspective drawing. In the first type, the *vertical edges* of the object are parallel to the picture plane

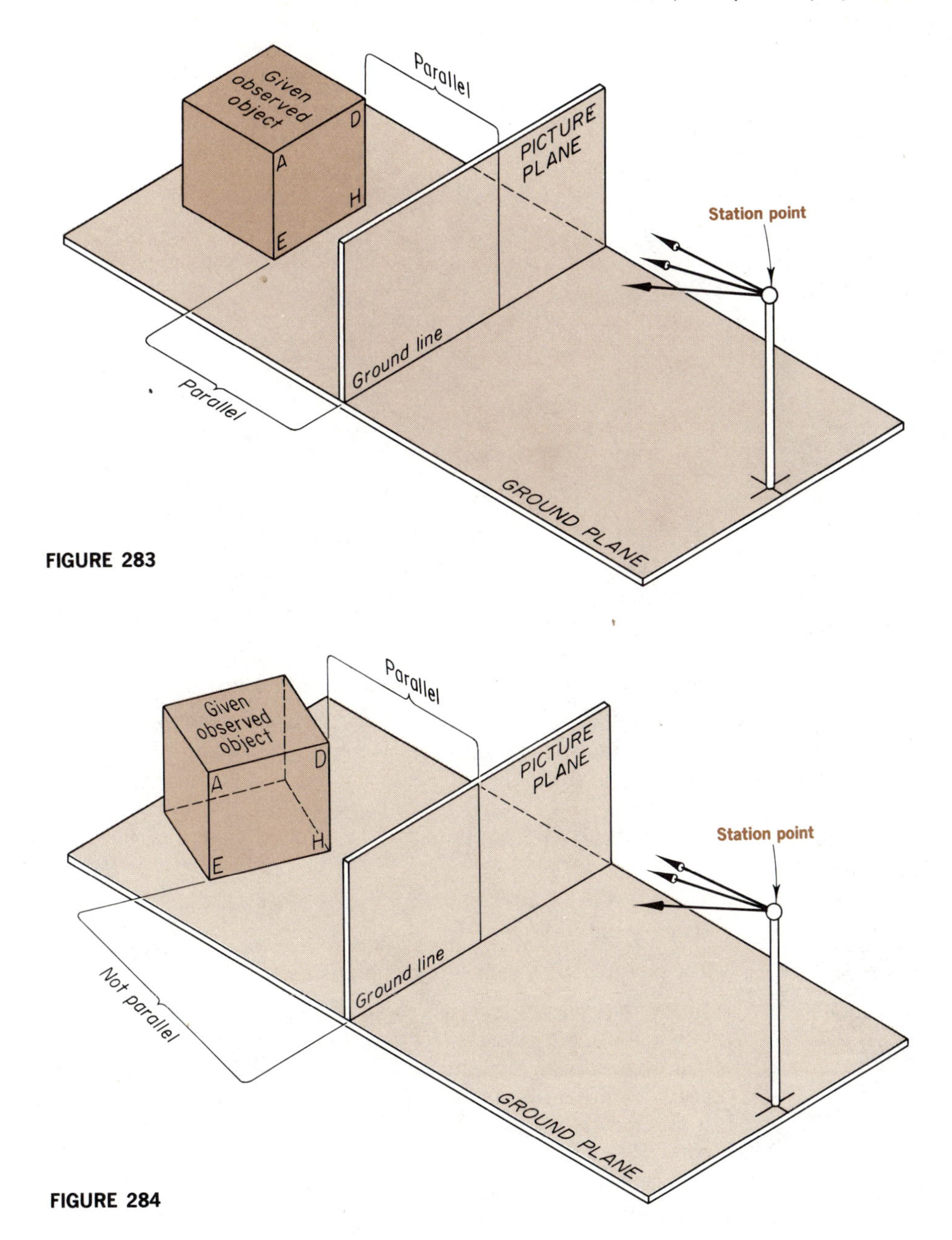

FIGURE 283

FIGURE 284

but the *faces* of the object are *not* (see Fig. 284). This results in a *two-point perspective,* which will be illustrated later.

In the second type of angular perspective drawing, *neither* the vertical edges *nor* the faces of the object are parallel to the picture plane. This results in a *three-point* perspective drawing (see Fig. 285).

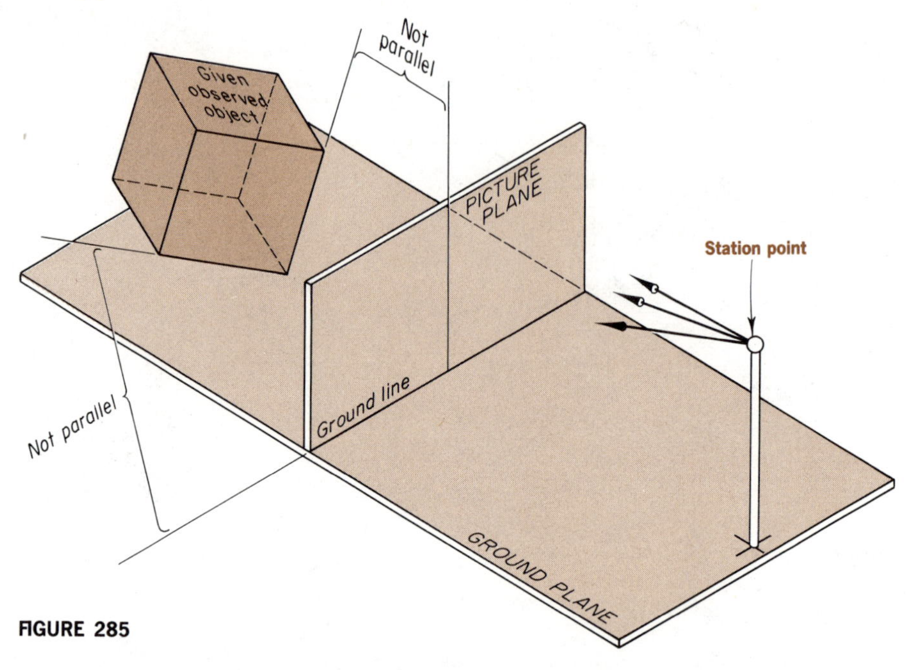

FIGURE 285

D. PERSPECTIVE DRAWING—THE METHOD OF DIRECT PROJECTION

In direct projection, the perspective projection is determined by the *piercing points* of the observer's radial lines of sight and the picture plane (see Figs. 281 and 286).

Orthographic Example: Development of a perspective drawing by direct projection (Fig. 286).

Given Data: The horizontal and left profile (or right profile) views of a cube and a station point S_P.

Required: The perspective projection of the given cube.

Construction Program

a. In Fig. 286, the left and right views of a cube are given. Use either view to solve the problem. (We have used the left profile view.) Note that the horizontal projection s_{p_1} of the station point is in front of the edge view of the picture plane (view #1) at a distance D. Distance D is also seen as the distance from the edge view of the picture plane (view #3) to the left profile projection s_{p_3} of the station point. These projections (s_{p_1} and s_{p_3}) *overlap* in the picture plane (view #2). (Carefully distinguish between radial lines of sight and projectors.)

b. From s_{p_1} construct radial lines to points a_1, b_1, c_1, and d_1 on the top face of the cube and to points e_1, f_1, g_1, and h_1 on the bottom face, or base.

c. Note the piercing points of the radial lines of sight and the edge view of the picture plane (view #1). These points are indicated as a_{1_p}, b_{1_p}, c_{1_p}, d_{1_p}, e_{1_p}, f_{1_p}, g_{1_p}, and h_{1_p}.

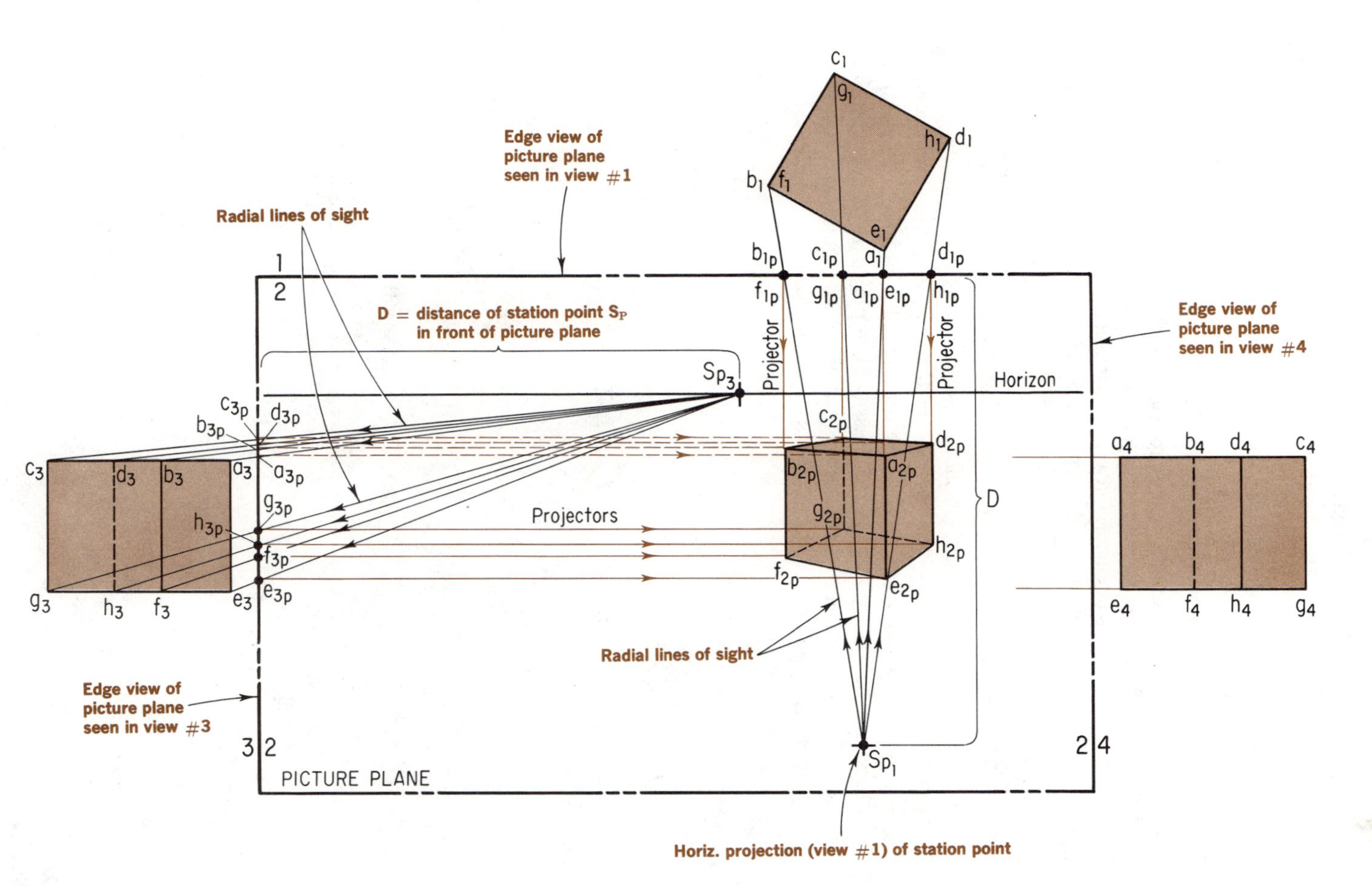
Edge view of picture plane seen in view #1
Radial lines of sight
D = distance of station point S_P in front of picture plane
Edge view of picture plane seen in view #4
Horizon
Projector
Projector
Projectors
D
Radial lines of sight
Edge view of picture plane seen in view #3
PICTURE PLANE
Horiz. projection (view #1) of station point
S_{p_3}
S_{p_1}
c_1 g_1 h_1 d_1 b_1 f_1 e_1 a_1
b_{1p} c_{1p} d_{1p} f_{1p} g_{1p} a_{1p} e_{1p} h_{1p}
c_{2p} d_{2p} b_{2p} a_{2p} g_{2p} h_{2p} f_{2p} e_{2p}
c_{3p} d_{3p} b_{3p} a_{3p} g_{3p} h_{3p} f_{3p} e_{3p}
c_3 d_3 b_3 a_3
g_3 h_3 f_3 e_3
a_4 b_4 d_4 c_4
e_4 f_4 h_4 g_4
1 2
3 2
2 4

FIGURE 286

d. Project the above piercing points to view #2.

e. From s_{p_3}, construct radial lines of sight to each point on the cube in view #3. Note where these radial lines of sight pierce the edge view of the picture plane. They are indicated as a_{3_p}, b_{3_p}, c_{3_p}, d_{3_p}, e_{3_p}, f_{3_p}, g_{3_p}, and h_{3_p}.

f. Project these piercing points from view #3 to view #2.

g. The intersection of the projectors from view #1 and the projectors from view #3 determine points a_{2p}, b_{2p}, c_{2p}, d_{2p}, e_{2p}, f_{2p}, g_{2p}, and h_{2_p} in the picture plane (view #2). Connect the points with straight lines. This is the required perspective drawing of the given cube. (The visibility can be determined by inspection.)

E. PERSPECTIVE DRAWING—THE METHOD OF VANISHING POINTS

There is a more direct and simpler construction available for developing a perspective drawing—the use of *vanishing points.* A vanishing point is the point at which a horizontal line theoretically intersects the horizon. (This can be illustrated by the classic example of a person standing between two railroad tracks, looking down along them and seeing an *apparent* intersection of these tracks at the horizon.)

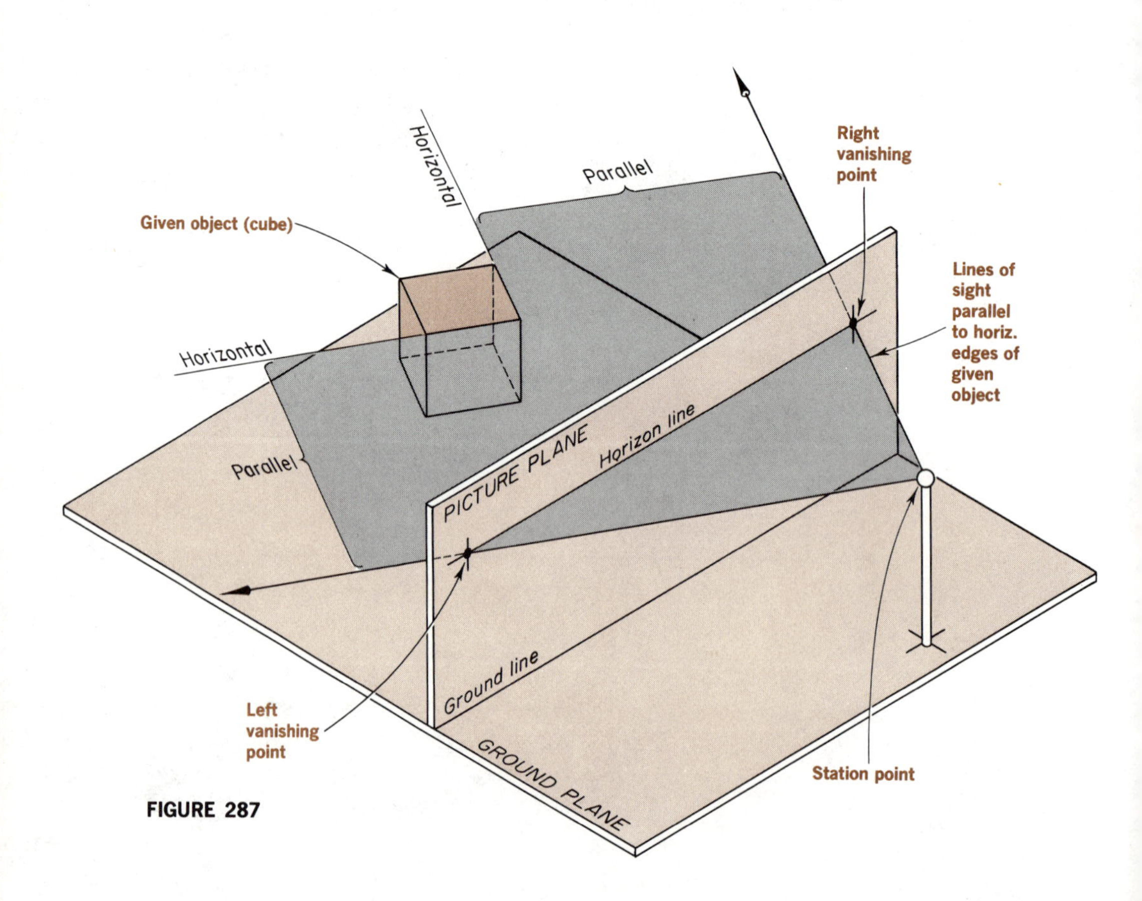

FIGURE 287

Referring to Fig. 287, note that to establish the vanishing points of a given object, construct lines of sight from the station point parallel to two horizontal edges of the object. The intersection of these lines of sight with the *horizon line* determines the left and right vanishing points (L_{VP} and R_{VP}, respectively).

Figure 288 shows the complete development of a perspective drawing of a cube, by the use of vanishing points.

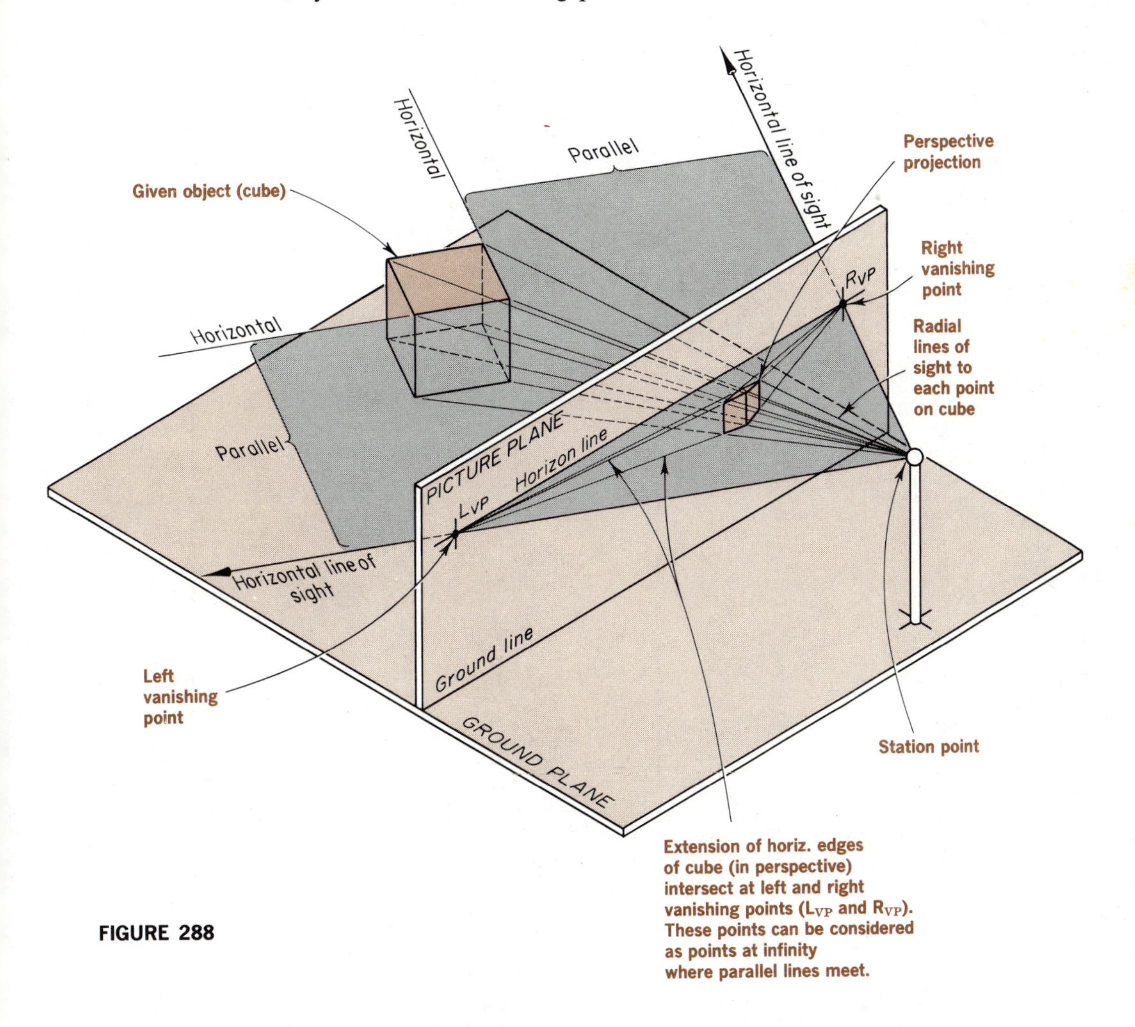

FIGURE 288

Orthographic Example: Development of a perspective drawing by the use of vanishing points (Fig. 289).

Given Data: Horizontal and profile projections (either left or right) of a cube, the horizontal projection of the station point s_{p_1}, and the horizon line.

Required: The perspective drawing of the given cube.

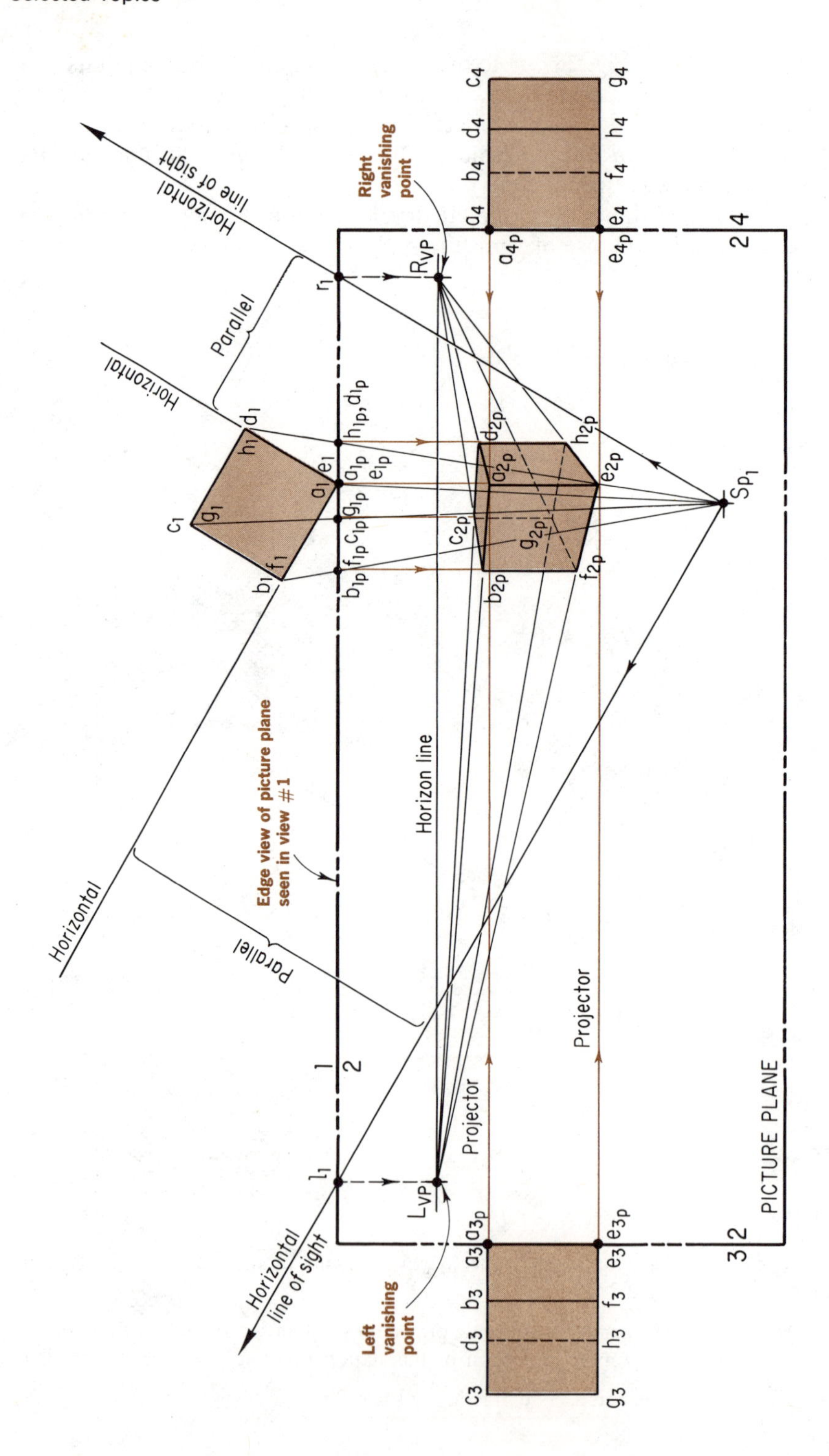

Horizontal line of sight
Horizontal
Parallel
Right vanishing point
Edge view of picture plane seen in view #1
Horizon line
Projector
Projector
PICTURE PLANE
Left vanishing point
Horizontal line of sight
Horizontal
Parallel

FIGURE 289

Construction Program:

a. The projections of the cube have been oriented so that vertical edge AE touches the picture plane (see views #1, #3, and #4). From station point s_{p_1}, construct lines parallel to edges a_1b_1 and a_1d_1 of the cube in view #1. These lines intersect the edge of the picture plane at points l_1 and r_1, respectively.

b. Project points l_1 and r_1 to the horizon line in view #2 and determine the left and right vanishing points (L_{VP} and R_{VP}, respectively).

c. From station point s_{p_1}, construct radial lines of sight to points a_1, b_1, c_1, and d_1 and to points e_1, f_1, g_1, and h_1 on the cube. These radial lines of sight intersect the edge of the picture plane in view #1 at points a_{1_p}, b_{1_p}, c_{1_p}, d_{1_p}, e_{1_p}, f_{1_p}, g_{1_p}, and h_{1_p}.

d. Project these piercing points to view #2.

e. From either view #3 or #4 (we have used #3 in our example), project points a_3 and e_3 to view #2. The intersection of these projectors and those from points a_1 and e_1 in view #1 determines points a_{2_p} and e_{2_p} in the picture plane (view #2).

f. From points a_{2_p} and e_{2_p} in view #2, construct converging straight lines to the left and right vanishing points. These lines represent (in perspective) horizontal lines that intersect at the vanishing points.

g. From view #1, project points a_{1_p}, b_{1_p}, c_{1_p}, and d_{1_p} and points e_{1_p}, f_{1_p}, g_{1_p}, and h_{1_p} to view #2. This locates points b_{2_p}, c_{2_p}, and d_{2_p} on their respective lines moving to the left and right vanishing points, and also locates points f_{2_p}, g_{2_p}, and h_{2_p} on their respective lines moving to each vanishing point. Connect these points with straight lines. The result is the required perspective drawing of the given cube. (The visibility can be determined by inspection.)

Principles of Descriptive Geometry Applied to Practical Problems

102 Introduction

This chapter is devoted to problems for solution that deal with various practical applications of three-dimensional descriptive geometry principles and procedures. Certain simplifying assumptions have been made in many of the problem presentations so that they can fit into a textbook format. A number of these problems are suitable for project-type efforts that encourage team work, class discussion, and consultation with your instructor, as well as individual creative approaches in their solution. However, whatever approach is adopted, you are encouraged to exercise independent judgment in the selection and sequential application of the specific principles of descriptive geometry to attain correct solutions. This means, in many instances, that there may be more than one way of attacking a problem to arrive at a correct solution.

The "Practical Application Problems" in this chapter are set up on cross-section paper in the same manner as the "Practice Problems" in the preceding parts of the book, with one important exception: the problems here are set up on cross-section paper having 5 × 5 squares to the *centimeter*.[1] (See the example of Practical

[1] The specific paper used in setting up problems in this chapter is the K. & E. 5 × 5 to the centimeter, 18 × 24 cm paper, No. 46 1612. However, any equivalent paper with centimeter and millimeter squares (grid) is satisfactory for the problem setups.

Problem setup of given data below.) The reason for doing this is that we are going to use some aspects of the system of measurement known as the "International System of Units," which is based on the metric system.

PRACTICAL PROBLEM SETUP OF GIVEN DATA

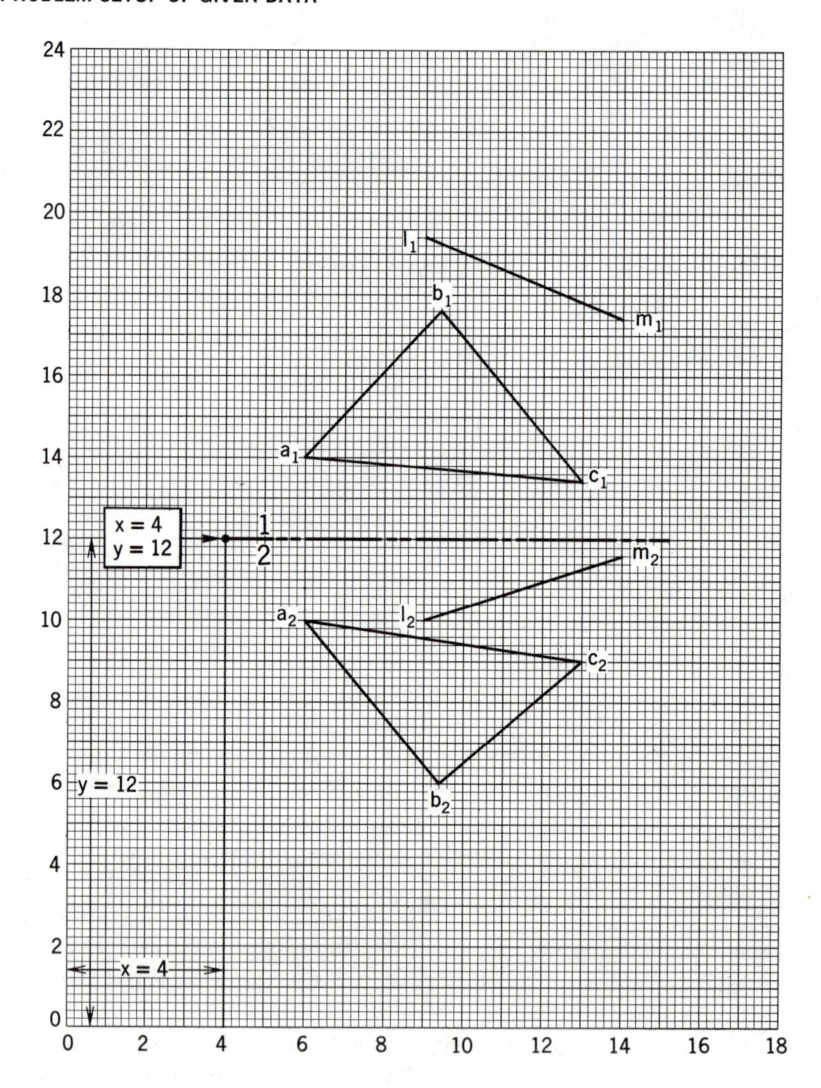

102 International system of units (SI)

The International System of Units is officially designated as "SI." This system is rapidly being adopted throughout the world of science, engineering, architecture, and industry. It is predicted that the SI system of units (measurements) will replace all other systems in the not so distant future, and that it will have a positive influence on all facets of our lives in which measurements play an important and critical role. We are presently in the transitional phase of changing our thinking from our traditional system of measurement to the SI system.

The SI system has the advantage of specifying only one unit for each physical quantity: for example, the *meter* for length is m; the *kilogram* for mass is kg; the *second* for time is s; and the *radian* for plane angles is rad. From these units all other units are derived.[2]

BASE UNITS

Quantity	Unit	SI Symbol
length	meter[a]	m
mass	kilogram	kg
time	second	s
electric current	ampere	A
thermodynamic temperature	kelvin	K
amount of substance	mole	mol
luminous intensity	candela	cd

[a] The British spelling is metre. There is as yet no worldwide agreement on how to spell this word in the English language.

SUPPLEMENTARY UNITS

Quantity	Unit	SI Symbol
plane angle	radian	rad
solid angle	steradian	sr
acceleration	meters per second squared	m/s^2
angular acceleration	radians per second squared	rad/s^2
angular velocity	radian per second	rad/s
area	square meters	m^2
force	pounds	lbf

CONVERSION FACTORS

1 mile = 1.6093 kilometers
1 kilometer = 0.6214 mile
1 foot = 30.48 centimeters
1 meter = 3.281 feet

1 meter = 39.37 inches
1 inch = 2.540 centimeters
1 kilogram = 2.2046 pounds
1 pound = 0.4535 kilograms

---

[2] An excellent detailed explanation of the entire SI system of units and measurements is presented in the "Metric Practice Guide," No. E-380-74 published by the American Society for Testing Materials, 1916 Race Street, Philadelphia, Pennsylvania 19103. This guide includes complete tables for conversion from the U.S.-British system to the SI system.

In keeping with the philosophy of the SI system, plane angles in this chapter will be specified only in degrees such as 10.25°, 42.5°, and the like, *not* in degrees and minutes. For your convenience, a compilation of some SI general units (Base Units and Supplementary Units) that are used in science, engineering, and architecture is given with a table of conversion factors:

The "Practical Application Problems" in this chapter will introduce you to some of the features of the SI system. The purpose is to make you familiar with some of the SI notation and to have this experience serve as a partial transition from the U.S.–British system to the International SI system of measurements. It will be possible for you to use most of your drawing equipment in solving the problems in this chapter except your inch-based scale. All linear SI measurements can be made by transferring distances with dividers to the metric grid and reading values according to the specified scales. Or, you may purchase a metric scale and use it directly.

103 Approach to practical problem solving

In setting up actual problems for geometric solution in industry, several approaches can be taken. Normally the key known information or data are accurately laid out on paper where many of the fine details of the specific "hardware" components of the problem are omitted in the initial analysis. For example, if we are interested in solving a problem concerning the clearance (or interference) between two cables in an airplane control system, normally the initial setup (Phase I Analysis) starts with the known locations and positions of the centerlines of the

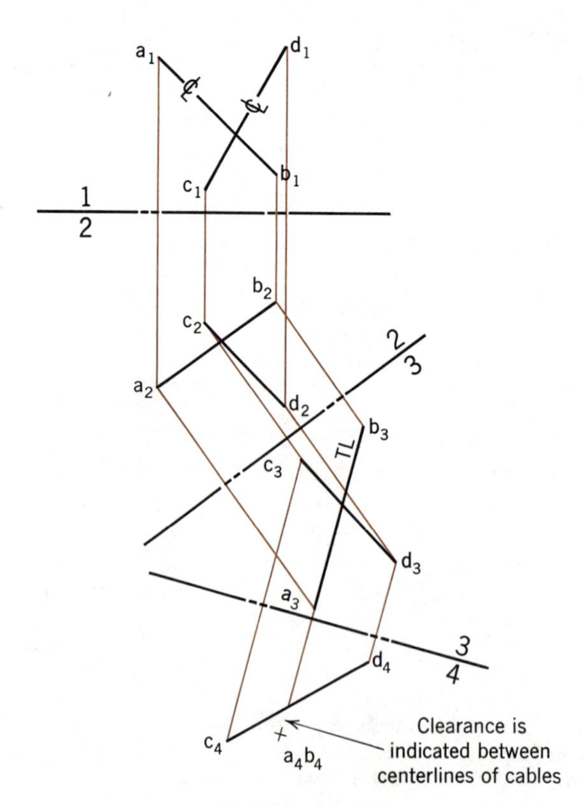

FIGURE 290

Phase I analysis

cables while the diameters and sag of the cables are ignored. (See Fig. 290, Phase I Analysis.)

In the Phase I Analysis, shown in Fig. 290, we see that the two cables clear each other. However, the Phase II Analysis (see Fig. 291) with the cable diameters in place, shows us that the cables, in fact, interfere with each other and one or both

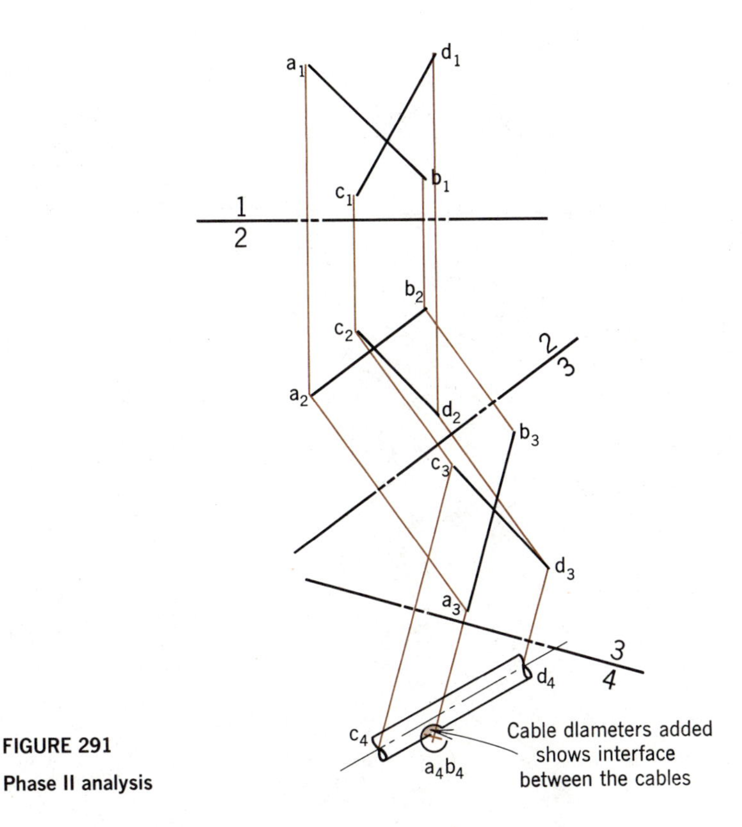

FIGURE 291
Phase II analysis

may be moved to attain the desired clearance. This leads to Phase III Analysis, which would apply the necessary redesign requirements (relocation of one or both cables) to arrive at the desired clearance (see Fig. 292).

A related part of the cable problem solution would also involve the redesign of the pulley brackets, which support the cables at the desired clearance positions. This would be considered Phase IV of the analysis and design process. But here the problem can become more complex since, in an engineering system, every component in the system is closely related to all other components in the system within given spatial or geometric constraints. Hence, it is possible that several other phases in the analysis and design process would have to take place before an acceptable solution to the problem was established.

In solving the problems in this chapter you should try to utilize approaches that are economic in terms of the number of steps used in their solution. An elegant solution is one that goes to the heart of the problems in the most efficient manner. Do not inhibit your imagination.

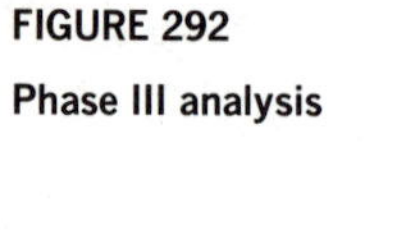

FIGURE 292
Phase III analysis

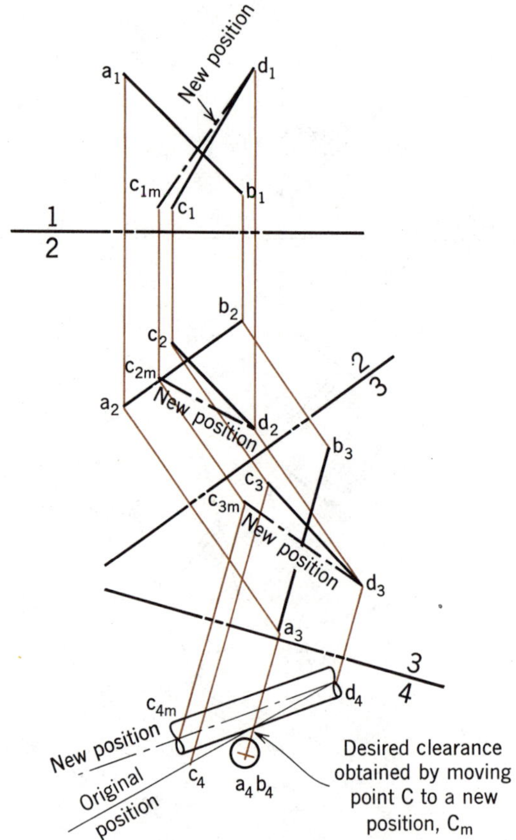

PRACTICAL APPLICATION PROBLEMS

PROBLEM 1 (LASER COMMUNICATIONS TRANSMISSION)

1. Given Data: An experiment in communications transmission involves the use of lasers and includes three relay mirrors at points *A, B,* and *C* located at various mountain-top elevations. *A* is at 1000 m, *B* is at 850 m and *C* is at 900 m. (All elevations are referenced to the 1000-m Datum Elevation.) (Scale: 1 cm = 50 m.)

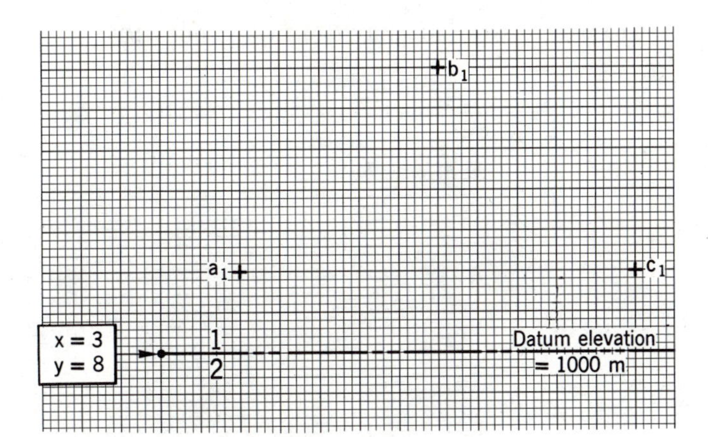

Required:

A. Actual transmitting distances between each mirror.

B. Slope angle of each transmitting distance; *A* to *B*, *B* to *C*, and *C* to *A*.

C. Slope of the plane formed by mirror locations *A*, *B*., and *C*.

D. Label and show all construction.

Ans.: TL $AB = 385$m Slope $AB = 23°$ Slope of Plane $ABC = 24°$
$BC = 360$m $BC = 9°$
$CA = 510$m $CA = 11.5°$

PROBLEM 2 (LIGHTHOUSE BUOY PROBLEM)

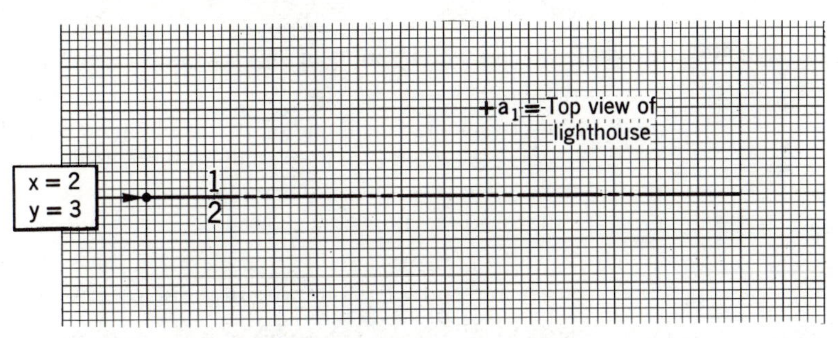

2. Given Data: From a lighthouse 26 m above the water (near the Bay of Fundy) a buoy is observed on a bearing of N45° W and a depression angle (negative slope) of 15°. Two hours later the same buoy is observed on a bearing of N45° E and a depression angle of 30°. In the meantime the tide has dropped 12 m.

Required:

1. Direction of current.
   (*Ans.:* N78.5° E.)
2. Velocity of current in meters per second.
   (*Ans.:* 0.023 m/s)

(NOTE: Use two sheets of paper taped together along their long edges for this problem. Use a scale: 1 cm = 10 m.)

PROBLEM 3 (POWER LINE—TELEPHONE WIRE PROBLEM)

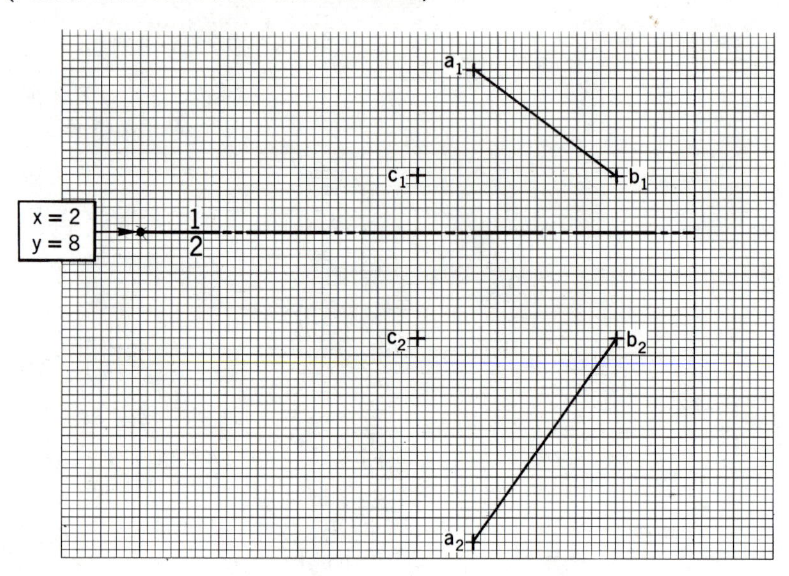

3. Given Data: A telephone wire, line *AB*, crosses the right of way of a high tension power line running through point *C*, rising 15° from the horizontal and bearing N60° E.

The allowable distance (clearance) between the two wires at the nearest point is 4.0 m.

Required:

A. Is the present location of the telephone wire safe, and by what margin?

B. If unsafe, change the direction of the telephone wire *AB* as follows: Keeping the point *A* in its present location, how far must point *B* be moved *due east* and at the *same level* to make the new location conform to the allowable clearance.

C. Draw the horizontal and frontal elevation views of the high tension power line and telephone wire in the safe location.

(NOTE: North is at the top of the cross-section paper. Neglect thickness and sag of the two wires. Use a scale 1 cm = 1 m.)

PROBLEM 4 (GEOLOGY-BOREHOLE-TUNNEL PROBLEM)

4. Given Data: To determine data concerning a specific geological formation under the earth, two boreholes are drilled, one at *A* and one at *B* on the earth's surface.

The borehole starting at *A* is vertical and intersects the headwall (upper surface) of the geological formation at point *X*, 20 m below the surface of the earth.

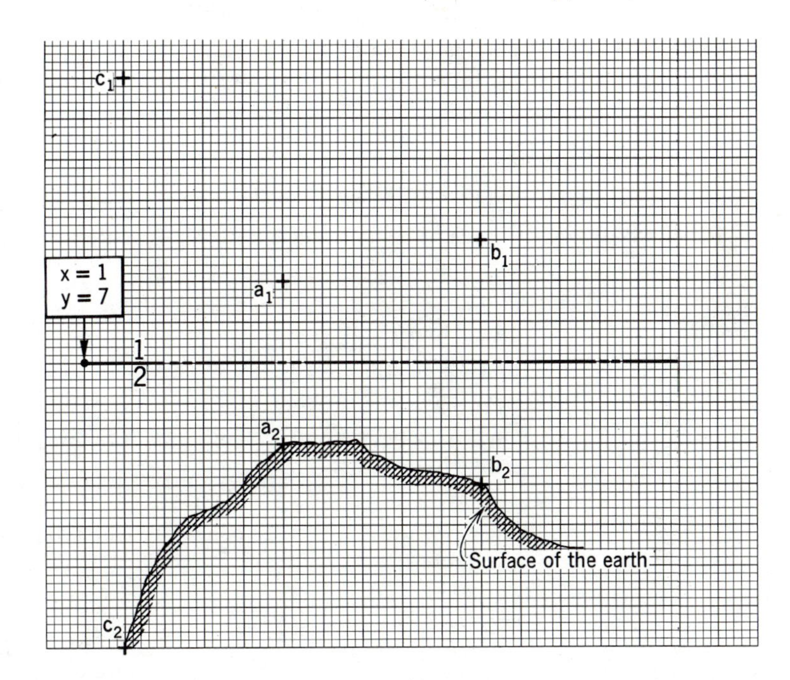

The borehole starting at *B*, with a bearing of N45° W and a slope of 60°, meets the underground geological formation at point *Y*, 30 m from the surface of the earth (measured along the line of the borehole).

Points *X* and *Y* establish a line on the headwall of the geological formation.

From the point *C* (also on the surface of the earth and on a side of a hill) an exploratory tunnel is constructed to the midpoint of line *XY* on the headwall.

Required:

A. Determine the true length, bearing (from *X*), and slope of line *XY*. (*Ans.:* TL = 47 m; bearing = N62° E; slope = -21°)

B. Determine the true length, bearing (from *C*), and slope of the exploratory tunnel.
(*Ans.:* TL = 76 m; bearing = S57° E; slope = + 17°)
(Use a scale of 1 cm = 10 m.)

PROBLEM 5 (WEATHER BALLOON PROBLEM)

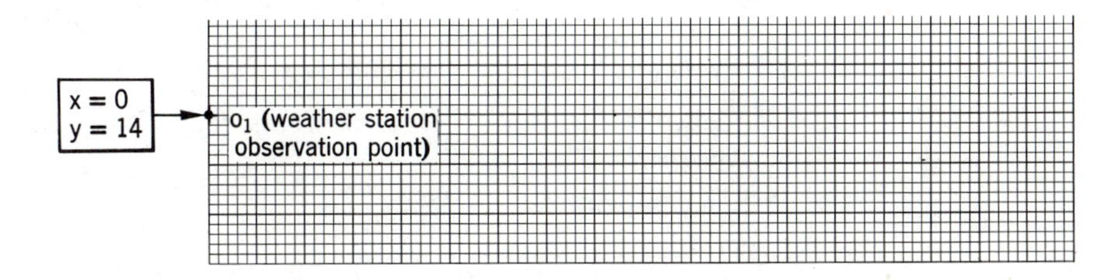

5. Given Data: A balloon is released from a weather station at point *O*. After 120 s is allowed for the balloon to reach stable air currents, it is observed on bearing of N45° E, at an elevation angle of 45°, and at a distance of 570 m from the observer at *O*. 240 s later another observation is made, when the balloon is seen on bearing N75° E, at an elevation angle of 20°, and at a distance of 1780 m from the observer at point *O*.

Required:

A. Altitude of balloon above observer at first observation.
(*Ans.:* 400 m)

B. Altitude of balloon at second observation.
(*Ans.:* 610 m)

C. Direction of wind.
(*Ans.:* From S81° W)

D. Wind velocity in meters per second. (That is velocity of balloon relative to the ground.)
(*Ans.:* 5.6 m/s)

E. Rate of vertical rise of balloon in meters per second between the two observations.
(*Ans.:* 0.87 m/s)
(Use scale of 1 cm = 100 m.)

PROBLEM 6 (SKI LIFT CABLEWAY PROBLEM)

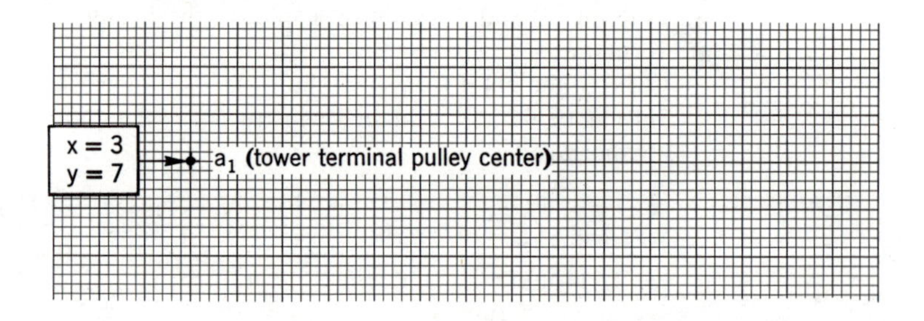

6. Given Data: A cableway for a ski lift is to be built up the side of a mountain. The tower terminal pulley center, *A*, is 300 m above sea level. The ski gondola will first climb N45° E at 32° with the horizontal, to an elevation of 680 m, where the cable will be suspended from a pulley attached to a tower anchored to the rock. At this point, the direction of the cable changes to N22° E and rises to the upper terminal pulley, elevation 1190 m, at an angle of 25° with the horizontal.

Required:

1. Total length of the cable if terminal wheels (pulleys) are 7 m in diameter. (*Ans.:* 3782 m)
2. Horizontal projection of cableway.

(NOTE: Neglect sag in cable. Use a scale of 1 cm = 100 m.)

PROBLEM 7 (AIRPLANE SPOTTER PROBLEM)

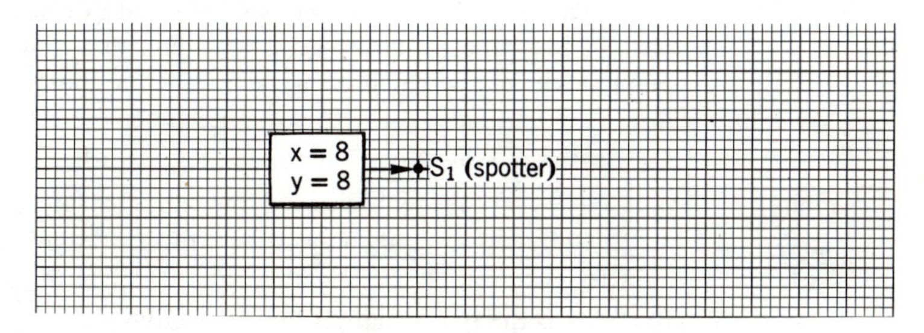

7. Given Data: An airplane spotter (at point S) sights an airplane at a bearing of N30° W at an elevation of 20° and a range of 7800 m. Exactly 60 s later the same airplane, flying a straight course, is spotted at a bearing of N60° E at an elevation angle of 10° and a range of 8300 m.

Required:

A. Bearing of the airplane's course.
(*Ans.:* S78.6° E)

B. Speed of the airplane relative to the ground in kilometers per hour.
(*Ans.:* 642 km/hr)

C. Actual airspeed of the airplane in kilometers per hour.
(*Ans.:* 654 km/hr)

D. Rate of climb or descent of the airplane in meters per second.
(*Ans.:* Rate of climb = 31.6 m/s)

(NOTE: Use a scale of 1 cm = 1000 m. Range is equal to the actual distance from spotter to airplane.)

PROBLEM 8 (FLUID FLOW PROBLEM)

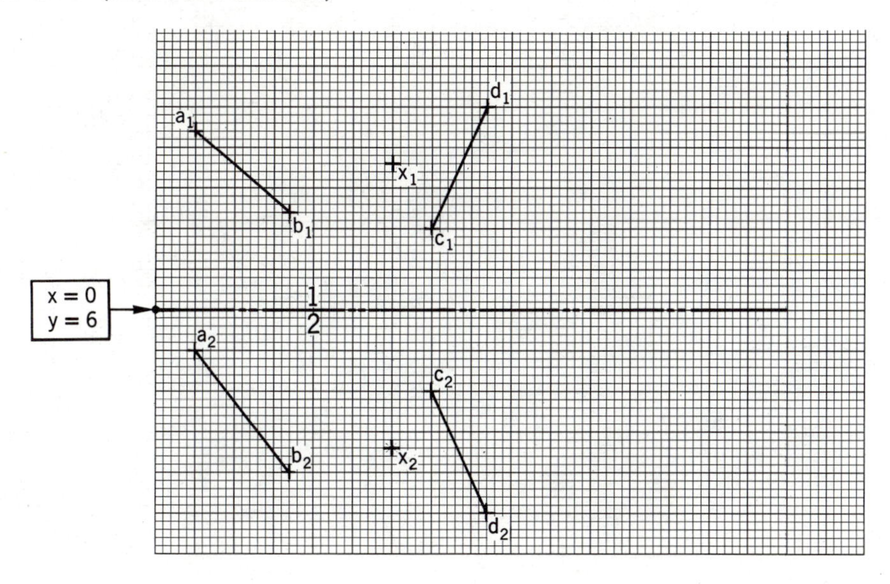

8. Given Data: Horizontal and frontal projections of lines *AB* and *CD* and a point *X*. Lines *AB* and *CD* establish the positions and directions of the centerlines of two pipes in an experimental fluid flow system. Point *X* is a point at which a flowmeter is to be installed to measure the amount of fluid flow that takes place in a straight line connecting pipe between pipes *AB* and *CD*.

Required:

A. Determine the centerline of the straight line pipe that passes through point *X* and connects with the centerlines of pipes *AB* and *CD*.

B. What is the true length of the connecting pipe? (*Ans.:* TL = 11.2 m)

C. What is the bearing and slope of the connecting pipe? (*Ans.:* Bearing = N85.5° E; slope = 10°. Down from *AB*.)

D. Show the horizontal and frontal projections of the centerline of the connecting pipe. (Check the accuracy of your work by comparing the frontal elevation measurements with auxiliary elevation measurements.)

(NOTE: Use a scale of 1 cm = 2 m. Exercise extreme accuracy in the layout and solution of this problem, since it is self-checking in nature when the horizontal and frontal projections of the connecting pipe are determined.)

PROBLEM 9 (LEM ANTENNA-HATCH PROBLEM)

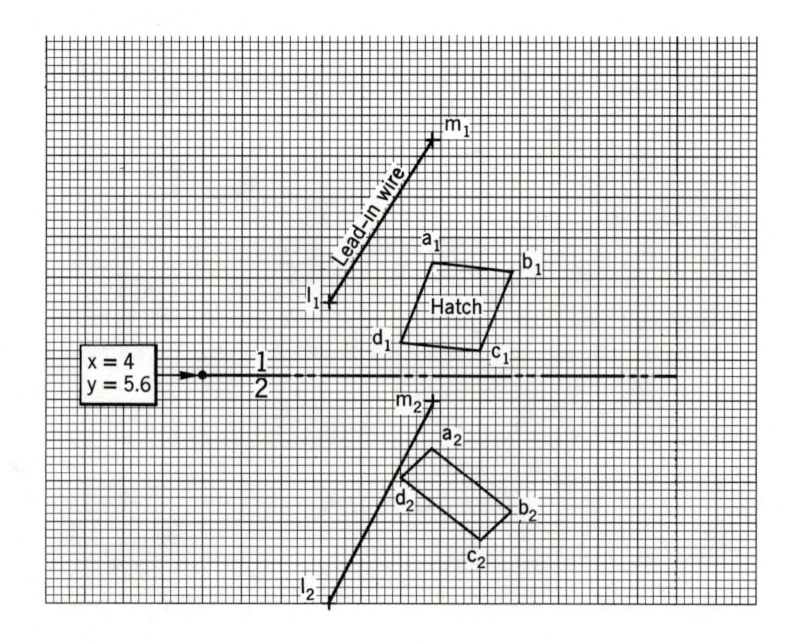

9. Given Data: Horizontal and frontal projections of a lead-in antenna wire attached at points *L* and *M* and the outline, *ABCD*, of the hatch on a lunar excursion module (LEM).

Required: In this given preliminary design layout for the entry-exit hatch of the LEM, the antenna lead-in wire interferes with the opening and closing of the hatch. The position of the lead-in wire can be changed but only at point *M* in a vertical direction (angle of lead-in wire will change) and by keeping point *L* fixed.

Determine a new location of Point *M* such that the lead-in wire will clear the hatch with a minimum of 400 mm. How far, vertically, does point *M* have to move to satisfy this condition?

Show the new position of the lead-in wire in all views.
(*Ans.:* 1450 mm or 1.45 m)

(NOTE: Use scale of 1 cm = 500 mm.)

PROBLEM 10 (PULLEY BELT PROBLEM)

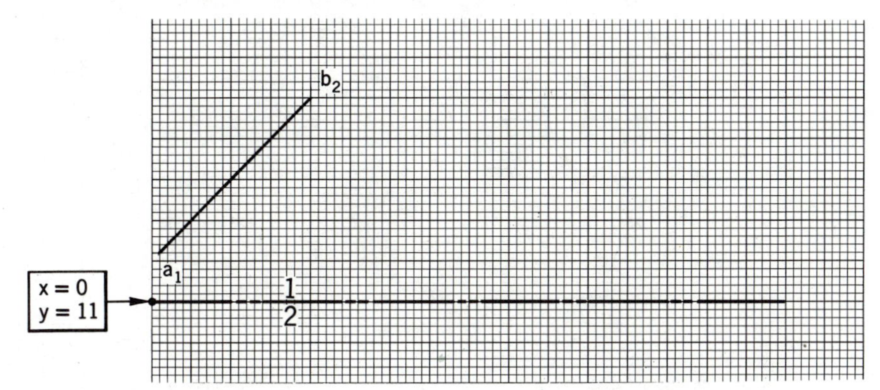

10. Given Data: Two horizontal parallel shafts are 4.45 m apart on center. The first shaft has its centerline (*AB*) 1.30 m above an 0.250-m-thick floor slab. The second shaft has its centerline 2.05 m below the floor slab. A 2.05-m-diameter pulley on the first shaft clears by 0.250 m the back of an 0.40-m-thick wall and drives a 2.75-m-diameter pulley on the second shaft on the opposite side of the wall by means of a belt 0.500 m wide.

Required:

1. Draw the front elevation view of the hole that must be cut in the wall to clear the belt 0.250 m all around.
2. Calculate the volume, in cubic meters (m^3), of material that must be cut out of the wall.
   (*Ans.:* 3.20 m^3)

Note:

A. Use a scale of 2 cm = 1.00 m.

B. Let the horizontal projection plane represent the top of the floor and let the frontal projection plane represent the front of the wall.

C. Neglect the thickness and sag of the belt.

D. Omit the front elevation view of the shaft, pulleys, and belt.

PROBLEM 11 (DERRICK PROBLEM I)

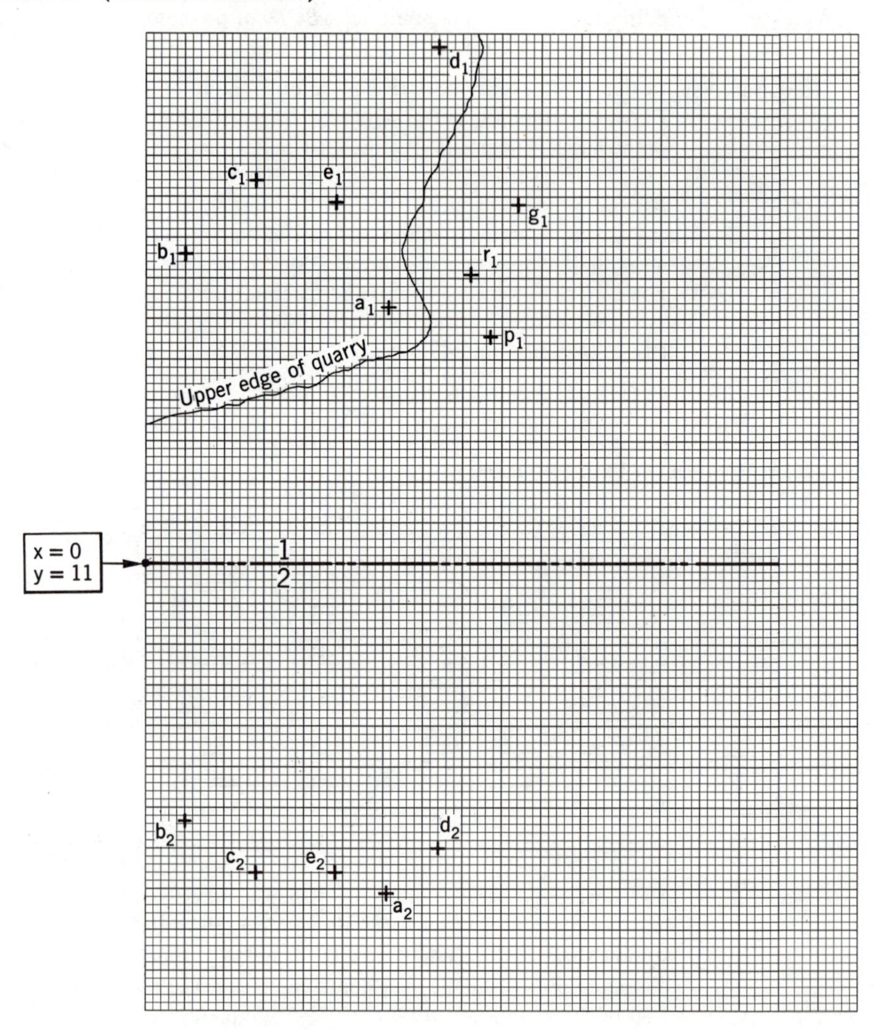

11. Given Data: At the edge of a Vermont granite quarry a derrick mast, 25 m high, is erected at postion E and held vertical by guy cables from its top to anchor points A, B, C, and D on the ground. A boom, 25 m long, is attached to the mast at point E and is supported and maneuvered by a cable from the top of the mast to its free end. The boom never goes below a horizontal position during its operation. (Neglect weight and sag of all cables.)

The derrick boom lifts a block of granite, weighing 1000 kg from one position to another at the bottom of the quarry.

Required:

1. What is the magnitude and type of load carried by each cable and the mast when the granite block is lifted and held at boom position G?

NOTE: Use a space scale of 1 cm = 5.0 m and a vector scale of 1 cm = 200 kg. (Use two sheets of cross-section paper, taped along their long edges, to solve the problem.)

(*Ans. for Position G:* Load in mast = 2280 kg compression,
Load in cable to anchor *B* = 615 kg tension,
Load in cable to anchor *C* = 1380 kg tension,
No loads in cables going to anchors *A* and *D*.

PROBLEM 12 (DERRICK PROBLEM II)

12. Given Data: Same as given for Problem 11, Derrick Problem I, including scales.

Required: What is the magnitude and type of load carried by each cable and the mast when the granite block is lifted and held at boom position *P*?

(NOTE: Use two sheets of cross-section paper taped along their long edges to solve this problem.)

(*Ans. for Position P:* Load in mast = 3990 kg compression,
Load in cable to anchor *B* = 2230 kg tension,
Load in cable to anchor *D* = 1980 kg tension,
No loads in cables going to anchors *A* and *C*.)

PROBLEM 13 (DERRICK PROBLEM III)

13. Given Data: Same as that given for Problem 11, Derrick Problem I, including scales.

Required: What is the magnitude and type of load carried by each cable and the mast when the granite block is lifted and held at boom position *R*?

(NOTE: Use two sheets of cross-section paper taped along their long edges to solve this problem.)

(*Ans.:* Load in mast = 2480 kg compression,
Load in cable to anchor *B* = 1740 kg tension,
Load in cable to anchor *D* = 1320 kg tension,
No loads in cables going to anchors *A* and *C*.)

PROBLEM 14 (SATELLITE CAMERA SHROUD PROBLEM)

14. Given Data: The Ranger 7 moon probe contained cameras that obtained the first close-up photography of the moon. These cameras were enclosed and protected in an aluminum shroud with a conical surface. The shroud was opened in such a way that the camera's fields of view would not be obstructed. This opening was established by cutting the conical surface of the shroud along two flat planes, as is shown in the given front elevation view.

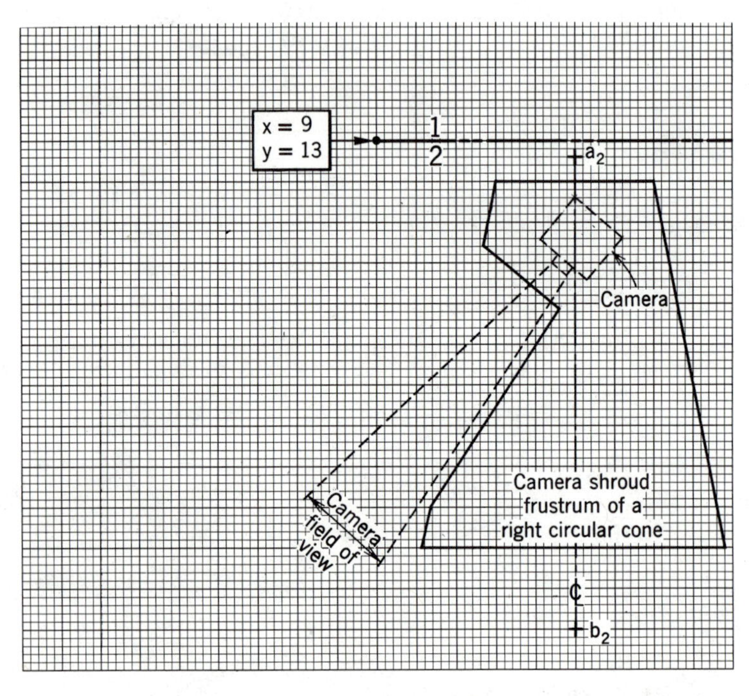

Required: Draw the horizontal and left-side projection views of the shroud showing the opening in both views.

PROBLEM 15 (SATELLITE CAMERA SHROUD SURFACE DEVELOPMENT PROBLEM)

15. Given Data: Same as that given for Problem 14 (Satellite Camera Shroud Problem)

Required: Using the information obtained in the solution of Problem 14, develop the surface of the camera shroud, including the cutout necessary to provide the necessary opening for the camera's unobstructed field of view.

Develop the surface so as to minimize the length of joint (seam) that has to be riveted.

PROBLEM 16 (HOUSE-SHED ROOF-RAFTER PROBLEM)

16. Given Data: A pictorial view of a small cottage and an attached tool shed. The front wall of the cottage measures 6.8 m wide by 3.4 m high. The side of the cottage is 4.2 m deep. The ridgepole, located symetrically, is 4.2 m long and 5 m above the ground floor.

The attached tool shed measures 1.8 m wide and 2.6 m deep, it is 2.6 m high on its attached side and 2.2 m high on the outside.

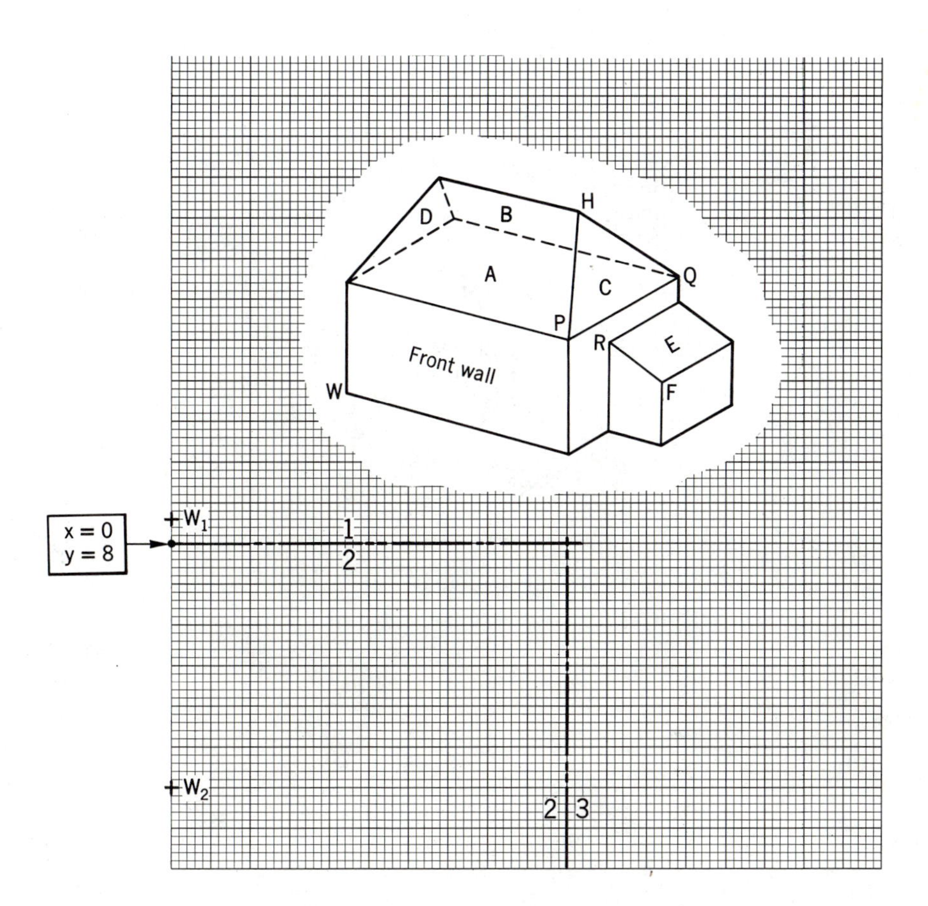

Required:

1. Draw the front, top, and right-side views of the house and the attached tool shed.
2. Determine the total roof surface area to be covered by sheet metal roofing.
   (*Ans.:* 43.6 m^2)
3. Find the true lengths and slopes of hip rafters *RF and HQ*.
   (*Ans.:* *RF*, TL = 1.7 m; slope = 15°,
   HQ, TL = 3.0 m; slope = 34.5°.)
4. Find the true slopes of roofs *A*, *B*, *C*, *D*, and *E*.
   (*Ans.:* Slopes of *A* and *B* = 40°,
   Slopes of *C* and *D* = 50.5°,
   Slope of *E* = 15°.)

(NOTE: Use a Scale of 1 cm = 1.0 m.)

PROBLEM 17 (PICTORIAL VIEW PROBLEM)

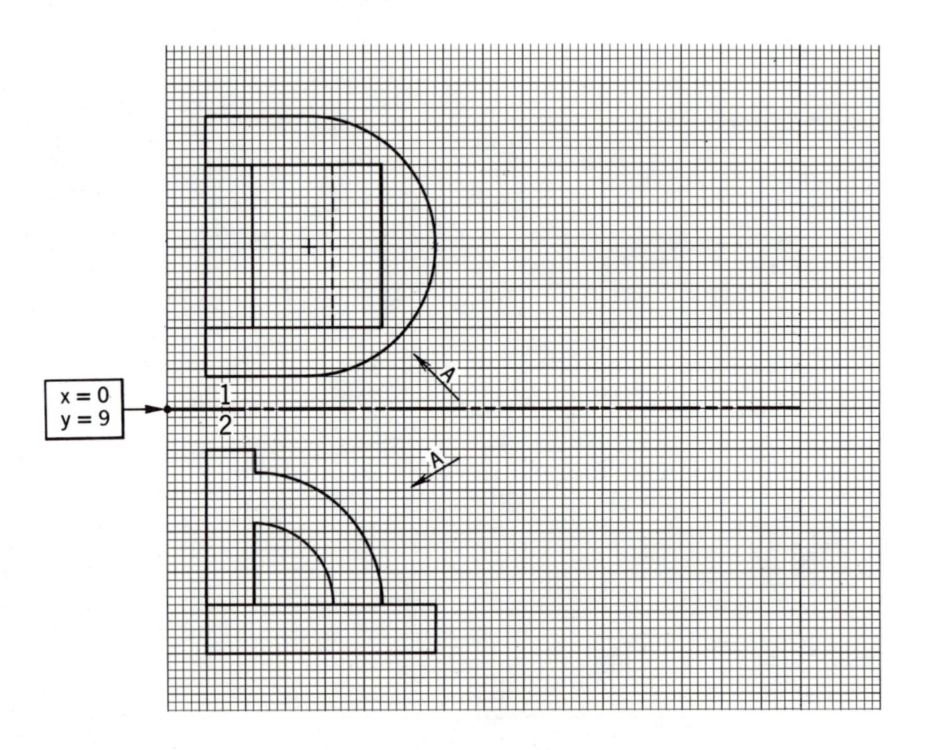

17. Given Data: Horizontal and frontal projections of a solid cast aluminum guide support, used in an experimental materials testing project, and a line of sight direction indicated by '*A*'.

Required: Draw the auxiliary view of the guide support as seen from the given line of sight direction '*A*'.

PROBLEM 18 (AIRPLANE FLIGHT PROBLEM)

18. Given Data: An airplane, in the air at point *P*, must fly directly (straight line flight) to the nearest point 1000 m above an inclined section of a roadway having a centerline *AB*.

Required:

1. Actual distance of flight from starting point *P* to destination point. (*Ans.:* 1240 m)
2. Bearing of flight path. (*Ans.:* N51.5° W)
3. Angle of climb. (*Ans.:* 19°)

(NOTE: Use a scale of 1 cm = 200 m.)

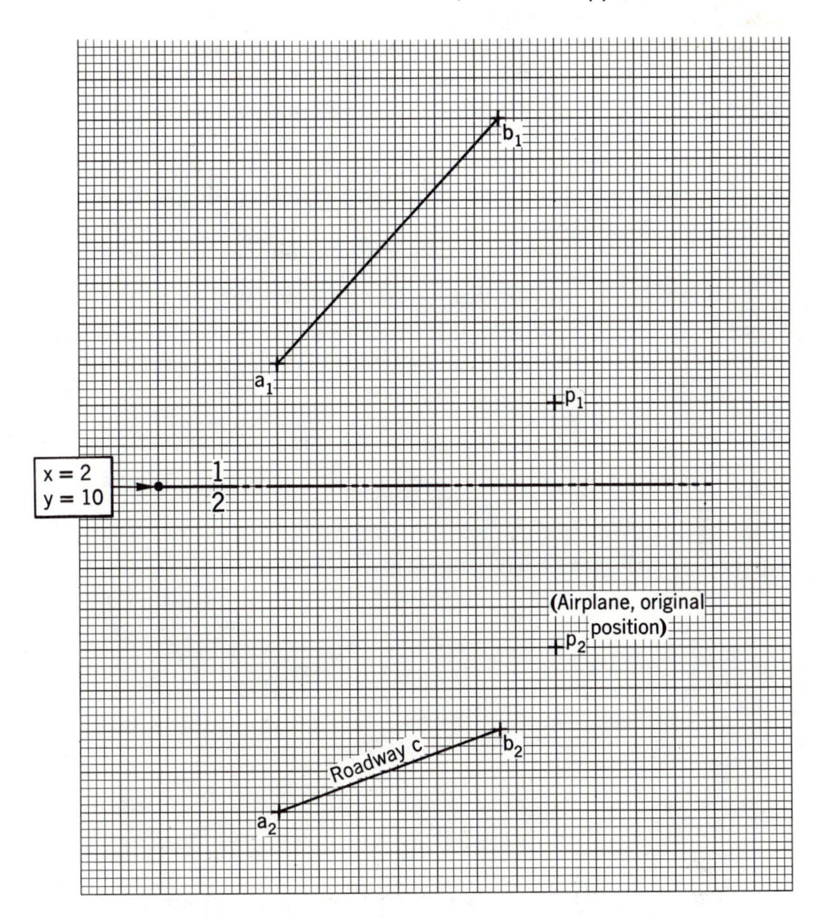

PROBLEM 19 (AIRPLANE FLIGHT PLAN PROBLEM)

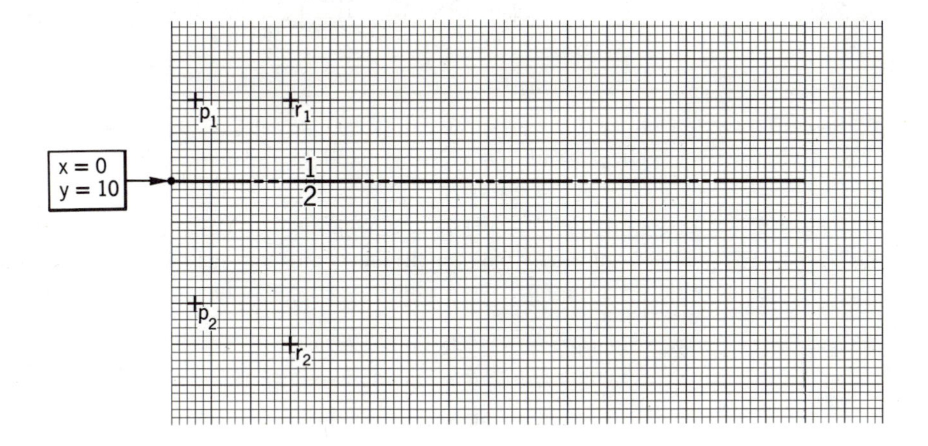

19. Given Data: An airplane (code name Alpha), at point *P*, having an altitude of 10,000 m, flies N45° E at a climbing angle of 20° for a distance of 35,000 m. At the 35,000-m point it changes its course and flies due East, maintaining its new altitude. Another airplane (code name Beta), at point *R*, having an altitude of 5000 m, flies due East for a distance of 35,000 m. At the 35,000-m point it changes its course to due North, climbing at an angle of 30°.

(Assume changes in course for each airplane to be instantaneous.)

Required:

1. Are the flight paths of the airplanes on a collision course?
2. Assuming that airplane Alpha is flying at a constant speed of 600 km/h and airplane Beta is flying at a constant speed of 480 km/h, how far apart (line of sight distance) are they after they have been flying 6 min from their initial positions?
(*Ans.:* 16,000 m)

(NOTE: Use a scale of 1 cm = 5000 m.)

PROBLEM 20 (MINING-GEOLOGY PROBLEM)

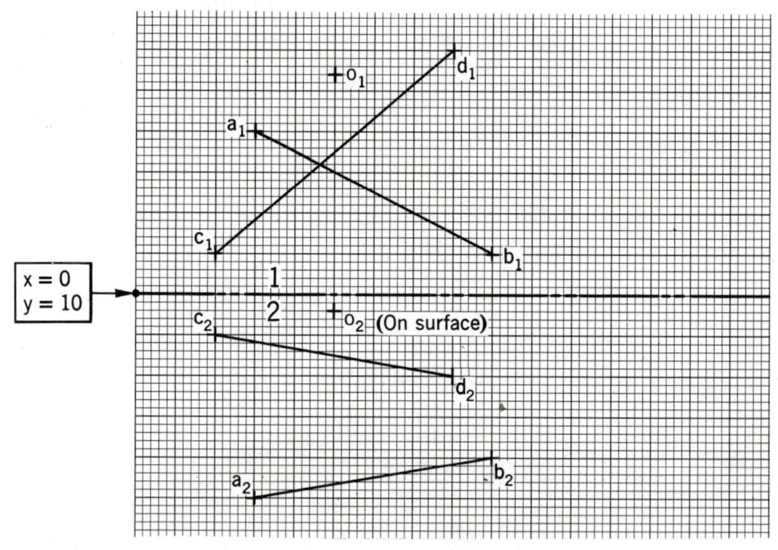

20. Given Data: It is desired to connect two mine tunnels, whose centerlines are represented by *AB* and *CD*, by a single straight line shaft from an entrance *O* on the surface of the earth.

Required: As a mining engineer–geologist, using the principles of descriptive geometry, show how this straight line connecting shaft is determined and what is its bearing from *O*, percent grade, and its length from *O* to *AB* and *O* to *CD*.

Ans.: Bearing = S31° E, grade = 82%,
Length *O* to *AB* = 610 m, length *O* to *CD* = 200 m.)

(NOTE: Use a scale of 1 cm = 100 m.)

PROBLEM 21 (ELECTRIC POWER LINE PROBLEM)

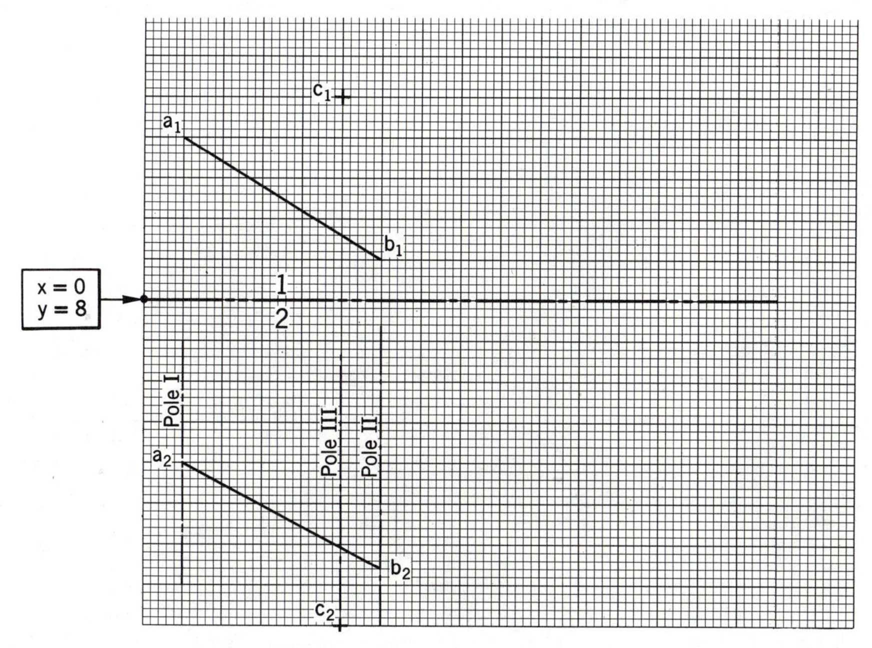

21. Given Data: An electric power line AB, attached at point A on Pole I and at point B on pole II; and another power line whose one end is connected to pole III at point C. (All poles are vertical.)

Required: To run the power line connected at point C, on pole III, to a connecting point D on pole I such that a minimum clearance of 1.5 m is maintained between the two power lines.

1. How far up pole I above point A must connection D be made to assure the minimum required clearance between the two power lines?
   (*Ans.:* = 3.8 m)
2. How long is the power line between connecting points C and D.?
   (*Ans.:* = 8.9 m)
3. Show all views of both power lines.

   (NOTE: Use a scale of 1 cm = 1.0 m.)

PROBLEM 22 (CULVERT DESIGN DATA PROBLEM)

22. Given Data: You are a member of a design group in a firm of consulting civil engineers that has been invited to bid on the design and construction of a retaining wall through which a suspended culvert is to pass.

The general layout and data that determine the slope of the wall, grade, and direction of the culvert have been established by a field

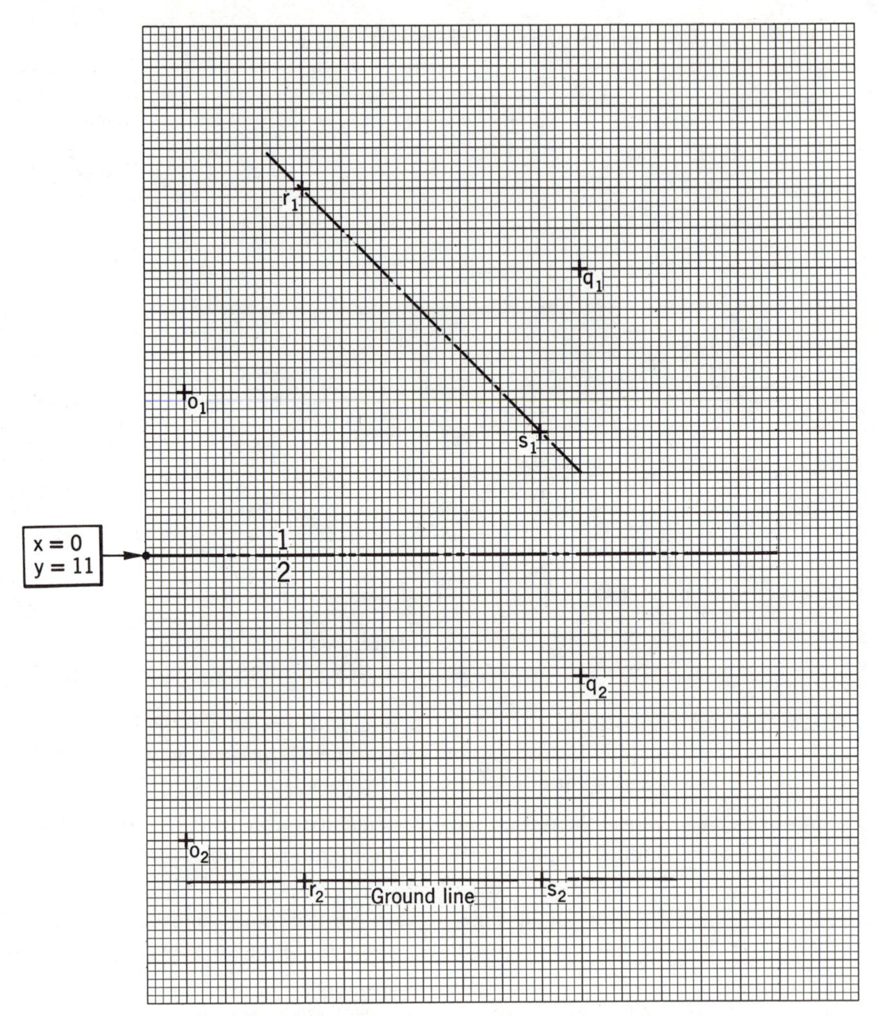

survey party and preliminary design decisions. Three points, *Q*, *R*, and *S*, have been established on a hillside where the retaining wall will be constructed. These points locate the plane of the retaining wall. Points *R* and *S* are on the line of intersection of the wall and the level ground; point *Q* is a third point on the hillside in the plane of the wall. The direction of the culvert is N75° E and its design grade is 15%.

Additional design data are required before the design group can complete the design and establish full construction details for the retaining wall. The part of the culvert in which you are interested begins at a point *O*, which is 5 m above the ground where the centerline of a section of suspended pipe ends and the culvert begins.

Required: The following additional design data:

1. Determine where the centerline of the culvert pierces the retaining wall. Show this point in all views.

2. What is the distance from point O to the retaining wall measured along the centerline of the culvert?
(*Ans.:* 42 m)
3. What is the slope of the wall in degrees?
(*Ans.:* 54.5°)
4. How far up the wall does the centerline of the culvert pierce, measured along the wall perpendicular to the level ground line?
(*Ans.:* 13 m)
5. What is the true size of the angle between the centerline of the culvert and the plane of the retaining wall?
(*Ans.:* 37°)

(NOTE: Use two sheets of paper taped along their long edges for this problem. Use a scale of 1 cm = 5.0 m.)

PROBLEM 23 (GAS PIPE LINE PROBLEM I)

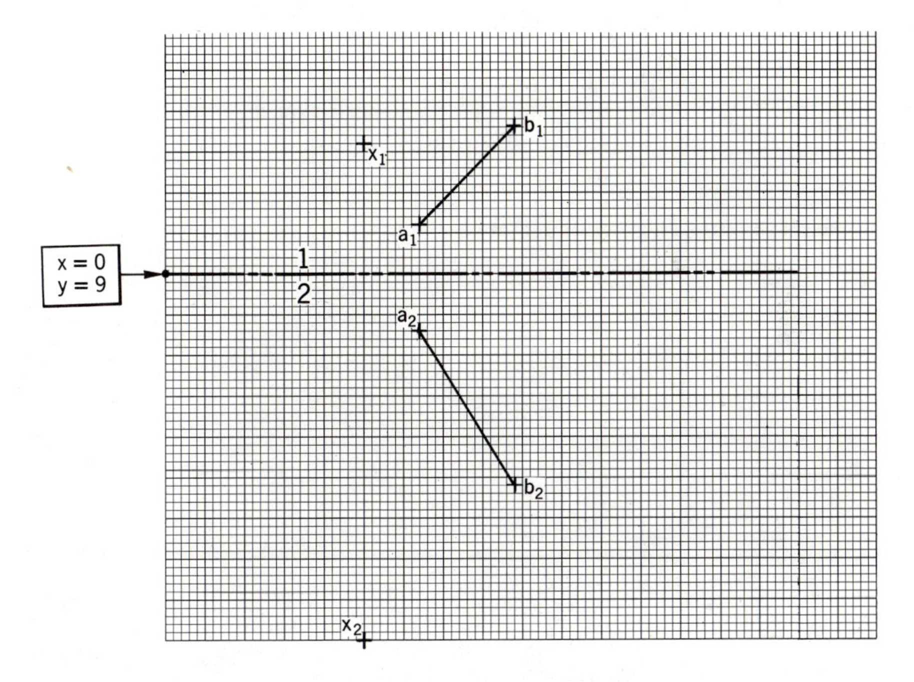

23. Given Data: A pipeline for natural gas is being laid through the hilly country of southwestern Pennsylvania. Pipelines for oil are usually laid as close to level grades as possible, to minimize the size and number of pumping stations. Grades in pipelines for gases, however, are not quite so critical and, therefore, the most advantageous routes for these lines are closely related to inherent property condemnations for rights of way. Hence, gas pipeline routes are designed to use the most direct way possible, no matter what the terrain, and normally may be laid up and down grade as necessary.

The line AB establishes the direction and position (not the length) of the centerline of a pipeline that has been laid. A particularly difficult terrain has been encountered near the vicinity of the B end of the pipe centerline. Not only is there a deep gulley to cross but there is also a massive rock formation at the bottom of the gulley which necessitates a deviation in the direction of the continuing pipeline to minimize the blasting and removal of rock. Point X is a planned pipe-support point through which the continuing pipeline must pass.

Required:

1. Show in all views how a curved section of pipe, with an 8-m-radius curve, can be installed to connect the pipe section whose centerline is AB to a continuing pipe section centerline CD that has a direction of due West. The connecting points are at the points of tagency of the curved connecting pipe to pipe directions AB and CD?
2. What is the percent grade of CD? (*Ans.:* 109%)
3. What is the length of the connecting curved section? (*Ans.:* 11.7 m)

(NOTE: Use a scale of 1 cm = 2.0 m. Use the given lines as centerlines of the pipes and neglect the diameter of the pipes and the details of the connections.

PROBLEM 24 (GAS PIPELINE PROBLEM II)

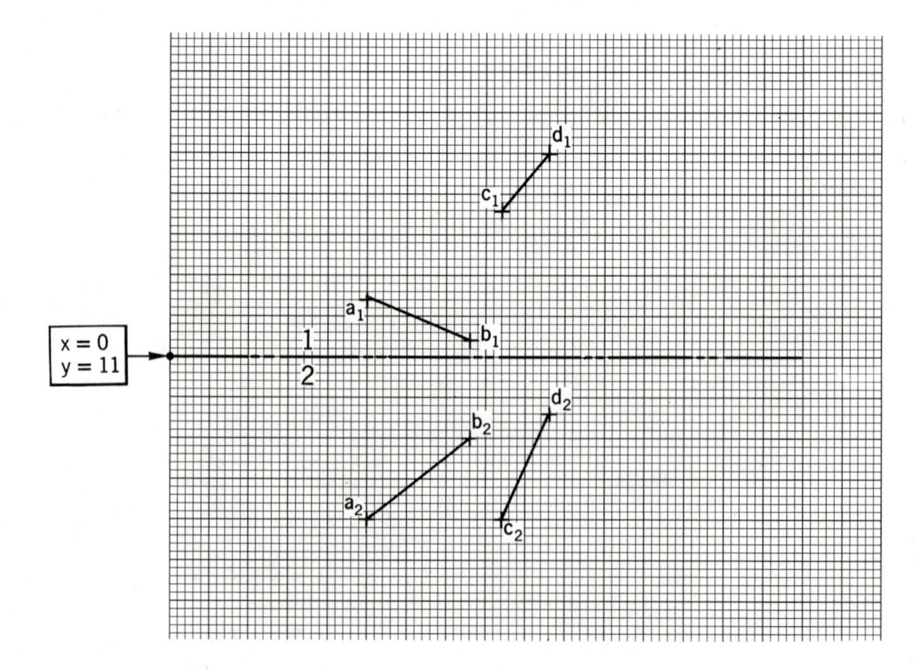

24. Given Data: Two natural gas pipelines whose centerlines and lengths are represented by line *AB* and *CD* are to be connected. The most economical way of physically connecting these pipes is to make the connection along the shortest straight-line distance between the pipes. However, the right of way for the connecting pipe is in question because of pending litigation by the owner of the land through which the pipe is to be laid. Because of the uncertainty of the outcome of the litigation, the engineers who are designing the connecting pipe have been instructed to design two versions of the route of this pipe as part of the contingency plans of the company that is installing the pipeline.

Required: As one of the engineers assigned to this job you have been given the responsibility to determine the following design information.

1. (Design I) Determine the true length and percent grade of the centerline of the shortest possible straight line connecting pipe, assuming that the given pipes, *AB* and *CD*, may be extended in length.
2. (Design II) Determine the true length, bearing, and percent grade of the centerline of the shortest possible straight line connecting pipe, assuming that the given pipes, *AB* and *CD*, may *not* be extended.
3. Which design uses the least length of pipe to make the required connection?

(NOTE: Use a scale of 1 cm = 200 m.)

PROBLEM 25 (GAS LINE PROBLEM III)

25. Given Data: Same as for Problem 24 (Gas Pipeline Problem II)

Required: The bearing of and angles that the shortest straight-line connecting pipe makes with the pipes *AB* and *CD* when they cannot be extended. (*Ans.:* Bearing = N37° E; Angle with *AB* = 90°; Angle with *CD* = 105.5°.)

PROBLEM 26 (AIRPLANE CONTROL CABLE PROBLEM, PROJECT)

26. Given Data: An aluminum alloy gusset plate in an airplane fuselage, whose plane is defined by *ABCD*, has been riveted into position. The control cable system is designed so that one of the cables, *MN* (points *M* and *N* are tangent points on the cable pulleys), controlling the rudder passes through a grommet hole in the gusset plate *ABCD*. The diameter of the hole is 32 mm.

The designed strength requirements of the gusset plate specify a minimum allowable edge distance of 27 mm from the grommet hole to any edge of the gusset plate.

Required:

1. Determine the actual edge distances of the present design.
   (*Ans.:* Distance to *AB* = 56 mm, to *BC* = 24 mm,
   Distance to *CD* = 26 mm, to *DA* = 24 mm.)

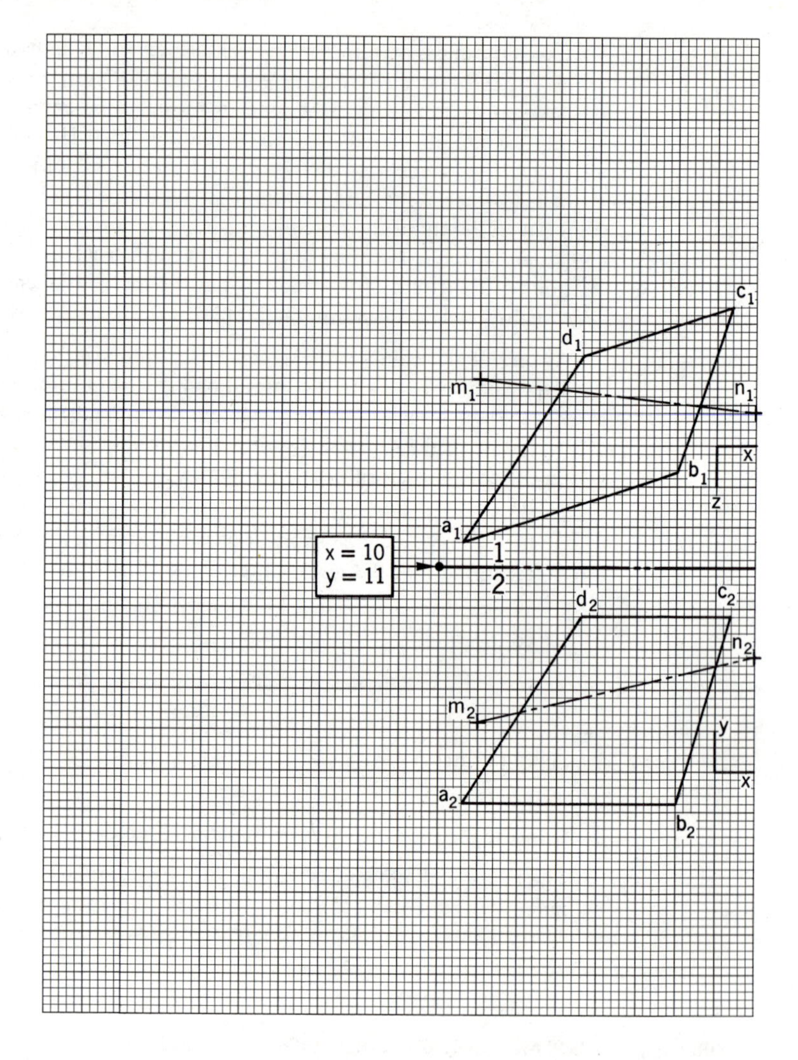

2. If any of the edge distances are less than the 27 mm minimum allowable, the control cable configuration and the location of the grommet hole in the gusset plate must be changed. The control cable configuration can be redesigned by changing the location of point M (while point N must remain fixed). What is the minimum distance that point M must be moved, in the XYZ directions (while keeping it parallel to the plane of the gusset plate $ABCD$) to satisfy the required minimum edge distance?
(*Ans.:* $x = 16$ mm, $y = 36$ mm, $z = 34$ mm)
3. How far must the center of the grommet hole be moved to obtain the minimum allowable edge distances?
(*Ans.:* 23 mm)

(NOTE: Use a scale of 1 cm = 20 mm.)

PROBLEM 27 (HOUSE-GARAGE ROOF INTERSECTION PROBLEM)

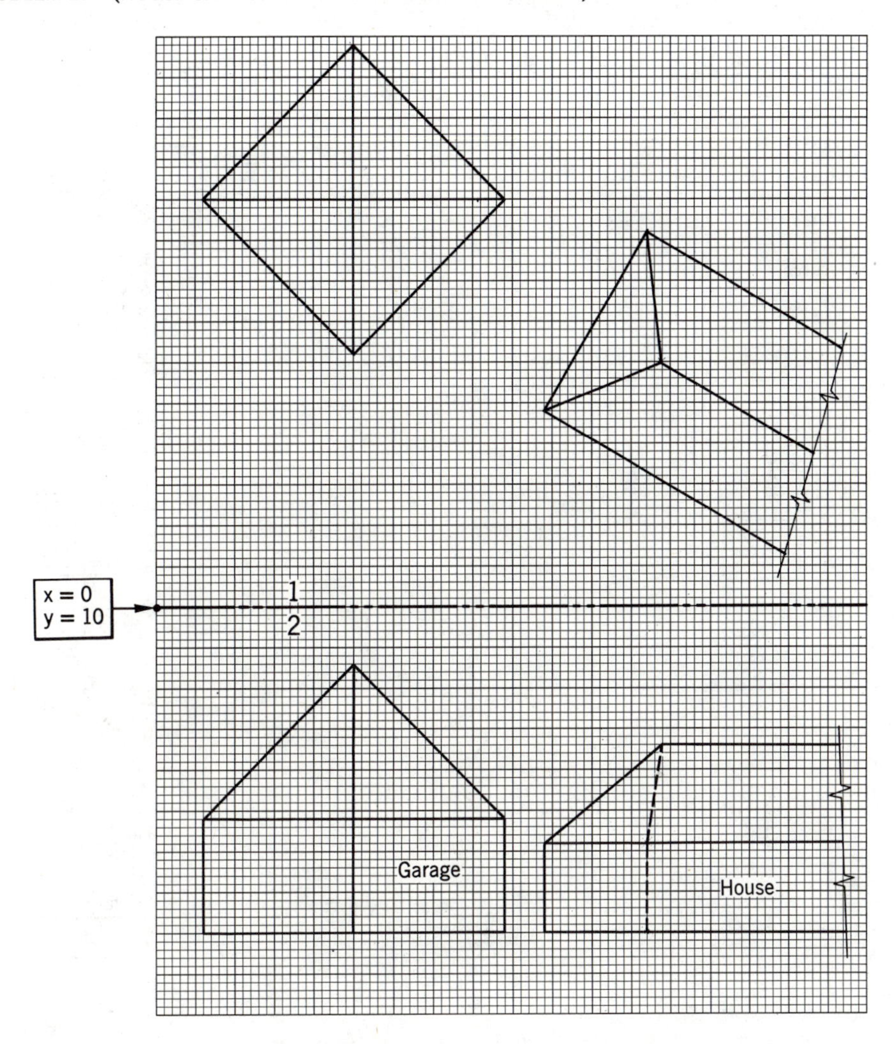

27. Given Data: The plan and elevation views of a detached house and garage. The house is to be extended to meet and be attached to the garage, keeping the existing roof design and slopes of both buildings.

Required: Determine the plan and elevation views of the intersection of the roof of the garage and the extended house.

PROBLEM 28 (SHADE-SHADOW PROBLEM)

28. Given Data: An architect plans to do a series of renderings for a house he has designed. He needs information about the shade and shadow charac-

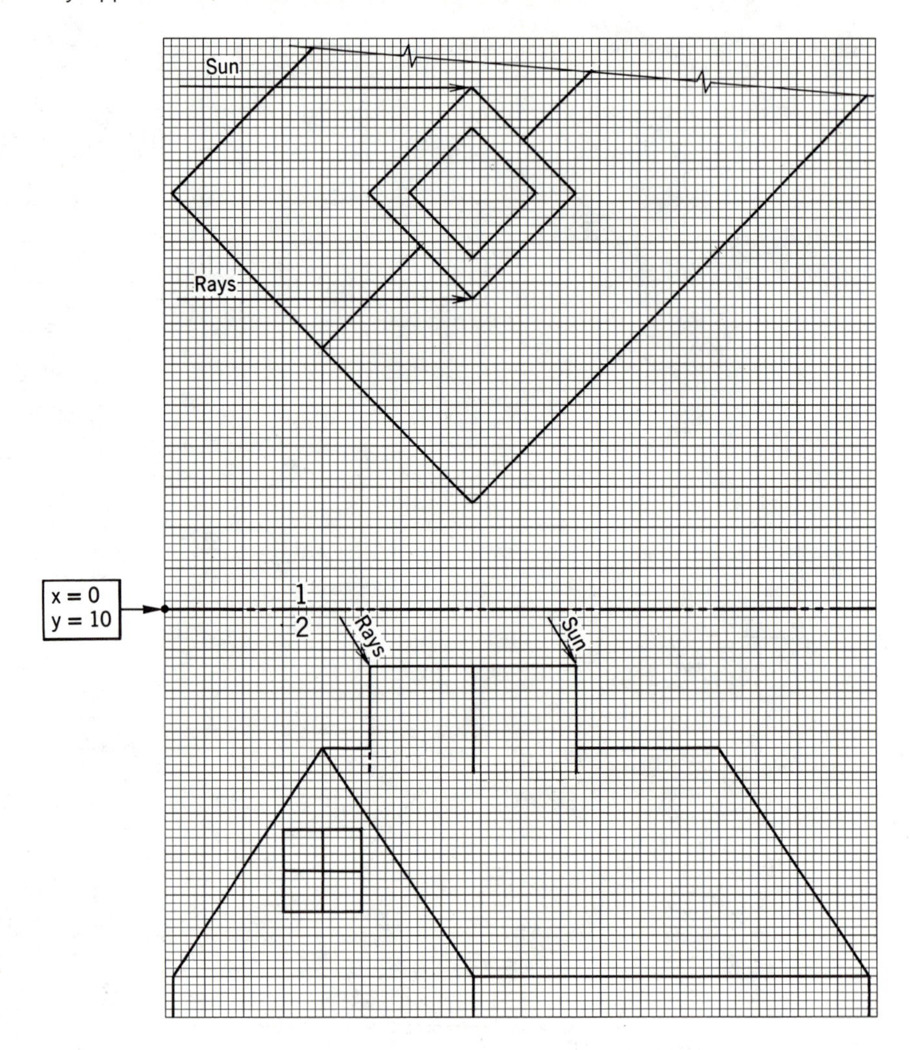

teristics of the chimney on the roof of the house when the sun is in a particular location in the sky. You are given the partial plan and elevation views of the house and the direction of the rays of the sun.

Required:

1. Determine the line of intersection of the chimney with the roof in the plan, front, and right-side elevation views.
2. Determine in plan, front, and right-side elevation views, the shade and shadow details of the chimney on the roof as determined by the given sun-rays direction.

   (NOTE: Use two sheets of cross-section paper taped along their long edges to solve this problem.)

PROBLEM 29 (SHADE-SHADOW PERSPECTIVE PROBLEM)

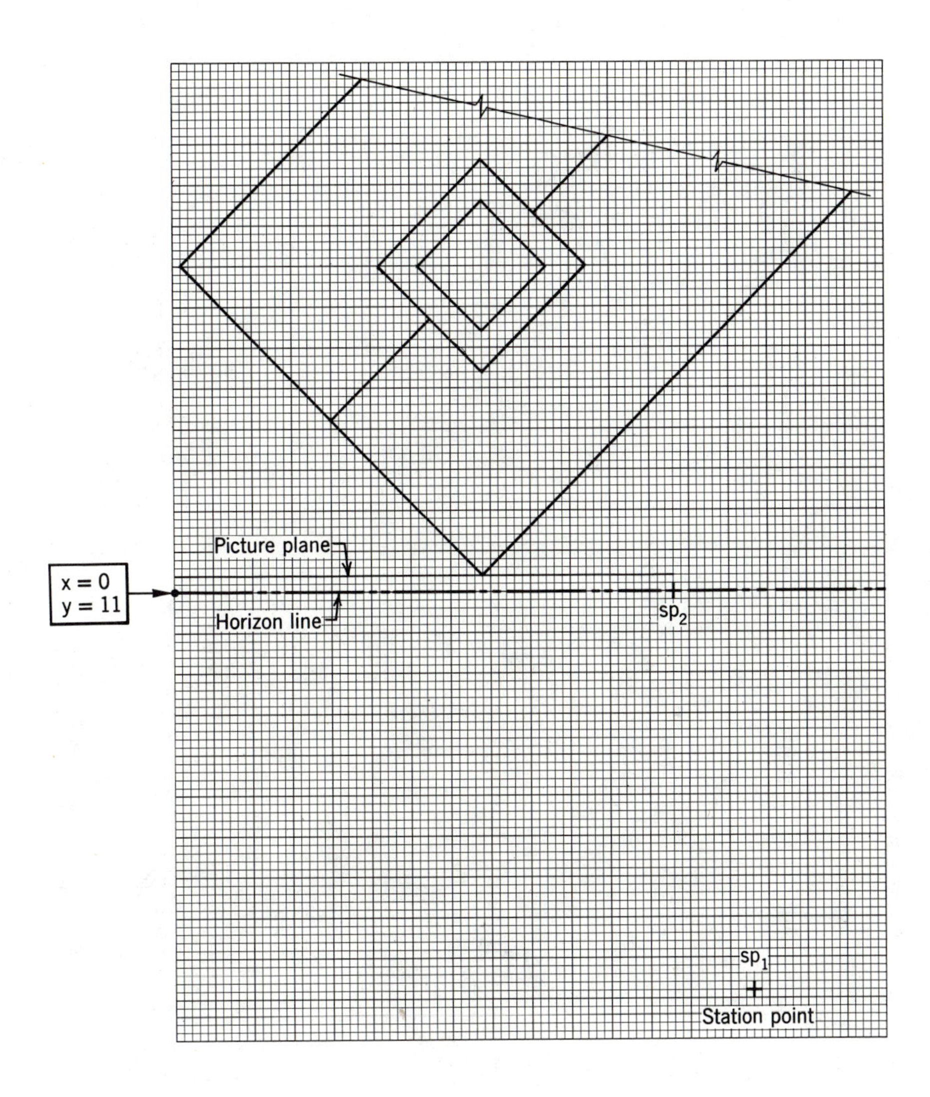

29. Given Data: Use data obtained in the solution of Problem 28 (Shade-Shadow Problem) and the given picture plane, horizon line, and station point.

Required: Construct a perspective drawing of the chimney and its shadow on the roof of the house given in Problem 28.

(NOTE: Use two sheets of cross-section paper taped along their long edges.)

PROBLEM 30 (SPEEDBOAT RUDDER CONTROL-BELL CRANK PROBLEM)

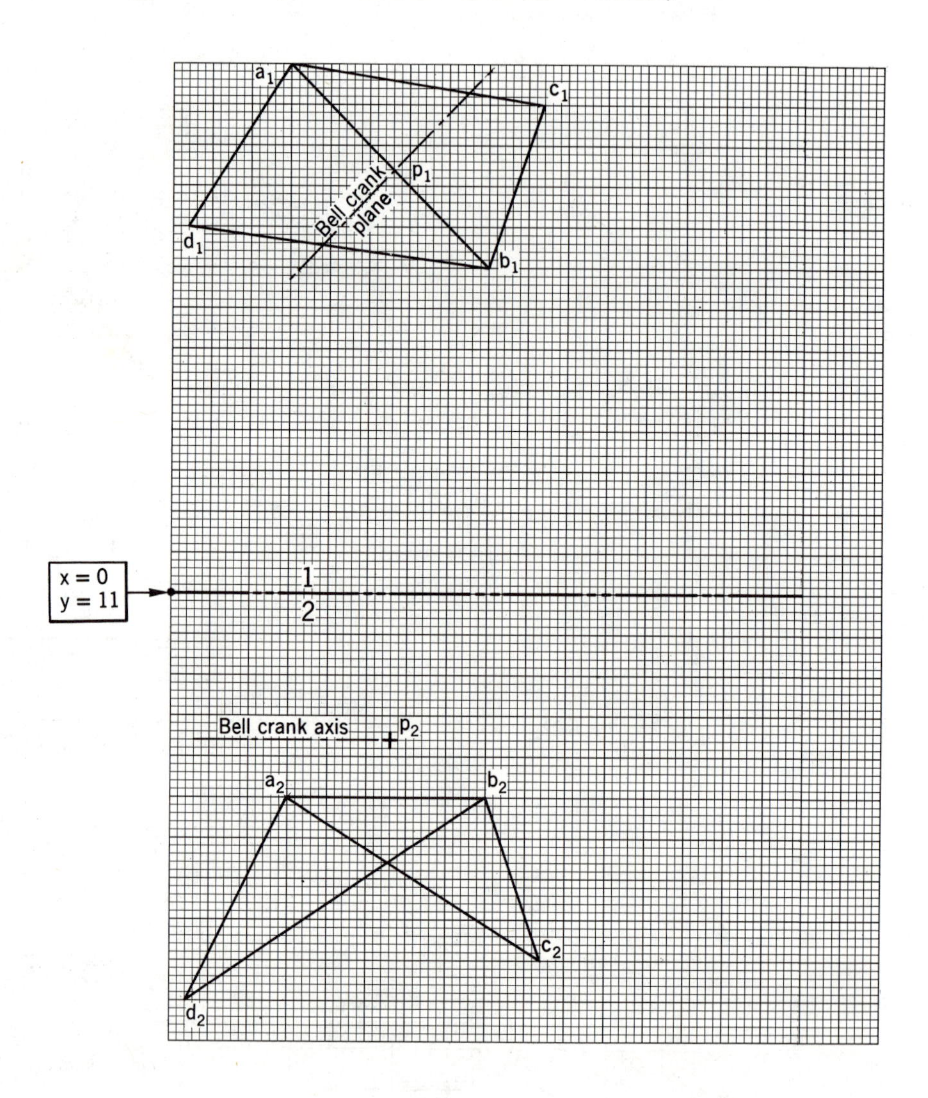

30. Given Data: Two structural members, whose planes are defined by *ABC* and *ABD*, in a speedboat powered by an outboard engine that is controlled directionally (rudder effect) from the front of the boat by a steering wheel connected to the engine frame by flexible steel cables. Parallel to and directly above the line of intersection (*AB*) of *ABC* and *ABD* is the axis of the rudder control bell crank, which is attached to the structural member by a bracket fixing the axis through point *P*.

One arm of the bell crank, 700 mm long, is in the horizontal position while the other arm, 500 mm long, makes a fixed angle of 130°, counterclockwise, with the horizontal arm.

Required:

1. Determine the number of degrees of rotation of the bell crank that is possible when it is in its extreme position (when it touches the structural members.
   (*Ans.:* 193°)
2. Show the extreme positions of the bell crank in all views.

   (NOTE: Represent the bell crank arms by single straight lines when solving this problem. Use a scale of 1 cm = 200 mm.)

Appendix

Sample Descriptive Geometry Test #I (Solutions)

1a. Solutions:
Projection plane #1, RL 1–2: *frontal proj. plane.*
Projection plane #2, RL 1–2: *horiz. proj. plane.*
Projection plane #3, RL 1–3: *horiz. proj. plane.*
Projection plane #1, RL 1–3: *aux. elev. proj. plane.*
Projection plane #5, RL 3–5: *aux. elev. proj. plane.*
Projection plane #2, RL 2–4: *aux. inclined proj. plane.*

1b. Solution:

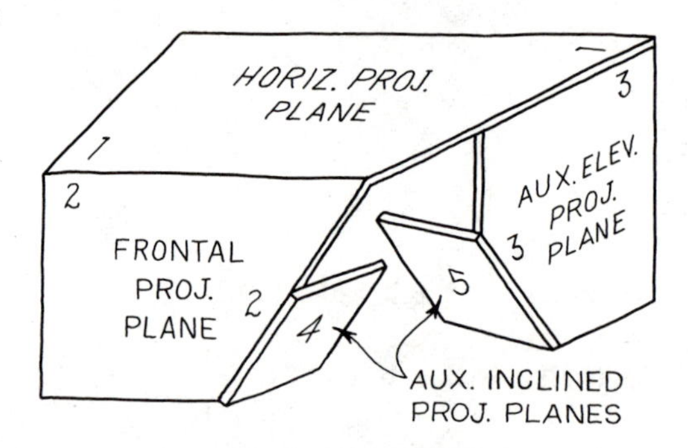

2. Solution: *c* **3.** Solution: *c* **4.** Solution: *c*

5. Solution: *A view in which at least one of the lines appears as a true length.*

6. Solution:

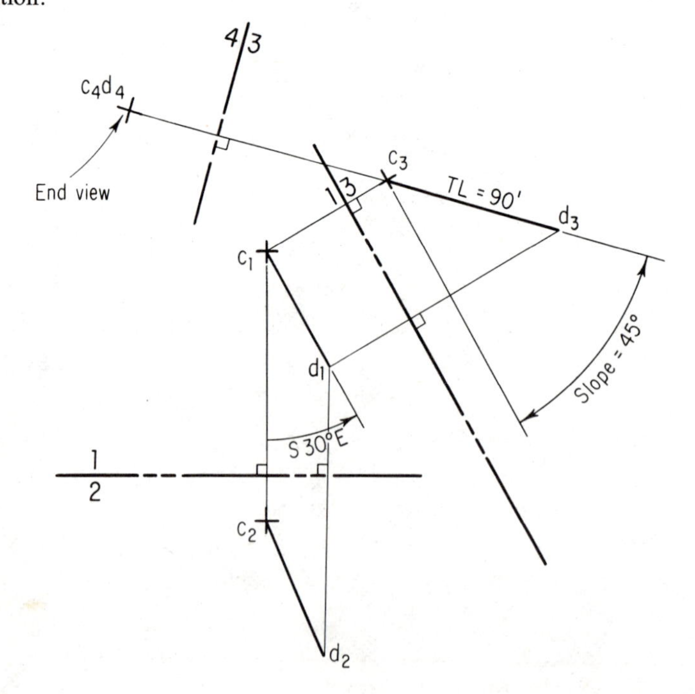

Sample Descriptive Geometry Test #II (Solutions)

1. Solution:

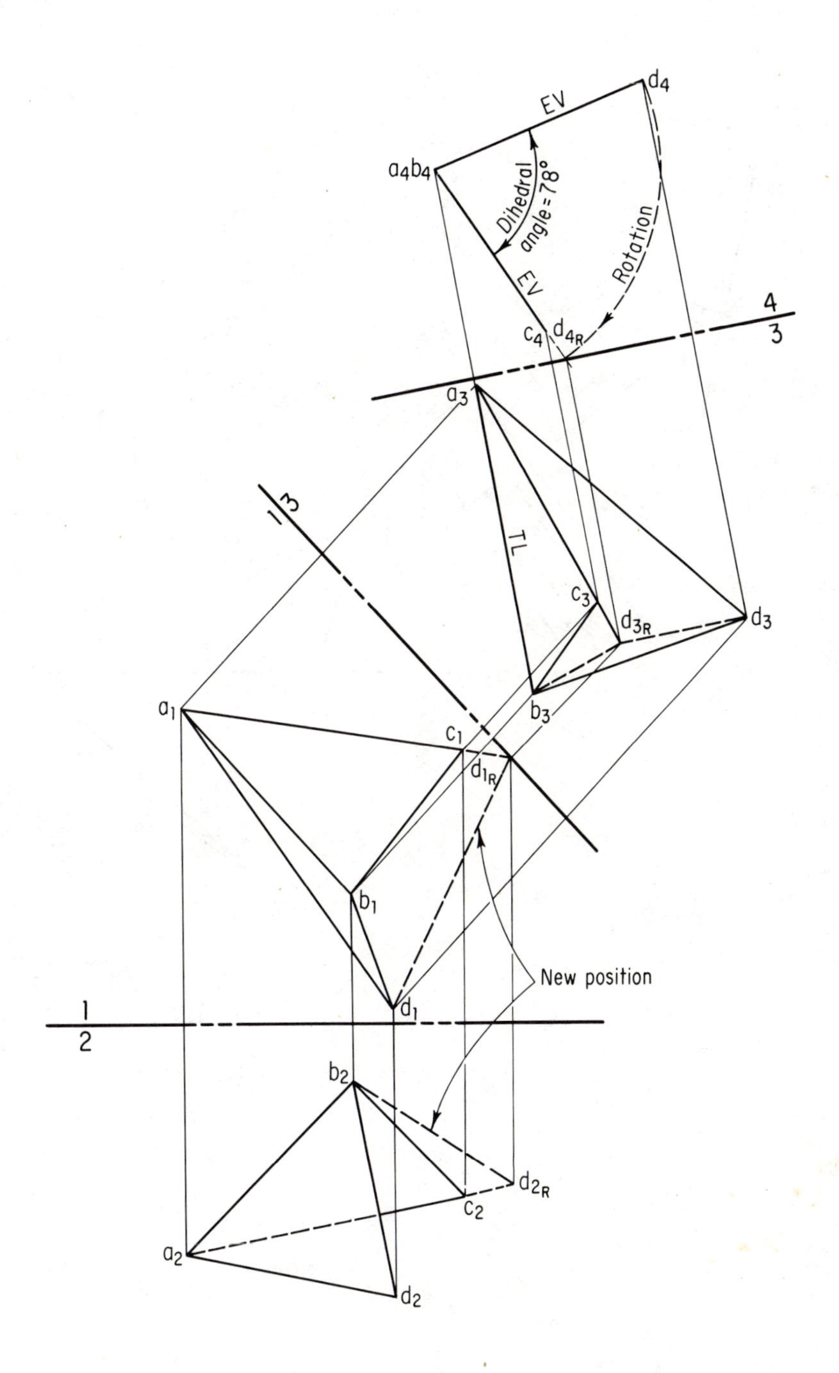

2. Solution:

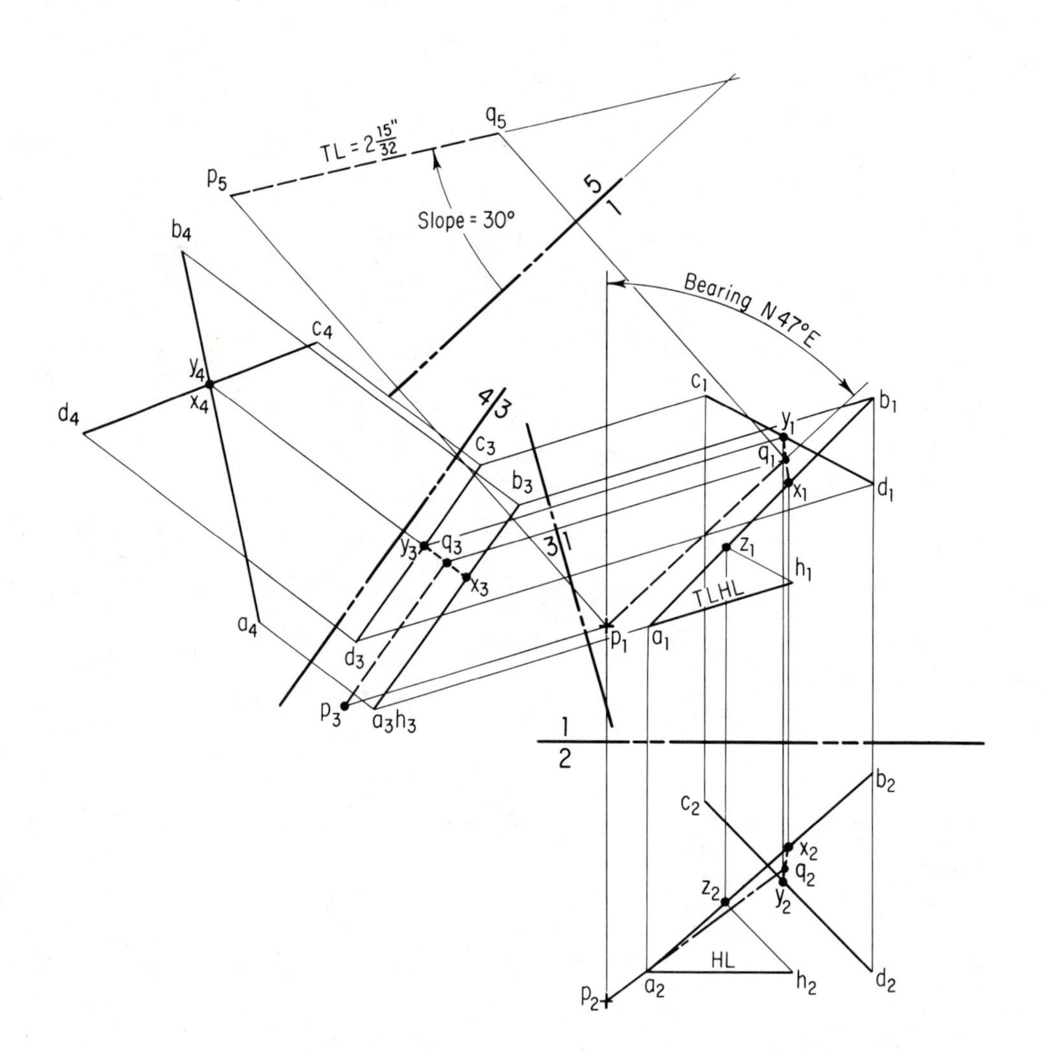

q5
TL = 2 15"/32
p5
5
1
Slope = 30°
b4
Bearing N 47°E
c4
y4
x4
c1
b1
d4
4
3
y1
c3
q1
b3
x1
d1
y3
q3
3
1
z1
x3
h1
TLHL
a4
p1
a1
d3
p3
a3h3
1
2
b2
c2
x2
q2
z2
y2
HL
p2
a2
h2
d2

3. Solution:

(Cutting plane method)

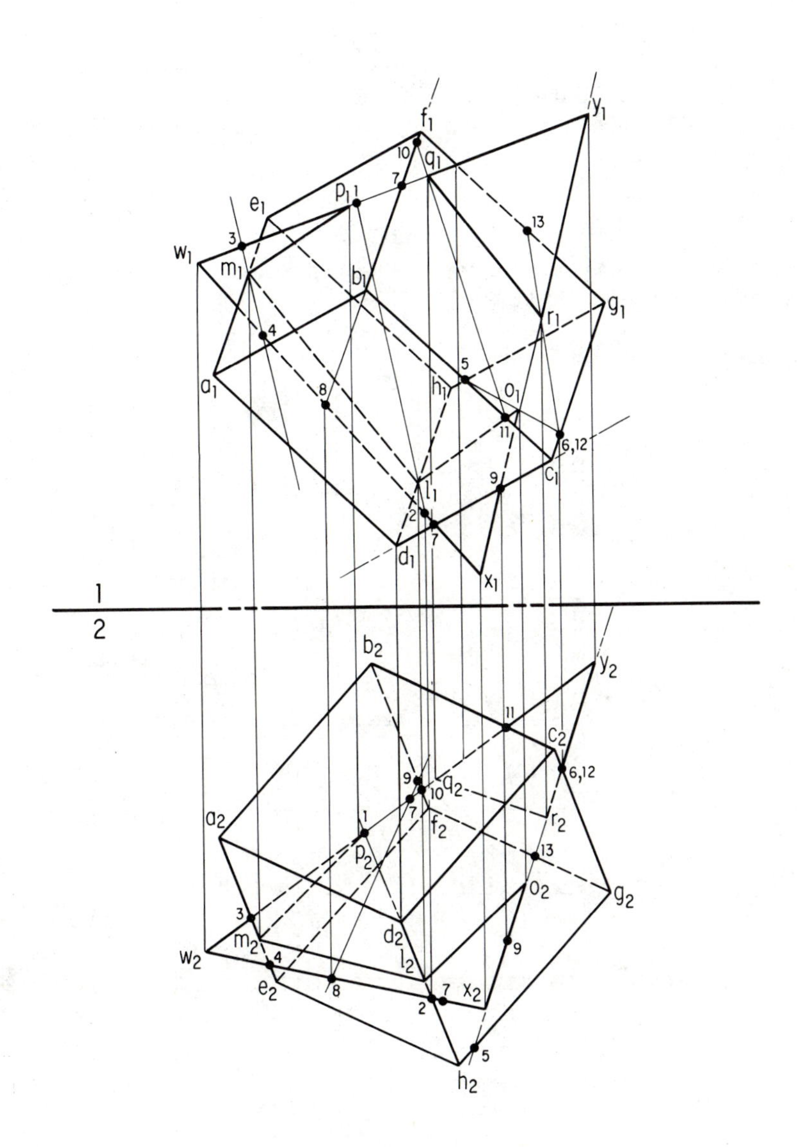

Sample Descriptive Geometry Final Examination

TIME LIMIT: 3 HOURS

NOTE: Reproduce examination problems on $8\frac{1}{2}$ in. × 10 in. cross-section paper having 8 in. × 8 in. divisions to the inch, horizontally and vertically. All construction should be shown and fully labeled. (Solutions to Final Examination problems appear in the Appendix.)

1. GIVEN DATA: Horizontal and frontal projections of a tetrahedron with faces *ABD*, *ACD*, *BCD*, and *ABC*.

REQUIRED: Using the rotation procedure find, measure, and note the dihedral angle between faces *ABC* and *BCD*.

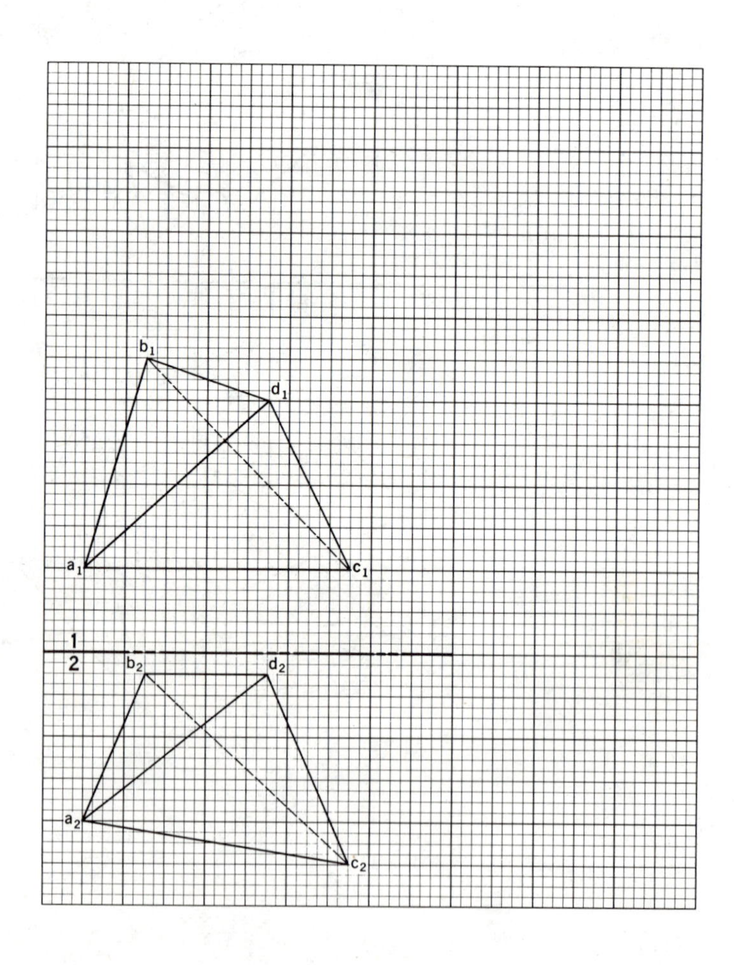

2. GIVEN DATA: Horizontal and frontal projections of a point (light source *L*), an unlimited plane defined by lines *WX* and *YZ*, and a triangular prism.

REQUIRED: The horizontal and frontal projections of the cast shadow of the prism on the given plane.

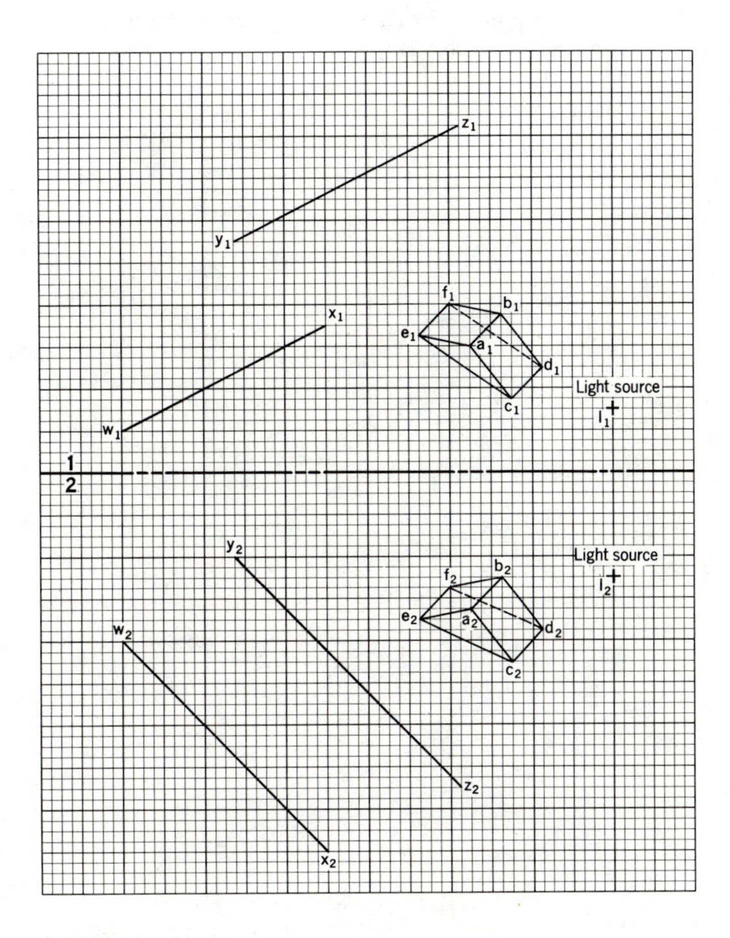

3. GIVEN DATA: Horizontal and frontal projections of a space diagram that illustrates a tripod (the legs of which are *OA*, *OB*, and *OC*) carrying a load of $F_1 = 160$ lb and $F_2 = 100$ lb at the apex. (Lines of action and senses of these forces are given.)

REQUIRED: Using the edge view method, graphically determine the magnitude and type of load carried by each leg of the tripod in order for it to maintain a condition of equilibrium. Summarize all data in view #1 of the space diagram. (SUGGESTION: Use a vector scale in which 1 in. = 80 lb.)

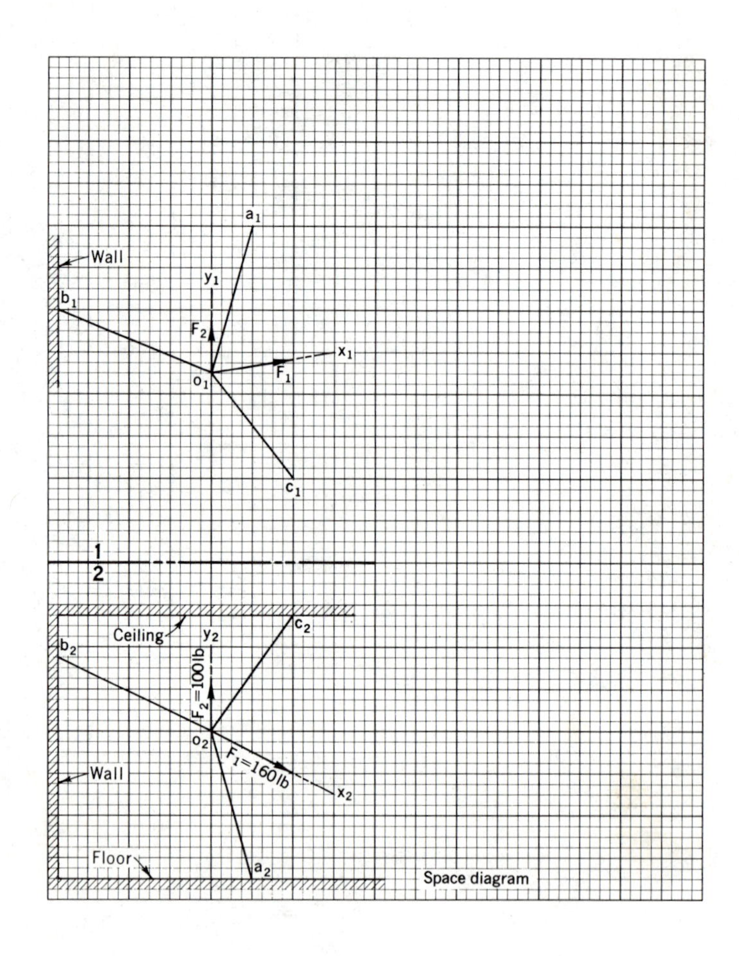

4. GIVEN DATA: Horizontal and frontal projections of two intersecting oblique cones, the bases of which lie in different planes.

REQUIRED: Draw the horizontal and frontal projections of the line of intersection between the given cones, showing the complete visibility of the lines of intersection and of the cones.

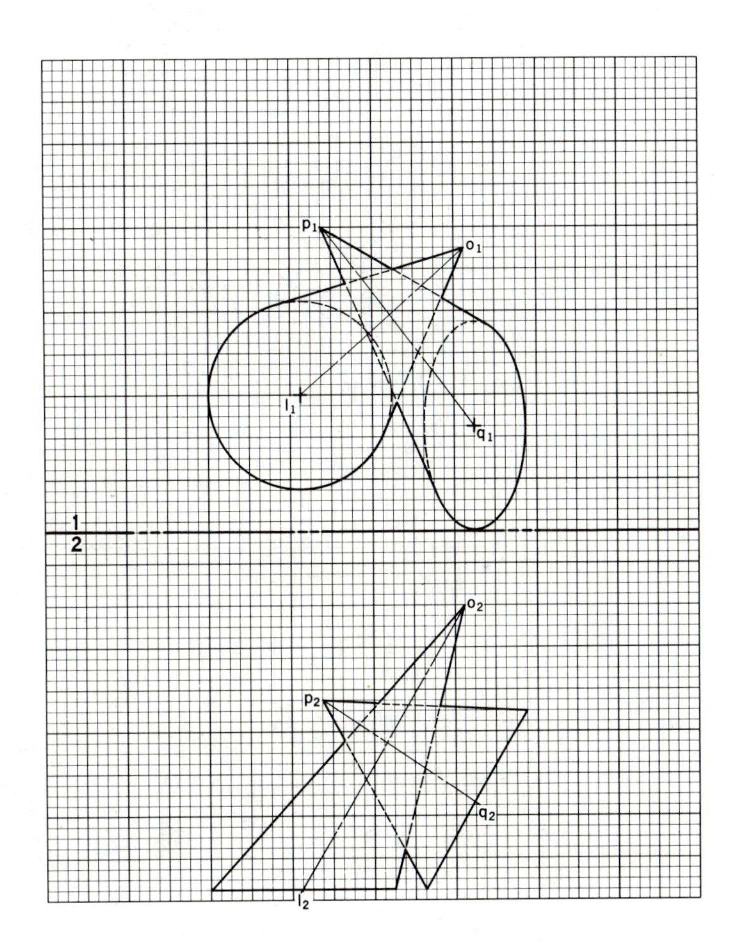

Sample Descriptive Geometry Final Examination (Solutions)

1. Solution:

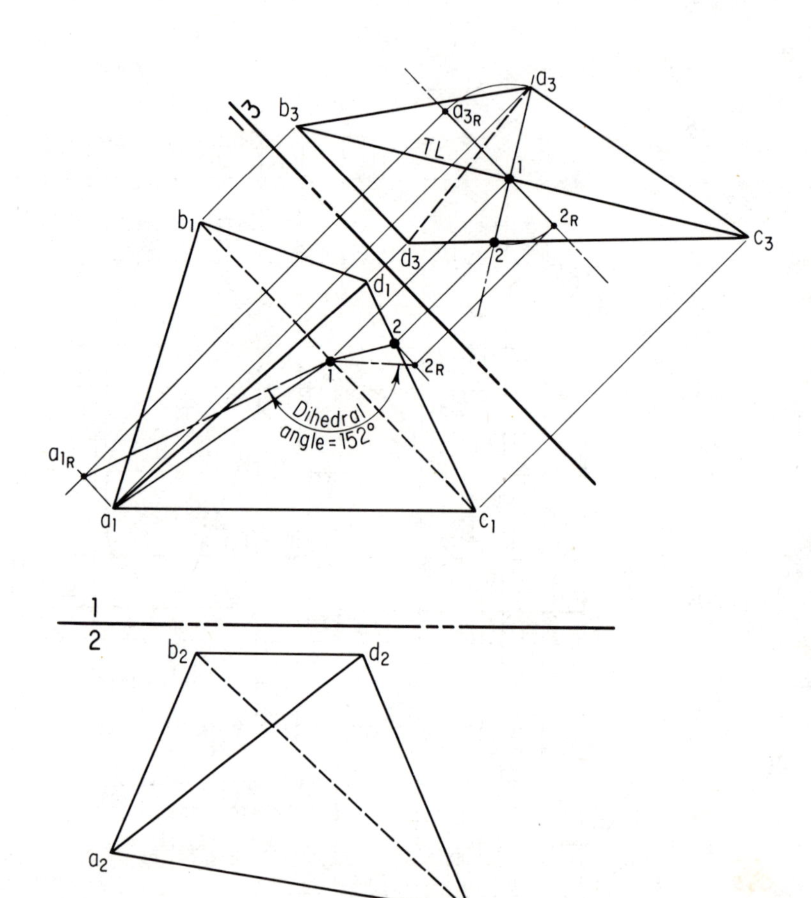

2. Solution:

(Cutting plane method—piercing point of a line and a plane)

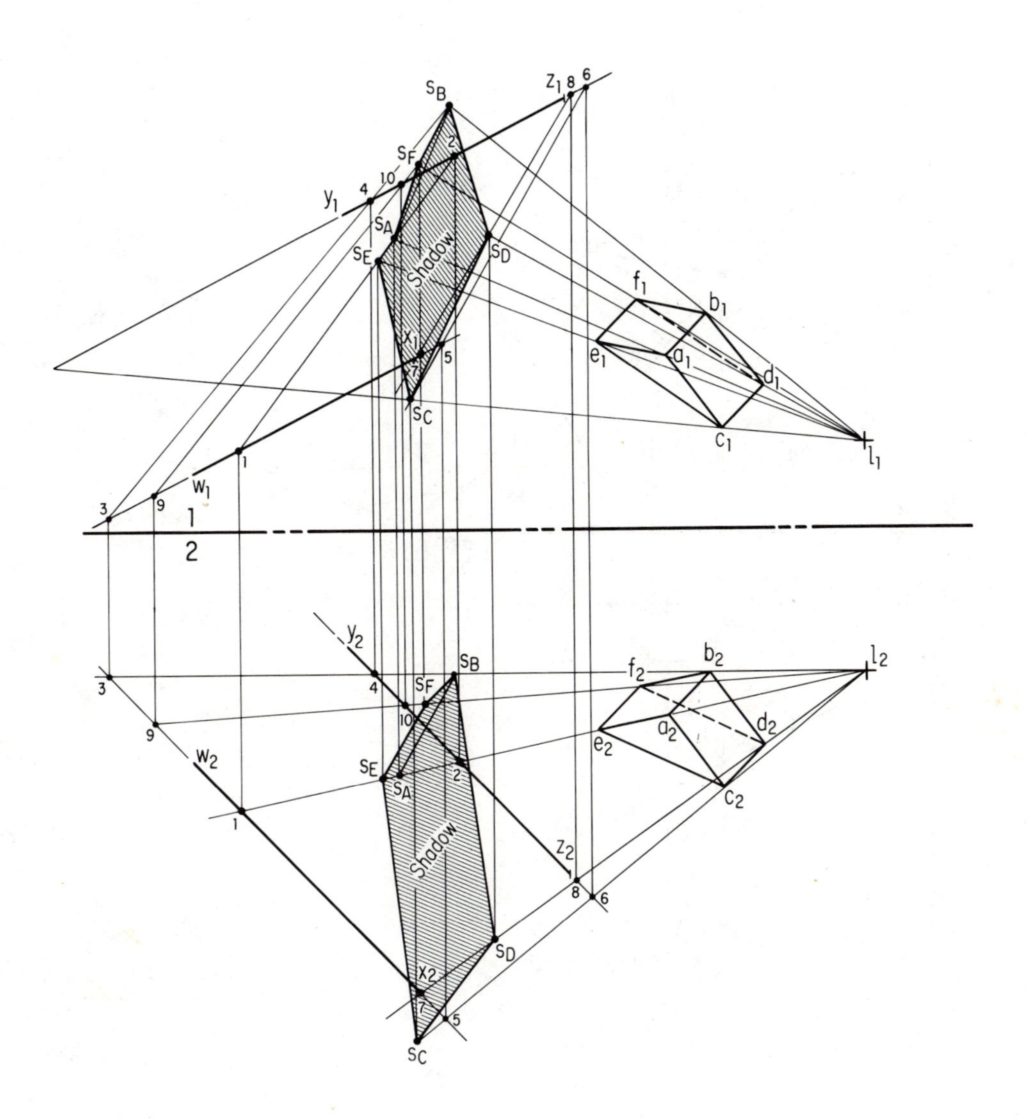

3. Solution:

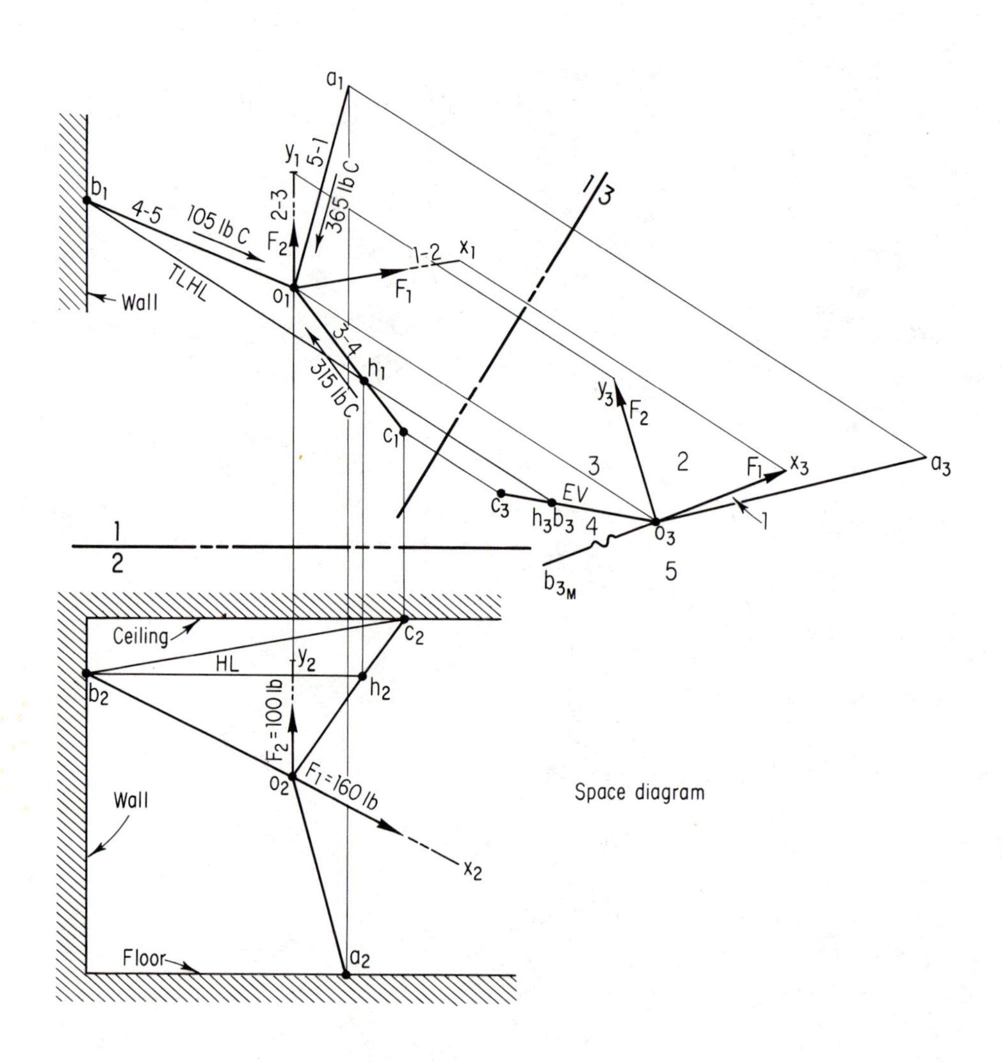

a1
y1
5-1
365 lb C
2-3
b1
4-5
105 lb C
F2
1-2
x1
TLHL
o1
F1
Wall
3-4
315 lb C
h1
c1
1/3
y3
F2
3
2
F1
x3
a3
c3
EV
h3b3
4
o3
1
1
2
b3M
5
Ceiling
c2
HL
y2
h2
b2
F2 = 100 lb
o2
F1 = 160 lb
Wall
x2
Floor
a2
Space diagram

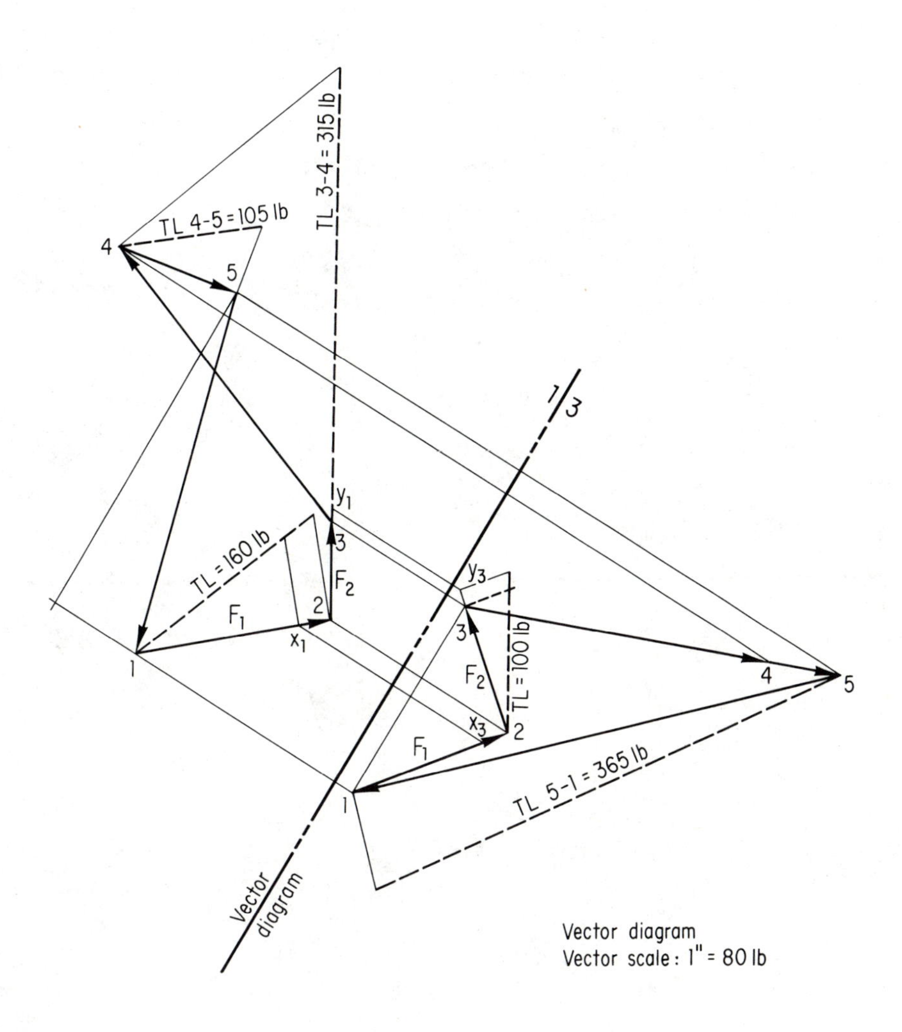

Vector diagram
Vector scale: 1" = 80 lb

4. Solution:

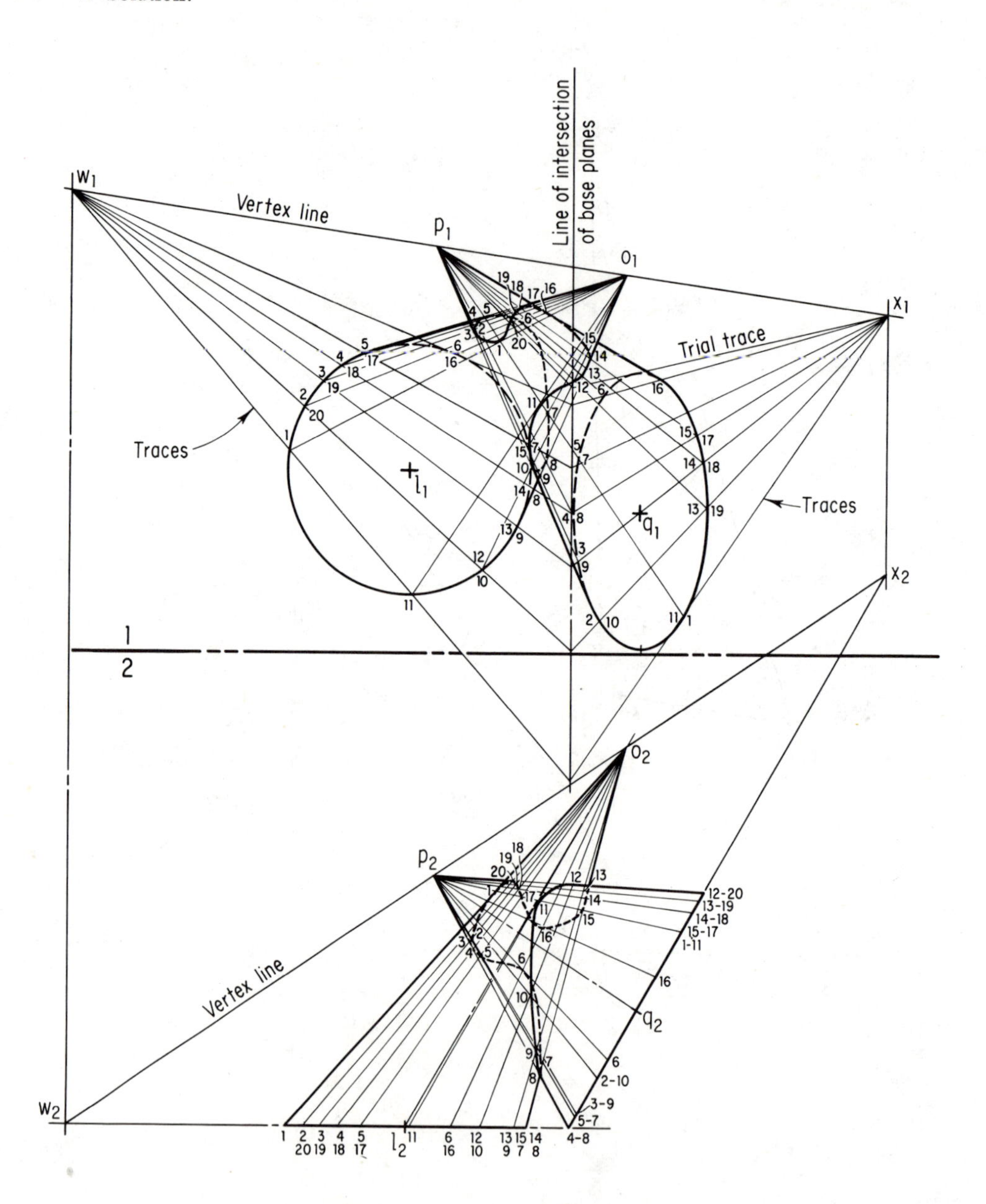

Line of intersection of base planes
Vertex line
Trial trace
Traces
Traces
Vertex line

Bibliography

O. P. Arvesen, *Beskrivende Geometri* (Oslo: H. Aschehoug, 1948), 211 pp.

C. E. Douglass and A. L. Hoag, *Descriptive Geometry* (New York: Holt, Rinehart & Winston, 1962), 208 pp.

Hiram E. Grant, *Practical Descriptive Geometry*, 2nd ed. (New York: McGraw-Hill, 1952), 253 pp.

F. A. Heacock, *Graphic Methods for Solving Problems* (Princeton, N. J.: Princeton University Press, 1952), 113 pp.

F. Hohenberg, *Konstruktive Geometrie für Techniker* (Berlin: Springer-Verlag, 1956), 272 pp.

H. B. Howe, *Descriptive Geometry* (New York: Ronald Press, 1951), 332 pp.

E. G. Pare, R. O. Loving, and I. L. Hill, *Descriptive Geometry* (New York: Macmillan, 1965), 368 pp.

Steve M. Slaby, *Engineering Descriptive Geometry*, College Outline Series, 6th ed. (New York: Barnes & Noble, 1965), 360 pp.

Frank M. Warner and Matthew McNeary, *Applied Descriptive Geometry*, 5th ed. (New York: McGraw-Hill, 1959), 243 pp.

Earle F. Watts and John T. Rule, *Descriptive Geometry* (Englewood Cliffs, N. J.: Prentice-Hall, 1946), 300 pp.

B. Leighton Wellman, *Technical Descriptive Geometry*, 2nd ed. (New York: McGraw-Hill, 1957), 640 pp.

INDEX

M

N

O

P

R

S

T

U

V

W

Z

T

U

V

W

Z

INCH – MILLIMETER SCALES

MILLIMETERS

CENTIMETERS

1 2 3 4 5 6 7 8 9 10 11 12 13 14 15

16th's TO INCH

10th's TO INCH

INCHES

1 2 3 4 5 6